Beyond the Synapse

Formation of synapses and the changes in their connections during life are the basis for learning and memory and recovery from brain disease or injury. Much interest has been focussed on how synapses function at the molecular level, while the cell–cell interactions controlling their formation and function receive far less attention. This book expands the scope of inquiry beyond the synaptic cleft to provide a comprehensive insight into how intercellular signaling enables neurons to communicate beyond the synapse, and to interact with other cells in the brain to alter synaptic connections appropriately. These are chapters devoted to consideration of glia, brain cells which have thus far been ignored in the majority of studies of learning and memory. Writing for academic researchers and professionals, contributors to this book reveal that there is much to learning and memory that lies beyond the synapse.

R. DOUGLAS FIELDS has worked at the NIH, where he now runs the Nervous System Development and Plasticity section, since 1987. Originally a marine biologist, Dr Fields has always based his primary research interests around the synapse. He is currently Editor-in-Chief of *Neuron Glia Biology*.

Beyond the Synapse
Cell–Cell Signaling in Synaptic Plasticity

Edited by
R. Douglas Fields

CAMBRIDGE UNIVERSITY PRESS
Cambridge, New York, Melbourne, Madrid, Cape Town,
Singapore, São Paulo, Delhi, Mexico City

Cambridge University Press
The Edinburgh Building, Cambridge CB2 8RU, UK

Published in the United States of America by Cambridge University Press, New York

www.cambridge.org
Information on this title: www.cambridge.org/9781107411562

© Cambridge University Press 2008

This publication is in copyright. Subject to statutory exception
and to the provisions of relevant collective licensing agreements,
no reproduction of any part may take place without the written
permission of Cambridge University Press.

First published 2008
First paperback edition 2012

A catalogue record for this publication is available from the British Library

Library of Congress Cataloguing in Publication Data
Beyond the synapse: cell-cell signaling in synaptic plasticity/[edited by] R. Douglas Fields.
   p. ; cm.
   Includes bibliographical references and index.
   ISBN 978-0-521-86914-0 (hardback)
   1. Synapses.   2. Neuroplasticity.   3. Cellular signal transduction.   I. Fields, R. Douglas.
   [DNLM: 1. Synapses–physiology.   2. Synaptic Transmission.   3. Brain Diseases–pathology.
   4. Memory–physiology.   5. Neuronal Plasticity.   WL 102.8 B573 2008]
   QP364.B49 2008
   612.8–dc22
                                                                                                2008020690

ISBN 978-0-521-86914-0 Hardback
ISBN 978-1-107-41156-2 Paperback

Cambridge University Press has no responsibility for the persistence or
accuracy of URLs for external or third-party internet websites referred to in
this publication, and does not guarantee that any content on such websites is,
or will remain, accurate or appropriate.

Contents

List of contributors — page vii

Introduction: Beyond the synapse — 1
R. DOUGLAS FIELDS

I SPANNING SCALES OF NEURAL PLASTICITY — 5

1. Memory beyond the synapse — 7
   STEVEN P. R. ROSE

2. Between synapses and behavior: functional circuitry of the hippocampus — 14
   HOWARD EICHENBAUM

3. Widening the lens: looking beyond the synapse for experience-driven brain plasticity — 22
   JULIE A. MARKHAM, AARON W. GROSSMAN, AND WILLIAM T. GREENOUGH

4. Activity-dependent myelination — 36
   R. DOUGLAS FIELDS

5. Bipolar disorder: involvement of signaling cascades and AMPA receptor trafficking at synapses — 43
   JING DU, JORGE QUIROZ, PEIXIONG YUAN, CARLOS ZARATE JR., AND HUSSEINI K. MANJI

II NOVELTY, STRESS, AND HORMONES IN PLASTICITY — 57

6. Sleep-dependent memory consolidation and reconsolidation — 59
   MATTHEW P. WALKER AND ROBERT STICKGOLD

7. Consolidation and reconsolidation of Pavlovian fear-conditioning: roles for intracellular signaling and extracellular modulation in memory storage — 72
   CHRISTOPHER CAIN, JACEK DEBIEC, AND JOSEPH E. LEDOUX

8. Emotional and cognitive reinforcement of rat hippocampal long-term potentiation by different learning paradigms — 89
   VOLKER KORZ AND JULIETTA U. FREY

9. Estrogen and hippocampal synaptic plasticity — 97
   MICHAEL FOY, MICHAEL BAUDRY, AND RICHARD F. THOMPSON

10. Steroid-induced hippocampal synaptic plasticity: sex differences and similarities — 111
    RUSSELL D. ROMEO, ELIZABETH M. WATERS, AND BRUCE S. MCEWEN

III CELL–CELL SIGNALING MOLECULES IN SYNAPTIC PLASTICITY — 121

11. MHC class I in activity-dependent structural and functional plasticity — 123
    LISA M. BOULANGER

12. Cytokine induction of neuronal receptor trafficking: relevance to synaptic function and excitotoxicity — 130
    DMITRI LEONOUDAKIS, STEVEN P. BRAITHWAITE, MICHAEL S. BEATTIE, AND ERIC C. BEATTIE

13	Neurotrophin signaling among neurons and glia during formation of synapses	142	17	Diffusible hydrogen peroxide generated by synaptic activity inhibits axonal dopamine release in striatum	181

13 Neurotrophin signaling among neurons and glia during formation of synapses   142
SARINA B. ELMARIAH,
ETHAN G. HUGHES,
EUN JOO OH, AND
RITA J. BALICE-GORDON

14 Regulation of neurogenesis by neurotrophins: implications in hippocampus-dependent memory   153
BAI LU AND JAY H. CHANG

15 Focal adhesion-like processes underlie induction of long-term potentiation in the Schaffer collateral–CA1 region of the hippocampus   160
RICHARD G. LeBARON,
RUBEN V. HERNANDEZ,
MARY M. NAVARRO, JAMES E. ORFILA,
LISA R. CURRY, AND JOE L. MARTINEZ JR

16 Signaling to the nucleus in long-term memory   169
OLENA BUKALO AND R. DOUGLAS FIELDS

IV NON-TRADITIONAL TRANSMITTERS AND GLIA   179

17 Diffusible hydrogen peroxide generated by synaptic activity inhibits axonal dopamine release in striatum   181
MARAT V. AVSHALUMOV, JYOTI C. PATEL,
LI BAO, DUNCAN G. MACGREGOR,
ZSUZSANNA SIDLÓ AND
MARGARET E. RICE

18 D-serine as a putative glial neurotransmitter   193
ASIF K. MUSTAFA, PAUL M. KIM, AND
SOLOMON H. SNYDER

19 A dialogue between glia and neurons in the retina: modulation of neuronal excitability   201
ERIC A. NEWMAN

20 Metabotropic glutamate receptors as a target for astrocytic control of inhibitory synaptic transmission in the hippocampus   210
WING-SONG LIU, QIWU XU,
JIAN KANG, AND MAIKEN NEDERGAARD

References   219
Index   304

Contributors

AVSHALUMOV, MARAT V.
Department of Physiology, Neuroscience, and Neurosurgery
New York University School of Medicine
550 First Avenue
New York, NY 10016

BALICE-GORDON, RITA J.
Department of Neuroscience
University of Pennsylvania School of Medicine
Philadelphia, PA 19104–6074

BAO, LI
Department of Physiology, Neuroscience, and Neurosurgery
New York University School of Medicine
550 First Avenue
New York, NY 10016

BAUDRY, MICHAEL
Department of Biological Sciences
University of Southern California
3641 Watt Way
Los Angeles, CA 90089–2520

BEATTIE, ERIC C.
Neurosciences Program
California Pacific Medical Center Research Institute
475 Brannan Street, Suite 220
San Francisco, CA 94107

BEATTIE, MICHAEL S.
Brain and Spinal Injury Center
University of California
San Francisco, CA 94110

BUKALO, OLENA
National Institutes of Health, NICHD
Bldg. 35, Room 2A214, MSC 3713
35 Lincoln Drive
Bethesda, MD 20892

BOULANGER, LISA M.
Department of Biological Sciences
University of California, San Diego
9500 Gilman Drive
La Jolla, CA 92093

BRAITHWAITE, STEVEN P.
Neurodegeneration Research
Wyeth Research
Princeton, NJ 08543

CAIN, CHRISTOPHER
Center for Neural Science
New York University
4 Washington Place
New York, NY 10003

CHANG, JAY H.
Section on Neural Development and Plasticity
NICHD, NIH
Building 35, Rm. 1c86914
35 Convent Dr., MSC 3714
Bethesda, MD 20892–3714

CURRY, LISA R.
Department of Biology
University of Texas at San Antonio
One UTSA Circle
San Antonio, Texas 78249

DEBIEC, JACEK
Department of Psychiatry
New York University School of Medicine
Center for Neural Science
New York University
4 Washington Pl., Room 809
New York, NY 10003

DU, JING
Laboratory of Molecular Pathophysiology
National Institute of Mental Health
Bethesda, MD 20892

EICHENBAUM, HOWARD
Center for Memory and Brain
Boston University
Department of Psychology
64 Cummington Street
Boston MA 02215

ELMARIAH, SARINA B.
Department of Neuroscience
University of Pennsylvania School of Medicine
215 Stemmler Hall
Philadelphia, PA 19104-6074

FIELDS, R. DOUGLAS
Nervous System Development and
Plasticity Section
National Institutes of Health, NICHD
Bldg. 35, Room 2A211, MSC 3713
35 Lincoln Drive
Bethesda, MD 20892

FOY, MICHAEL R.
Department of Psychology
Loyola Marymount University
1 LMU Drive
Los Angeles, CA 90045

FREY, JULIETTA U.
Leibniz Institute for Neurobiology
Department of Neurobiology
Brenneckstrasse 6
39108
Magdeburg
Germany

GREENOUGH, WILLIAM T.
Beckman Institute
University of Illinois
405 N. Mathews Ave.
Urbana, IL 61801

GROSSMAN, AARON W.
Beckman Institute
University of Illinois at Urbana-Champaign
405 M. Mathews Ave
Urbana, IL 61801

HERNANDEZ, RUBEN V.
Department of Psychology
San Diego State University
5500 Campanile Drive
San Diego, CA 92182

HUGHES, ETHAN G.
Department of Neuroscience
University of Pennsylvania School of Medicine
215 Stemmler Hall
Philadelphia, PA 19104-6074

KANG, JIAN
Center for Aging and Developmental Biology
Dept. of Neurosurgery
University of Rochester Medical Center
601 Elmwood Avenue
Box 645, KMRB1.9915
Rochester, NY 14642

KIM, PAUL M.
Department of Pharmacology
Johns Hopkins University School of Medicine
725 North Wolfe Street
Baltimore, MD 21205

KORZ, VOLKER
Department of Neurophysiology,
Leibniz-Institute for Neurobiology
Brenneckestr. 6,
D-39118 Magdeburg
Germany

LEBARON, RICHARD G.
Department of Biology
University of Texas at San Antonio
One UTSA Circle
San Antonio, Texas 78249

LEDOUX, JOSEPH E.
New York University
Center for Neural Science
6 Washington Place
New York, NY 10003

LEONOUDAKIS, DMITRI
Neurosciences Program
California Pacific Medical Center Research Institute
475 Brannan St, Suite 220
San Francisco, CA 94107

LIU, WING-SONG
Center for Aging and Developmental Biology
Dept. of Neurosurgery
University of Rochester Medical Center
601 Elmwood Avenue
Box 645, KMRB1.9915
Rochester, NY 14642

LU, BAI
Section on Neural Development and Plasticity
NICHD, NIH
Building 35, Rm. 1c86914
35 Convent Dr., MSC 3714
Bethesda, MD 20892–3714

MANJI, HUSSEINI K.
Laboratory of Molecular Pathophysiology
Mood and Anxiety Disorders Program
National Institute of Mental Health
Building 35, Room 1C-912
Bethesda, MD 20892

MACGREGOR, DUNCAN G.
Division of Neuroscience and Biomedical Systems
Institute of Biological and Life Sciences
University of Glasgow
University Avenue
Glasgow G12 8QQ, UK

MARKHAM, JULIE A.
Maryland Psychiatric Research Center
University of Maryland School of Medicine
P.O. Box 21247
Baltimore, MD 21228

MARTINEZ, JOE L. JR
Department of Biology
University of Texas at San Antonio
One UTSA Circle
San Antonio, Texas 78249

MCEWEN, BRUCE S.
The Rockefeller University
1230 York Avenue
New York, NY 10021

MUSTAFA, ASIF K.
The Solomon H. Snyder Department of Neuroscience
Johns Hopkins University School of Medicine
725 North Wolfe Street
Baltimore, MD 21205

NAVARRO, MARY M.
Department of Biology
University of Texas at San Antonio
One UTSA Circle
San Antonio, Texas 78249

NEDERGAARD, MAIKEN
Center for Aging and Developmental Biology
Dept. of Neurosurgery
University of Rochester Medical Center
601 Elmwood Avenue
Box 645, KMRB1.9915
Rochester, NY 14642

NEWMAN, ERIC A.
Department of Neuroscience
University of Minnesota
6–145 Jackson Hall
321 Church Street SE
Minneapolis, MN 55455

OH, EUN JOO
Department of Neuroscience
University of Pennsylvania School of Medicine
215 Stemmler Hall
Philadelphia, PA 19104-6074

ORFILA, JAMES E.
Department of Biology
University of Texas at San Antonio
One UTSA Circle
San Antonio, Texas 78249

PATEL, JYOTI C.
Department of Physiology, Neuroscience, and Neurosurgery
New York University School of Medicine
550 First Avenue
New York, NY 10016

QUIROZ, JORGE
CNS and Pain Therapeutic Area
Johnson and Johnson Pharmaceutical
Research and Development
1125 Trenton-Harbourton Road
Titusville, NJ 08560

RICE, MARGARET E.
Departments of Physiology & Neuroscience and Neurosurgery
New York University School of Medicine
550 First Avenue
New York, NY 10016 USA

ROMEO, RUSSELL D.
Laboratory of Neuroendocrinology
The Rockefeller University
Box 165
New York, NY 10021

ROSE, STEVEN P.R.
Dept of Biological Sciences
The Open University
Milton Keynes MK7 6AA, UK

SIDLÓ, ZSUZSANNA
Department of Physiology, Neuroscience, and Neurosurgery
New York University School of Medicine
550 First Avenue
New York, NY 10016

SNYDER, SOLOMON H.
Department of Psychiatry and Behavioral Sciences
Johns Hopkins University School of Medicine
725 North Wolfe Street
Baltimore, MD 21205

STICKGOLD, ROBERT
Harvard Medical School
Department of Psychiatry
BIDMC/E-FD 861
330 Brookline Ave
Boston, MA 02139

THOMPSON, RICHARD F.
University of Southern California
3641 Watt Way, HNB 522
Los Angeles, CA 90089, USA

XU, QIWU
Center for Aging and Developmental Biology
Dept. of Neurosurgery
University of Rochester Medical Center
601 Elmwood Avenue
Box 645, KMRB1.9915
Rochester, NY 14642

WALKER, MATTHEW P.
Sleep and Neuroimaging Laboratory
Department of Psychology and Helen Wills Neuroscience Institute
Tolman Hall, Room 333
University of California
Berkeley, CA 94720

WATERS, ELIZABETH M.
The Rockefeller University
1230 York Avenue
New York, NY 10021

YUAN, PEIXIONG
Laboratory of Molecular Pathophysiology
National Institute of Mental Health
Bethesda, MD 20892

ZARATE, CARLOS, JR.
Laboratory of Molecular Pathophysiology
Mood and Anxiety Disorders Program
National Institute of Mental Health
Building 35, Room 1C-912
Bethesda, MD 20892

Introduction: Beyond the synapse

R. Douglas Fields

Powerful new imaging techniques can now reveal synapses changing structure and track neurotransmitter receptors shuttling into and out of the synaptic membrane. Even as synaptic plasticity is studied at a finer scale than could be imagined previously, it is important to remember that learning and memory are behaviors, not molecules, cells, or synapses. Synaptic plasticity is central, but in widening the scope of consideration, the contributors to this book reveal that there is much to learning and memory which lies beyond the synapse.

All cells in the body communicate with one another, and they all change physiologically from past experience – an elemental form of learning. Many mechanisms of intercellular communication, intercellular messenger systems, cell adhesion molecules, growth factors, interactions with the immune system cells, and gene expression, for example, are relevant to how neural circuits change their structure and function from experience. Moreover, system- and circuit-level properties of nervous system operation are fundamental to understanding the behaviors of learning and memory.

Steven Rose begins with the Hebb hypothesis, the fulcrum which has leveraged the great mass of research on learning and memory in modern times, but, as he makes clear in his chapter, the Hebb hypothesis is a fulcrum, not an endpoint. His chapter also reminds us that memory is a process, not a single event like the sudden jump in voltage of a synaptic potential in a slice of brain triggered by the flick of a switch. Memory has multiple phases, spanning from milliseconds to weeks, and it results from a sequence of very different cellular processes. Many engage machinery well outside the synapse to reach into the nucleus of the cell. Transcription factors, cell adhesion molecules, and growth factors all come into play over the course of hours or days in learning the simplest task: a chick avoiding selecting a distasteful bead after a single experience with it.

Much like the work of Rose and his colleagues on learning in the chick, the involvement of cell adhesion molecules and extracellular matrix molecules in long-term potentiation (LTP), is the focus of the chapter by Richard LeBaron and colleagues. Here, the investigators show that the specialized cell surface molecules and the associated intracellular signaling enzymes of focal adhesions in non-neuronal cells provide an important foundation for plasticity in synaptic connections between neurons in the hippocampus.

Lisa Boulanger's research concerns molecules mediating cell–cell signaling in the immune system, the major histocompatibility complex (MHC) class 1 proteins, in the context of structural and functional plasticity in the nervous system. This work is an example of how broadening thinking beyond the synaptic cleft may lead to fundamental insights into how the brain changes structure and function through experience. Molecular mechanisms of cell–cell communication used by other cells in the body are likely to be adopted by neurons in several aspects of synaptic communication and plasticity. Also, work such as this, at the intersection between the immune and nervous systems, may further illuminate how immune responses and exposure to disease may affect development and plasticity of neural circuits.

Information processing in the hippocampus underlying episodic memory is examined in the chapter by Howard Eichenbaum. "I woke, put on my pants, ran down the stairs, and ate eggs while reading the paper in the kitchen." Without this sequential coherence, fragmentary impressions of events – no matter how indelibly recorded in the altered strengths of individual synaptic connections – could not yield a useful memory any more than the frames of a filmstrip sliced into fragments could yield a meaningful sequence. Although the entire record may be preserved, without the vital episodic connections, the record cannot provide a comprehensible memory. What are the cognitive processes and circuitry that preserves the sequence of records in an episodic memory? Eichenbaum argues that three cognitive processes supported by the hippocampus: associative representation; sequential organization; and relational networking, must be integrated to provide a coherent episodic memory.

William Greenough and colleagues approach the subject of memory from the opposite direction of most: beginning from the behavior and tracing to the roots of cellular changes induced by experience. His research shows that there are indeed changes in neurons and neuronal structure with experience, but the cellular changes in the brain sculpted by experience are hardly limited to neurons. Glia of many different types (and blood vessels) are altered by functional experience, along with changes in synapse number, synapse morphology, and neurogenesis. One particularly intriguing finding is that the myelin insulation on axons changes in animals brought up in impoverished or enriched environments, a subject that I consider briefly in a separate chapter. Operating well beyond the synapse, changes in myelin may be an underappreciated form of nervous system plasticity, affecting information-processing in the brain by regulating the speed of conduction in neural circuits, and thus the degree of temporal summation at synapses.

Several other chapters are devoted to consideration of glia, brain cells which have been ignored in the majority of studies of learning and memory. Glia remain absent from all computational neuroscience and that will remain so for some time to come. At present our knowledge of glia and their interactions with neurons is simply too limited to incorporate into a quantitative theory of brain function in plasticity, yet these cells have a powerful influence on synaptic and neuronal function. Dmitri Leonoudakis and colleagues consider the involvement of cytokines released from astrocytes in regulating α-amino-3-hydroxy-5-methyl-4-isoxazolepropionic acid (AMPA) receptor trafficking in central nervous system (CNS) neurons. Sarina Elmariah and colleagues examine the involvement of neurotrophin signaling among neurons and glia during synaptogenesis. Nedergaard and colleagues have pioneered studies on the involvement of astrocytes in hippocampal synaptic transmission, and the chapter by Liu and colleagues considers how inhibitory synaptic transmission is regulated through calcium-dependent release of glutamate from astrocytes. Eric Newman's pioneering studies of glia in regulating information-processing in the retina show that calcium signaling, glutamate, and purinergic signaling between neurons and glia contribute to light-evoked responses in the retina. The retina, being the most accessible part of the CNS, is a window into how circuitry in the brain operates, and through that window we see glia as a vital part of the mechanism.

Behavioral states of arousal are fundamental to learning, and Korz and Frey combine behavioral experiments with cellular electrophysiology to untangle the hormonal effects of stress and novelty on learning. Cain, Debiec, and LeDoux examine consolidation and reconsolidation of Pavlovian fear conditioning, and isolate in detail the mechanisms of emotional memory storage at a molecular level, with important implications for treating fear disorders in humans. The relationship between memory consolidation and reconsolidation emerges from studies of fear conditioning, with a wealth of information on the circuitry, receptors, ion channels, intracellular signaling cascades mobilized to consolidate short-term memories into long-term memories, and reconsolidate the memory once it is retrieved. Neurotrophin signaling, nitric oxide signaling, and neuromodulators and hormones as well as regulation of gene transcription are considered in this comprehensive chapter of fear conditioning memory.

The phases of memory consolidation extend through cycles of sleep and wakefulness, and in their chapter, Walker and Stickgold review the largely mysterious role of sleep in supporting memory consolidation and reconsolidation. Neuroimaging studies show that a task learned in the daytime re-emerges in the hippocampus during slow wave sleep. Cellular studies show that sleep can contribute as much to changes in synaptic connectivity as visual experience does to visual cortical neurons. Gene array studies show that sleep is hardly an idle period in brain function, transcription of approximately 100 genes is increased during sleep. This includes many of the same immediate early genes associated with memory formation.

Sex hormones are powerful agents driving behavior, and their effects on the brain are dramatic. Changes in synapse number and morphology are encountered during the phases of hormonal cycles, and male and female hormones have different effects on cognition and neuronal protection and plasticity. Foy, Baudry, and Thompson consider the role of estrogen in hippocampal synaptic plasticity, and Romeo, Waters, and McEwen examine sex differences and similarities in steroid-induced hipocampal synaptic plasticity.

The role of neurotrophins in neurogenesis during development is an active area of research, and Lu and Chang consider the involvement of neurotrophins in neurogenesis associated with learning and memory in the hippocampus. Neurogenesis and neurotrophins are regulated by many extrinsic factors, including, intriguingly, voluntary exercise. Although it is now clear that thousands of new neurons are born in the adult hippocampus each day, how they are encorporated into circuits supporting learning and memory remains a difficult link to make. In their chapter Lu and Chang present evidence that the original circuitry may be changed after learning by adding new neurons to replace existing neurons in the circuit.

In addition to the classical neurotransmitters, non-traditional transmitters are being isolated which have unique properties that regulate neuronal function differently from classic neurotransmitters, and more intriguingly, encompass other non-neuronal cells in the signaling. D-serine is a glial transmitter released from astrocytes, which can act on synapses by regulating N-methyl-D-aspartate (NMDA) receptor function through the glycine binding site on the NMDA receptor. Mustafa, Kim, and Snyder review research showing that D-serine is an endogenous co-agonist of the NMDA receptor. The actions of D-serine at synapses in the hippocampus and cerebellum can put the regal receptor most closely associated with synaptic plasticity, the NMDA receptor, under direct control of perisynaptic glia.

There is now wide recognition that diffusible gases, such as nitric oxide and carbon monoxide, are important intercellular messengers, regulating synaptic function, and communicating between neurons and non-neuronal cells. The chapter by Avshulumov et al. considers a less well-known intercellular messenger, hydrogen peroxide, in regulating dopamine release in the striatum. Hydrogen peroxide works together with the traditional neurotransmitters, glutamate, and gamma-aminobutyric acid (GABA), to regulate dopamine release from dorsal striatum. Their chapter also considers reactive oxygen species in neuron survival and the important role of glia in releasing antioxidants for neuroprotection. These non-traditional signaling molecules expand beyond the synapse and interact with non-neuronal cells to modulate neurons in normal and pathological conditions.

In their chapter, Du et al. link bipolar disorder to impaired intracellular signaling and AMPA receptor trafficking at synapses. Calcium-signaling and several protein kinase pathways involved in learning are also implicated in bipolar disorder through effects on AMPA receptor trafficking. Many mood stabilizers and antidepressants modulate synaptic plasticity in association with hormonal effects and complex intracellular signaling networks.

Learning does not require gene transcription or translation into protein, but long-term memory does. Without activating transcription of new genes in the nucleus, short-term memories will quickly fade. But how do signals reach the nucleus to activate transcription of the necessary genes to make memories permanent? Bukalo and I consider this question, where our research indicates that the widely assumed requirement for a synapse-to-nucleus signaling molecule to activate gene transcription for late-phase LTP is not necessary. Our work showing that action potentials are the critical factor, again, shows the importance of thinking beyond the synapse in understanding the mechanisms of memory.

Part I
Spanning scales of neural plasticity

1 · Memory beyond the synapse

Steven P. R. Rose

Let us assume then that the persistence or repetition of a reverberatory activity (or "trace") tends to induce lasting cellular changes that add to its stability. The assumption can be precisely stated as follows: When an axon of cell A is near enough to excite a cell B and repeatedly or persistently takes part in firing it, some growth process or metabolic change takes place in one or both cells such that A's efficiency, as one of the cells firing B, is increased.

The most obvious and I believe much the most probable suggestion concerning the way in which one cell could become more capable of firing another is that synaptic knobs develop and increase the area of contact between the afferent axon and efferent [cell body]. There is certainly no direct evidence that this is so ... There are several considerations, however, that make the growth of synaptic knobs a plausible perception.

(Hebb, 1949, pp. 62–63)

HEBB SYNAPSES

When Donald Hebb (1949) formulated this now famous proposition in his classic book *The Organization of Behavior* it was no more than a challenging hypothesis. For sure, it was not entirely original. Earlier versions may be traced back to Tanzi (1893) and even pre-date the discovery of synapses themselves, as hints can be found in Sechenov and even Descartes. However, it is Hebb's formulation that now appears in all the standard textbooks, and what was when he advanced it a mere speculative idea has become, at least for neuroscientists working at the molecular and cellular level, an item of faith, the paradigm within which for the most part our experimental questions are set and our results interpreted.

Put at its simplest, the Hebbian paradigm states that when learning occurs there is a change in neural connectivity, brought about by the modulation of particular synaptic strengths. This modulation may be transient, in which case memory for the learned experience is not retained (short-term memory), or it may be converted into some more permanent modification (consolidation; long-term memory). Implicit in this model is the corollary that recalling a memory of the learned experience requires a retracing of the same neural pathways, via the restructured synapses. But are all forms of memory similar? Hebb did not concern himself with the varied taxonomies of memory that are now common in the literature. Thus there is a key distinction between procedural memory (knowing *how*) and declarative (knowing *that*; itself divisible between semantic and episodic or autobiographical memory). It is easy to distinguish between these forms in humans, much harder in animals, although Clayton (2004), working with scrub jays, has made valiant attempts. As an animal can only tell us it has learned by performing some task, are any observed synaptic changes related to the performance, or the memory on which that performance is based? Are synaptic changes associated with procedural memory similar to those for declarative memory, differing only in which synapses are involved, or are different biochemistries and mechanisms involved?

However these questions are answered, the Hebbian view of memory makes it a special case of synaptic plasticity – though it is important to recognize that plasticity is a term more often adduced than inspected, as it has several meanings. Any release of neurotransmitter from a synapse involves a temporary alteration in its composition, morphology, and physiology: millisecond plastic changes after which the synapse may revert to its prior form. Longer-term and more stable changes occur during development (including apoptosis) as a result of experience and injury. All these are forms of plasticity. So too, are the continual making and breaking of synaptic connections which may be observed in time-lapse studies of living brains even in adults (Purves, 1988). Training an animal to acquire some new skill – learning – may involve any or all of these forms of plasticity, and there is a conceptual distinction to be drawn between identifying a synaptic change that occurs as a consequence or correlate of such training and a change which in some sense is part of the representation of that memory within the brain's encoding systems.

So, an important question within the Hebbian paradigm is whether the synaptic modulation that occurs during memory consolidation is of the same type, and involves the same molecular and physiological processes, as occur during the formation of more or less stable synaptic connections during development. It is how we answer this question that determines, for instance, whether we regard the well-documented changes in hippocampal synaptic physiology that occur after injection of a train of high-frequency pulses, long-term potentiation, as a mechanism of memory, or merely a model system for the study of synaptic plasticity (Bliss, Collingridge & Morris, 2004).

THE CHICK AS A MODEL SYSTEM

These questions have been studied in a variety of animal models ranging from molluscs to mammals. In this chapter, however, the focus is on my own studies in a simple learning model in the young chick. First, I briefly review past work which, seemingly, leads to a straightforward endorsement of the Hebbian paradigm. Training chicks on a one-trial passive avoidance task results in a molecular cascade occurring in a defined region of the chick forebrain, which culminates in seemingly lasting changes in synaptic morphology, biochemistry, and physiology. (A fuller review of the earlier results may be found in Rose, 2000). I will also discuss recent data concerning the role of the amyloid precursor protein (APP) in this cascade, and its potential relevance therefore to Alzheimer's disease. However, as is often the case in neuroscience, more detailed analysis, especially of the processes involved in recall of already learned memory, produces paradoxical results, which suggest a much more labile and dynamic model of memory beyond the synaptic level.

The one-trial passive avoidance task, introduced by Cherkin (1969), is based on the young chick's tendency to peck at small, bright objects such as beads. It has the merit of being rapid and sharply timed (chicks peck a bead within 10 seconds), enabling us to distinguish between the molecular correlates of the immediate training experience (the taste and sight of the bead, the motor act of pecking) and the downstream events associated with memory consolidation. In the standard version of the task in our laboratory, day-old chicks are held in pairs in small pens, pre-trained by being offered a small, dry white bead, and those which peck are trained with a larger (4 mm diameter) chrome or coloured bead coated with the distasteful methylanthranilate (MeA) (Lossner and Rose, 1983). Chicks that peck such a bead show a disgust reaction (backing away, shaking their heads, and wiping their bills) and will avoid a similar but dry bead for at least 48 hours subsequently. However, they continue to discriminate, as shown by pecking at control beads of other colours. Chicks trained on the bitter bead are matched with controls which have pecked at a water-coated or dry bead, and which peck the dry bead on test. Generally some 80% of chicks in any hatch group may be trained successfully and tested on this protocol. Each chick is usually trained and tested only once.

We have used two approaches to identify the cellular sequelae of pecking the bitter bead. In the correlative approach, appropriate brain regions may be dissected from trained and control birds at specific post-training times, and tissue is processed for biochemical, immunocytochemical, autoradiographic, or microscopic analysis. In the interventive approach all the birds are trained on MeA and injected either before or after training with drug, antibody or antisense, or vehicle to explore the possible enhancing or amnestic effect of the agent. Chicks which peck the previously distasteful bead on test are considered to be amnesic for the training. As this pecking response requires a positive and accurate act by the bird, it also controls for effects of the agent on attentional, visual, and motor processes.

The sharply timed nature of the learning experience, together with a combination of these experimental strategies, has enabled us to identify a biochemical cascade associated with memory consolidation in the minutes to hours following training. Thus, a change in some biochemical marker at a specific post-training time, occurring in trained compared with control chicks, might imply its direct engagement in memory expression at that time. Alternatively, it might indicate the mobilization of that marker as part of a sequence leading to the synthesis of a molecule, or cellular reorganization, required for expression of memory. A similar argument applies to the timing of the onset of amnesia after intracerebral drug injection.

Two regions of the chick brain are specifically involved in the biochemical responses to the learning experience. These are the intermediate medial mesopallium (IMMP; previously called the intermediate medial hyperstriatum ventrale), an association "cortical" area, and the medial striatum (MS; previously called lobus parolfactorius, a basal ganglia homologue) (Avian Brain Nomenclature Consortium, 2005). The chick brain is strongly lateralized (Andrew, 1999; Rogers & Deng, 1999), and many, though not all of the molecular events we have observed are confined to the left IMMP.

THE TEMPORAL CASCADE: THE FIRST HOUR

During training, and in the five minutes which follow, there is enhanced release of glutamate in the IMMP (Daisley & Rose, 2001). Over the same time period there is also an increase in potassium-stimulated calcium concentration in synaptoneurosomes isolated from the IMMP (Salinska et al., 1999). Within the succeeding 40 minutes, although we cannot assign them a precise temporal dependency, we have found: increases in NMDA-stimulated calcium flux in synaptoneurosomes (Salinska et al., 1999); in ligand binding to the NMDA-glutamate receptor (Steele, Stewart & Rose, 1995) and of phosphorylation of the presynaptic membrane protein B50/GAP43 (Ali, Bullock & Rose, 1988) coupled with a translocation of cytosolic protein kinase C (PKC) to the membrane (Burchuladze, Potter & Rose, 1990). There is increased release of the putative retrograde messenger arachidonic acid, in tissue prisms prepared 30–75 minutes post-training, though the onset time for amnesia if the arachidonic acid synthesis is blocked with phospholipase A_2 inhibitors is delayed until 75 minutes (Holscher & Rose, 1994; Clements & Rose, 1996). Intervention studies with MK801 (Burchuladze & Rose, 1992), the N-type calcium channel blocker ω-Ω conotoxin GVIA (Clements, Rose & Tiunova, 1995) and PKC inhibitors (Burchuladze, Potter & Rose, 1990), injected into the IMMP either just before or just after training, all produce amnesia with an onset time of 30 minutes to one hour. $GABA_A$ agonists are also amnestic at this time. So, too, is nitroarginine, which blocks synthesis of the putative retrograde messenger nitric oxide (NO) (Holscher & Rose, 1993; Rickard, Ng & Gibbs, 1998). Other laboratories have found an involvement of a variety of protein kinases, notably protein kinase A (PKA), over this period (Serrano, Rodriguez, Bennett & Rosenzweig, 1995).

Thus it would appear that the training experience generates a sequence of rapid synaptic transients which provide a temporary "hold" for the memory – the phases categorized as short- and intermediate-term memory by Gibbs and Ng (1977; see also Patterson et al., 1988). As well as forming the brain substrate of the remembered avoidance over this period, these transients must serve two other functions. They must initiate the sequence of pre- and post-synaptic intracellular processes which will in due course result in the lasting synaptic changes presumed to underlie long-term memory, and they must also serve to "tag" relevant active synapses, perhaps via membrane phosphorylations, so as to indicate those synapses later to be more lastingly modified.

THE TEMPORAL CASCADE: ONE TO EIGHT HOURS

A key step in the intracellular cascade must be the link between synapse and nucleus. Calcium is clearly a major player here, and that intracellular calcium signaling may be important is indicated by our recent observation that within 10 minutes post-training there is also a mobilization of synaptoneurosomal ryanodine-sensitive calcium stores (Salinska, Bourne & Rose, 2001), whilst dantrolene, which blocks calcium release from these stores, injected 30 minutes before or 30 minutes after training, produces amnesia by three hours post-training. Synaptoneurosomes are largely pre-synaptic, though they contain resealed post-synaptic (dendritic) elements as well, so it is not possible to distinguish whether the mobilized calcium stores are located at one, the other, or both sides of the cleft.

That activation of a number of transcription factors must be among the next steps in the process is clear from the elucidation of a role for cAMP response element-binding protein (CREB) in several mammalian learning paradigms. We, however, have focussed on the role of immediate early genes, c-fos and c-jun, both of which show increased expression in the hour after training

(Anokhin et al., 1991). Further evidence as to the necessity of fos expression for longer-term memory is provided by the observation that antisense to c-fos, given six or more hours before training, blocks its synthesis (Mileusnic, Anokhin & Rose, 1996) and chicks become amnesic within three hours after training.

One of the few universal findings in studies of biochemical processes in memory formation is that long(er)-term memory is protein synthesis-dependent (Davies & Squire, 1984). Passive avoidance training is no exception, and anisomycin injected into the IMMP either before or up to some 60 minutes post-training results in amnesia for the avoidance. If the anisomycin is injected before training, amnesia sets in by the end of the first post-training hour, leading to the suggestion that beyond this period memory is protein-synthesis independent (Gibbs & Ng, 1977). However, the earlier view that beyond this time a protein synthesis-independent, long-term memory has been established is no longer tenable. Whilst anisomycin injections two and three hours after training are without effect on memory, injections given four or five hours post-training are amnestic in animals tested at 24 hours (Freeman & Rose, 1995). Thus there is a second, downstream, wave of training-related protein synthesis that we interpret as being the period during which late genes are activated and structural proteins are synthesized.

Although much attention within the learning and memory community is directed toward the roles of the many transcription factors involved in the early phases of memory formation, we have focussed on identifying the later gene products, and in particular the cell adhesion molecules (CAMs), transmembrane molecules, whose glycosylated extracellular domains may bind either homophilically or heterophilically, providing a mechanism for associating pre- and post-synaptic membranes. Their potential role in synaptic plasticity has long been emphasized by Edelman (1985). As well as their adherent properties, they have a second role, in transmembrane signaling. Two, in particular, are required for longer-term memory: NCAM and NgCAM/ L1 (Scholey et al., 1993; Scholey et al., 1995).

Specific blocking of neural cell adhesion molecule (NCAM) synthesis with antisense, injected over the 24-hour post-hatching period before the birds are trained, does not prevent the chicks learning the avoidance, but amnesia sets in within three hours (Mileusnic, Lancashire & Rose, 1999). However, interference with the functioning of already-synthesized CAM molecules is also amnestic. Thus, if antibodies which bind to the extracellular domains of either NCAM or L1 are injected into the intermediate medial mesopallium (IMMP) at five to six hours post-training, chicks show amnesia when tested at 24 hours (Scholey et al., 1993, 1995), a time at which the antibodies themselves are no longer detectable in the brain. Antibodies to NCAM are not amnestic if injected at other times, but antibodies to L1, injected 30 minutes before training, are also amnestic when the chicks are tested at 24 hours. The extracellular domains of L1 include fibronectin and immunoglobulin regions, and using recombinant fragments to these regions we found that blocking the immunoglobulin domain at -30 minutes, but not at $+5.5$ hours, resulted in amnesia, whereas by contrast blocking the fibronectin domain at $+5.5$ hours but not at -30 minutes resulted in amnesia (Scholey et al., 1995). This biochemical version of a double dissociation experiment led us to postulate that it was the cell-signaling function of L1, mediated via the immunoglobulin domain, which was engaged in the early phases of memory formation, whilst the fibronectin domains of NCAM and L1 were required in the de-adherence or re-adherence processes at the later time-point. It is presumably at this time, five to eight hours downstream of the training event, while their epitopes on the external domains are open to attack, that antibody binding can occur and hence amnesia results.

Demonstrating a role of the CAMs in memory formation led us to think about the possible involvement of another adhesion molecule, APP. APP is a rapidly turned-over protein, whose extracellular domains have been implicated in a variety of functions, including neurite outgrowth and synaptic plasticity. Improper processing of APP results in the cleavage of the 42-amino acid Abeta fragment, which accumulates in the plaques characteristic of Alzheimer's disease, and as is well-known, memory loss is a characteristic early feature of the disease. A monoclonal antibody to the C-terminal of APP, injected prior to training, results in the rapid onset of amnesia. So, too, does down-regulating APP levels by injection of antisense (Mileusnic, Lancashire & Rose, 2000). More significantly, we have shown that it is possible to rescue the memory lost by either antibody or antisense injection by administering a small peptide, the palindromic sequence tripeptide Arg-Glu-Arg (RER), homologous to part of the growth-promoting domain of APP. The peptide also acts as a cognitive enhancer in weak versions of the training task. RER binds displaceably to two membrane proteins, of molecular weights 66–69 kD and 110 kD, present in both chick and human neuronal membranes (Mileusnic, Lancashire & Rose, 2000, 2004), and our working hypothesis is that it substitutes for APP in the transmembrane signaling required to activate the internal cascade leading to synaptic modulation. The potential therapeutic role of this peptide is currently under intense study.

STRUCTURAL ENDPOINTS?

The longer-term consequence of this cascade is thus the modification of synaptic connectivity, detectable biochemically in terms of changes in the configuration and distribution of NCAM, among other synaptic markers. The presumed endpoint for memory storage is modulation of synaptic connectivity, by altering synaptic number or relocating or structurally modifying existing synapses and dendritic spines, or both. Stewart and colleagues have been able to show changes in both pre- and post-synaptic elements. Thus, 24 hours after training there is increased dendritic spine density in projection neurons of the IMMP (Patel, Rose & Stewart, 1988) and, at the same time, changes in the numbers and dimensions of synaptic junctions, presynaptic boutons, and synaptic vesicle number in both IMMP and MS.

SIGNALING FACTORS BEYOND THE SYNAPSE

Having pecked a bead coated in MeA, chicks avoid a similar but dry bead for at least 24 hours subsequently. However, if the aversant is made less strong by, for instance, using a 10% solution of MeA in alcohol, the birds peck and display a disgust reaction, but will avoid similar beads for only six to nine hours subsequently (Sandi & Rose, 1994a; *see also* Burne & Rose, 1997). Although in so far as we have compared them, weak training initiates a similar set of synaptic transients to those produced in the strong version of the task, these are apparently not sufficient to result in gene expression, as CAM synthesis does not occur. Our assumption is that the temporal relationship between the fading of the memory trace for the weak training beyond six hours and the wave of glycoprotein synthesis that occurs at this time with strong training is not fortuitous (Rose, 2000). However, there are many factors which may affect the salience of this "weak-learning" experience, and which result in memory being retained as for the strong learning.

Chicks are normally held in their pens in pairs, as this diminishes stress. If they are trained on 10% MeA, and then separated, stress levels increase, and retention persists for 24 hours. The normal training procedure is indeed stressful, as is shown by the fact that for five to ten minutes after training chicks on the strong, but not the weak, version of the task there is an increase in plasma corticosterone levels (Sandi & Rose, 1997). Further, if corticosterone is injected into the IMMP just before or just after weak training, retention is also enhanced (Sandi & Rose, 1994a). The enhancing effects of stress may be blocked by injection of antagonists of glucocorticoid receptors into the IMMP, which is rich in such receptors. Blockade of these receptors is also amnestic for strong training (Sandi & Rose, 1994b), as is inhibition of peripheral corticosterone synthesis with metyrapone or aminoglutethimide (Loscertales, Rose & Sandi, 1997). As might be anticipated, the effects of corticosterone are dose-dependent in the classic inverted-U form: 1 μg injected prior to weak training enhances retention, whereas higher doses do not; 1–5 μg injected prior to strong training diminishes retention (Sandi & Rose, 1997). Similar enhancing effects are also apparent with neurosteroids such as dehydroepiandosterone (DHEA) (Migues, Johnson & Rose, 2001). Neurotrophins also affect the salience of weak training. Recombinant brain-derived neurotrophic factor (BDNF), but not nerve growth factor (NGF) or neurotrophin 3 (NT-3), injected just before or just after weak training, will enhance 24-hour retention. Reciprocally, antibodies to BDNF are amnestic for strong training, amnesia setting in within three hours (Johnston & Rose, 2001).

These findings are of both theoretical and practical relevance. First, they remind us that although, especially under the influence of the neurophysiological observations of synaptic interactions during LTP, cellular theories of memory formation are heavily based on Hebbian models, memory is not just a pre- or post-synaptic event. Rather, whether any particular experience is learned or not depends on a much wider array of neural and peripheral factors, humoral and perhaps also immunological (*see* McGaugh, 1989; Damasio, 1994). The entire animal is thus involved in any learning experience. Second, together with the observations on APP described above, they may point the way toward developing effective agents for therapeutic intervention in conditions of memory deficit.

MEMORY BEYOND THE IMMP

I have so far focussed on the sequence of biochemical events occurring in the chick IMMP consequent on passive avoidance training, and I have argued that the cascade we have identified, leading as it does to measurable morphologic changes in synaptic connectivity, is a necessary part of memory consolidation. Does this, however, mean that the IMMP contains some lasting representation of the association between bead and bitter taste, the elusive engram? A combination of electrophysiologic and lesioning experiments that have been conducted in parallel with those described here makes clear that this is not simply the case (Rose, 2000). Within the hours after training biochemical changes occur in brain regions other than the left IMMP, including the right IMMP and MS, and the memory trace, if such it is, becomes both fragmented and redistributed. The IMMP seems to retain some aspects of the memory including colour discrimination, whereas others, related perhaps to the size and shape of the bead, may be located to the MS (Patterson & Rose 1992; Barber et al., 1999). Again this points to the conclusion that learning and memory formation and retention engage not simply a discrete neuronal ensemble in a small brain region, but a much wider set of spatially and temporally dynamic processes, linked and given coherence by some form of binding mechanism (Rose, 2004).

REACTIVATING MEMORY

The evidence adduced so far suggests that although synapses are modified as a result of training, in accord with the Hebbian hypothesis, this modulation involves more than just intersynaptic signals, but engages wider systemic properties, growth factors, systemic and neuro-hormones, and neurotrophins. Furthermore, over time the memory trace becomes distributed, rather than localized to a simple neural network. There is a yet further complexity to be added. Within the simple Hebbian paradigm, the memory trace, once established, is permanent. However, there have been persistent reports in the literature suggesting that even well-established memories may be rendered labile and susceptible to amnestic agents if they are reactivated by giving an animal a reminder (Sara, 2001). In particular, it has been shown in a number of species and learning paradigms that if anisomycin is administered around the time of the reminder, the animal is rendered amnesic for the task. However, a debate has ensued as to whether this is a lasting or merely transient amnesia (Nader, 2003); that is, does "reconsolidation" recapitulate consolidation, or does the amnestic agent merely transiently block access to the memory? In our hands, both the temporal dynamics and kinetics of the amnesia after administration of anisomycin coupled with a reminder are different. Notably, the effects are transient, and lower doses of the inhibitor are required to produce it (Anokhin, Tiunova & Rose, 2003). Furthermore, whilst a reminder resulted in increases in 2-doexyglucose uptake into both IMMP and MS, as is the case following initial training, immediate early gene expression is enhanced only in the MS (Salinska, Bourne & Rose, 2004). One suggestion is that reactivating a memory might not require the full biochemical cascade triggered by training, but involves only local, synaptic, protein synthesis. It is relevant in this context that blocking axonal flow around the time of training with colchicine results in transient amnesia, there is no such effect following a reminder.

CONCLUSIONS

What are the lessons from these experiments? A biochemical cascade leading to modulation of synaptic connectivity is a necessary consequence of exposing an animal to a learning situation, and that this cascade is required for memory consolidation is clear. Our experiments have mapped this cascade in a specific and simple learning task in the young chick. We cannot say from this that we have identified a universal mechanism, even for all forms of learning in a single species. However, the similarities both in molecular processes and temporal dynamics between this cascade and those observed in other tasks and vertebrates (Izquierdo & Medina, 1997) are encouraging. It would be nice to be able to say that what is true for *Gallus gallus domesticus* is also true for *Homo sapiens sapiens*, in which case it may well be that our discovery of the memory-enhancing effects of short APP peptides may well lead to a novel rational therapy for Alzheimer's disease. But I doubt that this is the full story; the myriad of neuromodulators already known to affect both memory acquisition and recall is bound to increase, and will differ from task to task and from species to species. However, if synapses are to be modulated, the mechanisms and dynamics will be determined by biochemical constraints – how long it takes to activate late genes and to transport newly synthesized proteins to the synapse, for instance. And these constraints will largely transcend species differences.

But another more general lesson is that Hebbianism is not sufficient. Whilst it may well describe the processes involved in the early stages of memory formation, it cannot account for the ways in which, over periods which vary from hours to weeks and even months in some species, the putative memory traces are disassembled and redistributed. Nor can it account for the renewed lability of memory following a reminder, which suggests that each time a memory is recalled, it is in a molecular as well as a psychological sense, reconstructed. Further, it is important to recognize that in practice nearly all the molecular and cellular studies of memory study the processes involved in learning and consolidation, not in the recall of that memory. We have no idea how recall occurs, whether it is the neural processes engaged in scanning the putative memory store to recover the instruction not to peck at an attractive bead, or the name of the person we are talking to. Nor do we understand how chicks – or humans – derive a coherent image from such distributed cues – the so-called binding problem. And even if we were to explain this process in neural terms, I contend, no amount of sophisticated technology will enable us to read off, from the properties of whatever ensemble of neurons and their synapses, in whatever degree of detail, the actual *content* of that memory. In this sense, memories are found, not only beyond the synapse, but beyond the brain itself, in the complex web of interacting processes that in humans we call a mind (Rose, 2005).

ACKNOWLEDGEMENTS

The experiments described in this chapter are based on many years of collaboration with colleagues in the Brain and Behaviour Research Group, and especially my long-term co-workers Radmila Mileusnic and Chris Lancashire, and diverse sources of funding, including the Medical Research Council, Wellcome Trust, and Royal Society; none of these are required to take responsibility for the conclusions I draw from them, however.

2 · Between synapses and behavior: functional circuitry of the hippocampus

Howard Eichenbaum

Synaptic plasticity almost surely is a key fundamental mechanism for alterations in neural connectivity which underlie memory. Many of the chapters in this book provide substantial insights into the molecular and cellular events that constitute synaptic plasticity and some directly relate cellular plasticity mechanisms to the behavioral expression of memory. However, a full understanding of memory requires considerations between the level of the synapse and the level of overt behavior. Synaptic plasticity does not exert its effects directly on behavior, but rather mediates the modification of functional circuitry that encodes and retrieves memories that guide behavior. Here, I will outline some of the efforts in my laboratory which aimed to reveal the neural coding scheme and elemental information-processing functions accomplished by the hippocampus in the service of episodic memory.

The hippocampus plays a critical role in episodic memory, our capacity for recollection of unique personal experiences (Aggleton & Brown, 1999; Eichenbaum & Cohen, 2001; Squire, Stark & Clark, 2004). Tulving (2002) defined episodic memory by the subjective experience of conscious recollection, making it difficult to establish animal models that could provide platforms for neurobiological investigation. However, an examination of the content of episodic memories suggests features of this kind of memory that might be characterized in animals as well as humans. Consider the following typical example of episodic memory. When asked what we did upon awakening this morning, we would likely recall a rich body of information about the circumstances of awakening ("I woke up in my bed to the sound of an alarm clock"), and about the flow of events in the entire experience ("I woke up, put on my pants, ran down the stairs, and ate eggs while reading the paper in the kitchen"). In addition, the remembarer might relate specific events of that episode to others ("The newspaper article explained the results of elections I had read about yesterday"). The content of vivid episodic memories such as this example are thus characterized by three fundamental features: the recall of specific events as items and actions in the context in which they occurred; the organization of unique episodes as sequences of events; and the remembering of related events and episodes which provide the broader meaning of the episode for our lives.

The following considerations are guided by a framework for conceptualizing these features of episodic memory in terms of three elemental cognitive processes supported by the hippocampus: *associative representation*, *sequential organization*, and *relational networking*. I will first briefly explain what is meant by each of these types of information processing. Then I will outline some of the properties of hippocampal circuitry and plasticity that may support them. This will be followed by a review of our experimental studies exploring the role of the hippocampus in these functions and characterizing the neural representations within the hippocampus that may support these processes. My aim is to bridge between our understanding of synaptic plasticity and the neural circuitry that supports cognitive functions in the service of episodic memory. Finally, I will return to the issue of synaptic plasticity, and begin to make the case for the critical role of synaptic plasticity in each of the processing functions I have attributed to hippocampal circuitry.

COGNITIVE PROCESSES THAT UNDERLIE EPISODIC MEMORY

A central theme of this chapter is that the content of episodic memories (see Figure 2.1) may be deconstructed into three main elements, each corresponding to a particular kind of "binding" operation that might be supported by synaptic plasticity within the hippocampus.

First, episodic memories are composed of discrete events, each of which involves a set of relevant objects

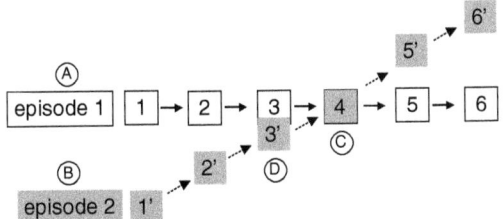

Figure 2.1. A schematic diagram of a simple relational network composed of two episodic memories (**A** and **B**). Each is construed as a sequence of elements (1–6) that represent the association of an event and the place where it occurred. **C** is an element that contains the same features in both episodes. **D** is an element that contains only some of common information.

and people, their actions, and the place and other contextual features that characterize where and when the event occurred. In the example given above, the subject wakes up in bed, walks down the stairs, and eats in the kitchen. I will suggest that each of these discrete events is composed as an *associative representation* encoded by single hippocampal neurons that bind together a broad variety of cues and contextual information which constitute that event.

Second, episodic memories are composed from representations of the series of events that constitute entire experience. The morning experience described above is composed of a sequence of waking, dressing, running, and eating events that unfold in a mental replay that characterizes the vivid recollection of that experience. I will suggest that ensembles of hippocampal neurons bind together the *sequential organization* of event representations that compose a unique episodic memory.

Third, episodic memories are not isolated, but instead have the capacity to evoke memories of related events and episodes. In the example above, reading about the explanation of an election evoked the memory of a related event where the remberer had first heard about that election. I will suggest that our ability to remember related experiences is supported by the establishment of a *relational network*, and that such networks are composed by binding the representations of related episodes through their common events. Furthermore, I will suggest that relational networks allow us to compare and contrast memories across related experiences, giving rise to the flexibility of expression that is characteristic of conscious memory (Cohen, 1984).

HIPPOCAMPAL PLASTICITY AND CIRCUITRY

Note that I have suggested that each of these three elemental cognitive processes is the outcome of a "binding" of features of a memory. This perspective suggests how even the most elaborate qualities of episodic memory may ultimately be supported by relatively simple synaptic "binding" operations that alter the connectivity of cells within hippocampal circuits. Here I will focus on a synaptic plasticity mechanism and two circuit properties that are characteristic of the hippocampus: *long-term potentiation, convergent afferents, recurrent connections*. Notably, these properties exist in cortical areas as well as the hippocampus (e.g. Bear, 1996a), suggesting that the functional mechanisms described here may contribute to memory processing within widespread areas of the cerebral cortex as well as the hippocampus (e.g. Frankland et al., 2004; Maviel et al., 2004). The special role of these mechanisms within the hippocampus may be derived from the high diversity of convergent inputs, the exceptionally strong recurrency within hippocampal circuitry, and the rapidity of long-term potentiation (LTP) that is especially prominent within hippocampal circuitry.

As outlined in some detail in other chapters of this book and elsewhere, the hippocampus is noted for the rapid development and prevalence of a form of synaptic plasticity known as *long-term potentiation* (LTP) (Bliss & Collingridge, 1993). In particular, a form of LTP that is dependent on N-methyl-D-aspartate (NMDA) receptors has been strongly linked to memory (Martin, Grimwood & Morris, 2000) and to the memory-associated firing properties of hippocampal neurons (Shapiro & Eichenbaum, 1999). A central question, then, is how LTP is put to use in modifying hippocampal circuitry in order to form episodic memories (see Figure 2.2). I will suggest that LTP within specific elements of hippocampal circuitry may underlie each of the three cognitive processes introduced above.

First, principle cells of the hippocampus have as their main afferents considerable high-level perceptual information about all manner of attended stimuli and spatial cues initially filtered and integrated by multiple cortical processing streams. The hippocampus receives these *convergent afferents* from virtually all cortical association areas and these inputs are widely distributed on to the cell population in multiple subdivisions of the

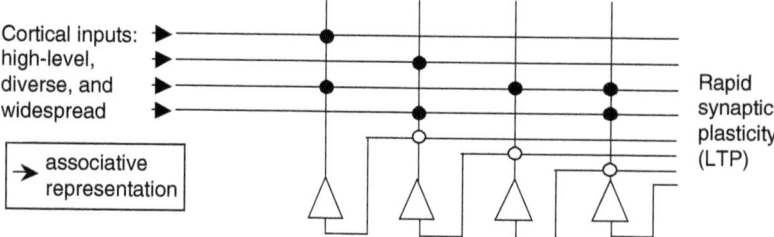

Figure 2.2. Schematic diagram of circuitry of the principal cells in hippocampal area CA3 which might mediate properties of relational networks. Two major pathways are shown: one pathway involving inputs from diverse neocortical association areas that send widespread projections on to the dendrites of pyramidal cells and the other pathway involving recurrent projections from pyramidal cells projecting sparsely on to other pyramidal cells in the same network. Connections of both of these pathways are subject to activity-dependent rapid synaptic plasticity. LTP, long-term potentiation.

hippocampus (Amaral & Witter, 1995). These convergent afferents are good candidates for supporting *associative representations* that bind a broad variety of inputs which constitute discrete events. It is well known that simultaneous activation of multiple afferents to CA3 principal cells leads to LTP of weakly as well as strongly activating inputs. This associative LTP supports pattern completion, so that subsequent presentation of any part of the representation would fire the cell, constituting retrieval of the entire association (for review, see Nakazawa et al., 2004). In addition, recurrent connections may support pattern completion by spreading the activation of some elements to all elements that compose a previously activated CA3 network (McNaughton & Morris, 1987; Treves & Rolls, 1994).

Second, several recent computational models have emphasized the *sequential organization* of memory representations supported by the hippocampus, and in particular area CA3. The principle neurons of area CA3 send considerable projections to other CA3 principle cells. These *recurrent connections* are broad across the CA3 population, sparse, and involve mainly excitatory glutamatergic synapses (Amaral & Witter, 1989, 1995; Treves & Rolls, 1994). Levy (1989, 1996) proposed that unique characteristics of hippocampal area CA3, specifically the sparseness and largely excitatory nature of its recurrent connectivity, combined with rapid synaptic plasticity, inherently produces asymmetric connections that can represent sequences of information from a single patterned input, and may spontaneously reproduce learned sequences. Thus, according to these models, when temporally patterned inputs reach the hippocampus, a rapid LTP mechanism enhances connections between cells that fire in sequence. The likelihood of a reciprocal (backward) connection of forward connected cells is very low owing to the overall sparse connectivity. Therefore the enhancement of recurrent connections is mostly unidirectional, leading to an asymmetry of the enhanced connectivity. When partial inputs are reproduced, the network is more likely to complete the sequence of the full initial input pattern. In addition, when sequences are repeated (practiced) just a few times cells representing neighboring sequential events that converge on background cells that enhance sustained activity, providing a context that bridges the firings of neurons representing salient sequential events (Levy, 1989, 1996; Wallenstein & Hasselmo, 1997; Wallenstein, Eichenbaum & Hasselmo, 1998).

Third, a simple and effective way to organize multiple the episodic memories is to encode common features of related experiences into the same representational elements. The same computational models that emphasize temporal organization in episodic memory representations provide a mechanism for linking distinct memory representations into a *relational network*. These networks include cells that receive no external inputs but develop firing patterns that are regularly associated with a particular sequence or with overlapping sequences (Levy, 1996; Sohal and Hasselmo, 1998; Wallenstein, Eichenbaum & Hasselmo, 1998; Hasselmo

& Eichenbaum, 2005). In the situation where episodes are repeated, these cells provide a local temporal context in which items within a particular sequence are linked together. When these links incorporate events that are unique to a particular episode, they can assist the network in disambiguating successive patterns in overlapping but distinct sequences. At the same time, when the links are activated similarly by separate episodes that share a series of overlapping features, they can allow the association of discontiguous episodes that share those features. Thus the same network properties that support encoding episodes as sequences of events also contain means to link and disambiguate related episodes (see Agster, Fortin & Eichenbaum, 2002 for an experimental test).

These considerations suggest that well-known properties of hippocampal circuitry and plasticity might support each of the elemental cognitive processes of episodic memory introduced above. The following sections consider the evidence from neuropsychologic and physiologic studies about each of these elemental cognitive processes and consider whether these circuit and plasticity properties may play a role.

ASSOCIATIVE REPRESENTATIONS

There is a large body of evidence that the hippocampus encodes associations among stimuli, actions, and places that compose discrete events. Many studies have shown that hippocampal neurons encode an animal's location within its environment, and some view this as the principle function of hippocampal populations (Muller et al., 1999; Best, White & Minai, 2001). In addition, however, many other studies have shown that hippocampal neurons also fire associated with the ongoing behavior and the context of events as well as the animal's location (Eichenbaum et al., 1999). In the most direct examination of this issue, Wood, Dudchenko and Eichenbaum (1999) directly compared spatial and non-spatial coding by hippocampal neurons by training animals to perform the same memory judgments at many locations in the environment. Rats performed a task in which they had to recognize any of nine olfactory cues placed in any of nine locations. Because the location of the discriminative stimuli was varied systematically, cellular activity related to the stimuli and behavior could be dissociated from that related to the animal's location. A large subset of hippocampal neurons fired only in association with a particular combination of the odor, the place where it was sampled, and the match–non-match status of the odor. In a remarkably similar study the coding properties of hippocampal neurons in humans, Ekstrom et al. (2003) recorded in subjects as they played a taxi driver game, searching for passengers picked up and dropped off at various locations in a virtual reality town. They observed that many of these cells fired selectively, associated with specific combinations of a place and the view of a particular scene or a particular goal.

Other recent studies highlight the associative coding of events and places by hippocampal neurons. In one study rats were trained on an auditory fear conditioning task (Moita et al., 2003). Prior to fear conditioning, few hippocampal cells were activated by an auditory stimulus. Following pairings of tone presentations and shocks, many cells fired briskly to the tone when the animal was in a particular place where the cell fired above baseline. Other studies have reported that hippocampal neuronal activity that encodes specific salient objects in the context of a particular environment as rats are engaged in foraging (Gothard et al., 1996; Rivard et al., 2004) and escape behavior (Hollup et al., 2001) in open fields. Another recent study examined the firing properties of hippocampal neurons in monkeys performing a task where they rapidly learned new scene-location associations (Wirth et al., 2003). As the monkeys acquired a new response to a location in the scene, neurons in the hippocampus changed their firing patterns to become selective to particular scenes. These scene-location associations persist even long after learning is completed (Yanike, Wirth & Suzuki, 2004). The combinations of studies described above indicate that, in rats, monkeys, and humans, a prevalent property of hippocampal firing patterns involves the representation of unique associations of stimuli, their significance, specific behaviors, and the places where these events occur.

Functional imaging studies on humans also support the notion that the hippocampus is activated during the encoding or retrieval of associations among many elements of a memory, a characteristic of context-rich episodic memories (for review see Cohen et al., 1999; Eldridge et al., 2000; Maguire, 2001; Addis et al., 2004). In studies that directly examined whether the hippocampus is activated when we remember items in their context, Henke et al., (1997) observed greater hippocampal activation when subjects associated a person with a house, as compared to making independent judgments about the person and house, and others have found selective hippocampal activation during recollection of the context of learning in formal tests of memory (see Davachi, Mitchell & Wagner, 2003; Ranganath et al., 2003). The coding of associations extends beyond item and context associations so that the hippocampus is also selectively activated during the encoding or retrieval of verbal (Davachi & Wagner, 2002; Giovanello, Schnyer & Verfaellie, 2003) and face–name associations (Small et al., 2001; Zeineh et al., 2003; Sperling et al., 2003).

In animals, there is also substantial evidence that selective hippocampal damage results in deficits in forming a memory for the context or location where items were once experienced (reviewed in Mumby, 2001). In one recent study, rats were initially exposed to two objects in particular places in one of two environmental chambers (Mumby et al., 2002). In subsequent recognition testing, the place of the object or the context was changed. Normal rats increased their exploration of objects that were moved to new places or put in novel contexts. By contrast, rats with hippocampal damage failed to recognize objects when either the place or context was changed (see also Eacott & Norman, 2004).

SEQUENTIAL ORGANIZATION

A property of episodic memory prominent in computational modeling involves the organization of an episode as a sequence of events that unfolds over time. Studies on animals show that the representation of memories by the hippocampus incorporates not only items that must be remembered, but also the events that precede and follow. For example, Honey, Eatt and Good (1998) provided a simple demonstration of the importance of temporal order in hippocampal processing, reporting that hippocampal lesions disrupted animals' normal orienting response when a pair of stimuli are presented in the opposite order of previous exposures.

We investigated the specific role of the hippocampus in remembering the order of a series of events in unique experiences by developing a behavioral protocol that assesses memory for episodes composed of a unique sequence of olfactory stimuli (Fortin, Agster & Eichenbaum, 2002; see also Kesner, Gilbert & Barua, 2002). Memory for the sequential order of odor events was directly compared with recognition of the odors in the list independent of memory for their order. On each trial rats were presented with a series of five odors, selected randomly from a large pool of common household scents. Memory for each series was subsequently probed using a choice test where the animal was reinforced for selecting the earlier of two of the odors that had appeared in the series. In later sessions we also tested whether the rats could identify the odors in the list independent of their order, by rewarding the selection of a novel odor against one that had appeared in the series. Normal rats performed both tasks well. Rats with hippocampal lesions could recognize items that had appeared in the series but were severely impaired in judging their sequential order.

How do hippocampal neuronal populations represent the sequences of events that compose distinct episodes? A common observation across many different behavioral protocols is that different hippocampal neurons become activated during every event that composes each experience, including during simple behaviors such as foraging for food (see Muller, Kubie & Ranck, 1987) as well as learning related behaviors directed at relevant stimuli that have to be remembered in studies that involve classical conditioning, discrimination learning, and non-matching or matching to sample tasks to tests and a variety of maze tasks (see Hampson, Heyser & Deadwyler, 1993; for review, see Eichenbaum et al., 1999). In each of these paradigms, animals are repeatedly presented with specific stimuli and rewards, and execute appropriate cognitive judgments and conditioned behaviors. Corresponding to each of these regular events, many hippocampal cells show time-locked activations associated with each sequential event. Also, as described above, many of these cells show striking specificities corresponding to particular combinations of stimuli, behaviors, and the spatial location of the event. Thus, hippocampal population activity may be characterized as a sequence of firings representing the step-by-step events in each behavioral episode.

Furthermore, these sequential codings may be envisioned to represent the series of events and their places that compose a meaningful episode, and the information contained in these representations both distinguishes and links related episodes. Recent studies on the spatial firing patterns of hippocampal neurons provide compelling data consistent with this characterization. In one study, rats were trained on the classic spatial alternation task in a modified T-maze (Wood *et al.*, 2000; see also Frank, Brown & Wilson, 2000; Ferbinteanu & Shapiro, 2003). Performance on this task requires that the animal distinguish left-turn and right-turn episodes and that it remember the immediately preceding episode to guide the choice on the current trial, and in that way, the task is similar in demands to those of episodic memory. If hippocampal neurons encode each sequential behavioral event and its locus within one type of episode, then most cells should fire only when the rat is performing within either the left-turn or the right-turn type of episode. This should be particularly evident when the rat is on the "stem" of the maze, when the rat traverses the same locations on both types of trials. Indeed, a large proportion of cells that fired when the rat was on the maze stem fired differentially on left-turn versus right-turn trials. The majority of cells showed strong selectivity, some firing at over ten times the rate on one trial type, suggesting they were part of the representations of only one type of episode. Other cells fired substantially on both trial types, potentially providing a link between left-turn and right-turn representations by the common places traversed on both trial types.

LINKING MEMORIES WITHIN RELATIONAL NETWORKS

The third elemental cognitive process proposed here involves the linking of episodic memories into relational networks in order to abstract the common features among related memories and to mediate flexible memory expression (Eichenbaum *et al.*, 1999).

Do hippocampal neurons extract the common features among related episodes? In virtually all the studies described above, a subset of hippocampal neurons encoded features that are common among different experiences – these representations could provide links between distinct memories. In the study by Moita and colleagues (2003) of auditory fear conditioning, whereas some cells only fired to a tone when the animal was in a particular place, others fired in association with the tone wherever it was presented across trials. In the study by Wood, Dudchenko and Eichenbaum (1999) on odor recognition memory, whereas some cells showed striking associative coding of odors, their match–non-match status, and places, other cells fired in association with one of those features across different trials. Some cells fired during a particular phase of the approach towards any stimulus cup. Others fired differentially as the rat sampled a particular odor, regardless of its location or match–non-match status. Other cells fired only when the rat sampled the odor at a particular place, regardless of the odor or its status. Yet other cells fired differentially associated with the match and non-match status of the odor, regardless of the odor or where it was sampled. Similarly, in the study by Ekstrom and colleagues (2003) on humans performing a virtual navigation task, whereas some hippocampal neurons fired associated with combinations of views, goals, and places, other cells fired when subjects viewed particular scenes, occupied particular locations, or had particular goals in finding passengers or locations for drop-off. Also, in the study by Rivard and colleagues (2004) of rats exploring objects in open fields, whereas some cells fired selectively associated with an object in one environment, others fired associated with the same object across environments. In studies that have recorded hippocampal neuronal activity as rats perform alternation tasks in a T-maze (Wood *et al.*, 2000; Frank, Brown & Wilson, 2000; Ferbintineau & Shapiro, 2003), whereas many cells distinguish overalapping actions and locations on the maze, some cells capture the common places and events between the different types of episodes.

The notion that these cells might reflect the linking of important features across experiences and the abstraction of common information was highlighted in recent studies on monkeys and humans. Hampson *et al.* (2004) trained monkeys on matching to sample problems, and then probed the nature of the representation of stimuli by recording from hippocampal cells when the animals were shown novel stimuli that shared features with the trained cues. They found many hippocampal neurons that encoded meaningful categories of stimulus features and appeared to employ these representations to recognize the same features across many situations. Kreiman, Kock and Fried (2000a) characterized hippocampal firing patterns in humans during presentations of a variety of visual stimuli. They reported a substantial number of hippocampal neurons that fired when the subject viewed specific categories of material, such as faces, famous people, animals, scenes, and houses, across many exemplars of each. A subsequent study showed that these neurons are activated when a subject simply imagines its optimal stimulus, supporting a role for hippocampal networks in recollection of specific memories (Krieman, Kock & Fried, 2000b). This combination of findings across species provides compelling evidence for the notion that some hippocampal cells represent common features among the various episodes that could serve to link memories obtained in separate experiences.

Studies of the effects of hippocampal damage have suggested how these representations of common events between distinct experiences are used in memory. In particular, some studies have focussed directly on the learning of multiple related problems and their integration into networks of memory that support flexible, inferential judgments. One study compared the ability of normal rats and rats with selective damage to the hippocampus on their ability to learn a set of odor problems and to interleave the representations of these problems in support of novel inferential judgments (Bunsey & Eichenbaum, 1996). Animals were initially trained on two sets of overlapping odor paired associates (e.g. A goes with B, B goes with C). Then the rats were given probe tests to determine if they could infer the relationships between items that were only indirectly associated through the common elements (A goes with C?). Normal rats learned the paired associates and showed strong transitivity in the probe tests. Rats with selective hippocampal lesions also learned the pairs over several trials but were severely impaired in the probes, showing no evidence of transitivity.

In another experiment, rats learned a hierarchical series of overlapping odor choice judgments (e.g. $A > B$, $B > C$, $C > D$, $D > E$) and were then probed on the relationship between indirectly related items ($B > D$?). Normal rats learned the series and showed robust transitive inference on the probe tests. Rats with hippocampal damage also learned each of the initial premises but failed to show transitivity (Dusek & Eichenbaum, 1997). The combined findings from these studies show that rats with hippocampal damage can learn even complex associations, such as those embodied in the odor paired-associates and conditional discriminations. But, without a hippocampus, they do not interleave the distinct experiences according to their overlapping elements to form a relational network that supports inferential and flexible expression of their memories (see also Buckmaster et al., 2004).

Complementary evidence on the role of the hippocampus in networking of memories comes from two recent studies indicating that the hippocampus is selectively activated when humans make inferential memory judgments. In one study, subjects initially learned to associate each of two faces with a house and, separately, learned to associate pairs of faces (Preston et al., 2004). Then, during brain scanning, the subjects were tested on their ability to judge whether two faces who were each associated with the same house were therefore indirectly associated with each other, and on whether they could remember trained face pairs. The hippocampus was selectively activated during performance of the inferential judgment about indirectly related faces as compared to during memory for trained face–house or face–face pairings. In the other study, subjects learned a series of choice judgments between pairs of visual patterns that contained overlapping elements, just as in the studies on rats and monkeys, and as a control they also learned a set of non-overlapping choice judgments (Heckers et al., 2004). The hippocampus was selectively activated during transitive judgments as compared to novel non-transitive judgments.

Combined, these findings suggest that the hippocampus contributes to the networking of distinct memories. The characterizations of neuronal firing patterns suggest that the hippocampus can abstract the common features that are shared across related experiences. Correspondingly, the results on the transitive inference paradigm indicate that the hippocampus plays a critical role in linking related memories according to their common features, and this binding results in a network that can support inferences between items in memory that are only indirectly related. Such findings provide a framework for understanding how a networking of memories can underlie the flexibility of expression in this memory system.

CONCLUSIONS: IS SYNAPTIC PLASTICITY THE BASIS FOR THE DISTINCT PROCESSING FUNCTIONS OF HIPPOCAMPAL CIRCUITRY?

Several studies have suggested that NMDA receptor-dependent plasticity is critical to each of the binding operations that support the above described cognitive processes. With regard to the encoding of events as items in their context, Day, Langston and Morris (2003) reported that NMDA receptor-dependent synaptic plasticity plays a critical role in the association of items and the place where they are experienced in episodic memory. They designed a task in which rats were initially allowed to find different-flavored rewards at specific locations in an open platform, and then tested their memory for the location of those events by providing an additional flavored reward associated with one of the locations. After a single exposure to the flavor in a particular location, animals could return to the location where a flavor had previously been consumed when cued by the flavor alone. By contrast, inactivation of the hippocampus or blockade of NMDA receptors prevented encoding of the flavor–place association.

Other recent studies have supported this observation and suggested that the representation of items in their context may be localized within hippocampal circuitry. Selective lesions of CA3, and not CA1, retarded the composition of a contextual representation (Lee & Kesner, 2004) and lesions of CA3, and not CA1, interfered with odor–place and object–place association (Gilbert & Kesner, 2003). Also, in a study that attempted to localize the role of synaptic plasticity in contextual processing within hippocampal circuitry, Lee and Kesner (2002) reported that blockade of NMDA receptors in CA3, and not CA1, impaired learning of a novel spatial organization. These data are fragmentary, but suggest a distinction in the role of NMDA receptors in associative representation within CA3 versus CA1 (Kesner, Gilbert & Wallenstein, 2000). Conversely, NMDA receptors, and particularly those in CA1, may play a critical role in the sequential organization of memories. Consistent with this notion, CA1, and not CA3, is critical to memory performance that requires bridging periods of time in each trial episode, including memory for the order of odors or places visited, temporal pattern separation, and trace odor–place memory (Kesner, Lee & Gilbert, 2004; Kesner, Hunsaker & Gilbert, 2005), and pharmacological blockade of NMDA receptors in CA1, but not CA3, impaired maintenance of memories over short periods (Lee & Kesner, 2002). Furthermore, in mice lacking NMDA-dependent LTP selectively in CA1, the temporal coordination of CA1 neurons, reflected in the extent to which CA1 ensemble activity reflects sequential movements through a maze, is severely diminished (McHugh *et al.*, 1996). That the NMDA-knockout is localized to CA1 in these animals, suggests the temporal coordination arises in CA1 itself. These observations have led us to speculate that NMDA receptors in CA1 may play a differential role in memory for the order of events in unique experiences in contrast to the role of NMDA receptors of CA3 in concurrent associations.

Studies on hippocampal synaptic plasticity have not yet been applied to an examination of hippocampal function in relational networking. It is not clear whether this relational networking is accomplished within CA3 processing that associates items in their unique contexts, or CA1 processing that links events within unique episodes, or some combination of these putative processing stages. Yet this question is eminently tractable with available techniques.

Here I have outlined a framework for thinking about the role of synaptic plasticity in the functional circuitry of the hippocampus. I have suggested that the mechanisms of synaptic plasticity provide means of binding information within hippocampal ensembles in distinct ways that support three elemental forms of information-processing that underlie episodic memory. Although the framework suggested here is speculative, such an integration of the continuity of levels of structure–function relations provides the best possibility of a full understanding of how memory works.

3 · Widening the lens: looking beyond the synapse for experience-driven brain plasticity

Julie A. Markham, Aaron W. Grossman, and William T. Greenough

INTRODUCTION

For centuries, scientists have studied the inheritance of human physical traits, focussing more recently on modern molecular genetics. The inheritance pattern of blood groups ABO, for example, was one of the early physical traits to be described (Landsteiner, 1900). It soon became apparent that blood type was determined by multiple alleles at a single gene locus, and more recent studies have revealed polymorphisms associated with each blood type (Yamamoto et al., 1990). We now understand that the molecular basis of this physical trait's inheritance pattern is genetically determined, and occurs largely independently of external influence.

The inheritance of human behavioral traits, however, is much more complex. For instance, in monozygotic (or identical) twins, the concordance rate for schizophrenia is 50%, and it is 17% in dizygotic (or fraternal) twins, indicating the clear contribution of both genetic and environmental factors (reviewed in Tsuang, 2000). Similarly, a polymorphism in the serotonin transporter gene modulates the onset of major depression in response to stressful life events (Caspi et al., 2003). It has been suggested that interaction between an individual's genetic makeup and his or her environment may account for individual differences not only in susceptibility to psychopathology, but also in things such as personality characteristics and neural and behavioral responses to the normal aging process (see Grossman et al., 2003; Caspi, Roberts & Shiner, 2005; Ryff and Singer, 2005).

This dynamic interplay between the environment (nurture) and brain biology (nature) is being called upon to further the understanding of complex psychopathologic disorders that are characterized by abnormal brain morphology or functional activation, or both; even those that are known to have a strong genetic component. The temptation to understand behavior in terms of either nature or nurture has therefore become considerably less compelling in recent decades, in part because of the recognition that experience induces measurable morphologic changes in cells of the brain – both neurons and glia – and in their interactions with one another.

Alterations in gene expression, neurochemistry, and physiological properties of these same cells are often found to accompany environment-induced reorganization of brain connectivity, as are measurable adjustments in behavior that are adaptive in the context of the new environment.

Elucidation of experience-induced alterations in brain morphology is facilitated by conducting studies in a laboratory setting where environmental conditions may be systematically varied. The complex environment paradigm, pioneered by Donald Hebb and his students (Hebb, 1949; Forgays and Forgays, 1952; Hymovitch, 1952), involves housing a group of animals, typically rats or mice, together in a large cage containing numerous toys, such as balls, tunnels, and ladders, which are changed daily to provide a continuously stimulating environment. This complex environment is sometimes referred to as an "enriched condition," or "EC," although it is important to note that these animals are only enriched relative to rats housed in the laboratory in isolation (in an "impoverished condition;" IC) or to rats housed in a group but without toys ("social condition;" SC), and not when compared with animals living in the wild. As Hebb's work first demonstrated, animals exposed to EC have long been known to demonstrate superior performance on a variety of tests measuring higher-order cognitive ability, such as the Hebb–Williams maze (Mohammed, Jonsson & Archer, 1986; Galani et al., 1997), the Morris water maze (Whishaw, Zaborowski & Kolb, 1984; Mohammed et al., 1990), and the radial arm maze (Galani, Courtureau & Kelche, 1998).

The first use of the EC paradigm as a tool to study experience-induced morphological plasticity in the brain was by Bennett et al. (1964), who reported that rats raised in EC possessed heavier and thicker cerebral cortices than did IC rats. Because the differences were largest in the visual cortex, much of the subsequent work was aimed at determining the underlying morphology contributing to the gross size difference in this region (see Diamond, Krech & Rosenzweig, 1964; Diamond et al., 1966; Volkmar and Greenough, 1972). Although not a ubiquitous phenomenon, morphological plasticity after exposure to EC has now been demonstrated in a number of other brain regions involved in the processing of and/or response to environmental stimuli, including the auditory cortex (Greenough, Volkmar & Juraska, 1973), primary somatosensory cortex (Coq and Xerri, 1998), hippocampus and entorhinal cortex (Fiala, Joyce & Greenough, 1978; Moser et al., 1997; Rampon et al., 2000), amygdala (Nikolaev et al., 2002), basal ganglia (Comery, Shah & Greenough, 1995; Comery et al., 1996), and cerebellar cortex (Greenough et al., 1986). Initially, alterations in neuronal structure were the focus of investigation; however, more recently it has become clear that other nervous system components, such as macroglial cells and cerebrovasculature, also exhibit robust plasticity in response to experience.

Morphological plasticity in the brain that occurs in response to an increase in environmental complexity appears to reflect adaptations of brain substrates to the demands and opportunities provided by the new experience. These opportunities may include relatively typical forms of learning and memory as well as behavioral adjustments associated with fundamental processes such as sensory, motor, and cognitive processing. Some important points concerning how this conclusion has been reached deserve mention: it was suggested, for example (following the initial reports of increased cortical thickness in EC rats), that the observed plasticity might potentially be attributed – instead of to learning-dependent processes – to changes in overall body growth, to increased social contact, to differences in stress that might occur as a result of differential housing, or to other hormonal or metabolic responses to behavioral manipulations. Our laboratory and others have addressed the potential confounds of stress hormones or other nonspecific hormonal or metabolic changes accompanied by EC exposure by employing paradigms in which the effects of learning are expected to be restricted to specific brain regions, and to compare plasticity in these regions to other "control" regions. This approach has shown plastic anatomical effects of training to be concentrated on the side of the brain to which training is delivered compared to the opposite "untrained hemisphere" (Chang and Greenough, 1982; Greenough, Larson & Withers, 1985) and to distinguish between activity-induced and learning-induced plasticity (Black et al., 1990; Anderson, Alcantara & Greenough, 1996; Kleim et al., 1996; Kleim et al., 1998; and discussed below). Furthermore, specificity of plasticity is even seen in subpopulations of neurons and synapses within the same brain region (Withers and Greenough, 1989; Kleim et al., 1998); for example, EC causes dendritic hypertrophy of cerebellar Purkinje cells but not granule cells (Floeter and Greenough, 1979).

Because: (1) both social and nonspecific hormonal or metabolic factors may be ruled out; (2) direct interaction with the environment and not just visual exposure is required to induce morphological plasticity in the brain (Ferchmin and Bennett, 1975); and (3) the brain undergoes similar remodeling in response to both training paradigms and EC exposure, occurring specifically in regions associated with the processing of environmental or task-related information, learning and memory remain the most plausible if not the only plausible explanation of brain plasticity observed after complex experience. For more comprehensive reviews on the specificity of brain plasticity, the reader is referred to Grossman et al. (2002) and to Markham and Greenough (2004).

NEURONAL PLASTICITY

Synapse number

Synaptogenesis in response to exposure to a complex environment has been demonstrated many times. Animals raised in EC have greater dendritic arborization, increased dendritic spine density, and more synapses per neuron in a number of brain areas as compared with IC animals (reviewed by Greenough and Chang, 1988). For instance, both dendritic branching and the number of synapses per neuron in the visual cortex are greater in rats raised in EC as compared to either IC or SC rats, which were equivalent on these measures (Volkmar & Greenough, 1972; Turner & Greenough, 1985; Sirevaag & Greenough, 1987). Interestingly, the magnitude of these two effects is similar (20–25% increase in EC), which may indicate that synaptogenesis in the visual cortex in response to visual experience is the result of elongation or elaboration of dendritic branches, or both, upon which new synapses are formed, rather than merely an increase in the packing density of synapses along an existing length of dendrite. Dendritic elaboration as a result of EC also occurs in other neocortical areas (see Greenough, Volkmar & Juraska, 1973; Kolb et al., 2003) and in the dentate gyrus and area CA3 of the hippocampus, although interestingly the direction of the changes in dendritic arbor there has been found to vary by sex (Juraska et al., 1985; Juraska, Fitch & Washburne, 1989). Increased dendritic length may be detected after as few as four days in EC in the visual cortex (Wallace et al., 1992), and contributes to the greater thickness of the visual cortex among EC animals that was initially reported (Bennett et al., 1964). Increases in dendritic spine density in response to EC exposure have also been reported, though the degree to which spine density increases, if at all, has been variable (see Globus et al., 1973; Comery, Sha & Greenough, 1995; Rampon et al., 2000; Kolb et al., 2003). The experience-dependent addition of synapses is most pronounced in, but certainly not limited to, developing animals. Briones, Klintsova and Greenough (2004) have recently observed that rats exposed *as adults* to EC for either 30 or 60 days had significantly (and equivalently) more synapses per neuron in layer IV of the visual cortex than did IC animals of the same age, as revealed by electron microscopy. Increases in synapse number and dendritic branching of neurons have also been demonstrated to occur in response to EC in aging rats (Green, Greenough & Schlumpf, 1983; Greenough et al., 1986).

One question within the field of brain plasticity that has warranted further investigation is whether continued experience and brain activity are required for the *maintenance* of experience-incremented neuronal connectivity. The answer appears to depend on the age of the animal and whether the experience is enhanced or reduced, relative to the animal's normal sensory input. Briones, Klintsova and Greenough (2004), for example, enhanced the housing experience of adult rats by exposing them to EC for either 30 or 60 days, as described above. This experiment also included another important experimental group: some animals that had experienced EC for 30 days were then placed in individual housing (IC conditions) for a subsequent 30 days. These animals exhibited increased synapse

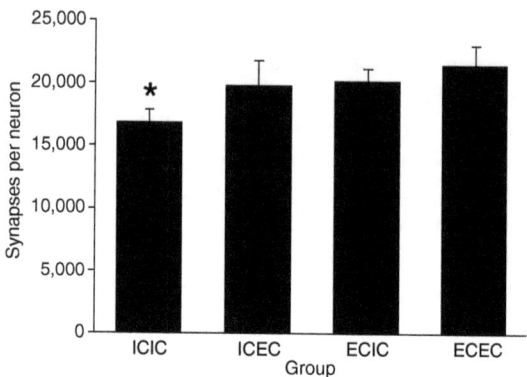

Figure 3.1. The EC-induced increase in the number of synapses per neuron in the adult rat visual cortex persists for at least 30 days after animals are removed from a complex environment. ICIC animals were individually caged (IC) for 60 days and were significantly different (*, $p < 0.05$) from each of the three other groups: ICEC animals (housed in IC for 30 days followed by EC housing for 30 days); ECIC animals (housed in EC for 30 days followed by IC housing for 30 days); ECEC animals (housed in EC for 60 days). Modified from Briones, Klintsova and Greenough (2004), with permission.

number to a statistically equivalent degree as those housed in EC for the full 60 days, reflecting the persistence of EC-induced synaptic plasticity (Figure 3.1) (Briones, Klintsova & Greenough, 2004). Similarly, animals raised in EC for 30 days beginning at weaning, followed by 30 days of IC housing conditions, had comparably greater dendritic arborization of visual cortical neurons (both stellate neurons of layer IV and pyramidal neurons of layer III) when compared to IC animals, as did animals that were examined immediately after 30 days of EC housing (Camel, Withers & Greenough, 1986). Together, these studies suggest that once an animal's sensory input has been enhanced the additional neuronal connectivity becomes a relatively stable component of the brain's new "wiring diagram."

Other researchers have studied the role of continued experience in brain plasticity by removing normal sensory input from the rodent somatosensory cortex, and have found that in younger animals this input is important for the normal development of this brain region, and in older animals is important for the maintenance of normal neuronal morphology. Trachtenberg et al. (2002), for example, repeatedly imaged dendritic spines on barrel cortex pyramidal neurons *in vivo* following sensory deprivation (whisker trimming). They found an increase in the proportion of spines that were transient (present for a single day or less) and a decreased proportion of spines observed to be stable over several days, suggesting that spine turnover is heavily influenced by sensory experience (Trachtenberg et al., 2002). Zuo et al. (2005) used similar repeated imaging techniques to study the effect of whisker trimming on developmental spine elimination from dendrites in layer I of mouse somatosensory cortex, reporting that in adolescent mice the rate of spine elimination was reduced, and that this effect was less pronounced in more mature animals. Another group recently studied the effects of sensory deprivation on the highly ordered dendritic orientation of neurons in layers III and IV of the barrel cortex of adult rats (Tailby et al., 2005). These authors found that in the absence of sensory input, dendrites on these cells (which normally extend into the center of the barrel) lost their orientation bias, and became directed away from the barrel center (Tailby et al., 2005). Clearly, the precise role of activity and experience in the maintenance or persistence of neural structure has yet to be determined and is heavily dependent on the age of the animal. It appears, however, from these studies in which sensory input was removed that some input threshold must reliably be met for normal synaptic plasticity to continue, just as it appears that continued enhancement of experience is *not* required for added synapses, once integrated into the "wiring diagram," to be maintained.

Training animals on complex motor tasks results in increases in synapse number similar to those observed after EC and suggests that the observed effects of EC on this measure are a result of the learning process and not merely increases in general activity levels. In a study designed to tease apart morphological changes associated with learning from those associated with general physical activity, a group of adult female rats that had been trained on a motor skill-learning task (using a challenging "acrobatic" course) were compared with animals allowed to exercise freely (on a treadmill) but with minimal opportunity for learning. The number of synapses per neuron in both motor and cerebellar cortices was greater in rats trained on the acrobat course when compared both to rats that exercised but did not engage in learning and to rats who were sedentary during the course of the experiment (Black et al., 1990; Kleim et al., 1996). These learning-induced changes in synapse number have been found to persist for at least four weeks after training has finished (Figure 3.2 (A))

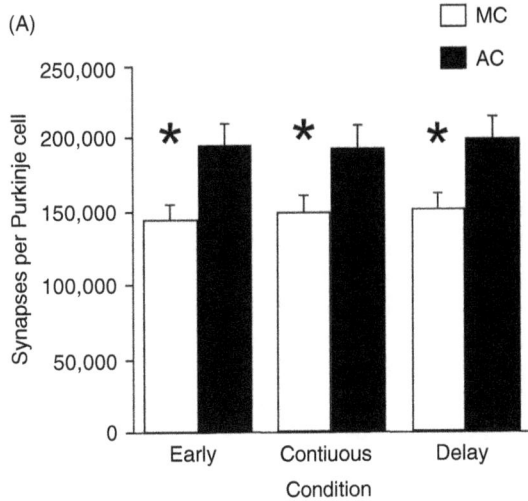

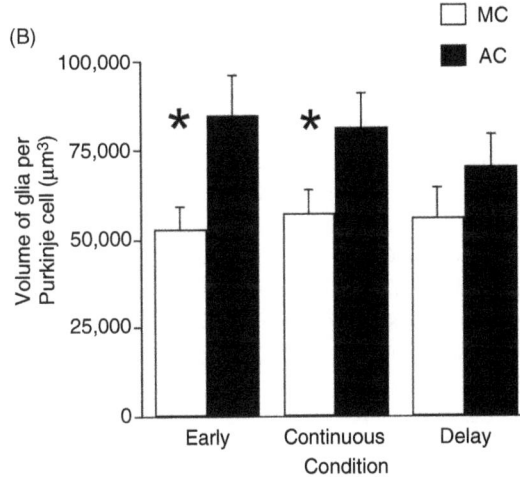

Figure 3.2. The increase in synapses per neuron in the motor cortex persists in the absence of continued training (A), whereas the glial response to motor skill learning appears to be less stable (B). AC (acrobat) rats were trained on a motor skill learning task, whereas MC (motor control) animals ran on a treadmill but were not given an opportunity for learning. Animals in the Early group participated in training (AC) or exercised (MC) for 10 days, animals in the Continuous group participated for 38 days, and animals in the Delay group participated for 10 days and then training (or exercise) was discontinued for the following 28 days before histological examination. (*, $p < 0.05$ for the comparison between the MC and AC animals of a particular group (Early, Continuous, or Delay).) Modified from Kleim et al. (1997), with permission.

(Kleim et al., 1997). Thus it appears that learning, and not merely neural activity, is required to induce synaptogenesis.

Using a skilled reaching task, Kleim and colleagues have continued to study the structural and functional correlates of motor learning. After training rats to reach food pellets through a thin slot in their cage using only their forelimbs, the authors found that the area of motor cortex controlling forelimb movements had expanded, and that in this region but not adjacent regions, the number of synapses per neuron had increased (Kleim et al.,1998; Kleim et al., 2002). Upon further testing they found that the addition of synapses precedes expansion of these motor maps, and that both occur during the late phase of training (Kleim et al., 2004). Although more research is needed to fully evaluate this hypothesis, the authors suggest that synaptogenesis and its functional correlates may play a role in the consolidation of motor learning rather than in the acquisition of motor skills.

Synapse morphology

In addition to inducing the formation of new synapses, experience may also modify the morphology of existing synapses, or, alternatively, induce the formation and/or loss of synapses exhibiting particular characteristics (reviewed by Greenough and Chang, 1988). Animals exposed to EC have larger synaptic components, both pre- and post-synaptic. For instance, the average size of the post-synaptic density (PSD), which is proportional to dendritic spine volume (Sorra & Harris, 2000), is increased by approximately 5–8% in the visual cortex after exposure of rats to EC for one month beginning around weaning (West & Greenough, 1972; Diamond et al., 1975; Sirevaag & Greenough, 1985; Turner & Greenough, 1985). In synaptic boutons, the cross-sectional area of vesicle aggregate profiles in layer IV of the visual cortex has been found to be approximately one-third greater in EC compared to IC rats (SC rats were intermediate) (Sirevaag & Greenough, 1987). The size of the synapse is thought to be related to its strength, and in support of this notion it has been found that in monocularly deprived cats, synapses innervated by fibers carrying information from the non-deprived eye are larger, both pre- and post-synaptically, than synapses that are innervated by fibers carrying information from the deprived eye (Tieman, 1984, 1985).

Not only the size but also the shape of dendritic spines, which are the primary location of excitatory synaptic input on to principal neurons in the neocortex and hippocampus, is important for their function, as shape is known to influence a spine's conductive properties (Sorra & Harris, 2000; Tsay & Yuste, 2004) and to affect biochemical compartmentalization (Noguchi et al., 2005). Reflective of the spine's relative maturational state, spine shape changes in similar ways (from the initial sessile shape, to exhibiting a clearly discernible head or neck, and finally to the large mushroom shape with a mature spine apparatus) over development (Galofre & Ferrer, 1987) and in response to EC (Sirevaag & Greenough, 1985) and long-term potentiation (LTP) (Chang & Greenough, 1984). For example, Comery et al. (1996) reported 60% greater density of multiple-headed dendritic spines on spiny neurons in the striatum of EC as compared to IC rats. On the pre-synaptic side, boutons in animals exposed to EC are more concave than those in IC rats, with SC rats being intermediate (Wesa et al., 1982). Additionally, Tieman (1985) showed that in monocularly deprived cats, synapses associated with the deprived eye are more pre-synaptically convex than those associated with the open eye. Finally, Dyson and Jones (1980) reported that synaptic contacts in the rat visual cortex become increasingly less convex with age. Recent advances in imaging techniques have improved visualization of dendritic spines, making it possible to assess parameters such as spine shape and relative spine volume repeatedly for the same spine during the induction of plasticity. Matsuzaki et al. (2004), for example, selectively stimulated individual spines and observed rapid spine enlargement, noting that this enlargement was transient in larger spines, was more persistent in smaller spines, and was associated with enhancement of AMPA-receptor-mediated currents. Therefore it appears that the size and shape of both pre- and post-synaptic components are indicative of the activation and maturational state of a synapse.

Recently it has become clear that other aspects of synaptic morphology are also sensitive to experience. For instance, perforated synapses (those whose PSD has enlarged and assumed a more complex shape such as a horseshoe or doughnut, resulting in apparent discontinuities in single-electron micrographs) are associated with synaptic plasticity in part because they are known to increase in the visual cortex across development and in response to EC (Greenough, West & DeVoogd, 1978; Jones & Calverley, 1991), in the motor cortex after training on a motor skill learning task (Jones, 1999), and in the hippocampus in response to kindling or LTP induction (Geinisman, Morrell & de Toledo-Morrell, 1990; Geinisman, de Toledo-Morrell & Morrell, 1991). Additionally, post-synaptic expression of AMPA receptors, the number of which is considered to be the major determinant of synaptic efficacy, was found to be a ubiquitous characteristic of perforated synapses in the hippocampus using immunogold electron microscopy (Ganeshina et al., 2004). In contrast, only a fraction (64%) of the non-perforated synapses examined expressed these receptors (Ganeshina et al., 2004). Experience also induces the formation of multiple synaptic boutons (MSBs; two post-synaptic contacts innervated by the same pre-synaptic varicosity) – specifically, it has been found that animals trained on a motor skill task had more MSBs in the cerebellum than animals that exercised without the opportunity for learning and than animals who were sedentary during the course of the experiment (Federmeier, Kleim & Greenough, 2002). Similarly, the number of MSBs per neuron that contacted both a dendritic spine and a dendritic shaft were greatly increased in layer IV of the visual cortex of rats exposed to EC for 60 days compared to either SC or IC controls (Jones et al., 1997). From these examples it is clear that the formation of novel dendritic contacts on to existing axonal boutons or varicosities is a common form of experience-driven synaptic plasticity, one that would seem to alter the efficacy of a pre-existing pathway rather than creating novel connections.

Although the steps leading to the formation of perforated synapses and MSBs, and their ultimate function, is less clear, some interesting hypotheses have been proposed. Carlin and Siekevitz (1983) advanced the model of the dividing synapse to account for the induction of perforated synapses – initially, the synaptic junction was thought to enlarge, develop a perforation, then split into two separate synaptic junctions within a single synaptic terminal, and finally, the spine itself would divide into two spines, each containing one synaptic junction. However, when Kristen Harris' group carefully examined synapses in CA1 of hippocampus using unbiased stereological techniques as applied to electron microscopy, they failed to observe even a single branched ("multiple-headed" spine) with the different "heads" in synaptic contact with the same pre-synaptic bouton (Sorra, Fiala & Harris, 1998), an intermediate stage that is predicted by the splitting hypothesis. Subsequently, they directly examined the issue of spine splitting by serially reconstructing synapses on hippocampal dendrites, and the surrounding neuropil, across development and in response to hippocampal LTP (Fiala, Allwardt & Harris, 2002). Their results indicate that the post-synaptic components of MSBs actually commonly arise from separate dendritic processes. When two post-synaptic components from the same dendrite contact the same pre-synaptic bouton, they do not appear to have derived originally from the same dendritic spine, because other stable structures – namely, mature axons – are found to pass between them. This topic is still hotly debated and represents an exciting avenue for future research in the neuro-anatomical correlates of plasticity.

Neurogenesis

Most neurons in the brain proliferate during gestation, and until recently, the notion that neurogenesis does not occur in the adult mammalian brain (outside of the olfactory bulb) was part of neuroscience dogma. Although there were earlier indications to the contrary (Altman, 1962, 1963; Kaplan, 1981), these were largely ignored until several key studies were published within the last decade. These studies confirmed the phenomenon of adult neurogenesis in the dentate gyrus of both rodents and primates; the question of whether neurogenesis occurs in the adult primate neocortex remains controversial (Eriksson et al., 1998; Gould et al., 1999a; Kornack & Rakic, 1999; Gould et al., 2001).

Although the number of neurons added to the adult brain is small in comparison both to total neuron number and to glial cell genesis, several environmental factors have been shown to influence this process. In general, stress – experienced both during development and during adulthood – and stress hormones, alcohol exposure, and the aging process decrease the number of new neurons added to the adult brain, whereas antidepressants, estrogen, exercise, and EC all increase it (Gould et al., 1997; Cameron, Tanapat & Gould, 1998; Kempermann, Kuhn & Gage, 1998; Tanapat et al., 1999; van Praag, Kempermann & Gage, 1999; Malberg et al., 2000; Brown et al., 2003; Mirescu, Peters & Gould, 2004). The mechanisms by which voluntary exercise and environmental complexity result in greater numbers of new neurons added to the dentate gyrus of the adult rodent appear to be different: exercise increases the rate of neurogenesis whereas EC exposure promotes the survival of new neurons (Kempermann, Kuhn & Gage, 1998; van Praag, Kempermann & Gage, 1999).

Recently, our laboratory has begun to investigate the interacting influences of voluntary exercise and aging on adult neurogenesis for the first time in a primate model. In collaboration with Judy Cameron's group at the University of Pittsburgh, young adult (aged 10–12 years) and mature adult (aged 15–17 years) female *Macaca fascicularis* monkeys were assigned to one of three treatment groups: *runners*, who ran on a treadmill for one hour a day, five days a week, for 24 weeks; *sedentary controls*, who sat on an immobile treadmill for an equivalent amount of time for 24 weeks; or *run-stops*, who exercised for 24 weeks exactly as the runners did, but then subsequently sat on an immobile treadmill for the allotted time for an additional 12 weeks. Preliminary results confirm that significant neurogenesis occurs in the dentate gyrus of the adult monkey. Additionally, our findings to date suggest that neurogenesis is increased in this area in young adult monkeys in response to exercise, but that the ability of exercise to increase neurogenesis in this region may be reduced with age.

As researchers explore potential functions of adult-generated hippocampal neurons, brain-derived neurotrophic factor (BDNF, which is known to be critical for use-dependent synaptic plasticity) is emerging as a common link between many of the factors known to impact adult neurogenesis. BDNF levels are increased by exercise, estrogen, and antidepressants (factors that increase neurogenesis in the adult dentate gyrus) and decreased by stress and aging (factors known to reduce neurogenesis) (reviewed by Cotman & Berchtold, 2002). Stress is a risk-factor for depression, and in animal models for depression, behavioral stress is often used to induce symptoms. Both exercise and antidepressants are capable of relieving behavioral correlates of depression, and BDNF is capable of alleviating behavioral symptoms in animal models of depression (Cotman & Berchtold, 2002) (e.g. Siuciak *et al.*, 1997; Shirayama *et al.*, 2002). Furthermore, antidepressant administration has been shown to block or attenuate both the stress-induced decrease of BDNF levels and neurogenesis in the dentate gyrus (Nibuya, Morinobu & Duman, 1995; Xu *et al.*, 2004). Interestingly, the time-course necessary for antidepressant administration to influence hippocampal neurogenesis is similar to that required for therapeutic benefit (i.e. chronic rather than acute) (Nibuya, Morinobu & Duman, 1995; Russo-Neustadt *et al.*, 2000). Finally, inhibiting hippocampal neurogenesis blocks the behavioral effects of antidepressant drug administration (Santarelli *et al.*, 2003). Thus regulation of BDNF levels and hippocampal neurogenesis may play a role in the behavioral effects of mood-stabilizing factors, such as antidepressants and exercise, in the adult brain.

Klintsova *et al.* (2004), studying the rat cerebellum and motor cortex, recently reported increased levels of BDNF and its receptor TrkB following complex motor skill training, though expression levels depended on the brain region and the duration of training. Increases in BDNF levels have also been found in the dentate gyrus and cerebral cortex of rats housed in EC (see Ickes *et al.*, 2000). Because EC rats exhibit superior learning and memory ability compared to IC rats, and because exercise and EC-exposure during adulthood increase neurogenesis in the dentate gyrus but not in the olfactory bulb (Brown *et al.*, 2003), a learning-specific role for neurons added to the adult brain has also been proposed. It has been shown that training on associative learning tasks that require the hippocampal formation, but not training on hippocampal-independent tasks, increases the number of new neurons in the dentate gyrus (Gould *et al.*, 1999b). Since exposure to EC is known to improve animals' performance on tests of learning and memory, it seems likely that EC-generated hippocampal neurons participate in the improved memory performance. Earlier this year, Rampon's group confirmed the benefit conferred on both memory performance and hippocampal neurogenesis by EC exposure, and furthermore reported that blocking adult neurogenesis (using the antimitotic agent methylazoxymethanol acetate (MAM)) abolished the EC-induced improvement in hippocampal-dependent memory (Bruel-Jungerman, Laroche & Rampon, 2005). Because EC appears to promote survival more than proliferation of new neurons, it may be the neurons born prior to the learning experience, and not those generated by the learning experience itself which are critical for memory performance. Mild irradiation, which inhibits adult neurogenesis, disrupts performance on the spatial (hippocampal-dependent) version of the Morris water maze (but was without effect on performance of the hippocampal-independent, visible platform version of the maze) when administered 4–28 days prior to maze training, but not when administered just prior or immediately after maze training (Snyder et al., 2005). This finding is perhaps not surprising in light of the fact that the brain must rely on past experiences to predict future ones. Thus cells may be added to the adult hippocampus in anticipation of their need to mediate the acquisition, storage, and/or consolidation of future memories.

PLASTICITY OF ASTROCYTES

Although the focus on experience-driven plasticity in the brain has traditionally been on altered morphology of neurons and synapses in particular, astrocytic glia also show robust changes in response to experience. Exposure to EC, which was originally designed to investigate the relationship between neuronal and behavioral changes, in fact was demonstrated in some early studies to induce changes in astrocytic morphology (Diamond, Krech & Rosenzweig, 1964; Szeligo and Leblond, 1977). However, limitations in quantification techniques available at the time resulted in inconsistent findings that made interpretation difficult. With the advent of improved quantification methods, in general referred to as unbiased stereological methodology, EC-induced increases in astrocytic cell size (hypertrophy) and number (hyperplasia) have been confirmed (Sirevaag & Greenough, 1987, 1991).

The hypertrophy of astrocytic processes in response to EC appears to be dependent both on duration of EC exposure and on the cortical layer (reviewed in Jones, 2002). In general, morphological plasticity of astrocytes in response to EC occurs on a timescale that is comparable to neuronal changes observed in this paradigm. For instance, just four days of EC housing during adolescence increased the surface density of glial fibrillary acidic protein-immunoreactive (GFAP-IR) astrocytic processes in layers II/III of the rat visual cortex (Jones, Hawrylak & Greenough, 1996), an exposure duration that also induced detectable dendritic growth in this same layer (Wallace et al., 1992). Additionally, after 30 days of differential housing, the astrocytic volume per neuron in the visual cortex was increased in EC rats by an amount comparable to the previously established increase in synapse number (Sirevaag & Greenough, 1985; Jones & Greenough, 1996). This may indicate that astrocytic hypertrophy in the visual cortex of the EC rat is driven by synapse formation, as is the case in the cerebellar cortex following motor skill learning (Anderson et al., 1994). Because exercise without skill learning does not induce synaptogenesis or astrocytic hypertrophy, and because astrocytic and synaptic changes in the cerebellar cortex are correlated on an animal-by-animal basis, increased astrocytic volume may be inferred to arise in association with learning-specific synaptogenesis and not merely to constitute a response to a general increase in neural activity (Anderson et al., 1994).

It is not just the morphology of astrocytes that is altered by experience; the relationship between astrocytes and neurons is also refined. In the neocortex, astrocytes cover pre- and post-synaptic elements of axodendritic synapses, but typically only partial covering is observed. The degree of synaptic ensheathement by fine astrocytic processes, as observed by electron microscopy, is increased in the visual cortex of EC rats (Figure 3.3) (Jones & Greenough, 1996), indicating that the function of demonstrated alterations in astrocytic processes in response to EC is related in some way to enhancing synaptic function. Clearly, there is an experience-dependent enhancement of astrocytic–synaptic communication, an important finding in light of the fact that peri-synaptic astrocytes modulate synaptic transmission in response to synaptically released neurotransmitters and in fact release neurotransmitter themselves (Oliet, Piet & Poulain, 2001; Zhang & Haydon, 2005). Additionally, astrocytes are involved in GABA and glutamate re-uptake and metabolism (Schousboe et al., 1992; Bezzi et al., 1999), and can conduct excitation via propagated Ca^{2+} waves which can directly influence neuronal activity (reviewed by Zhang & Haydon, 2005). In CA1 of the hippocampus, where nearly 60% of the synapses are directly contacted by astrocytic processes (Ventura & Harris, 1999), post-synaptic glutamate receptors are activated by glutamate spillover from neighboring terminals (in addition to glutamate released from the pre-synaptic terminal) (see Kullmann et al., 1999). In the cerebellum, Bergmann glial cells have appendages that ensheathe individual parallel fiber–Purkinje cell synapses. Many of these "glial microdomains" exhibit transient, isolated increases in Ca^{2+} concentrations in response to parallel

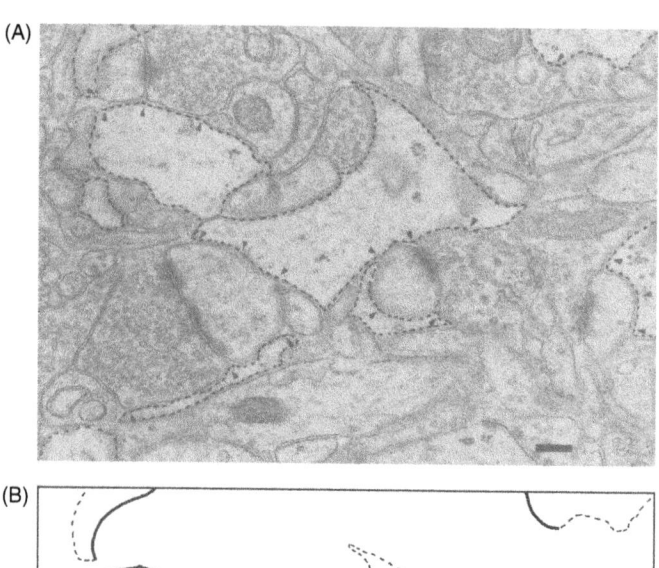

Figure 3.3. Astrocytic processes (dashed outline) within layer IV of the rat visual cortex as revealed by electron microscopy (A) and a tracing of these processes (B). Processes in direct apposition to synaptic elements are indicated (arrows, A; solid lines, B). Scale bar, 0.2 μm. Reprinted with permission from Jones and Greenough (1996).

fiber activation (Grosche et al., 1999). It has been suggested that these ionic responses could affect the synaptic efficacy of individual synapses or subpopulations of these synapses (Grosche, Kettenmann & Reichenbach, 2002). Astrocytic coverage of synapses may thus also serve to enhance input-specificity, though to our knowledge it has not been examined whether activity influences the degree of astrocytic ensheathement of synapses in the hippocampus or cerebellum.

As the role of astrocytes in synaptic function draws more attention (see Volterra, Magistretti & Haydon, 2002), more researchers are investigating the mechanisms by which glia can enhance synaptic strength. It has been demonstrated in the supraoptic nucleus of the hypothalamus, for example, that the degree of synaptic coverage by astrocytes is associated with the degree of glutamate clearance, which in turn influences glutamate concentration and diffusion at the synapse (Oliet, Piet & Poulain, 2001). Another recent study has shown that a protein produced by glia and released near synapses, tumor necrosis factor α, can trigger increased expression of post-synaptic AMPA receptors, thereby enhancing synaptic strength (Beattie et al., 2002). It seems that, whatever the mechanisms by which astrocytes enhance learning-dependent synaptic plasticity, these changes may be transient, as they appear to fade in the absence of continued behavioral or environmental stimulation. For example, when animals were first trained on a motor skill learning task for 10 days and then left idle for the following 28 days, synaptogenesis that had occurred during learning remained clearly evident in these "Delay" animals (see above Figure 3.2 (A)), whereas training-induced effects on astrocytes were reduced and no longer statistically significant (see above Figure 3.2 (B)) (Kleim et al., 1997). Uniquely positioned, both anatomically and functionally, to manage these diverse roles at the synapse, astrocytes clearly exhibit plasticity that, in concert with neuronal plasticity, may very well be crucial to the processes of learning and memory. It is tempting to speculate that astrocytic changes may be necessary to induce, but not to maintain, adaptive changes in the brain's "wiring diagram" in response to experience.

PLASTICITY OF OLIGODENDROCYTES AND MYELINATION

There is good evidence that both oligodendrocytes and the myelination process itself are sensitive to developmental experience. Early work in the optic nerve showed that visual deprivation resulted in reduced myelination (Gyllensten & Malmfors, 1963) and, conversely, that premature eyelid opening accelerated the onset of myelination (Tauber, Waehneldt & Neuhoff, 1980). It should be noted, however, that not all studies have found a relationship between early visual experience and myelination in the optic nerve (see Moore, Kalil & Richards, 1976; Fukui et al., 1991). Szeligo and Leblond (1977), who were the first to examine the influence of rearing environment on brain fiber tracts, found increases in oligodendrocytes in the visual cortex of EC rats. Subsequently, Sirevaag and Greenough (1987) also found the volume fraction of oligodendrocyte nuclei in the visual cortex to be greater among EC-raised rats, compared to their IC littermates. The influence of developmental experience on oligodendrocytes is not limited to the visual cortex; raising rats in EC increases the number of myelinated axons (as measured using electron microscopy) in the splenial portion of the corpus callosum, which contains axons of visual cortical neurons (Juraska & Kopcik, 1988). The positive effect of a complex rearing environment on the size of the corpus callosum has also been demonstrated in rhesus monkeys (Sanchez et al., 1998).

Despite studies indicating that myelination continues well into adulthood, in both rodents and humans (Yakovlev, 1967; Benes et al., 1994; Nunez et al., 2000), this process is still often treated as if it were a strictly developmental phenomenon. The question of whether myelination remains sensitive to experience during adulthood is interesting and largely unexplored. Recently, our laboratory has begun to investigate this. Preliminary data indicate that the splenial corpus callosum of rats exposed to EC during adulthood contains increased numbers of myelinated axons and that these changes persist for at least one month after the animal has been removed from the complex environment (Briones et al., 1999). Thus the ability of oligodendrocytes and the myelination process to respond to the demands of an animal's environment apparently extends beyond the developmental time-frame, and the experience-induced changes in the adult brain appear to be stable. At this time it is unknown whether the increased myelination in the EC rat is caused by an increase in the number of axons that a typical oligodendrocyte helps to myelinate or by an increase in the proliferation and/or survival of new oligodendrocytes. Although many questions remain to be answered, the discovery that adult myelination is regulated by experience provides a functional correlate to continued myelination in the brain across the lifespan.

Communication between oligodendrocytes and the axons they ensheathe has also been a focus of increased attention. This communication is known to be bi-directional – signals from axons help regulate myelination, and myelination may lead to local changes in axonal cytoarchictecture (e.g. de Waegh, Lee & Brady, 1992; Michailov et al., 2004). The protein neuregulin, for example, is released by axons and appears to play a role in signaling axonal size to myelinating cells, which may affect the thickness of the myelin sheath (Michailov et al., 2004). Neurotrophins, whose levels are sensitive to behavioral experience (Branchi, Francia & Alleva, 2004), also appear to be key mediators of myelination, acting either directly on myelinating cells or indirectly by initiating signaling cascades within axons (Hempstead & Salzer, 2002). Activating receptors for BDNF, for example, can enhance myelin formation in the peripheral nervous system whereas activating receptors for neurotrophin-3 (NT-3) inhibits myelination (Cosgaya, Chan & Shooter, 2002). Several activity-dependent axonal signals have also been proposed to initiate myelination, including adenosine triphosphate (ATP) and the neurotransmitters glutamate and aspartate (receptors for which are expressed by oligodendrocytes) (Butt & Tutton, 1992; Brady et al., 1999; Stevens & Fields, 2000; Stevens et al., 2002). It is clear that some form of communication between neurons and oligodendrocytes is being affected by experience, but the mechanisms by which an increase in environmental complexity triggers myelination remain uncertain at this time.

The respective roles of myelinated and unmyelinated fibers in information transfer are not yet understood. Unmyelinated fibers are found in substantially greater numbers in the corpus callosum than are myelinated fibers, and one possibility is that they allow utilization of a richer supply of information in cases where a rapid response is not essential. Myelinated axons, on the other hand, provide the framework for those more rapid responses, and the addition of new axons to this pool of myelinated fibers is a form of plasticity with a potential comparable to the addition or strengthening of synapses, with one additional very interesting feature: speed rather than strength of communication is enhanced. Typical myelinated fibers, which are generally larger in axonal diameter than non-myelinated axons, conduct at velocities 50–100 times faster than their non-myelinated counterparts (see Brinley, 1980). Assuming a velocity of 60 m/s for a large myelinated fiber, conduction of an action potential from one hemisphere to the other might require 1–2 m/s, whereas a typical unmyelinated fiber conducting at 1 m/s might take 50–100 ms to travel the same distance. During the difference in time of arrival between the myelinated and unmyelinated axonal input, a typical large cortical pyramidal neuron could have fired a burst of several action potentials in response to the earlier input. Hence it is hard to imagine that these two fiber types are working in synchrony on the same behavioral or thought processes. Moreover, if an unmyelinated fiber were to be recruited to the myelinated pool, as may occur during exposure to a complex environment, its input would subsequently arrive some 50–100 ms earlier than before – seemingly a substantial difference in a system that does not typically wait until all afferents have had their say. In short, where relatively long distance communication is involved (and for these purposes, communication between adjacent gyri in the human brain would be "long distance," with disparities of perhaps 30–40 ms between myelinated and unmyelinated fiber inputs) myelinated fibers should largely determine the early response, if there is one, and hence addition to the myelinated pool connecting brain regions involved in a particular task could qualitatively alter performance.

MITOCHONDRIAL PLASTICITY

To support experience-dependent plasticity of synapses; dendrites; neurons; and glia, mitochondria, the ATP-producing powerhouses of the cell, exhibit remarkable plasticity of their own as they must adapt to meet the increased metabolic demands. In the frontal cortex, for example, the relative volume fraction of mitochondria per dendrite and per axon terminal increases over development, with mitochondria per dendrite increasing more rapidly (Tamai, Ikari & Hayashi, 1983). Housing rats in a complex environment for 30 days also resulted in an increase in the volume fraction of mitochondria in the visual cortex, as well as the volume of mitochondria per neuron which increases even more by 60 days of EC exposure (Black, Zelazny & Greenough, 1991). Isaacs et al. (1992) subsequently found increases in mitochondrial volume per neuron following a motor skill learning task, but no increase after physical exercise (with relatively little learning), implying that mitochondrial proliferation is related to learning (possibly via synapse number changes) rather than to physical activity.

Despite these early studies, the role of mitochondria in the processes of synaptic plasticity has been considered only relatively recently. Mitochondria exist within the cell as dynamic organelles that appear to move primarily along microtubules within axonal and dendritic cytoplasm (Ligon & Steward, 2000). A recent paper by Li et al. (2004) supports the view that mitochondria are important for synaptic plasticity, as they respond dynamically to synaptic activation of cultured hippocampal neurons by redistributing into dendritic spines. Furthermore, genetic manipulations that deplete the dendrite of mitochondria also dramatically reduce the density of spines along these dendrites. The precise molecular processes involved in synaptic plasticity for which mitochondria provide metabolic support, are yet unknown. Schuman and Chan (2004) have suggested several functions: mitochondria may provide ATP for the ATPases that are charged with restoring ionic homeostasis following repeated synaptic depolarization, or they could help buffer the influx of Ca^{2+} from intracellular stores. The authors also suggest that synaptic and dendritic mitochondria may be critical for maintaining the metabolic support for processes such as dendritic protein synthesis, perhaps even serving as a form of "synaptic tag" to signal activated synapses.

PLASTICITY OF CEREBRO-VASCULATURE

In contrast to earlier studies (e.g. Diamond, Krech & Rosenzweig, 1964; Rowan & Maxwell, 1981), data from our own laboratory indicate that the brain's vasculature is quite responsive to experience. Animals raised in EC have larger and more elaborately branched capillaries in the visual cortex, compared with both IC- and SC-raised animals (Black, Sirevaag & Greenough 1987; Sirevaag et al., 1988; Black, Zelazny & Greenough, 1991). As described above, the EC paradigm cannot be used to distinguish the relative contribution of motor skill learning and increased motor activity to plasticity. Studies in which these contributions may be distinguished have shown that physical exercise induces angiogenesis in motor brain regions but that motor skill learning is required to induce synaptogenesis. The inference is that angiogenesis is driven more by the repeated performance of unskilled movements rather than by the learning process per se (Black et al., 1990). Also, the plasticity of cerebrovasculature in response to behavioral demands appears to be far greater than that of synapses – volume fraction of capillaries (which combines diameter and density effects) nearly doubles after EC exposure (Black, Sirevaag & Greenough 1987; Sirevaag et al., 1988). Additionally, the capacity of vasculature in the primary motor cortex to supply blood in response to increased demand was shown to be greater in exercised animals, using functional magnetic resonance imaging to specifically address blood flow (versus the more common BOLD signal arising from deoxyhemoglobin) and reserve capacity under load in anesthetized rats (Swain et al., 2003).

Recently, our laboratory has investigated the interacting influences of exercise and aging on cerebrovasculature. Capillary volume in the precentral gyrus was found to be increased in mature (aged 15–17 years) but not young (aged 10–12 years) female *Macaca fascicularis* monkeys who had exercised for one hour a day, five days a week for 24 weeks prior to tissue collection as compared to sedentary controls (Figure 3.4) (Rhyu et al., 2003).

Interestingly, capillary volume fraction in monkeys which had a 12-week period of inactivity following the 24 weeks of exercise had returned to inactive control levels, indicating that exercise-induced changes in cerebrovasculature are somewhat shortlived (Rhyu et al., 2003). Aside from this study, the persistence of experience-induced angiogenesis has not been well investigated; however, the magnitude of the effect and

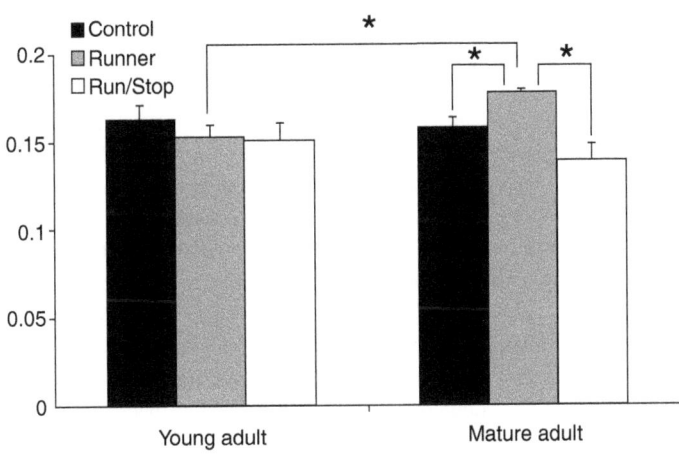

Figure 3.4. Exercise increases vascular volume fraction in the cortex of mature (aged 15–17 years) but not young (aged 10–12 years) monkeys. Among mature animals, runners (exercised over the course of 24 weeks) had greater capillary volume fraction compared with both sedentary control animals and to run/stop animals (that had a 12 week period of inactivity following 24 weeks of exercise), suggesting that continued physical activity is required for maintaining plasticity of cerebrovasculature. (*, $p < 0.05$) From Rhyu et al. (2003).

its activity-dependent nature suggest that the impact of experience on cerebrovasculature is likely to be considerably more shortlived than the quite stable changes in synaptic reorganization and myelination discussed above. Because like exercise, experimentally induced hypoxia also induces rapid angiogenesis (Harik, Hritz & LaManna, 1995), information concerning blood oxygen levels or a related metabolic demand, or both, may serve as physiological signals that trigger vascular proliferation. Although the precise signals are unknown, it is known that similar to synaptogenesis, angiogenesis in response to experience is greatest during development, also occurs during adulthood, and remains present, although diminished, during aging (Black, Polinsky & Greenough, 1989).

CONCLUDING REMARKS

Although there is much exciting work remaining to be done in the field of experience-induced morphological plasticity in the brain, especially in terms of non-neuronal components of the nervous system such as astrocytes, oligodendrocytes, and cerebrovasculature, some important observations may be drawn from the information available to date. The most general conclusion that can be confidently made from this body of work is that the brain is an extremely plastic organ, the structure of which is exquisitely sensitive to experience. A major function of the brain is thus to continuously reorganize itself, and what little we now know suggests that it does so in a way that is specifically tailored to result in behavior that is adaptive in the context of the individual's own unique environment. The nature of experience-driven plasticity is such that, while it has been demonstrated in many brain regions, in any given instance it is specific to those regions involved in processing the behaviourally relevant features of the environment. For instance, dramatic morphological changes occur in the visual cortex in response to a visually complex environment. On the other hand, learning a complex motor skill induces plasticity in motor areas of the brain such as the motor and cerebellar cortices. Such region-specific functional reorganization is not limited to experimental rodent models, as it has also been observed in monkeys (e.g. Recanzone et al., 1992) as well as humans, for example among string musicians (see Pantev et al., 2003) and those who have learned to read Braille (Pascual-Leone & Torres, 1993).

Experience-dependent plasticity is not limited to synapses or even to neurons. In fact, most if not every component of the nervous system exhibits robust, reproducible responses to experience. Thus, in addition to synaptogenesis, dendritic reorganization, and neurogenesis, other non-neuronal components are sensitive to experience, resulting, for example, in angiogenesis, increased myelination, mitochondrial plasticity, astrocytic hypertrophy, and increased astrocytic ensheathement of synapses. There are differences in both the type of experience that drive these changes and in their relative stability in the absence of the driving experience. For example, learning results in synaptogenesis, astrocytic hypertrophy, and survival of newly generated dentate gyrus neurons, whereas physical exercise without learning induces angiogenesis and dentate gyrus neurogenesis. Changes in synapses and myelination appear to be more stable or perhaps even permanent, possibly because they reflect reorganization of the brain's functional "wiring diagram," in comparison with the more generally acting transient effects on the brain's cerebrovasculature and astrocytes (changes in which appear, based on the available evidence, to be less robust than neuronal changes). The stability in morphological changes in synapses and myelination may occur because the brain must rely on past experiences to predict future ones. Therefore, organizational changes in these components of the nervous system at one point in time are very likely to be useful in the future. Another example in this regard is the addition of new cells in the dentate gyrus. As discussed above, whereas physical exercise increases the rate of neurogenesis, EC and, more specifically, learning, appear to enhance the survival of neurons generated at an earlier timepoint. Thus cells may be added to the adult hippocampus in anticipation of their need to mediate the acquisition, storage, and/or consolidation of future memories. From this brief review of the literature it may be inferred that experience induces multiple forms of plasticity in the brain that must be regulated, at least partially, by independent mechanisms. Finally, it should be emphasized that environmentally induced plasticity in the brain does not simply consist of changes in different classes of cells independently; rather, interactions between neurons and glia are also altered in ways that apparently more optimally meet behavioral demands. The greater ensheathement of synapses by astrocytes in response to a complex environment is one example of this, and we believe the future of the investigation of environmentally driven plasticity lies in understanding the integrative response of different brain elements to experience and in discovering the nature of their adaptive significance.

ACKNOWLEDGEMENTS

Portions of this work were supported by NIH grants HD37175, HD07333 (NICHD), AG10154 (NIA), MH35321 (NIMH), by the Illinois-Eastern Iowa District of the Kiwanis International Spastic Paralysis Research Foundation, and by the Retirement Research Foundation. The authors wish to thank the Beckman Visualization, Media, and Imaging Laboratory, Dr. James Churchill and Dr. Im Joo Rhyu for assistance with figures, and Dr. Willie Dong and Shawn Kohler for their thoughtful comments on the manuscript.

4 · Activity-dependent myelination

R. Douglas Fields

A central concept in synaptic plasticity is that synaptic inputs converging on the same post-synaptic neuron and firing simultaneously should be retained or strengthened. "Neurons that fire together wire together." However, the conduction time through axons from the pre-synaptic neurons is rarely considered. This is curious, since axon delays will determine the arrival time of action potentials to trigger synaptic transmission. The conduction time across distant regions of the brain may be significant, for example 30 ms across the corpus callosum connecting the two hemispheres (Swadlow, 1985). Extreme precision is necessary for coincident arrival and summation of synaptic potentials from converging inputs because the duration of a synaptic potential is only a few milliseconds. Although this is rarely considered in nervous system plasticity, the degree of temporal summation of converging inputs may increase the amplitude of synaptic responses to the same extent as physiological processes at the synapse regulating neurotransmission (long-term potentiation (LTP), for example).

Thus axonal conduction time is a critical variable in synaptic plasticity. Considering the large number of variables affecting conduction velocity and the lengthening of axons during development and growth of the body, genetic instruction alone would seem inadequate to specify the optimal conduction velocity in every axon. This implies a need for regulation of conduction velocity through an activity-dependent process. If the conduction time through axons is modifiable by an activity-dependent mechanism, this would represent another type of activity-dependent plasticity operating beyond the synapse (Fields, 2005).

Optimal function of a neural circuit would be disrupted if the spike arrival times are not highly synchronized. There are several instances described where conduction times through axons of different distances are adjusted to provide simultaneous arrival. Conduction through optic nerve axons from central regions of the retina is slower than from extrinsic regions (Rowe & Stone, 1976; Stanford, 1987). Similarly, conduction velocity in olivocerebellar projections to Purkinje cells in the cerebellum (Sugihara, Lang & Llinas, 1993) and the electromotor axons of certain electric fish (Waxman, Pappas & Bennett, 1972), are adjusted to compensate for differences in axon length to produce simultaneous arrival of impulses.

The primary mechanism regulating conduction velocity through vertebrate axons is myelin (Fields, 2008a). Myelin sheath thickness and internodal distance govern conduction velocity (Rushton, 1951; Smith & Koles, 1970; Waxman, 1972) and myelin may affect the localization of ion channels at the node and paranodal regions. Axon caliber is also a fundamental parameter in conduction velocity, but myelin can affect axon caliber through communication between myelinating glia and the axon. This raises the hypothesis that activity-dependent effects on myelin could regulate information-processing in the brain.

MYELIN

Myelin is formed by different types of cells in the peripheral nervous system (PNS) and the central nervous system (CNS) (Schwann cells and oligodendrocytes). These two cells have different embryologic origins and they respond to different signals throughout development in forming myelin. Myelination involves a complex orchestration of many highly regulated cell biologic processes, including cell recognition – indeed, subcellular recognition. Only axons of a neuron are myelinated, not the cell body or dendrites. Moreover, not all axons are myelinated; only those that must carry impulses at high velocities.

The sheath of myelin is elaborated and wrapped around the axon by extending a tongue of cytoplasm that spirals around the axon up to 150 times on larger fibers. This requires intricate neuron–glial communication to regulate cell motility, active synthesis of lipid membrane and cell adhesion molecules, and dynamic cytoskeletal changes to extrude the cytoplasm from between the consecutive layers of myelin to compact the layers of membrane tightly and seal the internodal regions from electrical current leaks. To enable salutatory conduction, the nodal and paranodal regions of myelinated axons are highly structured even at the ultrastructural level, where there is a precise arrangement of distinct ion channels and cell adhesion molecules at the node and paranode (Peles & Salzer, 2000).

Axon diameter is a primary determinant of conduction velocity, but myelinating glia may affect axon diameter. This is most evident at the node of Ranvier, where the axon diameter often changes abruptly in the nodal and paranodal regions. Axon diameter is reduced in myelin-deficient mutants (Cole et al., 1994), demonstrating an influence of myelinating glia (Schwann cells) on axon diameter. Oligodendrocytes also regulate regional expansion of axon caliber and local accumulation of neurofilaments. Myelination increases phosphorylation of axonal neurofilament-M and neurofilament-H, causing increased spacing between neurofilaments and expansion of the axon caliber (Hsieh et al., 1994; Yin et al., 1998). Myelin-associated glycoprotein (MAG) has been implicated in the signaling cascade controlling neurofilament phosphorylation and in turn the axon caliber (Yin et al., 1998; Lunn et al., 2002).

Myelin thickness is proportional to the diameter of the axon (Donaldson & Hoke, 1905; Friede, 1972), with maximal conduction velocity attained with a g-ratio (ratio of axon core diameter to total fiber diameter) of 0.65 (Rushton, 1951). Reduced neuregulin-1 (Nrg1) expression causes thinner myelination and reduced conduction velocity, whereas overexpression of Nrg1 in the axon causes hypermyelination (Michailov et al., 2004). The optic nerve of the cat is an example where myelin thickness has a predominant influence on conduction velocity. The fiber calibers are uniform, but five modes of conduction latencies correlate with the myelin thickness (Freeman, 1978). Thus, whilst on average myelin thickness to axon diameter displays the theoretical optimal ratio, there is wide variation around the optimum, with associated changes in conduction velocity among different fibers.

There is also an optimal spacing of nodes of Ranvier for maximal conduction velocity. Thus changes in internodal distance can increase or decrease conduction velocity. Axons in the barn owl auditory system operate as delay lines retarding the speed of action potentials to provide the correct localization of sound (Carr & Konishi, 1990). The conduction velocities of these fibers is as low as unmyelinated axons (3–5 m/s), and this is associated with an unusually close spacing of nodes of Ranvier inside the nucleus laminaris (60 µm) in contrast to 300 µm just outside the nucleus.

MYELIN IN COGNITION AND INFORMATION PROCESSING

Evidence for white matter involvement in information-processing and cognition has emerged from several neurologic and psychiatric disorders (Fields, 2008b). Some demyelinating disorders are associated with schizophrenic-like psychoses (Hyde, Ziegler & Weinberger, 1992), and progressive cognitive deterioration is one of the characteristics of multiple sclerosis (MS), a demyelinating disease of the CNS, (Kujala, Portin & Ruutiainen, 1997). Impaired memory or other cognitive function afflicts 40–60% of patients with MS.

Recently a surprising association between white matter abnormalities and several psychiatric conditions has been found by use of gene arrays and brain imaging. In an analysis of 6000 genes in the prefrontal cortex of schizophrenics, 89 genes were found to be abnormally regulated. Of these, 35 were genes involved in myelination (Hakak et al., 2001). Altered expression or polymorphisms of genes involved in myelination has been described as a molecular signature of schizophrenia (Hakak et al., 2001). The level of mRNA for the gene QKI, is decreased in several cortical regions and the hippocampus in schizophrenia subjects (Aberg et al., 2006a, 2006b; Haroutunian et al., 2006). QKI plays a fundamental role in oligodendrocyte differentiation and myelination. Polymorphisms of several other genes for myelin proteins, including MAG (Wan et al., 2005), CNP (Pierce et al., 2006), MOG (Zai et al., 2005), NRG1, ERBB4, and Olig2 are indicators of susceptibility to schizophrenia (Tkachev et al., 2003; Georgieva et al., 2006; McCullumsmith et al., 2007). This suggests that the physiology of myelinating glia and myelin may be altered in schizophrenia and in part underlie the cognitive defects in this disorder.

White matter abnormalities have been found in schizophrenic patients in brain imaging studies (Lim et al., 1988; Andreasen et al., 1994), and an autopsy study of schizophrenia patients revealed a 25–30% reduction in number of oligodendrocytes compared with control subjects (Hof et al., 2003). Both schizophrenia and bipolar disorder share some of the same symptoms, notably hallucinations and delusions during manic episodes for bipolar patients, and during depressive or psychotic episodes for schizophrenia patients.

Several other psychiatric disorders with pronounced cognitive defects involve alterations in white matter tracts or myelin genes. This includes autism (Akshoomoff, Pierce & Courchesne, 2002), Alzheimer's disease (Bartzokis, 2001, 2006), and depression (Tkachev et al., 2003; Aston, Jiang & Sokolov, 2005). Thus myelin abnormalities affect information processing and cognition.

MYELIN PLASTICITY

Plasticity of myelin is rarely considered outside the context of disease, but there is evidence that the concept of myelin as static may be incorrect. Myelin varies and it changes. Myelination begins in late stages of fetal development in most animals, but in humans and most other animals extensive myelination takes place in the early post-natal period. In humans, myelination continues through childhood and into early adult life (Giedd, 2004). Interestingly, this is the same period during which the cerebral cortex undergoes massive activity-dependent reorganization of synaptic connections, which are understood to modify functional circuits to optimize nervous system function in the environment experienced during rearing.

As the body grows, conduction velocity through axons must increase or there will be increasing delay in arrival of impulses. During the first two years after birth the conduction delay decreases in both sensory and motor pathways in the CNS and PNS, largely as a result of myelination. Remarkably, after two years, central somatosensory conduction delay and motor conduction delay remain constant, despite the substantial increase in body height (Eyre, Miller & Ramesh, 1991). Thus, conduction velocity increases during growth of the body proportionately to keep conduction delays fixed. This increased conduction velocity with axon growth in the CNS and increased myelination suggests that signals between axon and myelinating glia regulate conduction velocity in white matter tracts to minimize conduction delays.

A magnetic resonance imaging (MRI) method to assess the water fraction trapped between hydrophobic bilayers of the myelin sheath shows a steady increase in the myelin water fraction with age between 20 and 55 years (Flynn et al., 2003). This suggests continued myelin remodeling throughout adult life. Moreover, the myelin water fraction correlated linearly with the number of years of education.

New imaging techniques are reinforcing decades of evidence in the literature indicating that myelin may change according to environmental experience. The number of myelin-forming oligodendrocytes increases by 27–33% in the visual cortex of rats raised in enriched environments (Bennett et al., 1964; Szeligo & Leblond, 1977; Sirevaag & Greenough, 1987). An enriched environment is one that provides additional play objects and social interaction. The response of myelinating glia to environmental experience is not limited to the visual system or to rats. Enriched environments increase the number of myelinated axons in the corpus callosum of rats (Juraska & Kopcik, 1988; Markham & Greenough, 2004), and the corpus callsum increases in size in rhesus monkeys raised in enriched environments, and this increase correlates with improved performance in tests of cognitive function (Sanchez et al., 1996). Environmental effects on white matter extend beyond animal studies. Children suffering severe childhood neglect have 17% reduced corpus callosum area (Teicher et al., 2004).

Although the mechanisms are unknown, evidence that impulse activity can affect myelination has been in the literature since the 1960s, from experiments rearing mice in the dark (Gyllenstein & Malmfors, 1963) or opening the eye of neonatal rabbits prematurely (Tauber, Waehneldt & Neuhoff, 1980). Rearing animals in the dark reduces the number of myelinated axons in the optic nerve, and premature eye opening increases myelin protein expression. Electrical activity also promotes proliferation of oligodendrocyte progenitor cells in the optic nerve (Barres & Raff, 1993), and in cell culture experiments, increasing electrical firing of neurons or suppressing it with drugs, increases or decreases (respectively) the amount of myelin formed in cell cultures (Demerens et al., 1996), but the mechanism for the effect is unknown.

Learning complex skills, such as playing the piano, is accompanied by increased white matter development in the appropriate brain tracts involved in musical performance (Bengtsson et al., 2005). Importantly, this diffusion tensor imaging (DTI) study also found that the level of white matter structure increased proportionately to the number of hours each subject had practiced the instrument. This indicates that white matter structure changes in accordance with acquiring the skill rather than the level of musical performance being predetermined as a developmental limitation on white matter development. These responses of white matter to functional experience imply that the myelinating glia, which do not fire electrical impulses, must, in some way, "know" they are in a brain practicing the piano or in an environment that is enriched.

MYELINATING GLIA SENSE AND RESPOND TO NERVE IMPULSES

Glia are not electrically excitable, but it has become apparent in recent years that glia near synapses (astrocytes) can sense neurotransmitters released from the synaptic cleft (Fields & Stevens-Graham, 2002). These cells then communicate with other glia using diffusible signals, such as adenosine triphosphate (ATP) (Fields & Burnstock, 2006) and glutamate (Haydon, 2001). Myelinating glia, however, are located far from synapses, making it less obvious how they might sense impulse traffic in axons.

Optic nerve glia can respond to impulse activity in axons evoked by electrical stimulation or natural light stimulation of the retina (Orkand, Nicholls & Kuffler, 1966), by detecting the elevated concentration of potassium ions liberated by axons firing impulses. Voltage-sensitive dyes also reveal responses in the myelin sheath of CNS axons firing action potentials (Lev-Ram & Grinvald, 1986). In the PNS prolonged trains of impulse activity cause an influx of Ca^{2+} into the paranodal loops (Lev-Ram & Ellisman, 1995).

Action potentials can also cause release of other substances from axons, such as ATP (Fields & Burnstock, 2006) and glutamate (Kreigler & Chiu, 1993) to activate Ca^{2+} fluxes in Schwann cells and white matter glia. These neurotransmitters are released from axons firing action potentials through membrane transporters, ion channels, or synaptic vesicles (Kukley, Capetillo-Zarate & Deitrich, 2007; Ziskin et al., 2007). Clusters of synaptic vesicles are seen by electron microscopy accumulating at some nodes of Ranvier in white matter (Waxman, 1972), and functional synapses have been detected between axons and oligodendrocyte progenitor cells (Bergles, 2000).

Two mechanisms by which impulses regulate myelination by Schwann cells and oligodendroctyes have been identified from research using cell cultures equipped with electrodes to stimulate action potentials in axons of mouse DRG neurons. Electrical activity in neurons can alter expression of specific neuronal genes, depending on the frequency and pattern of impulse firing in the neuron (Itoh et al., 1997). These studies show that firing axons at the appropriate frequency to lower expression a cell adhesion molecule necessary for induction of myelin by oligodendrocytes (Barbin et al., 2004) and Schwann cells (Itoh et al., 1995), the cell adhesion molecule L1, inhibits myelin formation (Stevens, Tanner & Fields, 1998). Firing axons at a different frequency that does not affect the gene had no effect on myelination, suggesting that myelination may be regulated by the pattern of impulse activity in developing axons.

Diffusible substances released from axons firing bursts of action potentials have been identified that are detected by myelinating glia (Stevens & Fields, 2000), with subsequent effects on myelination. Studies in cell culture show that, at an early stage in development, release of ATP from axons firing impulses is followed by its degradation to adenosine. Adenosine then activates P1 purinergic receptors on oligodendrocyte progenitor cells, to inhibit cell proliferation and stimulate differentiation and maturation of oligodendrocyte progenitor cells, thus resulting in more myelinated axons (Stevens et al., 2002).

Later in development, after the progenitor cells have differentiated to oligodendrocytes, action potentials increase myelination through a different signaling process involving another glial cell, astrocytes (Ishibashi et al., 2006). ATP released by axons firing impulses activates membrane receptors for ATP (P2 receptors) on astrocytes, causing release the cytokine LIF (leukemia inhibitory factor). This in turn stimulates the formation of myelin by oligodendrocytes. Thus, electrical activity in premyelinated axons, acting through ATP (purinergic receptor) signaling, and astrocytes, can stimulate myelin formation in response to impulse activity. These three

MYELIN LIMITING SYNAPTIC PLASTICITY

Myelin directly controls axon sprouting and synapse formation. Several proteins in myelin, Nogo-A (Chen *et al.*, 1995; GrandPré *et al.*, 2000), MAG (McKerracher *et al.*, 1994), and OMgp (Wang *et al.*, 2002), cause the tips of growing axons to collapse and stop growth toward its target (Schwab & Thoenen, 1985). Originally studied in the context of axon regeneration after injury (Bregman *et al.*, 1995), the normal function of these growth-inhibiting proteins in myelin is now appreciated to be in suppressing axon sprouting after formation of appropriate connections in development.

In many neural circuits, myelination establishes the end of the critical period. Critical periods are developmental stages when functional activity may permanently alter development of specific sensory or cognitive abilities. The importance of visual stimulation in forming appropriate synaptic connections for sight (Hubel & Wiesel, 1962), and the necessity of hearing speech to develop language skills specific to our native language (Kuhl, Tsao & Liu, 2003) are well-known examples of critical periods. The critical period also describes youthful periods when the brain exhibits marked resilience in recovering from injury. Myelination is a significant, and sometimes dominant, determinant of critical periods, and thus an essential component in learning.

As suggested by the correlations between white matter and cognitive development, and the necessity of acquiring complex skills at an early age to reach optimal proficiency, myelination sets limits on the age at which specific brain connections may be modified by axon sprouting and synaptogenesis. In laboratory animals in which the gene for the Nogo-66 receptor is eliminated, the critical period for ocular dominance plasticity, normally 20–33 days after birth in mice, is extended well into adults (45–120 days) (McGee *et al,.* 2005).

In recovery from injury, myelination can have a dominant effect on the critical period. The North American opossum can recover from spinal cord injury if the cord is severed before 30 days of age. This is the age at which myelination is completed in the spinal cord of this marsupial (Ghooray & Martin, 1993). Genetic knock outs (see Woolf, 2003 for review) and antibody interference with myelin proteins such as Nogo (Bregman *et al.*, 1995) and MAG (McKerracher *et al.*, 1994) may improve regeneration of CNS axons and restore functional connections. Intriguingly, mRNA for Nogo is overexpressed in post-mortem samples of frontal cerebral cortices from individuals with schizophrenia, and a polymorphism containing a CAA insert in the 3'-untranslated region is more prevalent (Novak *et al.*, 2002).

ACTIVITY-DEPENDENT MYELINATION: UNANSWERED QUESTIONS

These new observations and hypotheses present several interesting questions and objectives for further investigation: Among these are: (1) defining precisely what myelin plasticity is. This will require examining the various mechanisms by which myelin can regulate conduction velocity in the context of activity-dependent plasticity. (2) Determining the signals, conditions, rules, and cellular mechanisms that regulate myelin to achieve optimal function. (3) Exploring the role of myelin in regulating synaptic plasticity. (4) Determining the causal relation underlying correlations between white matter and psychiatric or neuronal disorders affecting cognition.

Myelin plasticity. Is myelin plasticity strictly a change in the number of axons that become myelinated, or a change in the myelin sheath that would regulate impulse conduction speed? There is evidence for both processes. Current evidence best supports changes in the number of oligodendrocytes or number of myelinated axons with electrical stimulation or functional activity, but these changes are technically easier to detect than more subtle changes in morphology or composition of the myelin sheath, that could have substantial effects on conduction velocity.

Conduction velocity through axons may change. Interhemispheric conduction time through the corpus callosum is substantial, 30 ms or longer (Swadlow, Waxman & Geschwind, 1980). However, chronic recordings in rabbit show that the conduction velocity of 55% of callosal axons changes (increase or decrease) over one year, which is a substantial portion of the animal's lifespan (Swadlow, 1985).

If change in conduction velocity reflects changes in myelin structure, there are many possible mechanisms by which myelin could cause these changes. Few of them have been investigated in the context of activity-dependent myelination. Factors affecting impulse conduction velocity include: (1) the thickness of the myelin sheath; (2) the axon diameter; (3) the number and spacing of nodes of Ranvier; (4) the nodal structure and molecular composition of ion channels in the node and paranodal region.

The thickness of myelin sheath can have a dominant influence on conduction velocity. Considerable differences in g-ratio are seen between fibers, and even along the same fiber, and conduction velocity is not always constant along the entire length of an axon (Traub & Mendell, 1988; Baker & Stryker, 1990). In general, smaller fibers tend to have smaller than optimal g-ratios (0.34 for 0.2 micron diameter fibers) and this is increased to 0.7 for the largest fibers (Berthold, Nilsson & Rydmark, 1983).

Second, the g-ratio may change. The number of wraps of myelin increases as animals grow and their axons lengthen, suggesting that myelin production is a perpetual process (Berthold, Nilsson & Rydmark, 1983). The cross-sectional area of myelin sheath in gamma fibers in the lumbar nerve of cat increases by as much as 85% as the cat ages. Myelin thickness compensates for increased internodal distance during body growth, which would otherwise reduce conduction velocity (Schroder, Bohl & Brodda, 1978; Friede, Brzoska & Hartmann, 1985; Hara et al., 2003).

Experimentally increasing nodal distance, by lengthening the femur of a rat, increases myelin protein expression by 160% (Hara et al., 2003), suggesting a feedback signaling between changes in internodal distance and myelin thickness, to regulate conduction velocity.

Axon diameter is a critical determinant of conduction velocity. Larger fibers conduct impulses at higher speeds because the resistance to flow of electrical current through them is reduced as the caliber of the fibers increases. However, myelin can also regulate axon diameter. An abundant protein in myelin, MBP (myelin basic protein) is phosphorylated by repetitive impulse activity in optic nerve (Murray & Steck, 1984) and hippocampus (Atkins & Sweatt, 1999), through a signaling pathway involving PKC and reactive oxygen species. Phosphorylation of MBP by extracellualr signals regulates microtubule stability (Dryer et al., 1994) and changes actin cytoskeletal organization (Dryer & Benjamins, 1989). Phosphorylation of MBP by action potentials may regulate dynamic functions of myelin.

The node of Ranvier is the most complex cell–cell adhesive junction known. The structure of the node and the number of nodes along a fiber markedly influence conduction velocity. In the electric fish *Sternarchus*, abnormally large and electrically passive nodes add capacitance to *delay* the propagation of the impulse through the axon and modify the waveform of the voltage spike (Waxman, Pappas & Bennett, 1972). The shape and amplitude of the nerve impulse, as well as the firing frequency, and the speed of impulse propagation, are all critical properties in neural coding, information-processing, and synaptic function. Little is known of how the dimensions of the node or the morphology of the paranodal loops of myelin flanking the node affect conduction velocity, or how these parameters might change as a result of functional activity. Swelling of the paranodal region after repetitive activation has been observed, however (Wurtz & Ellisman, 1986), suggesting the possibility of activity-dependent structural dynamics at the node.

Myelinating glia have a major influence on the localization of ion channels into concentrated domains at the node during development (Dupree et al., 2004). The particular types of ion channels inserted into the membrane, and the spatial localization of different types of channels in the axon – most prominently, the concentration of sodium channels in the node and potassium channels clustered in the paranodal region – determine the excitability (voltage change required to initiate an impulse), frequency of impulse firing (singularly or in bursts), the refractory period (recovery time before another impulse can be fired), and velocity of impulse conduction. Both contact and diffusible factors from glial regulate ion channel localization in the axon membrane (Dupree et al., 2004).

Thus there are several possible mechanisms for myelinating glia to affect conduction velocity and considerable evidence for axon–glial communication and dynamics of these features of myelin. Whether these features of myelin are altered by impulse activity to promote optimal conduction and information transmission, has not yet received significant experimental attention.

Signals regulating activity-dependent myelination. The signals mediating activity-dependent communication between axons and oligodendrocytes, the developmental time course of activity-dependent effects, and the cellular mechanisms that would regulate myelin to optimize conduction velocity are only known at a superficial level at present.

Thus far, only two mechanisms have been described for activity-dependent regulation of myelination: changes in cell adhesion molecules on the axon (Stevens, Tanner & Fields, 1998), and purinergic signaling mediated by ATP release from axons (Stevens *et al.*, 2002; Ishibashi *et al.*, 2006). However, there are many other potential channels of communication between axons and myelinating glia, including neurotransmitters, ions (pH, K^+, Na^+), neuromodulators, growth factors, nitric oxide and other gasses, and even possible axon-glial synapses, which might inform glia of electrical activity in axons and regulate myelination accordingly.

The critical aspects of impulse activity in regulating myelination are poorly known. Research on gene expression shows that the frequency of impulse activity may be a crucial factor affecting myelination (Stevens, Tanner & Fields, 1998), but the duration of stimulation, frequency, phase with respect to firing in other axons, and the stimulus burst train parameters are poorly characterized in terms of communication between axon and glia and regulation of myelination.

Myelination involves a complex set of biological processes, including cell recognition, cell adhesion, migration of cells on axons, changes in glial cell morphology, differentiation, synthesis of glial membrane, and motility of individual subcellular glial extensions in wrapping axons. How impulse activity affects each of these is only beginning to be explored.

Current evidence best supports activity-dependent changes in myelin during early development or while specific tracts in the brain are still actively undergoing myelination. In the study of white matter structure in pianists, for example, the effects were only detected in regions of the brain that had not yet fully myelinated (Bengtsson *et al.*, 2005). Pianists who began practicing the instrument in their late teens, for example, only showed increased white matter structure in the forebrain region that was still undergoing myelination. Similarly, the effects of enriched environments on oligodendrocytes and myelin (Markham & Greenough, 2004), the effects of modulating visual input on optic nerve myelination (Gyllenstein & Malmfors, 1963), and the cellular or molecular mechanisms for increased myelin in cultures stimulated to fire action potentials chemically (Demerens *et al.*, 1996) or with electrical stimulation (Stevens *et al.*, 2002; Ishibashi *et al.*, 2006), are only observed during narrow developmental windows. It remains to be determined whether activity-dependent myelination is restricted to early life, and participates in sculpting the brain for optimal performance in the particular environment during rearing, or whether it extends throughout life. It is possible that changes in white matter structure are simply easier to detect earlier in life, but may continue into adulthood.

If conduction velocity is regulated by myelin to optimize performance through synchronization of impulse transmission rather than simply maximizing conduction velocity, how is optimal synchrony evaluated and communicated to myelinating glia? Some cellular or molecular integrator and comparator would be necessary to monitor the degree of synchrony in arrival of inputs converging on the same post-synaptic target. No specific mechanism has been proposed, but astrocytes have the anatomical features to sense activity at the node of Ranvier and synapses, and to regulate myelination by the release of LIF (Ishibashi *et al.*, 2006).

CONCLUSION

White matter plasticity appears to provide another cellular mechanism of learning complementing the well-studied mechanism of synaptic plasticity. This plasticity encompasses types of learning which includes acquisition of complex skills and abilities that require efficient interactions across distant cortical regions and prolonged practice to achieve. The involvement of white matter plasticity in cognition has important implications for early childhood experience on brain development and many psychiatric illnesses. Finally, activity-dependent myelination expands the scope of nervous system plasticity well beyond the synapse and even beyond neurons (Bullock *et al.*, 2005).

ACKNOWLEDGEMENTS

Portions of this chapter have been adapted from Fields (2008b).

5 · Bipolar disorder: involvement of signaling cascades and AMPA receptor trafficking at synapses

Jing Du, Jorge Quiroz, Peixiong Yuan, Carlos Zarate Jr., and Husseini K. Manji

INTRODUCTION

Bipolar disorder (BPD) is a common, severe, chronic, and often life-threatening illness with lifetime prevalence rates of 3.9% (Kessler et al., 2005) and estimated lifetime costs in the USA of between $24 and $42 billion (Simon, 2003). BPD has been identified by the Institute of Medicine as a priority area for national action to improve healthcare. A number of studies have reported on poor quality of life (Yatham et al., 2004) and low rates of remission (Goldberg & Harrow, 2004) for patients with BPD. Despite the devastating impact that BPD has on millions of people worldwide, little is known about its underlying etiology and neurobiology. Historically, the brain systems receiving the greatest attention in neurobiologic studies of mood disorders are the monoaminergic neurotransmitter systems, which are extensively distributed throughout the network of limbic, striatal, and prefrontal cortical neuronal circuits thought to support the behavioral and visceral manifestations of mood disorders (Drevets, 2001; Manji, Drevets & Charney, 2001; Nestler, Gould & Manji, 2002). In addition to the growing appreciation that investigations into the pathophysiology of complex mood disorders have primarily focussed on monoaminergic systems, there is growing concern that progress in developing truly novel and improved medications has been limited. Recognition of a clear need for better treatments, and the lack of significant advances in our ability to develop novel, improved, therapeutics for these devastating illnesses has led to investigation of the putative roles of intracellular signaling cascades and synaptic plasticity. Recent evidence demonstrating that impairments of neuroplasticity may underlie the pathophysiology of mood disorders, and that antidepressants and mood stabilizers exert major effects on intracellular signaling pathways that regulate cellular plasticity, has generated considerable excitement among the clinical neuroscience community, reshaping views about the neurobiologic underpinnings of these disorders (Manji & Lenox 2000; Manji, Drevets & Charney, 2001; D'Sa & Duman, 2002; Nestler et al., 2002; Young, 2002).

SIGNALING NETWORKS REGULATED BY MOOD STABILIZERS AND ANTIDEPRESSANTS AS KEY MODULATORS OF SYNAPTIC PLASTICITY

Recent research into the pathophysiology and treatment of mood disorders has moved from a focus on neurotransmitters and cell surface receptors to intracellular signaling cascades. Multi-component cellular signaling pathways interact at various levels, thereby forming complex signaling networks that allow cells to receive internal and external cues, process those cues, and respond to information (Figure 5.1). These signaling networks enable the integration of signals across multiple timepoints and generation of distinct outputs depending on input strength and duration. Moreover, these signaling networks regulate intricate feed-forward and feedback loops (Bourne & Nicoll, 1993; Bhalla & Iyengar, 1999; Weng, Bhalla & Iyengar, 1999). Given their widespread and crucial role in the integration and fine-tuning of physiologic processes, it is not surprising that abnormalities in signaling pathways have now been identified in a variety of human diseases.

Signaling pathways also represent major targets for a number of hormones, including glucocorticoids, thyroid hormones, and gonadal steroids (Manji, 1992; Speigel, 1998). These biochemical effects may play a role in mediating certain clinical manifestations of altered hormonal levels in mood disorder subjects (e.g. the frequent onset of BPD in puberty, triggering of episodes in the post-partum period, association of depression and potentially rapid cycling with hypothyroidism, association of thyroid dysfunction in relatives of patients with mood disorders and triggering of mood episodes in response to exogenous glucocorticoids).

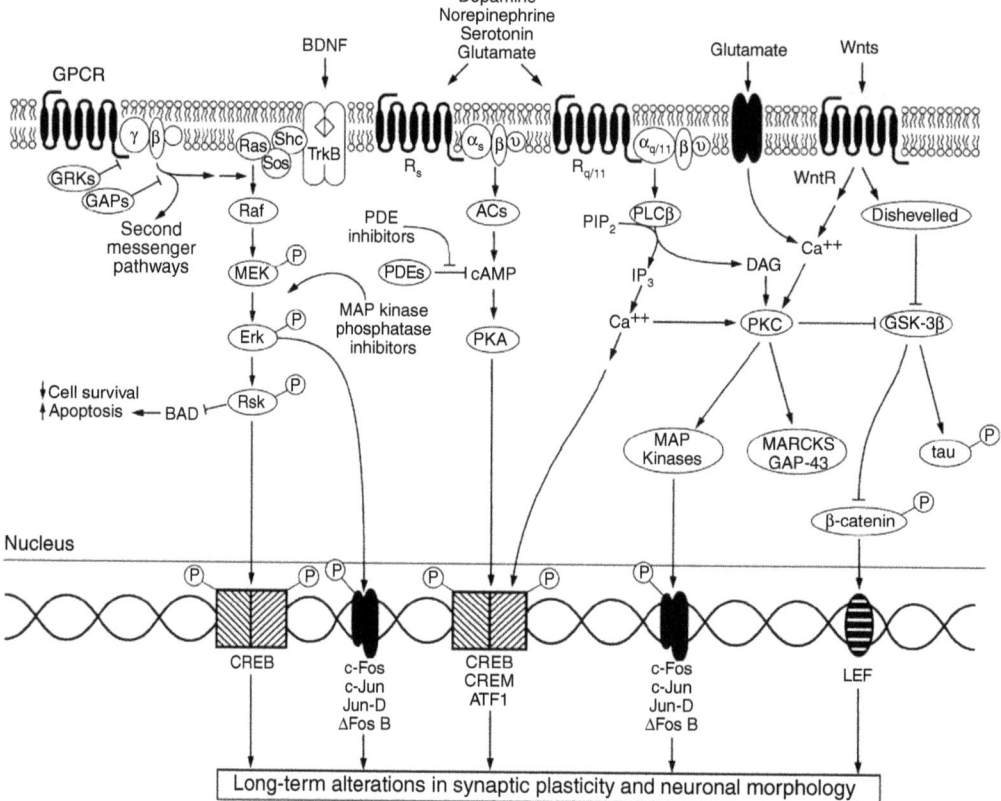

Figure 5.1 Major intracellular signaling pathways in bipolar disorder. The figure depicts some of the major intracellular signaling pathways involved in neural and behavioral plasticity. Cell surface receptors transduce extracellular signals such as neurotransmitters and neuropeptides into the interior of the cell. Most neurotransmitters and neuropeptides communicate with other cells by activating seven transmembrane spanning G protein-coupled receptors (GPCRs). As their name implies, GPCRs activate selected G proteins, which are composed of α and β γ subunits. Two families of proteins turn off the GPCR signal, and may therefore represent attractive targets for new medication development. G protein-coupled receptor kinases (GRKs) phosphorylate GPCRs and thereby uncouple them from their respective G proteins. GTPase activating proteins (GAPs, also called RGS or regulators of G protein signaling proteins) accelerate the G protein turn off reaction (an intrinsic GTPase activity). Two major signaling cascades activated by GPCRs are the cAMP generating second messenger system, and the phosphoinositide (PI) system. cAMP activates protein kinase A (PKA), a pathway which has been implicated in the therapeutic effects of antidepressants. Among the potential targets for the development of new antidepressants are certain phosphodiesterases (PDEs). PDEs catalyze the breakdown of cAMP; thus PDE inhibitors would be expected to sustain the cAMP signal, and may represent an antidepressant augmenting strategy. Activation of receptors coupled to PI hydrolysis results in the breakdown of phosphoinositide 4,5-biphosphate (PIP$_2$) into two second messengers – inositol 4,5-trisphosphate (IP3) and diacylglycerol (DAG). IP$_3$ mobilizes Ca^{2+} from intracellular stores, whereas DAG is an endogenous activator of protein kinase C (PKC), which is also directly activated by Ca^{2+}. PKC, PKA, and other Ca^{2+}-dependent kinases directly or indirectly activate several important transcription factors, including CREB, CREM, ATF-1, c-Fos, c-Jun, Jun-D, and ΔFos B. Endogenous growth factors such as brain-derived neurotrophic factor (BDNF) utilize different types of signaling pathways. BDNF binds to and activates its tyrosine kinase receptor (TrkB); this facilitates the recruitment of other proteins (SHC, SOS), which results in the activation of the ERK-MAP kinase cascade (via sequential activation of Ras, Raf, MEK, Erk, and Rsk). In addition to regulating several transcription factors, the ERK-MAP kinase casade, via Rsk, down-regulates BAD, a pro-apoptotic protein. Enhancement of the ERK-MAP kinase cascade may have effects similar to those of endogenous neurotrophic factors; one potential strategy is to utilize inhibitors of MAP kinase phosphatases (which would inhibit the turn-off reaction) as potential drugs with neurotrophic properties. In addition to utilizing GPCRs, many neurotransmitters (e.g. glutamate

Complex signaling networks may be especially important in the central nervous system (CNS), where they balance and integrate diverse neuronal signals, transmit them to effectors, thus forming the basis of a complex information-processing network (Bourne & Nicoll, 1993; Bhalla & Iyengar, 1999; Weng, Bhalla & Iyengar, 1999). The highly complex signaling networks may be one mechanism by which neurons acquire the flexibility for generating the wide range of responses observed in the nervous system. These pathways are undoubtedly involved in regulating diverse vegetative functions such as mood, appetite, and wakefulness therefore they are likely to be involved in the pathophysiology of BPD. We will now describe direct and indirect evidence supporting a role for signaling pathway abnormalities in the pathophysiology and treatment of BPD.

The Gs protein/cAMP generating signaling pathway

Several laboratories report abnormalities in G protein subunits in patients with BPD. Data from post-mortem brain studies on patients with BPD demonstrate increased levels of the stimulatory G protein (Gαs), accompanied by increases in post-receptor stimulated adenylyl cyclase (AC) activity (Young et al., 1993; Warsh, Young & Li, 2000). These observations are supported by the demonstration of increased agonist-activated (Emamghoreishi et al., 1997) GTPγS binding to G protein α subunits in the frontal cortical membranes of patients with BPD (Wang & Friedman, 1996). Several studies found elevated Gαs protein levels and mRNA levels in peripheral circulating cells in BPD, although the dependency of these levels on clinical mood state remains unclear (Manji et al., 1995; Lenox & Manji 1998; Mitchell et al., 1997; Spleiss et al., 1998; Wang & Friedman 1999; Warsh, Young & Li, 2000).

Recent studies from our laboratory and other laboratories has shown that basal protein kinase A (PKA) activity was decreased in hippocampal samples from animals treated chronically with lithium or valproate (Du et al., 2004a ; reviewed by Gould et al., 2004). In addition, Perez et al. (2000) looked further downstream in this signaling pathway and found higher levels of cAMP-stimulated phosphorylation of a $\sim$22 kDa protein (subsequently identified as Rap1) in platelets obtained from treated euthymic patients with BPD compared with healthy subjects (Perez et al., 2000).

The PKC signaling pathway

Protein kinase C (PKC) is a group of calcium and phospholipids-dependent enzymes. It comprises a family of closely related kinase subspecies and appears to play a critical role in the regulation of synaptic plasticity and various forms of learning and memory (Stabel & Parker, 1991; Nishizuka, 1992, 1995; Newton, 1995). PKC is one of the major intracellular signal mediators generated after external stimulation of cells via a variety of neurotransmitter receptors (including muscarinic M1, M3, and M5 receptors, noradrenergic α_1 receptors, metabotropic glutamatergic receptors, and serotonergic $5HT_{2A}$ receptors) that induce the hydrolysis of various membrane phospholipids.

Recent cumulative evidence indicated that alterations in PKC activity may play a significant role in mood disorders. Friedman et al. (1993) investigated PKC activity and PKC translocation in response to serotonin in platelets obtained from patients with BPD before and during lithium treatment. They report that the ratios of platelet membrane-bound to cytosolic PKC

Caption for Figure 5.1 (cont.) and GABA) produce their responses via ligand-gated ion channels. Although these responses are very rapid, they also bring about more stable changes via regulation of gene transcription. One pathway gaining increasing recent attention in adult mammalian neurobiology is the Wnt signaling pathway. Wnts are a group of glycoproteins active in development, but now known to play important roles in the mature brain. Binding of Wnts to the Wnt receptor (WntR) activates an intermediary protein, Disheveled, which regulates a glycogen synthase kinase (GSK-3β). GSK-3β exerts many cellular effects; it regulates cytoskeletal proteins, including tau, and also plays an important role in determining cell survival or cell death decisions. GSK-3β has recently been identified as a target for Li$^+$'s actions. GSK-3β also regulates phosphorylation of β-catenin, a protein that when dephosphorylated acts as a transcription factor at LEF (lymphoid enhancer factor) sites. CREB, cAMP response element binding protein; R_q and R_s, extracellular GPCRs coupled to stimulation or inhibition of adenylyl cyclases (ACs), respectively. Rq/11, GPCR coupled to activation of phospholipase C (PLC), MARCKS, myristoylated alanine rich C kinase substrate, a protein associated with several neuroplastic events. Modified and reproduced, with permission from Du et al. (2004).

activities were elevated in the manic subjects. In addition, serotonin-elicited platelet PKC translocation was found enhanced in those subjects. Wang and Friedman (1996) measured PKC isozyme levels, activity, and translocation in post-mortem brain tissue from patients with BPD. They reported increased PKC activity and translocation in brains of patients with BPD compared with control subjects, effects that were accompanied by elevated levels of selected PKC isozymes in the cortices of the patients with BPD.

Evidence from several investigative groups clearly demonstrates that lithium, at therapeutically relevant concentrations, exerts significant effects on the PKC signaling cascade. Data suggest that chronic lithium attenuates PKC activity, and downregulates the expression of PKC isozymes α and ε in the frontal cortex and hippocampus of patients with BPD (Manji & Lenox, 1999, 2000b). Chronic lithium also dramatically reduces the hippocampal levels of a major PKC substrate, myristoylated alanine-rich C kinase substrate (MARCKS), which has a role in regulating long-term neuroplastic events.

In order to validate the biochemical finding of PKC in attributing therapeutic relevance, it is noteworthy that on PKC α and ε isozymes and MARCKS protein the structurally dissimilar antimanic agent valproic acid (VPA) produces effects very similar to lithium (Manji & Lenox, 1999, 2000a). Following chronic lithium treatment, PKC activation was significantly reduced in rat brains, as measured by translocation of the cytoplasmic PKC to membrane-bound PKC. Lithium and VPA may bring about their effects on the PKC signaling pathway by distinct mechanisms, consistent with clinical observations that some patients show preferential response to one or other of the agents, and that additive therapeutic effects are often observed in patients when the two agents are co-administered. Recent studies have shown that lithium inhibited phorbol myristate acetate (PMA) induced PKC activity in prefrontal cortex, which may have an impact on learning and memory (Birnbaum et al., 2004).

In view of the pivotal role of the PKC signaling pathway in the regulation of neuronal excitability, neurotransmitter release, and long-term synaptic events (Conn & Sweatt, 1994; Hahn & Friedman, 1999), we postulated that the attenuation of PKC activity may play a role in the antimanic effects of lithium and VPA. To test this idea, we piloted a study in seven bipolar manic patients treated with tamoxifen, a non-steroidal anti-estrogen known to be a PKC inhibitor at higher concentrations (Couldwell et al., 1993). Tamoxifen showed antimanic efficacy in this study (Bebchuk et al., 2000). Owing to the small sample size, however, these study results are considered preliminary. In view of the preliminary data suggesting the involvement of the PKC signaling system in the pathophysiology of BPD, the results suggest that PKC inhibitors may be very useful in the treatment of mania, and warrant larger double-blind, placebo-controlled studies of tamoxifen and of novel selective PKC inhibitors.

GSK

The enzyme GSK-3 is a crucial kinase that functions as an intermediary in numerous intracellular signaling pathways. Recent research indicates the importance of this enzyme in BPD research (Chen et al., 1999; Gould & Manji, 2002). GSK-3 was directly inhibited by the mood stabilizer lithium in concentrations within therapeutic range ($\sim$1 mM) (Klein & Melton, 1996). GSK-3 has two nearly identical isoforms (slight variations) in mammals: the α and β isoforms. GSK-3 is unique among kinases because it is generally constitutively active, and therefore intracellular signals (e.g. the Wnt pathway, PI3 kinase pathway, protein kinase A, protein kinase C, among many others) inactivate this enzyme. A number of endogenous growth factors (e.g. nerve growth factor and brain-derived neurotrophic factor (BDNF)) use the PI 3-kinase signaling cascade as a major effector system. Thus, growth factors may bring about many of their neurotrophic and neuroprotective effects in part by inhibiting GSK-3. GSK-3 phosphorylates – and thereby inactivates – many transcription factors, and modulates the function of cytoskeletal proteins such as the Alzheimer's disease protein tau (a previous name for GSK-3 was tau kinase). Thus, inhibition of GSK-3 results in the release of this inhibition and the activation of multiple cellular targets.

Growing evidence suggests that GSK-3 plays an important role in regulating neuroplasticity and cellular resilience, including synapse formation and axonal growth. Studies suggest that changes in GSK-3-mediated mitogen activated protein (MAP)-1B (a cytoskeletal protein) phosphorylation are associated with the loss and/or unbundling of stable axonal microtubules (Lucas et al., 1998). Furthermore, GSK-3β inhibition results in accumulation of synapsin I, a protein involved in synaptic vesicle docking and release of growth cone-like areas (Lucas & Salinas, 1997). Recent data also have shown that VPA protects cells from endoplasmic reticulum (ER) stress-induced lipid accumulation and apoptosis by inhibiting GSK-3 (Kim et al., 2005).

Amphetamine-induced hyperactivity is the most established rodent model for mania; this hyperactivity has been attenuated by a number of mood stabilizers including lithium, anticonvulsants, and antipsychotics. Recently, Beaulieu et al. (2004) reported that dopamine-dependent activity increase in mice is mediated via a GSK-dependent mechanism and that both lithium and alternate GSK-3 inhibitors attenuate the hyperactivity in mice lacking dopamine transporter. We have also found that peripheral administration of a GSK-3 inhibitor decreases amphetamine-induced hyperactivity in rats (Gould et al., 2004)). Taken together, these data support the possibility that inhibition of GSK-3 may be involved in the antimanic effects of lithium.

ERK MAP kinase's pathways

Previous studies have demonstrated that lithium and VPA, at therapeutically relevant concentrations activate the extracellular signal-regulated kinase (ERK) MAP kinase cascade in human neuroblastoma SH-SY5Y cells and in critical limbic and limbic-related areas of rodent brain (Einat et al., 2003). MAP kinase is involved in neurotrophic factor signaling pathways, to promote cell survival and may play important roles in neuroprotection and neuroplasticity in mood disorders (reviewed by Du et al., 2003a).

Calcium signaling abnormalities

Calcium ions play a critical role in regulating the synthesis and release of neurotransmitters, neuronal excitability, and long-term neuroplastic events, and therefore it is not surprising that a number of studies investigated intracellular Ca^{2+} in peripheral cells in BPD. Intracellular Ca^{2+} signaling and homeostasis are maintained by an intricate array of processes acting in concert including, for example, inositol trisphosphate (IP_3)- and ryanodine-stimulated release of Ca^{2+} from ER storage pools, voltage- and ligand-gated ion channel-mediated Ca^{2+} influx, store-operated Ca^{2+} entry (SOCE), plasma membrane and sarcoplasmic/ER Ca^{2+}-ATPase pumps (PMCAs and SERCAs), and mitochondrial Ca^{2+} uptake, storage, and release (Pisani et al., 2004). Regardless of the complexity that intracellular calcium is regulated, impaired regulation of Ca^{2+} cascades is the most reproducible biological abnormality described in BPD research. For this reason, mechanisms involved in Ca^{2+} regulation are postulated to underlie aspects of the pathophysiology of BPD. To date, cumulative studies consistently revealed elevations in basal intracellular Ca^{2+} levels in platelets, lymphocytes, or neutrophils of patients with BPD (Emamghoreishi et al., 1997).

Post-mortem brain studies have also revealed changes that may reflect possible "signatures" of abnormal Ca^{2+} homeostasis in BPD. These include a marked blunting of G-protein activated phosphoinositide (PI) hydrolysis (Jope et al., 1996) and altered mRNA expression levels of two candidate proteins, which may have important roles in Ca^{2+} homeostasis, inositol monophosphatase (IMPase) type II (Yoon et al., 2001a), and a transient receptor potential channel, TRPM2 (TRPC7 in earlier nomenclature) (Yoon et al., 2001b), a ligand-gated plasma membrane ion channel which also mediates Ca^{2+} entry into cells. In addition, both lithium and VPA inhibit IMPase (Hallcher & Sherman, 1980), an effect suggested to diminish overactive signaling through PI-linked second messengers and, in turn, Ca^{2+} (Berridge, Downes & Hanley, 1982).

In summary, mood disorders associate with alterations in signaling network, which may subsequently regulate synaptic plasticity and cell resilience of neuronal circuitry associated with affective disorders. It is noteworthy that modulation of synaptic plasticity by signaling cascade has been extensively studied in glutamatergic systems – specifically, AMPA receptor trafficking.

AMPA RECEPTOR TRAFFICKING PLAYS CRITICAL ROLES IN THE REGULATION OF VARIOUS FORMS OF NEURAL PLASTICITY

Surprisingly, the potential role of the glutamatergic system in the pathophysiology and treatment of BPD is only recently undergoing earnest evaluation. Glutamate is the major excitatory synaptic neurotransmitter regulating numerous physiologic functions in the mammalian CNS, such as synaptic plasticity, learning, and memory, and representing a major neurotransmitter system in the circuitry thought to subserve many of the symptoms of severe, recurrent mood disorders (Drevets, 2001). Three major classes of ionotropic glutamate receptors are expressed throughout the mammalian CNS, including amino-3-hydroxy-5-methylisoxazole-4-propionic acid (AMPA), kainate, and N-methyl-D-aspartate (NMDA) receptors. AMPA receptors mediate the majority of excitatory synaptic transmission in the CNS; the AMPA R channel is composed of the combination of GluR1, GluR2, GluR3, and GluR4 subunits. The AMPA receptor is stimulated by the presence of glutamate and characteristically produces a fast excitatory synaptic signal that is responsible for the initial reaction to glutamate in the synapse.

The NMDA receptor is activated by glutamate, a co-agonist, namely glycine or D-serine and depolarization, to be activated to open and permit the entry of both Na^+ and Ca^{2+} (Figure 5.2). The NMDA receptor channel is composed of combinations of NR1, NR2A, NR2B, NR2C, NR2D, NR3A, and NR3B subunits (Figure 5.2). In fact, it is generally believed that it is the activation of the AMPA receptor that results in neuronal depolarization sufficient to liberate the Mg^{2+} cation from the NMDA receptor, thereby permitting its activation. NMDA receptors play a critical role in regulating synaptic plasticity (Malenka & Nicoll, 1999), including AMPA receptor trafficking. The best-studied forms of synaptic plasticity in the CNS are long-term potentiation (LTP) and long-term depression (LTD) of excitatory synaptic transmission. The molecular mechanisms of LTP and LTD are extensively characterized and proposed to represent cellular models of learning and memory (Malenka & Nicoll, 1999). During NMDA-receptor-dependent synaptic plasticity, Ca^{2+} influx through NMDA receptors can activate a wide variety of kinases and/or phosphatases that, in turn, modulate synaptic strength (Figure 5.2).

Emerging data suggest AMPA receptor trafficking, including receptor insertion, internalization, and delivery to synaptic sites, provides an elegant mechanism for activity-dependent regulation of synaptic strength. AMPA receptor subunits undergo constitutive endocytosis and exocytosis; however, the process is highly regulated with a variety of signal transduction cascades being capable of producing short- or long-term changes in synaptic surface expression of AMPA receptor subunits (Malenka & Nicoll, 1999). Indeed, although the mechanisms of LTP and LTD have not been completely elucidated, it is widely accepted that AMPA receptor trafficking is the key player in these phenomena.

REGULATION OF AMPA RECEPTOR TRAFFICKING BY SIGNALING CASCADES

Most vesicle trafficking requires the ordered coating of a donor membrane, budding, and fusion to form transport vesicles, transport by passive or active delivery along microtubule, and final fusion with the target membrane (Antonny & Schekman, 2001). AMPA receptors adopted this mechanism to be delivered to the neuronal membrane surface. Each subunit of AMPA receptors is composed of an N-terminal extracellular domain, membrane-spanning domain, and C-terminal intracellular domain (Bennett & Dingledine, 1995; Wo, Bian & Oswald, 1995). AMPA receptor trafficking is subunit-specific and regulated by phosphorylation of its C-terminal domain, and subsequently alteration of protein–protein interactions.

PKA pathway

The GluR1 subunit appears to govern the trafficking behavior of heteromeric GluR1 or GluR2 receptors, preventing constitutive exchange and conferring inducible delivery of the heteromer (Shi et al., 2001). Phosphorylation of GluR1 at the PKA site p845 facilitates the insertion of GluR1 onto the membrane and synapses, and is often associated with LTP (Lee et al., 2000). Dephosphorylation of the GluR1 by protein phosphatases (e.g. calcineurin and PP1) target GluR1 to recycling endosomes, where rephosphorylation by PKA may occur and the receptors will be reinserted onto the membrane (Ehlers, 2000). Phosphorylation of GluR1 at PKA site may be enhanced by a synapse-associated protein 97 (SAP97) or protein A kinase anchoring protein (AKAP79) complex that directs PKA to GluR1 via a PDZ domain interaction (Colledge et al., 2000).

A growing body of data suggest the importance of this PKA phosphorylation site of GluR1 in psychiatric

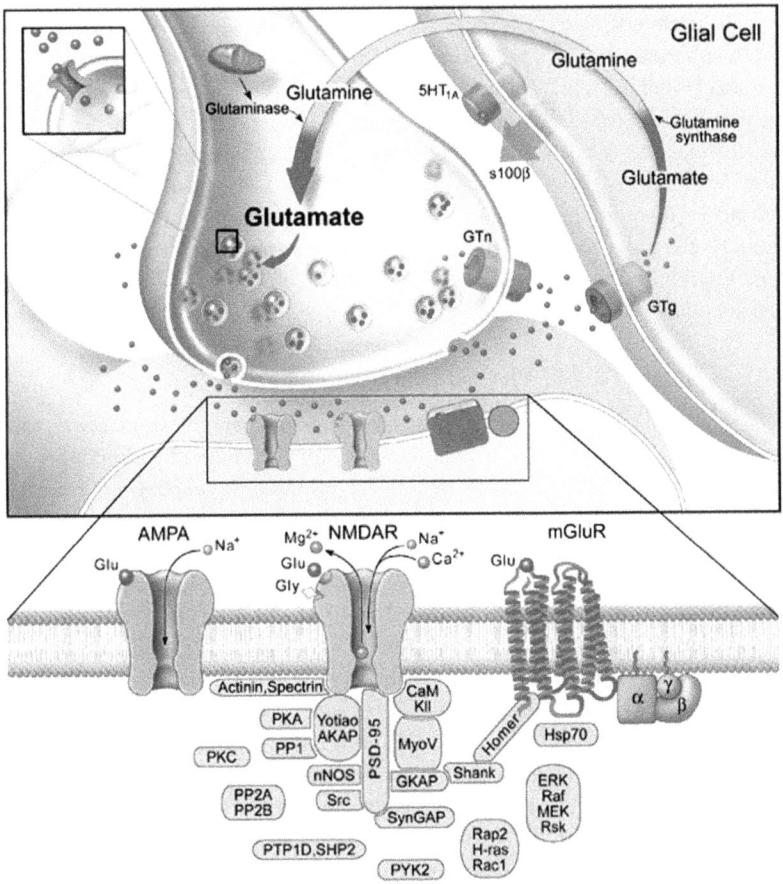

Figure 5.2 Glutamatergic system. This figure depicts the various regulatory processes involved in glutamatergic neurotransmission. The biosynthetic pathway for glutamate involves synthesis from glucose and the transamination of α-ketoglutarate; however, a small proportion of glutamate is formed more directly from glutamine by glutamine synthetase. The latter is actually synthesized in glia, and via an active process (requiring ATP) is transported to neurons where glutaminase is able to convert this precursor to glutamate. Furthermore, in astrocytes glutamine can undergo oxidation to yield α-ketoglutarate, which may also be transported to neurons and participate in glutamate synthesis. Glutamate is either metabolized, or sequestered and stored, into secretory vesicles by vesicular glutamate transporters (VGluT). Glutamate may then be released by a calcium-dependent-excitotoxic process. Once released from the pre-synaptic terminal, glutamate is able to bind to numerous excitatory amino acid (EAA) receptors, including both ionotropic (e.g. AMPA or NMDA) and metabotropic receptors. Pre-synaptic regulation of glutamate release occurs through metabotropic glutamate receptors (mGluR$_{2/3}$), which subserve the function of autoreceptors. However, these receptors are also located on the post-synaptic element. Glutamate has its action terminated in the synapse by reuptake mechanisms utilizing distinct GLU transporters (GLUTs) which exist not only on pre-synaptic nerve terminals, but also on astrocytes; indeed, current data suggests that astrocytic Glu uptake may be more important for clearing excess Glu, raising the possibility that astrocytic loss (as has been documented in mood disorders) may contribute to deleterious GLU signaling, but more so by astrocytes. It is now known that there are a number of important intracellular proteins that are able to alter the function of glutamate receptors (*see diagram*). Modified and reproduced with permission from Du *et al*. (2004).

disorders. Very recent studies also demonstrated that chronic administration of antidepressant agents enhances GluR1 membrane expression of GluR1 as well as phosphorylation of GluR1 at this PKA site (p845) (Martinez-Turrillas, Frechilla & Del Rio, 2002; Svenningsson et al., 2002). It has also been shown that dopamine, which is a neurotransmitter involved in psychostimulant effects, binds to D1 receptors and enhances surface expression of GluR1 receptors through activation of PKA at this site and AKAP 79 (Mangiavacchi & Wolf, 2004).

Calcium or calmodulin-dependent kinase II pathways

Numerous studies demonstrate that calmodulin-dependent kinase II (CaMKII) is required for the proper formation of LTP in slice preparations, and in regulating learning and memory in rodents (Fink & Meyer, 2002). In response to stimulation, CaMKII translocates to post-synaptic sites, where it has two major effects on AMPA receptor activity at the post-synaptic density during the formation of LTP (Fink & Meyer, 2002). First, the AMPA single conductance is directly increased by CaMKII at Ser831 of GluR1 subunit (Derkach, Barria & Soderlinh, 1999). Second, CaMKII is required for the delivery of AMPA receptor to the synapse, which is lacking AMPA receptors (Shi et al., 1999; Liao, Scannevin & Huganir, 2001; Shi et al., 2001). This enhancement of synaptic GluR1 level by activation of CaMKII requires an intact C-terminal domain of GluR1, and is possibly involved in interaction with SAP97 (Hayashi et al., 2000). Protein phosphatase 1 (PP1), an important modulator for learning and memory, may dephosphorylate the phosphorylation of GluR1 at p831 site by CaMKII (Genoux et al., 2002). It is noteworthy that this CaMKII site of GluR1 is phosphorylated after antidepressant treatment (Svenningsson et al., 2002) and CaMKII may also be involved in the pathophysiology of mood disorders (Du et al., 2004b).

ERK MAP kinase pathway

A recent study reported that small GTPases, Ras and Rap, are involved in AMPA receptor trafficking through a post-synaptic signaling mechanism. Ras mediates activity-evoked increase in GluR1 or GluR4 containing AMPA receptor surface expression at synapses via a pathway that requires p42/44 MAPK activation. In contrast, Rap mediates NMDA-dependent removal of synaptic GluR2/3-containing vesicles via a pathway that involves p38 MAP kinase. The regulation through Ras and Rap, which work as molecular switches, may in turn control the AMPA receptor level at synapses (Zhu et al., 2002).

PKC pathways

AMPA GluR2 receptors respond to secondary signals by constitutive receptor recycling. Phosphorylation of Ser880 on GluR2 provides a switch from receptor retention at the membrane by binding to ABP/GRIP, to receptor internalization by binding to PICK 1. Therefore, phosphorylation of GluR2 at Ser880 by PKC may release the AMPA receptor from the anchoring proteins and initiate the internalization of receptors (Matsuda et al., 2000; Xia et al., 2000; Kim et al., 2001a; Perez et al., 2001).

The mechanism for AMPA receptor trafficking is specific for brain region and type of neuron. For example, the endocytosis of AMPA receptors mediating LTD is triggered by very different signaling cascades in different cell types despite the fact that a conserved cell biological mechanism (i.e. clathrin- or dynamine-dependent endocytosis) appears to be involved. Specifically, in CA1 pyramidal cells, protein phosphatases seem to be involved in triggering LTD through dephosphorylation of GluR1. Phosphorylation of PKA site on GluR1 is associated with LTP (Ehlers, 2000). However, in midbrain dopamine cells, activation of PKA appears to trigger LTD and endocytosis of AMPA receptors (Gutlerner et al., 2002). Most importantly for the present discussion, AMPA receptor trafficking is highly regulated by the PKA, PKC, CaMKII and MAP kinase signaling cascades (reviewed by Du et al., 2004b). These are the very same signaling cascades upon which mood stabilizers and antidepressants exert major effects (Popoli et al., 2003; Gould et al., 2004; Payne et al., 2004). These observations led to an extensive series of studies, which clearly demonstrate that AMPA receptor trafficking is highly regulated by antidepressants and mood stabilizers (Du et al., 2003b; Gray et al., 2003).

MOOD STABILIZERS MODULATE SYNAPTIC PLASTICITY BY REGULATING AMPA RECEPTOR TRAFFICKING

In view of the critical role of AMPA receptor trafficking in regulating various forms of plasticity, it is important to determine if two structurally highly dissimilar antimanic agents, lithium and VPA, exert effects on AMPA receptor trafficking. Lithium, a monovalent cation, and VPA, an eight-carbon fatty acid, are the two most commonly used agents in the treatment of mania. Because lithium and VPA require several weeks to exert their therapeutic effects, it is widely believed that adaptive changes in intracellular signaling or cellular physiology, or both, underlie the beneficial effects. Interestingly, these two agents exert robust effects on the very same signaling pathways known to regulate AMPA receptor trafficking.

Recent experiments have investigated if lithium and VPA regulate synaptic plasticity and AMPA receptor trafficking in the hippocampus, a brain region presumed to be involved in the circuitry of mood disorders. This research indicates that the structurally highly dissimilar antimanic agents lithium and VPA have a common effect on downregulating AMPA GluR1 synaptic expression in the hippocampus after prolonged treatment with therapeutically relevant concentrations as assessed both *in vitro* and *in vivo* (Figure 5.3).

However, total protein levels of GluR1, and synaptotagmin remains unchanged after lithium and VPA treatment *in vitro* and *in vivo*. In cultured hippocampal neurons, lithium and VPA attenuate surface GluR1 expression after long-term treatment (Du et al., 2004a). Phosphorylation of a specific PKA site (GluRp845) is significantly attenuated by lithium and VPA treatment by 52% and 33%, respectively.

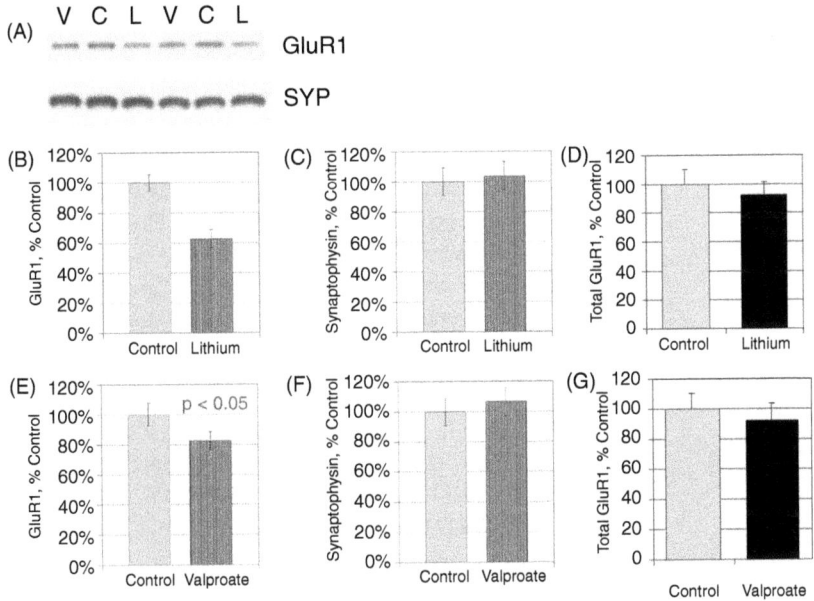

Figure 5.3 AMPA receptor subunit GluR1 was attenuated in synaptosomal preparation from long-term lithium and valproate treated animals. (A) Western blot analysis of hippocampal synaptosomal preparation from lithium or valproate treated animals with anti-GluR1 or anti-synaptophysin antibodies. (B) Quantification of GluR1 content in hippocampal synaptosome from lithium-treated ($n=5$) and control animals ($n=6$), (Student's t-test, $p<0.01$). (C) Quantification of synaptophysin protein in hippocampal synaptosomal preparation from lithium-treated and control animals. (D) Total GluR1 expression levels in hippocampal tissue homogenates from lithium-treated and control animals. (E) GluR1 protein content in hippocampal synaptosomes from valproate-treated ($n=8$) and control animals ($n=7$), (Studuent's t-test, $p<0.05$). (F) Synaptophysin protein levels in hippocampal synaptosomal preparation from valproate-treated and control animals. (G) Total GluR1 expression levels in hippocampal tissue homogenates from valproate-treated and control animals.

Sp-cAMP treatment reverses the attenuation of phosphorylation by lithium and VPA, and also recruits GluR1s back to the surface, suggesting that phosphorylation of GluRp845 is involved in the mechanism of GluR1 surface attenuation (Du et al., 2003a, 2003b, 2004, 2008; Gray et al., 2003). The therapeutic relevance of the finding is supported by experiments showing that an agent that provokes mania, the antidepressant imipramine, has an opposite effect as it upregulates AMPA synaptic strength in the hippocampus (Figure 5.4).

Further supporting these data are recent studies showing that AMPA receptor antagonists attenuate several "manic-like" behaviors produced by amphetamine administration. Thus, AMPA antagonists attenuate psychostimulant-induced development or expression of sensitization and hedonic behavior without affecting spontaneous locomotion. Additionally, some studies demonstrate that AMPA receptor antagonists reduce amphetamine- or cocaine-induced hyperactivity (Burns et al., 1994; Li et al., 1997; Mead & Stephens, 1998; Hotsenpiller, Giorgetti & Wolf, 2001; Backstrom & Hyytia, 2003; Tzschentke, 2003). The need to use caution in the appropriate application of animal models to complex neuropsychiatric disorders is well articulated, and, in fact, it is unlikely we will ever develop rodent models that display the full range of symptomatology as clinically expressed in humans (Nestler et al., 2002; Einat et al., 2003). However, one current model of mania, extensively used, and with reasonable heuristic value in the study of mood disorders, involves the use of psychostimulants in appropriate paradigms. Thus, psychostimulants like amphetamine and cocaine are known to induce manic-like symptoms in healthy volunteers, and to trigger frank manic episodes in individuals with BPD (Goodwin & Jamison, 1990). Thus, the best-established animals of mania utilize the administration of amphetamine or cocaine to produce hyperactivity, risk-taking behavior, and increased hedonic drive – all very important facets of the human clinical condition of mania. Moreover, these psychostimulant-induced behavioral changes are attenuated by the administration of chronic lithium in a therapeutically relevant timeframe. Thus, the fact that AMPA receptor antagonists are capable of attenuating psychostimulant-induced sensitization, hyperactivity, and hedonic behavior (Backstrom & Hyytia, 2003; Burns et al., 1994; Li et al., 1997; Mead & Stephens 1998; Hotsenpiller, Giorgetti & Wolf, 2001; Tzschentke, 2003) provides compelling behavioral support for our contention that AMPA receptors play important roles in regulating affective behavior. Taken together, the biochemical and behavioral studies investigating the effects of antimanic (lithium and VPA) and pro-manic (antidepressants, cocaine, amphetamine) agents on GluR1 strongly suggest that AMPA receptor trafficking is an important target in the pathogenesis and treatment of certain facets of BPD.

THE ROLE OF GLUTAMATERGIC SYSTEMS IN TREATMENT OF MOOD DISORDERS

The noradrenergic and serotonergic hypothesis of mood disorders, which was developed from the pharmacological effects of early drug development, no longer provides a satisfactory explanation of the mode of action of all antidepressant agents or of the underlying pathophysiology in depression. Several agents associated with glutamatergic systems demonstrated antidepressant efficacy in clinical studies. In the 1950s, D-cycloserine (DCS), a partial agonist at the NMDA receptor glycine site used as a part of multidrug antituberculosis treatment, was reported to have

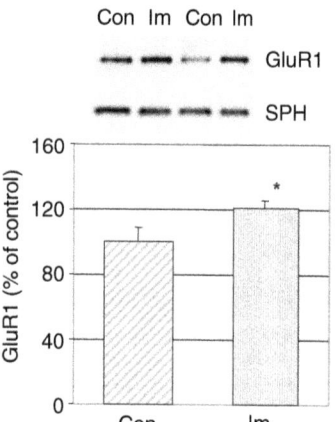

Figure 5.4 Antidepressant imipramine induces an increase on synaptic GluR1 in vivo. Rats were treated with imipramine (10 mg/kg, twice daily, intraperitoneal injection) for four weeks and hippocampal tissues were isolated and synaptosome preparations were obtained using Ficoll gradient assay. Equal amounts of proteins were separated by gel electrophoresis and analyzed with anti-GluR1 antibodies and anti-synaptophysin antibodies. The bands were analyzed with a Kodak imaging system. ($n = 8$; Student's t-test, $p < 0.05$) reprinted from Du et al. (2004).

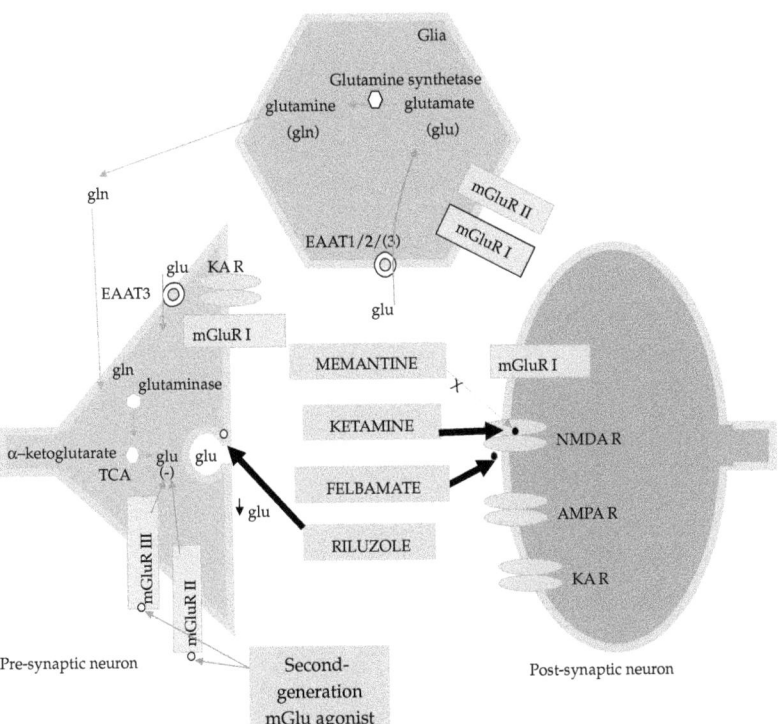

Figure 5.5 The glutamatergic receptor system: antiglutamatergic agents as putative treatments for mood disorders. (A) *Upper left*: Ionotropic receptors: activation of the AMPA receptor (AMPA R) by glutamate permits the depolarization of the membrane. When glutamate and glycine are present this depolarization results in the release of magnesium from the NMDA receptor (NMDA R) channel. Calcium also enters through the NMDA R pore. Interchange of cations additionally occurs via NMDA R and Kainate glutamate rceptor (KA R). *Upper right*: Metabotropic receptors: activation of group I metabotropic glutamate receptors (mGlu I), which are couped with a G protein (Gq/11), produces activation of phospholipase C- β (PLC- β). Activation of group II metabotropic glutamate receptors (mGlu II), which are coupled with Gi or Go, produce either inhibition of adenylyl cyclase (AC) or opening of potassium channels (not shown) respectively. *Bottom*: The subunit coposition of known receptor subtypes are shown in the figure. (B) Receptors that may represent a target for novel agents for the treatment of mood disorders. Glutamate is synthesized in neuron from α-ketoglutarate through the tricarboxylic acid cycle (TCA). After release, glutamate is reuptaken by glutamate transporters (EAAT1/2/3), shown in glia and a pre-synaptic neuron (EAAT 3). In the glia, glutamate is catabolized to glutamine (through the enzyme glutamine synthetase), diffuses to the neurons and is then metabolized back to glutamate (through the enzyme glutaminase). The different glutamate receptors and the presumed antiglutamatergic drug site of action are presented: **Memantine** is a non-competitive antagonist NMDA receptor. **Felbamate** is non-competitive NMDA receptor antagonist (glycine NR1 and glutamate NR2B), an AMPA receptor antagonist, an mGlu group I receptor antagonist, and a glutamate release inhibitor (acting through blockade of Ca^{++} and Na^+ voltage- dependent channels). **Riluzole** is a glutamate release inhibitor (acting through blockade of Ca^{++} and Na^+ voltage-dependent channels), a $GABA_A$ agonist, and probably an AMPA and KA antagonist. **Zinc**: is an endogenous modulator of ligand- and voltage-gated ion channels. The second-generation of mGlu groups II and III receptors agonists is also depicted. Reprinted from Du *et al.* (2004).

mood-elevating effects (Crane, 1959, 1961; Heresco-Levy & Javitt, 1998). Subsequent studies with DCS have reported cases of euphoria and amnestic effects in humans that resemble the effects of subperceptual doses of non-competitive NMDA antagonists (Krystal *et al.*, 1994). Early evidence that NMDA antagonism may be important for antidepressant effects in humans comes from the cases series and blinded trials conducted with the non-competitive NMDA antagonist amantadine. Amantadine has been shown to have antidepressant effects in patients with Parkinson's disease and in unipolar and bipolar patients (Parkes *et al.*, 1970; Rizzo &

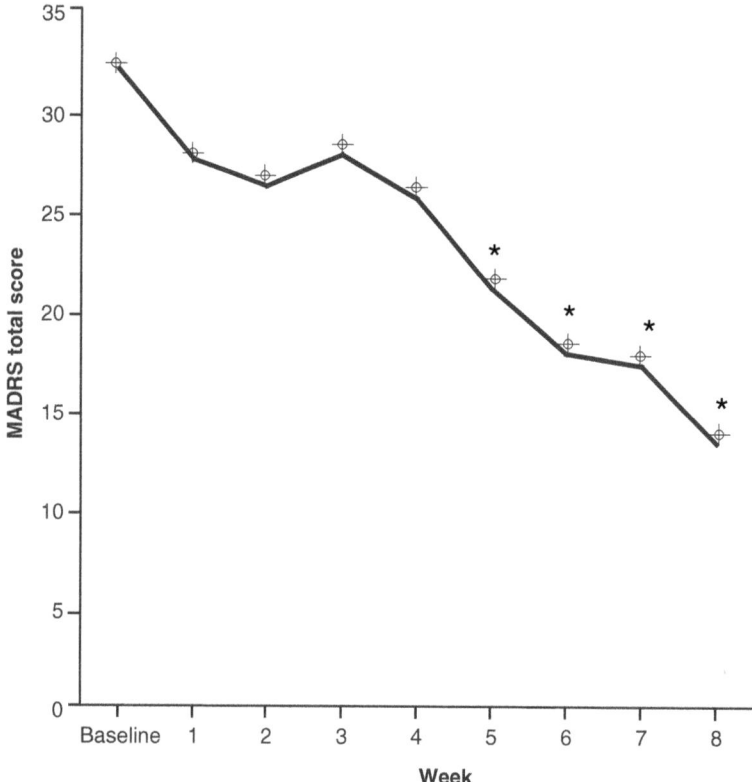

Figure 5.6 Mean change in MADRS total scores from baseline in patients with treatment-resistant bipolar depression who were treated with riluzole. After open treatment with lithium for a minimum period of four weeks, subjects who continued to have a Montgomery–Asberg Depression Rating Scale (MADRS) rating scale score >20, received riluzole (50–200 mg/day) for eight weeks. Fourteen bipolar depressed patients entered the study. The linear mixed models for total MADRS showed a significant treatment effect ($n = 14$; * $p < 0.05$ compared to baseline). No switch into hypomania or mania was observed. Overall, riluzole was well tolerated. Reprinted from Du et al. (2004).

Morselli, 1972; Vale, Espejel & Dominguez, 1971) (Figure 5.5). However, it is important to note that amantadine has indirect effects on other receptors (e.g. dopamine). Recently, Berman et al. (2000) reported the first placebo-controlled, double-blind trial assessing the treatment effects of a single dose of the NMDA receptor antagonist, ketamine, in seven patients with depression (Figure 5.5). Ketamine is a high-affinity NMDA receptor antagonist; it has less, but potentially relevant, affinity for the μ opiate receptors and weak antagonist activity for the dopamine receptor (Wong et al., 1996) (Figure 5.5). Perhaps the greatest evidence that glutamate modulation may be important in the pathophysiology of mood disorders comes from the clinical use of the anticonvulsant lamotrigine. Recently in several double-blind, placebo-controlled studies, lamotrigine was reported to be effective in bipolar depression (Calabrese et al., 1999). Although the exact mechanism of action of lamotrigine is unknown, inhibition of an excessive release of glutamate is postulated as a likely mechanism of action for this drug (Calabresi et al., 1996; Leach, Marden & Miller, 1986; Wang et al., 1996).

In addition, a recent study also suggested the potential of AMPA receptor modulator in the treatment of mood disorder that AMPA potentiator has an antidepressant effect in animal models (Zarate et al., 2003). As discussed in this review, mood stabilizers lithium, and VPA appear to attenuate glutamatergic function through multiple mechanisms. Repeated administration of lithium, the prototype mood stabilizer, may promote the uptake to glutamate from the synapse (Dixon & Hokin, 1998), attenuate the function of glutamate receptors (Nonaka, Hough & Chuang, 1998), and reduce the function of intracellular signaling cascades that are activated by the binding of glutamate to its receptors (Manji & Lenox, 1999) (Figure 5.6).

Given the pre-clinical and clinical data which support a role for the glutamatergic system in the pathophysiology of mood disorders, there are a number of therapeutic agents that act on this system which should be explored as potential treatments for mood disorders. These agents include riluzole, memantine, felbamate, and zinc. Riluzole, a neuroprotective agent with anticonvulsant properties, can easily cross the blood–brain barrier (Benavides et al., 1985). Riluzole inhibits the release of glutamate instead of producing NMDA antagonism and is one of the few antiglutamatergic agents available for use in humans. It is available for clinical use in severely debilitated medically ill patients (ALS) and it has a very favorable side effect profile; these characteristics make it an ideal candidate to test in patients with mood disorders (Zarate et al., 2003). A recent study from our laboratory has demonstrated riluzole in combination with lithium showed efficacy for bipolar depression (Figure 5.6) (Zarate et al., 2005) suggesting riluzole may indeed have antidepressant efficacy in subjects with bipolar depression. Memantine is a potent noncompetitive voltage-dependent NMDA antagonist that can easily cross the blood–brain barrier, with a receptor effect comparable to MK-801 (Bormann, 1989) and has been demonstrated to inhibit (Jope et al., 1996) MK-801 binding to human hippocampal NMDA receptors (Berger et al., 1994). Memantine is one of the few NMDA antagonists available for use in humans and is ideal for testing in mood disorders as it has been in clinical use for many years with minimal side effects (Kornhuber et al., 1994) and has a very favorable pharmacological profile (see Figure 5.5). Unfortunately, the low-to-moderate affinity NMDA antagonist memantine was found to be devoid of significant antidepressant effects in a recently conducted double-blind placebo-controlled trial (Zarate et al., 2006). Although no significant antidepressant effects were found with the low-to-moderate affinity NMDA antagonist memantine, this finding does not exclude the possibility that antidepressant or mood-stabilizing effects might occur by using higher-affinity NMDA antagonists. Felbamate (2-phenyl-1, 3-propanediol dicarbamate) is a unique broad-spectrum anticonvulsant (Brown & Aiken, 1998). Kass (1994) reported substantial improvement with felbamate, in a patient with severe BPD (see Figure 5.5). However, the use of felbamate has been limited by the unexpected increased risk of aplastic anemia. Zinc ($ZnSO_4$) is a very potent inhibitor of the NMDA receptor complex and plays an important role in a wide range of biochemical processes (Harrison & Gibbons, 1994), and might have a role in the treatment of mood disorders.

CONCLUDING REMARKS

For many years, the notion that the monoamine neurotransmitter systems are dysfunctional in bipolar disorder was supported by a considerable body of data, explaining why they have become a common target for pharmacological interventions. However, conceptual and experimental evidence of abnormalities in the regulation of signal transduction cascades and neuroplasticity has been found to be an integral phenomena that could more primarily underlie the pathophysiology of BPD. It is thus noteworthy that there is growing evidence from pre-clinical and clinical research that the glutamatergic system is involved in the pathophysiology and treatment of mood disorders. For a complex disease like BPD, it is not surprising that many molecules involved in network of signaling cascades and synaptic plasticity are regulated. The findings that mood stabilizers – in therapeutically meaningful paradigms – regulate AMPA receptors at synapses, opens new potential avenues for new drug development in regards to regulating glutamatergic synaptic strength in critical neuronal circuits. The development of new modulators of the glutamatergic system for the treatment of mood disorders may lead to improved therapeutics for these devastating disorders.

ACKNOWLEDGMENTS

The authors would like to thank Mrs. Patricia J. Williams and Mrs. Holly Giesen for their contribution in editing and organizing the references.

Part II
Novelty, stress, and hormones in plasticity

6 · Sleep-dependent memory consolidation and reconsolidation

Matthew P. Walker and Robert Stickgold

Molecular, cellular, and systems-level processes convert initial, labile memory representations into more permanent ones, available for continued reactivation and recall over extended periods of time. But these processes of memory consolidation and reconsolidation are not all-or-none phenomena, but a continuing series of biologic adjustments that enhance both the efficiency and utility of stored memories over time. In this chapter, we review the role of sleep in supporting these disparate but related processes.

INTRODUCTION

The question of "sleep-dependent memory consolidation" is a complex one. Each term in the phrase – sleep, dependent, memory, and consolidation – begs for clarification. For a start, the term "memory" covers a wide range of memory types, which differ in the kinds of information stored, the brain structures mediating this storage, and, in humans, whether the information is accessible to conscious awareness. There is no clear consensus at this time on how many such memory systems there are, and how they should be defined, either in terms of information content or brain structures involved in their storage (Squire & Zola-Morgan, 1996). The most widely accepted taxonomy divides human memories first into declarative and non-declarative, based on their accessibility to conscious recall, and then into finer and finer subdivisions of these basic categories (Figure 6.1 (A)) (Schacter & Tulving, 1994).

Similarly, the term "memory consolidation" refers to a poorly defined set of processes which take an initial, unstable memory representation and convert it into a form that is both more stable and more effective. At present it is unclear how memories are altered after initial encoding and no consensus as to which of the processes contributing to this alteration should be included under the umbrella of memory consolidation. When the term was first introduced (Muller & Pilzecker, 1900), it referred to an as yet unknown process which, over a period of minutes to hours, made recently formed memories resistant to degradation by, for example, electroconvulsive shock or from competing memories (Misanin, Miller & Lewis, 1968; Schneider & Sherman, 1968; McGaugh, 2000). But this notion of a single, relatively rapid process of memory consolidation has yielded to another one including phases of stabilization, enhancement, and integration, extending over hours to years.

More recently, the concept "memory reconsolidation" has resurfaced to describe yet another aspect of post-encoding memory modification (Nader, 2003). There is now evidence that when previously consolidated information is reactivated, either by returning an animal to an earlier learning environment or by having humans briefly rehearse or recall a previously learned task, the memory is destabilized, and returns to a labile state in which it is again susceptible to interference, thus necessitating *re*consolidation.

It now appears that many of the steps in this memory processing cascade occur preferentially or even exclusively during periods of sleep. Researchers are now focussing on identifying those types of memory, and, for each, the individual steps which show strong sleep-dependency. But, again, sleep does not refer to a simple unitary phenomenon; instead, it represents a complex array of brain states that differ in their physiology, chemistry, functional anatomy, and phenomenological experiences. Indeed, sleep has been broadly divided into rapid eye movement (REM) sleep and non-REM (NREM) sleep, which alternate across the night, in humans, in a 90-minute cycle (Figure 6.1 (B)). NREM sleep is further subdivided into NREM stages 1–4 (Rechtschaffen & Kales, 1968), with both NREM and REM appearing to differ in their contribution to sleep-dependent memory consolidation (Walker & Stickgold, 2004).

Before describing these systems of memory processing in greater detail, we would like to offer an

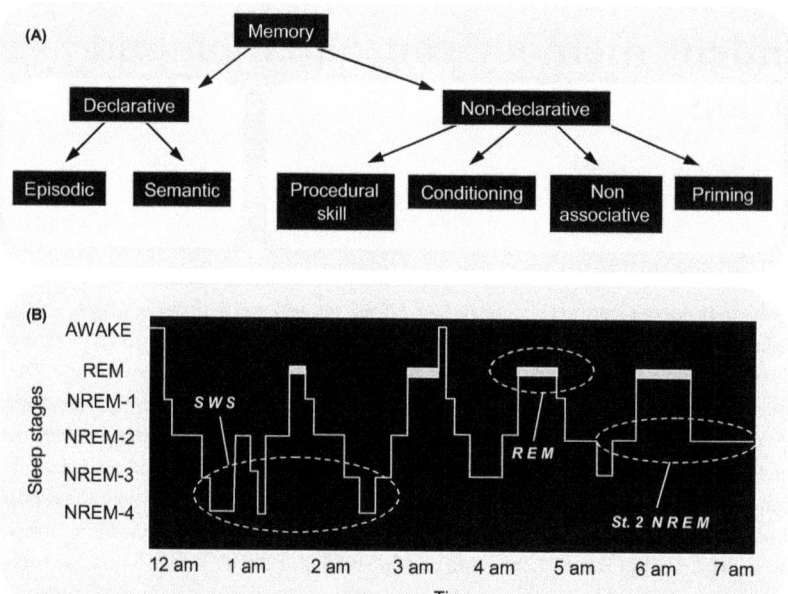

Figure 6.1. Forms of memory and stages of sleep. Neither memory (A) nor sleep (B) represent homogeneous phenomena. (A) Declarative memory includes consciously accessible memories of fact-based information (i.e. knowing "what"), and contains several subcategories, including episodic memory (memory for events in our past) and semantic memory (memory for general knowledge) (Tulving, 1985). In contrast, non-declarative memory includes all non-conscious memories, and includes subcategories such as conditioning, implicit memory, and procedural memory (i.e. knowing "how"). (B) In primates and felines, NREM sleep has been divided into sub-stages 1 through 4, corresponding to increasingly deeper states of sleep (Rechtschaffen & Kales, 1968). The deepest NREM stages, stages 3 and 4, are collectively referred to as "slow wave sleep" (SWS), based on a prevalence of low-frequency (0.5–4 Hz) cortical oscillations. Dramatic changes in brain electrophysiology, neurochemistry, and functional anatomy occur across these sleep stages, making them biologically distinct from the waking brain, and dissociable from one another. For example, SWS is characterized by a diminution in cholinergic activity and REM sleep by a suppression of release of norepinephrine from the locus coeruleus and serotonin from the Raphe nucleus.

overview of our perspective on memory consolidation and reconsolidation. First, both these processes occur over time automatically, outside of awareness and without intent. Furthermore, both are characterized by a range of mechanisms, from intracellular gene inductions to brain-wide, system-level reorganizations of memory representations. Second, although these processes occur automatically, they are nevertheless modulated by other factors. As a result, the multiple components of memory consolidation and reconsolidation form a coherent whole, which function to optimally integrate initially encoded memories into an organism's existing informational networks, and which continue to refine and remodel these memories in the context of ongoing experience, during wake and sleep. In short, memories do not simply form in the brain: they evolve.

STAGES OF MEMORY: CONSOLIDATION

The evolution of memories can be a long and complex process, occurring in several distinct stages (Figure 6.2). Whilst the initial encoding of a memory is a rapid (milliseconds; ms) process, its long-term maintenance requires processes that continue to modify it over hours to years (Muller & Pilzecker, 1900).

Even the original view of consolidation as stabilization is now in flux. When originally proposed in 1900 (Muller & Pilzecker, 1900), consolidation was

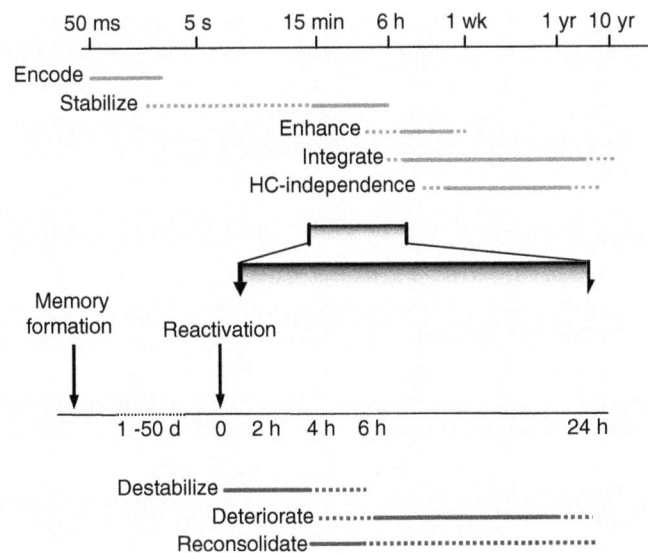

Figure 6.2. Time course of memory processes. (Top) Memory formation and consolidation – after the initial rapid encoding of a sensory experience, the neural representation of the memory may go through a number of automatic processes, independent of rehearsal, intent, or awareness. These can stabilize and enhance a memory, so it is resistant to interference and more effective to guiding behavior, and may also integrate the memory into larger associative networks. The latter process is thought to permit episodic memories to be recalled without hippocampal (HC) involvement. The extent to which such processes affect different memory systems is unclear. Note logarithmic time scale. (Bottom) Memory reactivation and reconsolidation – after stabilization is complete, reactivation of a memory can lead to its return to an unstable form. Normally, such memories appear to be reconsolidated after this destabilization, but if such reconsolidation is blocked, degradation of the memory can ensue.

defined as resistance to interference from competing memories. Animal studies in the middle of the twentieth century demonstrated that such consolidation was also required for the development of resistance to the more drastic actions of electroconvulsive shock (ECS) (Duncan, 1949) and protein synthesis inhibitors (Agranoff, Davis & Brink, 1965). More recently, the concept of memory consolidation has been simultaneously extended and challenged (Walker, 2005), and proposed to contain substages – namely stabilization and enhancement.

Stabilization: As with declarative memories, human procedural skill memories have also been shown to be disrupted by training on an alternate task within the first hours after training, suggesting that such learning initially requires a process of stabilization (Brashers-Krug, Shadmehr & Bizzi, 1996; Walker *et al.*, 2003). More importantly, these newer findings demonstrate that the time-course of stabilization may be functionally significant. When initially conceptualized, consolidation was considered to be an inexorable process which, once started, continued to completion except under the most severe of insults, such as ECS. But the finding that ecologically relevant stimuli (i.e. competing memories) may also interrupt consolidation suggests a functional role for this process, where a memory is consolidated unless other similar memories are formed shortly after the first memory. This could allow for the functional correction and/or updating of inadvertently or imprecisely formed memories before they are stabilized.

The recent failure of Caithness and others to reproduce these interference effects for motor adaptation learning (Caithness *et al.*, 2004) is perhaps not surprising, since it is still unclear what the stimulus characteristics are for an effective blockade of this early consolidation. Whilst they failed to observe interference with new learning on their tasks, it is still likely that even these memories remain sensitive to ECS and protein synthesis inhibitors for several hours (Muellbacher *et al.*, 2002), reflecting a required process of consolidation. Indeed, there is little objection to the view that conversion of initial memory traces into long-term memories requires protein synthesis (McGaugh, 2000).

Enhancement: Whilst memory consolidation clearly serves to stabilize memories, this is far from all that it does. For a start, periods of consolidation also appear capable of *enhancing* memories, defined as the improvement of behavioral performance, independent of further practice (Walker, 2005). Although these two phases of consolidation could reflect a single process, we believe that this is unlikely for at least two reasons. First, the consolidation process leading to enhancement of a motor sequence learning task continues for up to ten times as long as the earlier stabilizing phase (Walker *et al.*, 2003) (see Figure 6.2, top – note the logarithmic scale), and for a visual discrimination process continues over at least two to four days (Stickgold *et al.*, 2000). Second, whilst stabilization of this motor sequence task occurs over six hours of waking, the enhancement phase for both tasks occurs only during sleep, indicative of dissociable stages (Stickgold *et al.*, 2000; Walker *et al.*, 2002). Other post-encoding stages of memory processing which have been considered under the umbrella of consolidation include the integration of recently encoded memories into existing memory networks (memory integration or association) (Stickgold, 2002; Dumay & Gaskell, 2007), the development of hippocampal independence for declarative memories (McClelland, McNaughton & O'Reilly, 1995; Hasselmo, 1999), and even the active weakening of memory representations ("memory erasure") (Crick & Mitchison, 1983). Interestingly, all of these are thought to be facilitated by sleep.

We noted above that memories are not simply formed, but evolve over time. This time-course appears to serve two conceptually distinct functions. Some post-encoding stages of memory consolidation appear necessary simply to cope with the constraints of the brain. Thus, the molecular mechanisms which support rapid memory encoding (e.g. calcium influxes, unmasking of existing network connections) are inadequate for long-term maintenance of synaptic changes, whereas those processes which support long-term maintenance (e.g. protein synthesis) cannot be accomplished quickly enough to support rapid encoding (Kandel, 2001).

Similarly, network structures that can capture an episodic memory may be incapable of supporting dense network storage of memories (McClelland, McNaughton & O'Reilly, 1995). A recent review of offline memory reorganization has described many of these features in detail (Frankland & Bontempi, 2005).

In addition, processes of consolidation also serve to optimize behavior. As examples, they could: (1) automate behaviors, shifting representations from declarative to procedural systems and reducing frontal demands; (2) extract valuable details from complex episodic memories, so that we can, for example, recall the sum of five plus five without recalling an episodic memory of when and where it was learned; and (3) integrate information, so that we associate all the "addition facts" with one another, thereby developing semantic or associative networks. For both classes of memory consolidation processes, sleep is now thought to play an important role in meeting the demands of the organism.

STAGES OF MEMORY: RECONSOLIDATION

Memories may be retained for weeks to years, during which time they can be effectively recalled. But the mere act of memory recall can destabilize the memory and return it to a labile form, where it is again vulnerable to interference and degradation. Reconsolidation – the transformation of this now destabilized memory into a restabilized form is necessary if the memory is to be retained in the face of additional interference (Nader, 2003). Otherwise it can degrade relatively quickly (see Figure 6.2, bottom). (To be more precise, it is unclear whether such degradation actually weakens the memory trace or, instead, simply makes it inaccessible to recall mechanisms, but this distinction is irrelevant for the current review.)

Our understanding of memory reconsolidation is at a much earlier stage. Although originally reported in the 1960s (Misanin, Miller & Lewis, 1968; Schneider & Sherman, 1968), the details of memory reconsolidation have only recently come under intensive investigation (Nader, 2003). Its component processes, their time courses, and functions, are far less well-defined, and almost no attention has been paid to their possible dependence on wake–sleep states. Likewise, there has been little discussion of the significance or possible functions of these processes.

Conceptually, there are at least four processes that a consolidated memory can undergo: (1) reactivation, leading to (2) destabilization, which in turn leads to either (3) degradation or (4) reconsolidation. However, the time-courses of these individual steps, the mechanisms, and brain states which produce them, and even their biological functions, remain unclear.

While memory reactivation may presumably occur in a fraction of a second, the destabilizing effects of such reactivation appear to depend on longer periods of reactivation. Anywhere from 30 seconds to 10 minutes may be required to produce destabilization (defined by memory degradation after the prevention of reconsolidation) (Nader, Schafe & Le Doux, 2000; Suzuki et al., 2004; Krakauer, Ghez & Ghilardi, 2005), with longer times required when the intensity and duration of the initial training is increased. The duration of this destabilization appears to be of the order of five to six hours, after which the memory becomes reconsolidated and again resistant to interference (Przybyslawski, Roullet & Sara, 1999; Nader, Schafe & Le Doux, 2000; Gruest, Richer & Hars, 2004).

Once destabilized, and in the absence of subsequent reconsolidation, degradation of a memory has generally been considered a passive process, perhaps based on molecular turnover. Alternatively, it may be that the memory is not degraded at all, but its recall ability is lost. Regardless, the nature of this degradation remains unclear. Currently, degradation is defined behaviorally as diminished performance of the learned task or response. There is little data on the time-course over which this reduced efficacy, let alone its molecular correlates, develops. Following reactivation and blockade of reconsolidation, previously learned behaviors are still intact two to four hours later (Nader, Schafe & Le Doux, 2000; Debiec, Le Doux & Nader, 2002; Duvarci & Nader, 2004; Suzuki et al., 2004). This makes sense, since reconsolidation appears to take at least this long, and it would be counterproductive for memories to begin to degrade before reconsolidation normally has completed. By 24 hours after reactivation, any degradation of the memory appears to be complete; see also Myers & Davis, 2002).

Although early studies using ECS or administration of protein synthesis inhibitors were not able to suggest a practical purpose for reactivation and reconsolidation, other than preventing the inadvertent degradation of the memory, hints of these more complex mechanisms and functions come from studies showing that inhibitors of cholinergic (Boccia et al., 2004) and noradrenergic (Przybyslawski, Roullet & Sara, 1999) neuromodulation can also prevent reconsolidation. In addition, N-methyl-D-aspartate (NMDA) antagonists reportedly can block the destabilization associated with reactivation (Nader, Ben Mamou & Komorowski, 2004). Finally, even interference from training on competing tasks has now been shown to block reconsolidation (Walker et al., 2003). In light of these more recent findings, we propose that destabilization and reconsolidation of memories simply represent yet another sophisticated mechanism for remodeling and improving pre-existing memories.

In summary, there appear to be a number of stages of memory processing, which use distinct brain processes to perform unique functions ranging from stabilizing and enhancing new memories, to destabilizing them, allowing subsequent remodeling, refining, and even degrading of pre-existing information. With developments in understanding these processes, an intensive research effort has begun to ask what role sleep, and its specific stages, may play in these memory processes.

SLEEP-DEPENDENT MEMORY CONSOLIDATION

Perhaps the earliest reference to a relationship between sleep and memory is by the Roman rhetorician Quintillian, stating, "It is a curious fact, of which the reason is not obvious, that the interval of a single night will greatly increase the strength of the memory," and suggesting that "the power of recollection ... undergoes a process of ripening and maturing" (Quintillian, first century AD). Despite this remarkable insight describing what is now known as memory consolidation, it is only in the last 50 years, following the discovery of REM and NREM sleep, that researchers began testing the hypothesis that sleep actively participates in processes of learning and brain plasticity. We have recently reviewed the evidence for the critical role of sleep in memory consolidation (Walker & Stickgold, 2004), and only briefly summarize the behavioral literature here (for an opposing viewpoint, see Vertes, 2004).

Specific stages of sleep appear to be critical for discrete steps in the consolidation of various forms of memory, whilst for other steps, sleep appears unnecessary (for review see Walker, 2005). For example, *stabilization* of some forms of procedural motor memory may develop specifically in the absence of sleep, across three to six hours of waking (Brashers-Krug, Shadmehr & Bizzi, 1996; Muellbacher et al., 2002; Walker et al., 2003). In contrast, the *enhancement* of procedural sensory and motor memories has almost always been found to depend on overnight sleep, with equivalent periods of wake failing to produce any performance gains (Karni et al., 1994; Smith & MacNeill, 1994; Gais et al., 2000; Stickgold, James & Hobson, 2000; Stickgold et al., 2000; Atienza, Cantero & Dominguez-Marin, 2002; Fischer et al., 2002; Walker et al., 2002; Fenn, Nusbaum & Margoliash, 2003; Walker et al., 2003; Atienza, Cantero & Stickgold, 2004; Gaab et al., 2004; Huber et al., 2004; Kuriyama, Stickgold & Walker, 2004; Robertson, Pascual-Leone & Press, 2004) (see Figure 6.3 for examples; see also Robertson, Pascual-Leone & Press (2004) for sleep-independent enhancement). Such overnight enhancement has been seen for a variety of memory tasks, but individual tasks differ dramatically in the sleep stages or sleep characteristics required. Consolidation of motor skills have been connected to NREM sleep stages, stage 2 in some cases and slow-wave sleep (SWS) in others, as well as to specific physiological characteristics of NREM, (Smith & MacNeill, 1994; Fogel, Jacob & Smith, 2001; Walker et al., 2002; Huber et al., 2004; Robertson, Pascual-Leone & Press, 2004). In contrast, both SWS and REM sleep have been associated with consolidation enhancement of memory for a visual texture discrimination task (Karni et al., 1994; Gais et al., 2000; Stickgold et al., 2000), suggesting that consolidation of different forms of procedural skill memories require different types of sleep at different times of the night.

The evidence for sleep-dependent consolidation of declarative memories is less consistent, although more subtle features of what *type* of declarative memory is being learned may drive this inconsistency. First, it is difficult to separate processes of stabilization from processes of enhancement, since performance on declarative tasks generally deteriorates across both wake and sleep. But when performance after equivalent periods of wake and sleep is compared, early studies split over a role for sleep in memory consolidation (for review see Smith et al., 2001). More recently, however, several studies by Born and his colleagues have consistently shown actual performance enhancement on a paired-word associates test after early night sleep, rich in SWS (Gais & Born, 2004). These findings are striking in the face of earlier studies that showed no effect. But this discrepancy may well reflect the nature of the word pairs used. While older studies used unrelated word pairs, such as dog–leaf, Born has used related word pairs, such as dog–bone (Gais & Born, 2004). The nature of the learning task thus shifts from forming and retaining completely novel associations (dog–leaf) to the strengthening or tagging of well-formed associations (dog–bone) for recall at testing. As further support, they have shown that daytime training can trigger changes in characteristics of early-night SWS, with modifications reported in both the number of sleep spindles (Gais et al., 2002), and in the coherence of NREM slow-frequency electroencephalogram (EEG) oscillations. Adding to these findings, Peigneux and colleagues (Peigneux et al., 2004) have reported that overnight improvement on a hippocampally mediated spatial memory task is positively correlated with increased hippocampal activation during SWS, a finding that would also seem to argue in favor of actual memory enhancement.

Thus, the role of sleep in declarative memory consolidation, rather than being absolute, might depend on more specific aspects of the consolidation task. A further example of this subtlety pertains to the emotional nature of the material being learned. For example, Wagner and colleagues (Wagner, Gais & Born, 2001) have shown that late-night sleep, rich in REM sleep, selectively favors retention of previously learned declarative emotional texts relative to neutral texts. Similarly, Hu, Stylos-Allen and Walker (2006) have investigated the time-course of emotional and neutral episodic declarative memory consolidation across the day and overnight. Subjects performed an initial study session containing standardized emotional and neutral pictures, either in the evening or morning. Twelve hours later, after sleeping or waking, subjects performed a recognition test, discriminating between these original and novel pictures. Whilst memory recognition was consistently superior following sleep, consolidation effects were particularly strong for emotional rather than neutral stimuli, with recognition accuracy for emotional pictures improving by 42% overnight, relative to the 12-hour waking period. These

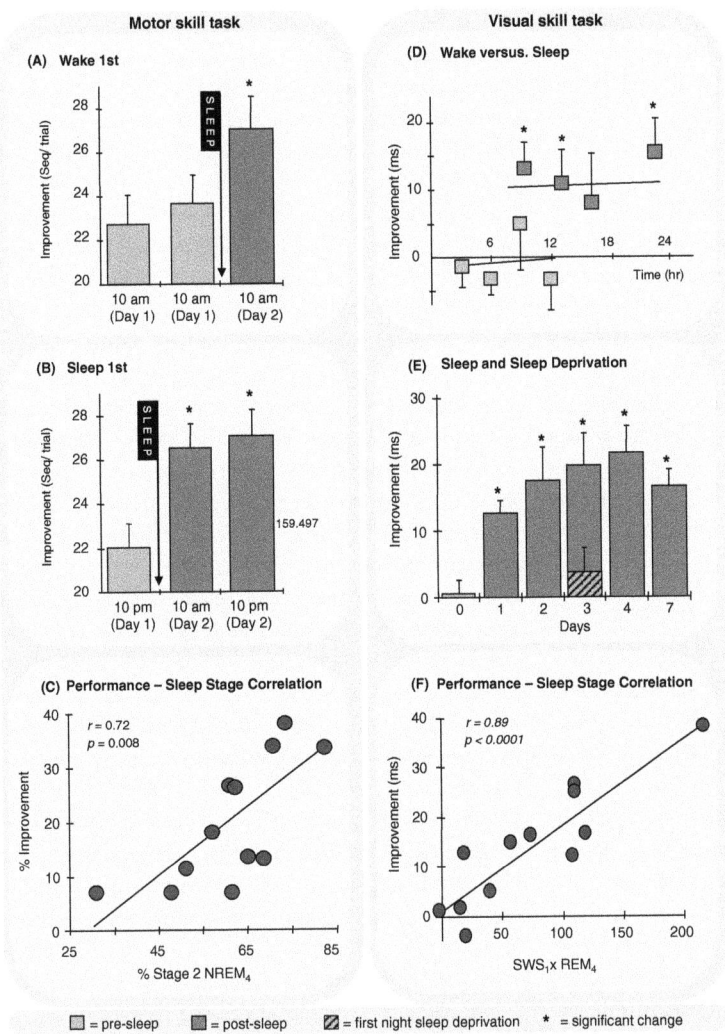

Figure 6.3. Sleep-dependent visual and motor skill learning in the human brain. (A–C) *Motor skill task*. (A) Wake 1st – subjects ($n = 15$) trained at 10 am (Day 1) showed no significant change in performance at re-test after 12 hours of waking (10 pm, Day 1). However, by the second re-test, after a night of sleep (10 am, Day 2) performance improved significantly. (B) Sleep 1st – after evening training (10 pm, Day 1), subjects ($n = 15$) showed significant improvements in speed just 12 hours after training following a night of sleep (10 am, Day 2) but expressed no further significant change after an additional 12 hours of waking (10 pm, Day 2). (C) The amount of overnight improvement on the motor skill task correlated with the percentage of stage 2 NREM sleep in the last quarter of the night (Stage 2 NREM$_4$). (From Walker et al., 2002.) (D–F) *Visual skill task*. Subjects were trained and then re-tested at a later time, with improvement (ms) in performance illustrated across time. Each subject was retested only once, and each point represents a separate group of subjects. (D) Wake versus sleep. Subjects trained and then retested on the same day ($n = 33$), after either 3, 6, 9, or 12 hours of subsequent waking (lower squares), showed no significant improvement as a consequence of the passage of waking time for any of the four time intervals. In contrast, subjects ($n = 39$) trained and then retested 8, 12, 15, or 23 hours later, after a night's sleep (upper squares), showed significant improvement. (E) Sleep and sleep deprivation. Subjects ($n = 89$) trained and re-tested one to seven days later (sold bars) continued to improve after the first night, without additional practice. Subjects ($n = 11$) sleep-deprived the first night after training showed no improvement (hatched bar), even after two nights of recovery sleep. (F) Overnight improvement was correlated with the percentage of slow-wave sleep (SWS) in the first quarter of the night (SWS$_1$) and REM sleep in the last quarter of the night (REM$_4$). (From Stickgold, James & Hobson, 2000; Stickgold et al., 2000.) * $p < 0.05$; error bars, SEM.

data indicate the selective facilitation of emotional declarative memory consolidation across sleep, rather than simply across time *per se*, resulting in the enhancement of memory retrieval.

SLEEP AND BRAIN PLASTICITY

Memory formation, in particular consolidation, depends on brain "plasticity" – a lasting structural or functional, or both, neural change in response to a stimulus (such as an experience). If sleep is to be considered a critical mediator of memory consolidation, then evidence of sleep-dependent plasticity would greatly strengthen this claim. Indeed, there is now a wealth of data describing sleep-dependent brain plasticity at a variety of different levels in both animals and humans, complementing evidence of sleep-dependent changes in behavior.

Neuroimaging studies

Several studies have investigated whether daytime training is capable of modifying functional brain activation during subsequent sleep. Based on animal studies, neuroimaging experiments have explored whether the signature pattern of brain activity elicited while practicing a memory task actually re-emerges, that is, is "replayed," during subsequent sleep. Using brain imaging, Maquet and colleagues (Maquet *et al.*, 2000) have shown that patterns of brain activity expressed during training on a sequential motor task reappear during subsequent REM sleep, while no such change in REM sleep brain activity occurs in subjects who received no daytime training. Furthermore, the extent of learning during daytime practice exhibits a positive relationship to the amount of reply during REM sleep (Peigneux *et al.*, 2003). As has been described animal studies (Datta, 2000), these findings suggest that it is not simply experiencing the task which modifies subsequent sleep physiology, but the process of learning itself. Similar findings have been reported by use of a virtual maze task. Daytime task learning is initially associated with hippocampal activity. Then, during post-training sleep, there was a re-emergence of hippocampal activation, this time specifically during SWS. Most compelling, however, the amount of SWS reactivation in the hippocampus was proportional to the amount of next-day task improvement, suggesting that this reactivation leads to offline memory improvement (Peigneux *et al.*, 2004). Such sleep-dependent replay may potentially modify synaptic connections established within specific brain networks during practice, strengthening some synaptic circuits while potentially weakening others in the endeavor of refining the memory.

A second approach, that more directly examines sleep-dependent plasticity, compares patterns of brain activation before and after a night of sleep. In contrast to measuring changes in functional activity *during* sleep, this technique aims to determine whether improved performance results from an overnight, sleep-dependent *re-structuring* of the neural representation of the memory. Using the sleep-dependent motor-skill task, Walker and colleagues have recently used functional magnetic resonance imaging (fMRI) to investigate differences between patterns of brain activation before and after sleep (Walker *et al.*, 2005a). After a night of sleep, and relative to an equivalent intervening time-period awake, increased activation was identified in motor control structures of the right primary motor cortex (Figure 6.4 (A)) and left cerebellum (Figure 6.4 (B)) – changes which allow more precise motor output (Ohyama *et al.*, 2003) and faster mapping of intention to key-press (Ungerleider, Doyon & Karni, 2002). There were also regions of increased activation in the medial prefrontal lobe and hippocampus (Figure 6.4 (C) and (D)), structures recently identified as supporting improved sequencing of motor movements (Koechlin *et al.*, 2000; Koechlin *et al.*, 2002; Poldrack & Rodriguez, 2003; Schendan *et al.*, 2003). In contrast, decreased post-sleep activity was identified bilaterally in the parietal cortices (Figure 6.4 (E)), possibly reflecting a reduced need for conscious spatial monitoring (Seitz *et al.*, 1990; Toni *et al.*, 1998; Muller *et al.*, 2002) owing to improved task automation (Kuriyama, Stickgold & Walker, 2004), together with regions of signal decrease throughout the limbic system (Figure 6.4 (F) to (G)), suggesting a decreased emotional task burden. In total, these results suggest that sleep-dependent motor learning is associated with a large-scale plastic reorganization of memory throughout several brain regions, allowing skilled motor movements to be executed more quickly, more accurately and more automatically after sleep. They may also signify a potential role for sleep in clinical rehabilitation following brain damage, such as stroke. Walker and colleagues have also used fMRI to

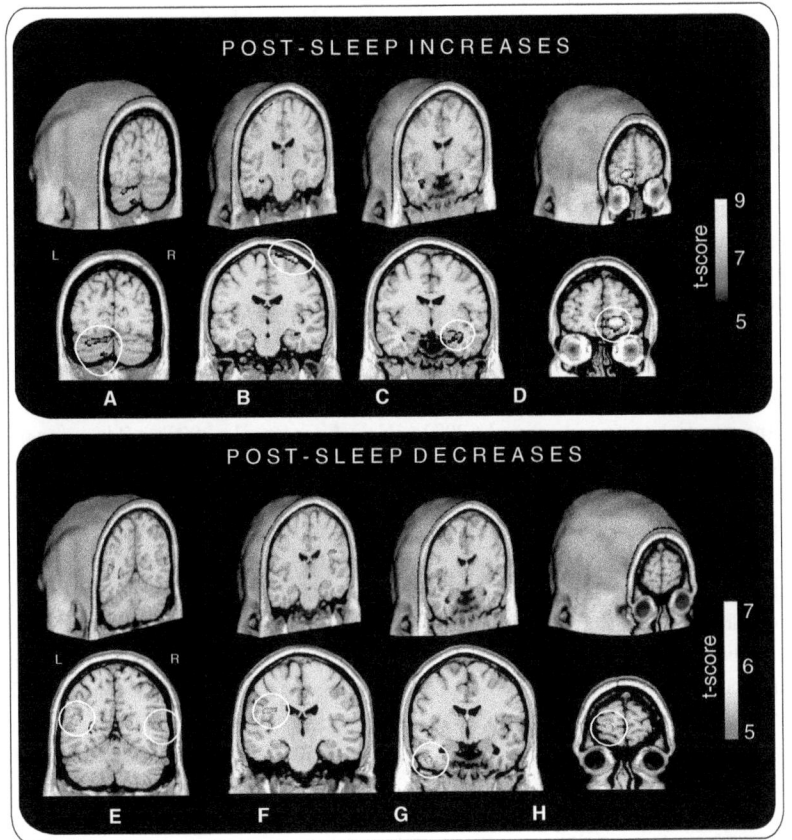

Figure 6.4. Sleep-dependent motor memory reorganization in the human brain. Subjects were trained on a sleep-dependent motor skill task and then tested 12 hours later, either after a night of sleep or following intervening waking, during a functional magnetic resonance imaging (fMRI) brain-scanning session. Scans after sleep and waking were compared (subtracted), resulting in regions showing increased fMRI activity post-sleep (small and circled areas; A–D) or decreased signal activity (small and circled; E–H) post-sleep, relative to post-waking. Activation patterns are displayed on three-dimensional rendered brains (top panel of each graphic), together with corresponding coronal sections (bottom panel of each graphic). After sleep, regions of increased activation were identified in the right primary motor cortex (A), the left cerebellum (B), the right hippocampus (C), and the right medial prefrontal cortex (D). Regions of decreased activity post-sleep were expressed bilaterally in the parietal lobes (E), together with the left insula cortex (F), left temporal pole (G), and left fronto-polar area (H), all regions of the extended limbic system. All data are displayed at a corrected threshold of $p < 0.05$. (From Walker et al., 2005b.)

investigate whether overnight reorganization similarly occurs in sensory-perceptual systems (Walker et al., 2005b) using the sleep-dependent visual texture discrimination task described earlier. Subjects were trained with or without intervening sleep. Relative to the condition without sleep, re-test after sleep was associated with significantly greater activation in an area of primary visual cortex corresponding to the visual target location. However, there were also several other regions of increased post-sleep activity, throughout both the ventral object recognition (inferior parietal and occipital-temporal junction) and dorsal object location (superior parietal lobe) pathways (Ungerleider & Haxby, 1994), together with corresponding decreases in the right temporal pole, a region involved in emotional visual processing. Thus, a night of sleep appears to reorganize the representation not only of procedural motor but visual skill memories as well, with greater activation throughout the visual processing pathways offering improved identification of both the stimulus form, and its location in space, and with signal decreases in the temporal pole reflecting a reduced emotional task burden resulting from the overnight learning benefits.

Maquet et al. (2003) have investigated the detrimental effects of sleep deprivation on underlying brain activity using a visuo-motor adaptation task. Subjects were trained on the task and re-tested three days later, with half the subjects deprived of sleep the first night. Control subjects, who slept all three nights, showed both enhanced behavioral performance at re-test, and a selective increase in activation in the superior temporal sulcus (a region involved in the evaluation of complex motion patterns) relative to subjects deprived of sleep the first night. In contrast, no such enhancement of either performance or brain activity was observed in these subjects, indicating that sleep deprivation had interfered with a latent process of plasticity and consolidation. This study offers an early indication that sleep deprivation not only disrupts consolidation, but the underlying neural mechanisms that support it.

Electrophysiological studies

Throughout the sleep cycle, both REM and NREM sleep stages contain numerous distinct electrophysiological events that contribute to each unique biologic state. Many of these electrical phenomena have been implicated in processes of plasticity, either potentiating or depressing synaptic connections (Benington & Frank, 2003). For example, it has been proposed that sleep spindles, seen most commonly during stage 2 NREM sleep, may provide brief trains of depolarizing inputs to targets in the neocortex, which are similar to spike trains used experimentally to induce long-term synaptic potentiation (Contreras, Destexhe & Steriade, 1997; Steriade, 1997; Steriade, 1999; Sejnowski & Destexhe, 2000). Indeed, Steriade and colleagues (Steriade, 2001) have shown that experimental trains of impulses similar to those produced by sleep spindles can produce lasting changes in the responsiveness of cortical neurons. Similarly, theta waves, seen in the hippocampus during REM sleep in both humans (Cantero et al., 2003) and other animals (Poe et al., 2000), greatly facilitate the induction of long-term potentiation (LTP) in the hippocampus, potentiation that is believed to be a physiologic mediator of memory development (Pavlides et al., 1988; Huerta & Lisman, 1995).

As noted earlier, phasic events during REM sleep, and ponto-geniculo-occipital (PGO) waves in particular, have been associated with learning. Sanford et al. (2001) demonstrated that fear conditioning in rats can increase the amplitude of elicited P-waves during REM sleep, suggesting again that they represent a homeostatically regulated component of a sleep-dependent mechanism of learning and plasticity (see Datta, 2000). These PGO waves occur in a phase-locked manner with theta wave activity during REM sleep (Karashima et al., 2002a; Karashima et al., 2002b). This is particularly interesting because experimental hippocampal stimulation at the peaks of theta waves facilitate LTP, whereas the same stimulation applied at the troughs of the theta waves instead leads to long-term depression of synaptic responses (Pavlides et al., 1988; Holscher, Anwyl & Rowan, 1997). These findings suggest that natural PGO activity during REM sleep may serve as an endogenous mediator of synaptic plasticity, based on its coincidence with theta wave oscillations, which, depending on its phase relationship, could either strengthen or weaken synaptic connections, both of which are necessary for efficient network plasticity.

Selective reactivation is seen not only from the human neuroimaging studies described above, but from more precise measurement of sleep-dependent network reactivation in the rat. Several groups have investigated the firing patterns of large networks of individual neurons across the wake–sleep cycle in a variety of cortical and subcortical regions of the rat brain. The signature firing patterns of these networks, expressed during waking performance of spatial tasks and novel experiences, are replayed during subsequent SWS and REM sleep, with replay during REM being at speeds similar to those seen during waking, but those in SWS being an order of magnitude faster in some, but not all, studies (Wilson & McNaughton, 1994; Skaggs & McNaughton, 1996; Poe et al., 2000; Louie & Wilson, 2001; Ribeiro et al., 2004). Dave and colleagues (Dave, Yu & Margoliash, 1998; Dave & Margoliash, 2000) have shown that waking patterns of pre-motor activity observed during song learning in the zebra finch are also replayed during sleep, with a temporal structure similar to that seen in waking.

Together, these data indicate that temporal patterns of network activity seen during waking experiences are consistently replayed during subsequent sleep across a broad spectrum of phylogeny. This replay of events is hypothesized to trigger distinct but complementary processes within reactivated neuronal ensembles. Ribeiro et al. (2004) suggested that SWS re-instantiates the memory representation through network reverberation, while subsequent REM sleep then potentiates the memory for subsequent post-sleep recall, through gene-induction mediated synaptic plasticity.

Cellular studies

Recently, a form of sleep-dependent plasticity at the cellular level has been elegantly demonstrated during early post-natal development of the cat visual system (Shaffery et al., 1998; Shaffery et al., 1999). Under normal circumstances, brief periods of monocular visual deprivation during critical periods of development lead to the remodeling of synaptic connectivity, with the deprived eye's inputs to cortical neurons being first functionally weakened and then anatomically diminished (Antonini & Stryker, 1993). Frank, Issa and Stryker (2001) have now shown that when six hours of monocular deprivation are followed by six hours of sleep, the size of the monocularity shift doubles. In contrast, if the cats are kept awake for these same six hours (in the dark, without input to either eye), a non-significant *reduction* in the size of the shift occurs. Thus, sleep may contribute as much towards developmental changes in synaptic connectivity as visual experience, presumably by enhancing the initial changes occurring during a prior period of monocular deprivation. In contrast, sleep-deprivation results in a loss of previously formed, experience-dependent synaptic changes.

Shaffery et al. (2002) have reported similar findings of sleep-dependent plasticity in the rat visual cortex, suggesting that REM sleep, in conjunction with visual experience, modulates the initial course of visual cortex maturation. In rats under 30 days of age, electrical stimulation produces increased excitability (potentiation) in specific layers of the visual cortex, while stimulation after this early developmental stage fails to produce such potentiation. Depriving rats of REM sleep during this period extends this window of plasticity by as much as seven days, suggesting that events occurring during REM sleep modulate the duration of this period of experience-dependent plasticity.

Molecular studies

At the molecular level, Smith, Tenn and Annett (1991) have shown that administration of protein synthesis inhibitors to rats during REM sleep windows thought to be critical for consolidation, prevents behavioral improvement following the sleep period. Such protein synthesis could reflect the activation of genetic cascades which produce key molecules for synaptic remodeling. Our understanding of such gene inductions during sleep is only beginning. Although several of the known "immediate early genes" (IEGs) are specifically downregulated during sleep (Cirelli & Tononi, 1998; Cirelli & Tononi, 2000a, 2000b), approximately 100 genes are specifically upregulated during sleep (Cirelli, Gutierrez & Tononi, 2004), almost the same number that are upregulated during wakefulness. Moreover, upregulation of these genes during sleep was seen only in the brain tissue.

This extensive upregulation of genes during sleep is seen even in the absence of any specific learning tasks being performed prior to sleep. If this upregulation were specifically related to consolidation of recent learning and memory formation, we would expect that such gene inductions would be seen specifically after training on tasks that undergo sleep-dependent consolidation. Indeed, such learning-specific upregulation has been observed. Ribeiro and colleagues (Ribeiro et al., 1999) found upregulation in rats of *zif-268*, a plasticity associated IEG, during REM sleep following exposure to a rich sensorimotor environment, but its downregulation during both SWS and REM sleep in the absence of such exposure (Figure 6.5). This rich environment effect may also be mimicked by brief electrical stimulation of the medial perforant pathway (Ribeiro et al., 2002), which normally carries signals from the cortex into the hippocampus. Unilateral stimulation results in a wave of *zif-268* expression during subsequent REM sleep, with expression seen predominantly in the ipsilateral amygdala, entorhinal, and auditory cortices during the first REM sleep episodes after LTP induction, but extending into somatosensory and other cerebral cortices during subsequent REM

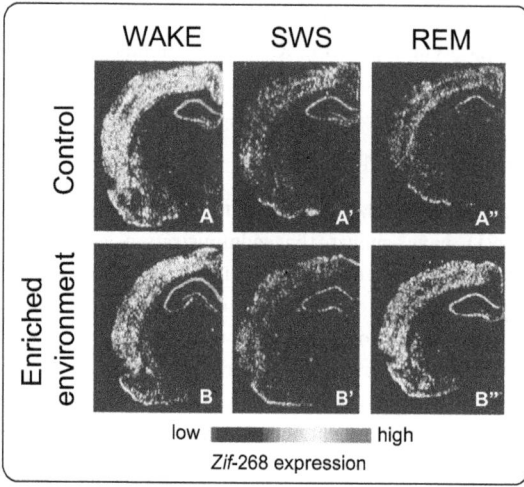

Figure 6.5. Experience-dependent upregulation of *zif-268* gene expression during waking, slow-wave sleep (SWS) and REM sleep states in the rat (Ribeiro *et al.*, 1999). Autoradiograms of frontal coronal brain sections whose gene expression levels best represent the means for each group studied. In controls, *zif-268* expression decreased from WAKE (A) to SWS (A') and REM (A"). In enriched environment animals, *zif-268* levels decreased from WAKE (B) to SWS (B'), but increased from the latter to REM (B"). This effect was particularly noticeable in the cerebral cortex and the hippocampus. (From Ribeiro *et al.*, 1999.)

periods (Ribeiro *et al.*, 2002). This provides additional molecular evidence for the existence of windows for increased neuronal plasticity during REM sleep periods following enriched waking experience, in agreement with both behavioral, physiological, and neuroimaging studies. In summary, learning and memory are dependent on mechanisms of brain plasticity, and such processes must mediate sleep-dependent learning and memory consolidation. Many examples of such plasticity during sleep have now been reported at several descriptive levels across numerous species, several of which have been specifically induced by learning experiences.

SLEEP-DEPENDENT MEMORY RECONSOLIDATION

Memory destabilization and reconsolidation may also be facilitated by sleep. Although there is little data that directly pertains to this question, we propose that both degradation and reconsolidation processes can, and in some circumstances must, occur during sleep. Indeed, most rodent studies of reconsolidation are carried out during the light (sleep) phase of the circadian cycle, and it is likely that animals in all of these studies slept between reactivation and subsequent measurements of reconsolidation. Thus existing evidence cannot distinguish between time-dependent and sleep-dependent reconsolidation.

Support for this comes from studies of procedural memory reconsolidation in humans (Walker *et al.*, 2003). After training on a finger-tapping motor sequence task, subjects showed normal overnight sleep-dependent gains in performance accuracy (see Walker *et al.*, 2002). But if subjects were taught a second, competing sequence immediately after learning the first, the second motor memory interfered with the initial motor memory, resulting in a blockade of normal overnight consolidation improvement for that original motor memory when tested the next morning. Thus, only that which was learned last, the second motor memory, was enhanced overnight. This demonstrates that immediately after learning, the original memory is more unstable and susceptible to interference. However, if a four- to six-hour period of time was inserted before learning of the second motor memory, rather than immediately afterward, it afforded the first memory immunity to the influence of interference. Therefore, >6 hours of time, and continuing across a night of sleep, the first memory, if given the chance, became stable.

Most interestingly, however, if the original memory was reactivated (through 90 seconds of rehearsal) after the night of sleep (Day 2), the memory appeared to be destabilized, returning to a labile state, as was the case soon after initial learning on Day 1. As a consequence, training on a second sequence after reactivation, despite having previously been consolidated successfully, led to interference and hence loss of the improvement seen across the first night when tested the following morning (Day 3). Thus, the simple act of recalling a consolidated memory offers the chance for remodeling, or in this case, disruption. Furthermore, these results suggest that the deterioration in performance seen after blockade of reconsolidation might be limited to the reversal of earlier sleep-dependent consolidation enhancement.

In a recent pilot study, we asked whether nonconscious or non-volitional reactivation of memory traces *during* sleep was sufficient to destabilize these memories and make them again susceptible to interference and degradation (unpublished results). We trained human subjects on the motor sequence task shortly before bed. Six hours into sleep, when sleep-dependent consolidation is thought to be in progress for this task (Walker et al., 2002), subjects were awakened and either trained on a new motor sequence or allowed to read a magazine for an equivalent period of time. They did not, however, practice, and hence intentionally reactivate, the initial memory as in earlier studies (Walker et al., 2003). Following an additional two hours of sleep, they were again awakened, and now retested on the original sequence learned the day before. Control subjects, who did not receive late-night interference training, showed a normal 21% enhancement of speed in the morning. In contrast, subjects who had received late-night interference training showed only 9% improvement, 59% less than the controls ($p = 0.03$; unpublished observations) – strikingly similar to the 57% deterioration seen after daytime reactivation and interference. Since previous studies have shown that similar training is resistant to interference six hours after training, it would suggest that non-conscious reactivation occurred *during* sleep, again destabilizing the memory, and allowing the subsequent remodeling. Whether such memory modification could occur during sleep (e.g. whilst dreaming) is unknown, and these are still preliminary findings, but the implications are striking. If extended, they suggest that, in addition to the offline process of consolidation that occurs for newly formed memories, there is also an offline, sleep-dependent process for the reactivation and potential modification of existing memories across the lifespan.

CONCLUSION

In summary, consolidation, reactivation, and reconsolidation may be viewed as components of a complex system of memory processing which modifies the strength, stability, form, and integrative connectivity of acquired information across the lifespan. For many of these processes, sleep, and often specific stages of sleep, appears to be either permissive or obligatory. Thus memories must be viewed as constantly evolving; an evolution that is controlled by a series of time and brain-state dependent processes occurring during wake and sleep states. By way of continued multidisciplinary research, and with a measured appreciation of the fundamental role that sleep plays in consolidating and reforming stored information, we can look forward to new advances in understanding memory, and treating its disorder.

ACKNOWLEDGEMENTS

This work was supported by grants from the National Institutes of Health (MH48832, MH65292, MH67754, and MH69985).

7 · Consolidation and reconsolidation of Pavlovian fear-conditioning: roles for intracellular signaling and extracellular modulation in memory storage

Christopher Cain, Jacek Debiec, and Joseph E. LeDoux

INTRODUCTION

Pavlovian fear conditioning in the rodent has been a leading paradigm for investigating the brain mechanisms of associative learning and memory. Since the identification of a simple neural circuit critical for fear conditioning, and the demonstration of synaptic plasticity within this circuit in response to learning, there has been an upsurge in research regarding the cellular and molecular mechanisms underlying memories established through fear learning. In this chapter, we focus on the molecular requirements of fear memory consolidation and reconsolidation: processes important for the stable storage of memory over time. We also consider contributions of both intracellular signaling cascades and extracellular modulators as they relate to these processes. This research has led to a fairly comprehensive, mechanistic model of emotional memory storage with important implications for treating fear disorders in humans.

Learning and memory are related but distinct processes. *Learning* refers to the acquisition of new information through experience. *Memory* refers to the storage of this information over time. Initially, memories are labile and subject to disruption, but within hours they begin to stabilize and persist. The initial labile state is called short-term memory (STM), the later persistent state is called long-term memory (LTM), and the process of transforming STM into LTM is called "consolidation" (Davis & Squire, 1984; McGaugh, 2000a). Without consolidation, newly acquired information is retained only transiently, but consolidated memories may persist for long periods, sometimes throughout the lifetime of the organism (Gale et al., 2004). Given that humans have a unique capacity to learn and remember it is not surprising that questions about how and when memory consolidation occurs have been important research topics in experimental psychology since its inception in the late nineteenth century (Ebbinghaus, 1885; Muller & Pilzecker, 1900).

Consolidation may be inferred by simple analysis of normal human learning and memory. Learning occurs almost continuously during waking experience yet only some of these experiences are consolidated to LTM. However, the strongest evidence for the existence of consolidation has come from studies showing that amnesia may result from various kinds of manipulations that follow learning. For instance, retrograde amnesia, a loss of memory for recent events, can be produced by a variety of manipulations, including hypothermia (Riccio, Hodges & Randall, 1968; Hamm, 1981; Blozovski & Buser, 1988), electroconvulsive shock (Duncan, 1949; Gerard, 1949; Luttges & McGaugh, 1967), head trauma (Dalrymple, 1995; Ahmed et al., 2000; Rees, 2003), seizures (Zeman, Boniface & Hodges, 1998; Pantoni, Lamassa & Inzitari, 2000), and anesthesia (Weissman, 1967; Gerlai & McNamara, 2000). Typically, in these studies, memory is intact shortly after the manipulation (STM is unaffected), but then declines within several hours (LTM is disrupted). This pattern of intact STM and impaired LTM is the key evidence supporting the notion that consolidation occurs.

What factors contribute to the transformation of STM into stable LTM? How are learning processes related to consolidation processes? Does consolidation occur more than once in the life of a memory? How are memories made to persist at the cellular level? Answers to these questions are beginning to emerge from studies of consolidation in reduced preparations and model organisms (Alberini et al., 1995; DeZazzo & Tully, 1995; Belvin & Yin, 1997; Abel & Kandel, 1998; Bailey et al., 2000; McGaugh, 2000a; Nader et al., 2000a; Carew & Sutton, 2001; Schafe et al., 2001; Korzus, 2003). Our focus in this chapter will be on mechanisms of consolidation as revealed through studies of mammals.

While consolidation has been studied using a variety of behavioral paradigms in mammals (Guzowski & McGaugh, 1997; Cartford, Gould & Bickford, 2004; LaLumiere, Nguyen & McGaugh, 2004; Krakauer & Shadmehr, 2006; Wang et al., 2005; Bahar, Dorfman & Dudai, 2004; Eisenberg et al., 2003), research on Pavlovian fear conditioning has been especially useful in relating memory at the level of behavior to underlying synaptic and molecular mechanisms that mediate consolidation. We therefore focus on studies of Pavlovian fear conditioning, beginning with an overview of consolidation and its relation to synaptic plasticity, and then turning to studies that have investigated consolidation using fear conditioning.

MEMORY CONSOLIDATION, RECONSOLIDATION, AND PROTEIN SYNTHESIS

Memory for a wide variety of experiences may persist for days to years, and evidence that certain manipulations may disrupt LTM, but leave STM intact, strongly suggest that memory consolidation is a process distinct from learning. Although there are many manipulations that selectively disrupt memory consolidation (see above), one manipulation in particular ties together all consolidation research: inhibition of protein synthesis (Davis & Squire, 1984). Protein synthesis inhibition is most commonly achieved by administration of drugs that impair mRNA translation, such as anisomycin (Grollman, 1967; Bull, Ferrera & Orrego, 1976), cyclohexamide (Bull, Ferrera & Orrego, 1976; Gibbs, Richdale & Ng, 1979) and emetine (Dunn, Gray & Iuvone, 1977; Oleinick, 1977). Administration of these drugs before, or shortly after, a controlled learning experience abolishes LTM for different types of learning across a diverse range of organisms (Quartermain & McEwen, 1970; Ng et al., 1991; Bailey et al., 1992; Rosenblum, Meiri & Dudai, 1993; Tully et al., 1994; Bartsch et al., 1995; Ghirardi, Montarolo & Kandel, 1995; Staubli, 1995; Bourtchouladze et al., 1998; Xia, Feng & Guo, 1998; Crow, Xue-Bian & Siddiqi, 1999; Schafe et al., 1999; Rose, 2000; Maren et al., 2003; Sangha et al., 2003a; Quevedo et al., 2004; Wang et al., 2005) (Figure 7.1).

The traditional view is that consolidation is a one-time event in the life of a memory. That is, once consolidated, the memory trace is stable and not subject to disruption by protein synthesis inhibition or other amnesia-inducing treatments (McGaugh, 1966; Squire & Alvarez, 1995; Sara, 2000). However, a recently revived area of study suggests that the traditional view is not completely correct since the retrieval or reactivation of a memory can return the trace to a labile state where it is again susceptible to disruption (Nader, Schafe & LeDoux, 2000a, 2000b; Duvarci & Nader, 2004). Such findings suggest that after reactivation memories are "reconsolidated" via protein synthesis-dependent processes.

In summary, the cellular assembly of new proteins appears to be a common requirement for memory stabilization. This stabilization appears to be required both for consolidation after learning and for reconsolidation after retrieval. As we will see below, these new proteins participate in the stabilization of memory by affecting the persistence of synaptic plasticity in brain regions that subserve learning and memory storage.

RELATIONSHIP BETWEEN MEMORY CONSOLIDATION AND SYNAPTIC PLASTICITY

Learning and memory at the level of behavior result from experience-dependent changes in the way neurons process stimuli. In 1894, Ramon y Cajal proposed that learning results from changes in the strength of neuronal synapses and that these changes may persist under certain circumstances resulting in LTM formation (Cajal, 1909). In an influential 1949 paper, Donald Hebb hypothesized that coincident firing of two interconnected neurons results in the strengthening of the synapse between them, that this strengthening may endure, and that synaptic plasticity of this type may represent the cellular basis of learning and memory (Hebb, 1949). These predictions proved to be prescient, as evidence for activity-dependent long-term changes in synaptic strength began to emerge in the latter half of the twentieth century. The first such evidence came from studies of the sea slug *Aplysia*. Kandel and colleagues demonstrated that habituation of the gill-withdrawal reflex resulted from synaptic changes in a circuit that processed sensory stimuli (Castellucci et al., 1970; Kupfermann et al., 1970; Pinsker et al., 1970; Carew & Kandel, 1973).

The first such evidence in mammals came in 1973 from researchers investigating synaptic strength in the

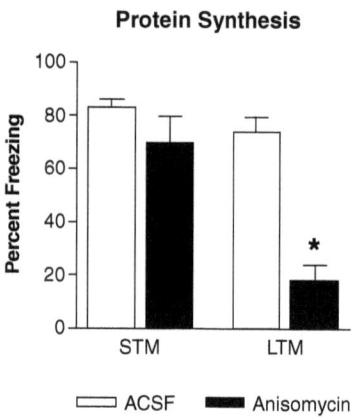
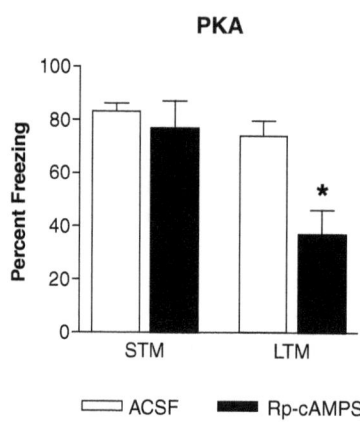
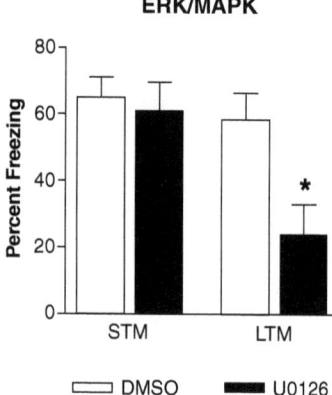

Figure 7.1. Inhibitors of protein synthesis (Anisomycin, *Left*), PKA signaling (Rp-cAMPS, *Middle*) or ERK/MAPK (U0126, *Right*) impair consolidation of auditory fear condition. In each case, STM is unaffected but LTM is severely disrupted. Drugs were infused intraventricularly or intra-LA (U0126) prior to fear conditioning and freezing behavior was assessed 30–60 minutes (STM test) or one day (LTM test) later. (From Schafe et al 1999; 2000).

rabbit hippocampal formation, a brain region strongly linked to declarative memory formation and spatial learning (Bliss & Lomo, 1973). These researchers demonstrated that brief pulses of electrical stimulation to hippocampal afferents resulted in synaptic strengthening. Further, this synaptic plasticity was persistent and was labeled "long-term potentiation" or LTP for short. Other researchers quickly recognized the potential importance of LTP to learning and memory processes, and there has been an explosion in synaptic plasticity research since this first report. LTP has since been demonstrated in nearly every brain region thought to subserve memory, including the cortex (Bear, 1996b; Feldman, Nicoll & Malenka, 1999), striatum (Centonze et al., 1999; Arbuthnott, Ingham & Wickens, 2000; Gubellini et al., 2004), amygdala (Maren, 1999; Huang, Martin & Kandel, 2000; Blair et al., 2001) and cerebellum (Thompson et al., 1997; Hansel, Linden & D'Angelo, 2001; Hartell, 2002). Recent research even confirms Hebb's more specific prediction that coincident firing of interconnected neurons results in LTP; action potentials generated in both pre- and post-synaptic neurons result in LTP only when their occurrence is correlated in time (spike-timing dependent plasticity; Bi & Poo, 2001; Stuart, 2001; Dan & Poo, 2004).

Subsequent studies of synaptic plasticity have identified another feature of LTP that makes it an attractive candidate mechanism for learning or memory. A single brief train of stimulation to afferents results in LTP that lasts only 60–90 minutes in both the hippocampus and amygdala (Frey, Huang & Kandel, 1993; Huang, Martin & Kandel, 2000). However, multiple trains spaced in time result in LTP that lasts at least three to six hours. Thus, the strength of the LTP induction protocol may determine the persistence of synaptic plasticity, and these transient and longlasting forms of LTP have been dubbed *early-* or E-LTP and *late-* or L-LTP, respectively. Memory may also be transient or persistent depending on the strength of the training experience and the discovery of E- and L-LTP have bolstered the argument that synaptic plasticity is the cellular basis of memory formation. The differentiation between E-LTP and L-LTP may be particularly important for studying the cellular basis of LTM consolidation. Indeed, protein synthesis inhibitors prevent L-LTP but not E-LTP (Frey, Huang & Kandel, 1993; Huang, Martin & Kandel, 2000) just as they disrupt LTM but not STM. Thus, plasticity mechanisms at the synapse may participate in memory consolidation by triggering protein synthesis and by incorporating these new proteins in ways that produce stable, long-term strengthening of synaptic connections.

CONSOLIDATION OF FEAR CONDITIONING

Pavlovian fear conditioning in the rodent has been intensely studied in recent decades for many reasons: (1) it is a simple form of associative learning usually resulting in LTM for the experience; (2) it is a mammalian defensive learning paradigm proven to be a valid model for the study of human fear and anxiety; and (3) a critical component of the neural circuitry mediating fear conditioning has been identified, making it possible to relate cellular processes to behavioral learning and memory. The bulk of research in this area involves auditory fear conditioning in rodents using behavioral freezing as the dependent measure, and this will be the focus of the following sections. Important contributions to our understanding of LTM have also been made using other learning paradigms and species (DeZazzo & Tully, 1995; Abel & Kandel, 1998; Davis & Shi, 1999; Carew & Sutton, 2001; McGaugh, McIntyre & Power, 2002; McGaugh & Roozendaal, 2002; Medina *et al.*, 2002; Fanselow & Poulos, 2004; Rosen, 2004).

Conditioned fear responses

In a typical auditory fear-conditioning experiment (Figure 7.2), rodents are presented with an emotionally neutral tone (conditioned stimulus; CS) that is paired in time with an aversive footshock (unconditioned stimulus; US). Before such a pairing, the rodent exhibits no defensive responses to the tone. After such a pairing, tone presentations elicit a cassette of defensive responses, including freezing (Blanchard & Blanchard, 1969; Fanselow & Bolles, 1979), autonomic reactions (Schneiderman *et al.*, 1974; Fitzgerald & Brackbill, 1976), neuroendocrine responses (Mason *et al.*, 1961; Korte *et al.*, 1992a, 1992b), as well as potentiation of somatic reflexes such as startle (Davis, 1986). Collectively these CS-elicited learned responses are referred to as "conditioned fear" responses. Fear responses elicited by the CS after pairing with the US indicate that associative learning takes place during conditioning, provided that similar responses fail to occur when the CS and the US are unpaired during conditioning (Rescorla, 1967). Conditioned fear responses may last for the lifetime of the animal (Gale *et al.*, 2004).

Neural circuitry

A large body of evidence suggests that the amygdala is a critical site of plasticity for fear conditioning (Fanselow & LeDoux, 1999; LeDoux, 2000; Maren & Quirk, 2004; Paré, Quirk & LeDoux, 2004). The amygdala is composed of a dozen or so nuclei (Pitkanen, Savander & LeDoux, 1997). Of these, three have been the focus of much of the research on fear conditioning: the lateral nucleus (LA), central nucleus (CE), and the basal nucleus (B) (Figure 7.3 (A)).

Neurons in the LA receive auditory (CS) and somatosensory (US) inputs from thalamic and cortical processing regions (LeDoux *et al.*, 1990; Turner & Herkenham, 1991; Amaral *et al.*, 1992; Mascagni, McDonald & Coleman, 1993; Romanski *et al.*, 1993; McDonald, 1998). Tone signals arrive in the LA as early as 12 ms after stimulus onset and the same neurons receive somatosensory inputs (Clugnet, LeDoux & Morrison, 1990; Bordi & LeDoux, 1992; Bordi *et al.*, 1993; Romanski *et al.*, 1993; Bordi & LeDoux, 1994a, 1994b). LA neurons in turn connect with CE both directly and indirectly via projections

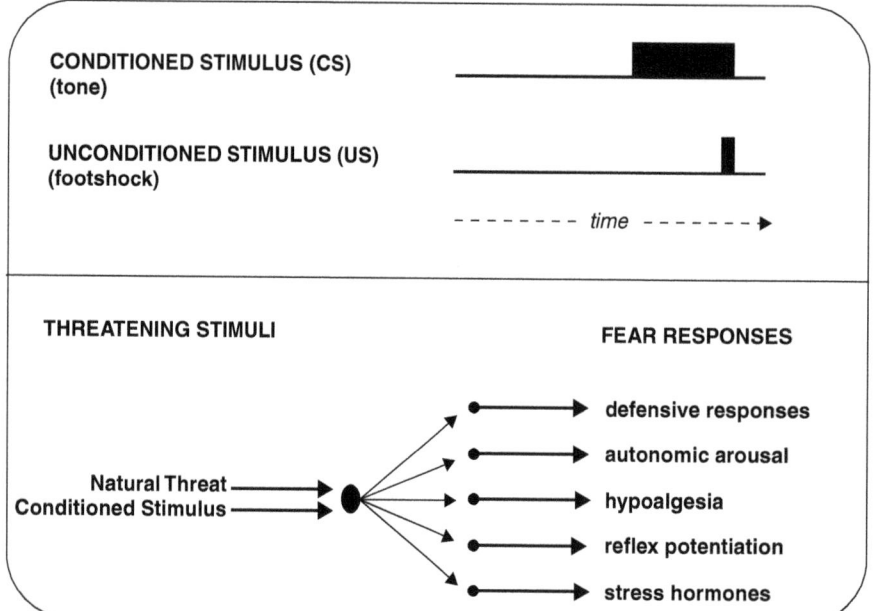

Figure 7.2. Pavlovian fear conditioning involves temporal pairings of an emotionally-neutral conditioned stimulus (CS), such as a tone, with an aversive unconditioned stimulus (US), such as a footshock (top). Prior to conditioning, the CS does not elicit fear responses. After conditioning, CS presentations elicit a number of fear responses (bottom). Figure recreated with permission from *Annual Reviews of Neuroscience* (LeDoux, 2000).

to B (Pitkanen, Savander & LeDoux, 1997; Pare & Smith, 1998).

The LA appears to be critical for the acquisition and storage of fear conditioning. Both electrolytic and excitotoxic lesions of the LA prevent acquisition and expression of fear conditioning (LeDoux *et al.*, 1990; Campeau & Davis, 1995; Amorapanth, LeDoux & Nader, 2000). Further, temporary inactivation of the LA with muscimol prevents CS-elicited freezing when given before training or testing, but not if given immediately after training (Muller *et al.*, 1997; Wilensky, Schafe & LeDoux, 1999). CE, via its projections to the hypothalamus and brainstem (LeDoux *et al.*, 1988; Bellgowan & Helmstetter, 1996; Davis, 1998; De Oca *et al.*, 1998), appears to mediate the expression of conditioned fear (Hitchcock & Davis, 1986; Kapp *et al.*, 1992; Amorapanth, LeDoux & Nader, 2000; Goosens & Maren, 2001; see also Koo, Han & Kim, 2004). The role of B in fear conditioning is somewhat controversial at present. Pre-training lesions of B have no effect on learning or expression (Amorapanth, LeDoux & Nader, 2000; Goosens & Maren, 2001; Nader *et al.*, 2001; Sotres-Bayon, Bush & LeDoux, 2004), but a recent study indicates that post-training lesions impair expression of learning (Anglada-Figueroa & Quirk, 2005). This suggests that B is not required for learning or expression of fear conditioning, but may participate in one or both of these processes under normal circumstances.

Taken together, anatomic, lesion and inactivation studies strongly suggest that LA participates in the learning and storage of fear conditioning while the B and CE participate at least in fear expression. Notably, LA manipulations that impair fear conditioning do not alter tone or shock sensitivity, the ability to freeze, or even non-associative learning processes such as shock-induced sensitization (Fanselow & LeDoux, 1999). Thus, the LA appears to be selectively involved in associative learning about emotionally significant stimuli.

Synaptic plasticity

Short latency (<15 ms) auditory-evoked unit responses of LA neurons are enhanced following behavioral fear conditioning (Quirk, Repa & LeDoux, 1995; Collins & Pare, 2000; Repa & LeDoux, 2001; Repa *et al.*, 2001)

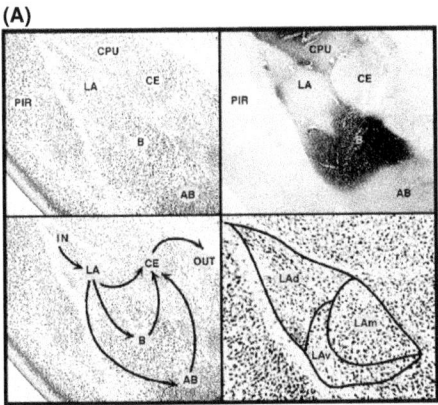
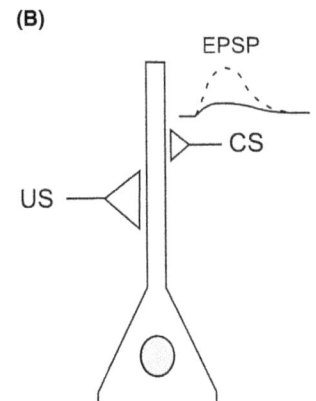

Figure 7.3. (A) At least 12 nuclei comprise the mammalian amygdala. Particularly important for fear conditioning are the lateral nucleus (LA), the basal nucleus (B), and the central nucleus (CE). Top panels show coronal slices taken from rat brain (*left*: Nissl stain, *right*: Acetycholinesterase stain) including LA, B and CE. The basic amygdala circuit mediating auditory fear conditioning is shown in the lower left panel. Sensory inputs enter through the LA, and LA connects to CE both directly and indirectly by way of B. CE is the major output nucleus mediating expression of conditioned responses. Bottom right: higher magnification image of the LA demonstrating that amygdala nuclei can also be divided into functional sub-regions called sub-nuclei. Pir = piriform cortex, AB = accessory basal nucleus, CPU = caudate putamen, LAd = dorsal subdivision of the LA, LAm = medial subdivision of the LA, LAv = ventral subdivision of the LA. Figure reprinted with permission from *Annual Reviews of Neuroscience* (LeDoux, 2000). (B) Schematic illustrating the basic cellular hypothesis of fear conditioning. Prior to conditioning, CS afferents synapse on dendrites of LA principle cells resulting in small EPSPs incapable of driving LA neurons, and downstream regions important for fear expression. When CS and US inputs are correlated in time, as during a CS–US pairing, synaptic plasticity occurs so that subsequent CS presentations elicit larger EPSPs that are capable of driving LA principle cells and downstream regions important for fear expression. Figures reprinted with permission from *Annual Reviews of Neuroscience* (LeDoux, 2000) and *Learning and Memory* (Blair et al., 2001).

(Figure 7.4). These changes are believed to reflect synaptic plasticity induced by the associative pairing of the tone and shock. To examine this, Rogan, Staubli and LeDoux (1997) assessed auditory CS-evoked field responses in the LA and CS-evoked fear behavior in freely behaving rats before, during, and after fear-conditioning. They found an increase in slope and amplitude of this response during fear-conditioning that paralleled fear behavior and persisted (Rogan, Staubli & LeDoux, 1997). Importantly, unpaired CS and US presentations produced no enhancement of this response or fear of the CS. These and other studies also highlight the thalamus→LA synapse as an important mediator of fear-conditioning. The earliest CS-evoked LA responses occur <15 ms after tone-onset, are modified by fear-conditioning and are likely driven by direct connections from the thalamus (Quirk, Repa & LeDoux, 1995; Rogan & LeDoux, 1995; Quirk, Armony & LeDoux, 1997; Rogan, Staubli & LeDoux, 1997). Consistent with this notion, LA excitatory post-synaptic potentials evoked by stimulation of thalamic afferents are selectively enhanced in brain slices taken from fear-conditioned rats (McKernan & Shinnick-Gallagher, 1997).

Fear-conditioning also results in synaptic plasticity in structures afferent to the LA (e.g. thalamus, cortex (Weinberger, 1995; Quirk, Armony & LeDoux, 1997)). However, these are unlikely to be essential for fear learning at the level of behavior for two reasons: (1) inactivation of the LA prevents fear learning and memory indicating that these structures alone cannot support learning (Muller *et al.*, 1997; Wilensky, Schafe & LeDoux, 1999) and (2) plasticity in these afferent structures appears to depend on LA function (Armony, Quirk & LeDoux, 1998; Maren, Yap & Goosens, 2001). Together these findings suggest that fear-conditioning changes the way a CS is processed in an emotional circuit involving LA, and this plasticity allows the CS to control expression of defensive responses after the aversive experience.

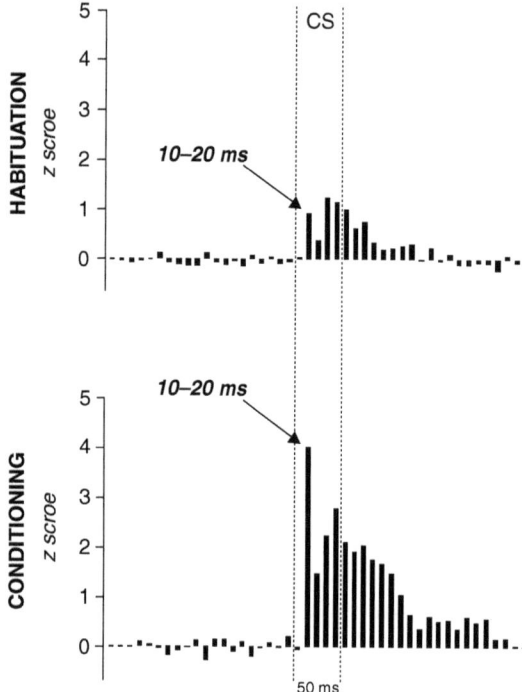

Figure 7.4. Tone-responsive neurons in the lateral amygdala show plasticity related to behavioral fear conditioning. Top, Peri-event time histogram (PETH, 10 ms bins) showing tone-elicited unit activity in the LA before conditioning (8 habituation tones). Bottom, PETH of same units recorded during fear conditioning (8 tone-shock pairings). Note that firing frequency increases especially during the earliest component of the response (10–20 ms). Based on Repa et al., 2001.

The demonstrated importance of the LA to learning and memory for fear-conditioning, coupled with discoveries of synaptic plasticity in LA, led to a cellular hypothesis of fear conditioning (Blair et al., 2001). Briefly, before auditory fear-conditioning, tone presentations result in weak depolarization of LA neurons and little to no activation of downstream brain areas mediating expression of defensive responses. However, when tone and shock stimuli are paired in time, neurons are strongly depolarized, resulting in initiation of an LTP-like process that strengthens the synapses between auditory afferents and LA neurons. After fear-conditioning, tone presentations result in strong depolarization of LA neurons and activation of downstream brain areas mediating expression of defensive responses (see Figure 7.3 (B)).

Molecular mechanisms of fear-conditioning and synaptic plasticity in LA

Armed with knowledge of a basic neural circuit and physiological mechanism for fear-conditioning, a multitude of laboratories have begun to unravel the detailed molecular mechanisms responsible for acquisition and consolidation of fear-conditioning. The majority of studies employ genetic and pharmacologic manipulations coupled with fear-conditioning to determine the function of specific molecules. Manipulations carefully timed with respect to training and testing allow researchers to distinguish between involvement in learning, STM, and LTM processes (Rodrigues, Schafe & LeDoux, 2004a). Related studies have also probed the molecular mechanisms of LTP, usually using in vitro brain slice preparations while stimulating thalamic auditory inputs (contained in the internal capsule) or cortical auditory inputs (contained in the external capsule) and recording in the LA. As with the behavioral studies, carefully timed manipulations help to distinguish between involvement in LTP induction, E-LTP, and L-LTP.

In discussing the molecular mechanisms of fear-conditioning and synaptic plasticity in LA, we will first consider receptors and ion channels that translate synaptic activity into intracellular responses, mainly by raising intracellular levels of calcium (Figure 7.5). Then we discuss the second messenger cascades and protein kinases that are activated by calcium. Next, we discuss gene expression and protein synthesis. Finally, we consider extracellular modulation of these basic processes.

Receptors and ion channels couple synaptic signals to intracellular cascades

The N-methyl-D-aspartate (NMDA) receptor (NMDAR) is the most widely studied molecule in relation to learning. Indeed, demonstrations of NMDAR involvement in associative learning, including fear conditioining, are now widespread (Muller, Joly & Lynch, 1988; Staubli et al., 1989; Falls, Miserendino & Davis, 1992; Davis & Klinger, 1994; Fanselow & Kim, 1994; Sakimura et al., 1995; Saucier et al., 1996; Tonegawa et al., 1996; Cain, 1997; Murphy & Glanzman, 1999; Nicoll & Malenka, 1999; Walker & Davis, 2000; Rodrigues, Schafe & LeDoux, 2001; Riedel, Platt &

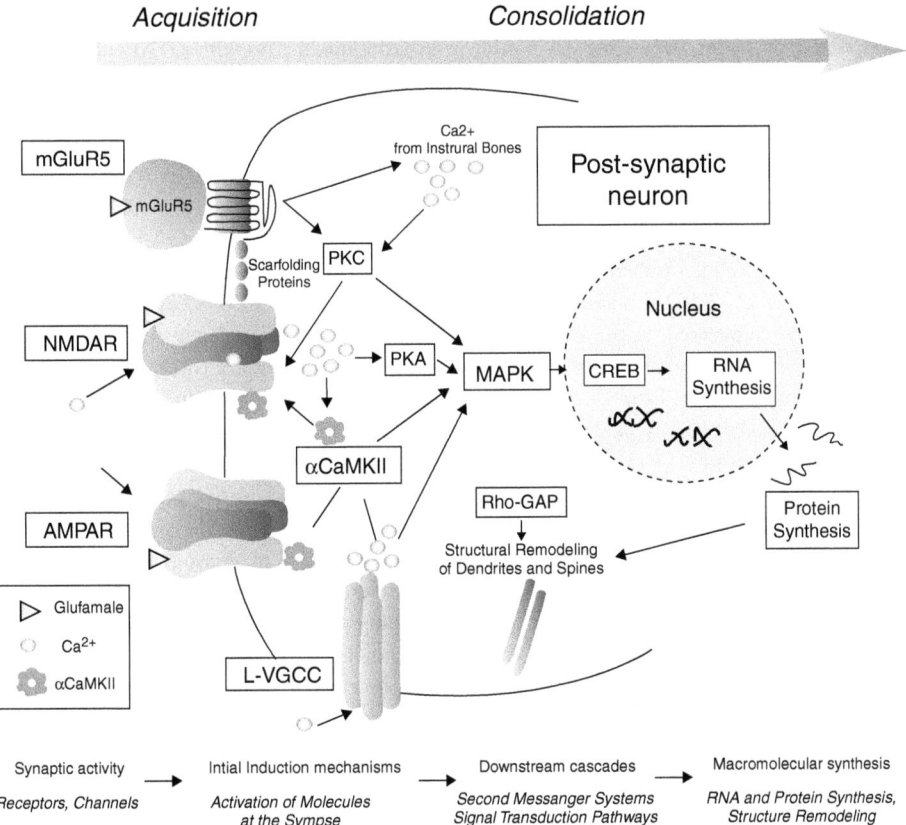

Figure 7.5. Lateral amygdala intracellular signaling pathways related to learning and memory for Pavlovian fear conditioning. Extracellular modulators, such as those mentioned in the text, are omitted from the present figure. Less is known about how these molecules interact with intracellular signaling pathways to influence fear conditioning and synaptic plasticity in the LA. Reprinted with permission from *Neuron* (Rodrigues *et al.*, 2004).

Michaeu, 2003; Goosens & Maren, 2004; Xia *et al.*, 2005). The NMDAR is an ionotropic glutamate receptor that passes Ca^{2+}. It is also an attractive candidate molecule for associative learning because it is a Hebbian "coincidence detector" that only opens when glutamate binds while the post-synaptic cell is depolarized (Mayer, Westbrook & Guthrie, 1984; Nowak *et al.*, 1984).

NMDARs are expressed at LA synapses and are found post-synaptic to auditory afferents (Farb, Aoki & LeDoux, 1995; Farb & LeDoux, 1997; Weisskopf & LeDoux, 1999). In the context of fear conditioning, it has been proposed that NMDARs on LA neurons open when glutamate is released from pre-synaptic auditory fibers while these neurons are depolarized by afferents signaling the shock (Blair *et al.*, 2001). When these conditions are met, during a CS–US pairing, Ca^{2+} can enter the post-synaptic cell and triggers a number of second-messenger cascades related to plasticity.

In support of this hypothesis, infusion of the NMDAR blocker (2R)-amino-5-phosphonovaleric acid (APV) into the LA before fear-conditioning prevents the acquisition of both STM and LTM (Miserendino *et al.*, 1990; Campeau, Miserendino & Davis, 1992; Fanselow & Kim, 1994; Lee & Kim, 1998; Bauer, Schafe & LeDoux, 2002). Post-training infusions of APV have no effect on LTM, indicating that the NMDAR may not be involved in consolidation processes. However, subsequent research found that APV impairs synaptic transmission in the LA (Li, Phillips & LeDoux, 1995; Maren *et al.*, 1996; Weisskopf & LeDoux, 1999) and also interferes with the expression of some amygdala-dependent defensive responses

(Maren et al., 1996; Lee & Kim, 1998; Lee et al., 2001a). Since NMDAR properties may differ depending on their subunit composition, several laboratories have attempted to circumvent this difficulty by using subtype-specific manipulations. For instance, LA infusions of ifenprodil, a drug that selectively blocks NR2B-containing NMDARs (Chenard & Menniti, 1999), prevents STM and LTM for fear-conditioning, but has no effect on fear expression or baseline LA transmission (Rodrigues et al., 2001; Bauer et al., 2002). Consistent with this finding, genetically engineered mice with NR2B overexpressed in the forebrain exhibit better STM and LTM for fear-conditioning (Tang et al., 1999). Finally, several *in vitro* studies of LA plasticity indicate that APV and ifenprodil disrupt LTP (Huang & Kandel, 1998; Bauer et al., 2002; Lee, Lee & Choi, 2002). Taken together, these findings strongly implicate the NDMAR in acquisition of associative fear-conditioning and induction of LA LTP.

Group I metabotropic (G-protein coupled) glutamate receptors (mGluRs) have also been implicated in fear-conditioning. This group includes mGluR1 and mGluR5 receptors, but mGluR5 may be particularly important. Glutamate binding to mGluR5 triggers activation of protein kinase C (PKC) and also release of Ca^{2+} from intracellular stores (Hermans & Challiss, 2001). Pre-training LA infusions of the mGluR5 antagonist MPEP impairs both STM and LTM for fear-conditioning, whereas post-training infusions have no effect on STM or its consolidation into LTM (Fendt & Schmid, 2002; Rodrigues et al., 2002). Thus, like the NMDAR, mGluR5 in the LA appears to participate in the learning, but not stabilization, of fear conditioning. Consistent with this, MPEP also blocks LA LTP induction *in vitro* (Fendt & Schmid, 2002; Rodrigues et al., 2002).

As noted above, elevations in cytoplasmic Ca^{2+} may be particularly important for triggering synaptic plasticity related to fear-conditioning. In addition to entry through NMDARs and release from intracellular stores, Ca^{2+} may also enter LA neurons during depolarization through voltage-gated calcium channels. Though there are many types of voltage-gated calcium channels, the L-type channel (LVGCC) may have a special role in fear-conditioning. LVGCCs are expressed on the soma and dendrites of neurons (Pinard, Mascagni & McDonald, 2005) and open in response to strong depolarization (Yuste & Tank, 1996; Magee & Johnston, 1997; Stuart et al., 1997; Johnston et al., 1999). LA infusion of LVGCC blockers prior to fear-conditioning prevents LTM but not STM, and pre-test infusions have no effect on expression of fear (Bauer, Schafe & LeDoux, 2002). This unique profile suggests that LVGCCs are not required for learning or STM, but are required for triggering consolidation processes that convert STM to LTM. LTP in the LA is also sensitive to LVGCC blockers (Huang & Kandel, 1998; Weisskopf, Bauer & LeDoux, 1999; Bauer, Schafe & LeDoux, 2002; Lee, Lee & Choi, 2002); however, these drugs appear to disrupt induction, so the relationship of this plasticity to LTM consolidation is unclear at present.

Second-messengers and protein kinases

The molecules mentioned in the previous section are important for transducing specific patterns of neural activity into activation of plasticity-related intracellular signaling cascades. Protein kinases, enzymes that catalyze the phosphorylation of specific target molecules and change their function, represent the next major step in fear memory formation. The second-messengers Ca^{2+} and cAMP are particularly important for activating appropriate kinases and there is now a large body of evidence implicating three specific protein kinases in fear-conditioning: alpha Ca^{2+}/ calmodulin-dependent protein kinase II (αCaMKII), protein kinase A (PKA), and mitogen-activated protein kinase (MAPK).

αCaMKII is present in dendrites and is activated by Ca^{2+} influx through both NMDARs and LVGCCs (Yasuda, Sabatini & Svoboda, 2003). Upon binding Ca^{2+}/calmodulin, αCaMKII autophosphorylates, translocates to the post-synaptic density, and becomes associated with the NR2B subunit of NMDARs (Strack et al., 1997; Strack & Colbran, 1998; Shen et al., 2000; Bayer et al., 2001; Yasuda, Sabatini & Svoboda, 2003). Autophosphorylation allows αCaMKII to remain active even after intracellular Ca^{2+} concentrations fall and, because of this property, it is sometimes called a "memory molecule" (Soderling, Chang & Brickey, 2001; Lisman, Schulman & Cline, 2002). LA infusions of the αCaMKII inhibitor (KN-62), given either before or immediately after training, result in impairments of both STM and LTM (Rodrigues et al., 2004b). A role for αCaMKII in STM and LTM for fear-conditioning is supported by genetic studies in mice where αCaMKII is disrupted in multiple brain regions, including the LA (Chen et al., 1994; Mayford et al., 1996; Frankland et al., 2001; Ohno et al., 2001; Bejar et al., 2002; Elgersma *et al.*, 2002; Miller *et al.*, 2002; Wang *et al.*, 2003a). One target of αCaMKII is the AMPA receptor (AMPAR) and αCaMKII may exert its effects on STM, at least partially, through regulating AMPAR function at the synapse (see below).

The cAMP/PKA pathway may play an important role in consolidation of fear conditioning. cAMP is produced by adenylate cyclase (AC) molecules that are stimulated by G-proteins or by Ca^{2+}. cAMP then activates PKA (Selcher *et al.*, 2002). Post-training infusion of PKA inhibitors, either into the ventricles or LA, impairs LTM but leaves STM intact (Bourtchouladze *et al.*, 1998; Schafe *et al.*, 1999; Schafe & LeDoux, 2000). Similarly, transgenic mice with impaired PKA signaling show normal STM but impaired LTM for fear-conditioning (Abel *et al.*, 1997). Disrupting PKA binding to the A-kinase anchoring protein in the LA also results in an identical behavioral profile (Moita *et al.*, 2002). Though PKA has many substrates, it may exert its effects on LTM by activating the MAPK pathway or by translocating to the nucleus and activating cyclicadenosine monophosphate responsive element-binding (CREB) -dependent transcription (see below).

The MAPK family, and especially extracellular-regulated kinase (ERK), has received considerable recent attention in learning or memory studies. There are at least two main reasons for this interest: (1) ERK/MAPK has been implicated in long-term plasticity in both vertebrates and invertebrates (English & Sweatt, 1997; Martin *et al.*, 1997; Thomas & Huganir, 2004) and (2) the ERK/MAPK pathway appears to be a point of convergence tying together most other intracellular molecules implicated in LTM formation. For instance, PKA and αCaMKII both converge on this pathway (Chen *et al.*, 1998a; Lin *et al.*, 2001; Adams & Sweatt, 2002; Wang *et al.*, 2004). Like PKA, activated ERK/MAPK may also translocate to the nucleus and phosphorylate the CREB transcription factor (Alberini *et al.*, 1995; Milner, Squire & Kandel, 1998; Silva *et al.*, 1998), initiating gene transcription and protein synthesis. In fear conditioning, pre-training LA infusions of ERK/MAPK inhibitors impair LTM but leave STM intact (Schafe *et al.*, 1999; Schafe & LeDoux, 2000) and systemic inhibitors of the MAPK pathway impair LTM when injected after training (Atkins *et al.*, 1998). Together, these results strongly suggest that ERK/MAPK in the LA selectively participates in consolidation of LTM for fear-conditioning.

These second-messengers and kinases also play important roles in LTP in the LA. Loading LA neurons with BAPTA – a Ca^{2+} chelator – blocks LTP induction (Huang & Kandel, 1998; Weisskopf, Bauer & LeDoux, 1999). Bath application of forskolin, an activator of AC that raises intracellular cAMP concentration, produces L-LTP on its own (Huang, Martin & Kandel, 2000). Consistent with a role in LTM, inhibitors of PKA and ERK/MAPK block L-LTP in the LA but not E-LTP (Huang, Martin & Kandel, 2000). One study examined αCaMKII antagonists in LA LTP and found impaired LTP within minutes (Rodrigues *et al.*, 2004b). These findings provide further correlative evidence that LA LTP-like processes underlie behavioral fear-conditioning, since both processes depend on the same second-messengers and protein kinases.

Transcription, translation, and expression of LTM
A major consequence of synaptic events strong enough to induce LTM processes is the translocation of PKA and ERK/MAPK to the nucleus. A common target of these kinases is CREB, a transcription factor that facilitates gene expression when phosphorylated (Stevens, 1994; Waltereit & Weller, 2003). Phosphorylation by PKA allows DNA binding while ERK/MAPK phosphorylation may be necessary for sustained activation (Yin & Tully, 1996; Kandel, 1997; Milner, Squire & Kandel, 1998; Silva *et al.*, 1998; Alberini, 1999; Lamprecht, 1999; Dolmetsch *et al.*, 2001). The result of CREB-mediated transcription is believed to be the generation of new mRNA, followed by translation of new proteins, and, ultimately, the creation of physical synaptic changes that are responsible for long-term plasticity and LTM.

Consistent with this model, impairing gene transcription or protein synthesis tends to interfere with LTM, but not STM, for fear-conditioning. Genetic disruption of CREB function in mice disrupts LTM, but not STM (Bourtchouladze *et al.*, 1994; Kida *et al.*, 2002), and viral-mediated overexpression of CREB in the LA enhances LTM for fear-conditioning (Josselyn *et al.*, 2001; Wallace *et al.*, 2004). Furthermore, blockade of all gene transcription in the LA with pre-training infusions of actinomycin-D prevents consolidation of fear-conditioning without affecting learning (Bailey *et al.*, 1999). Similarly, protein synthesis inhibitors infused into the LA either before or immediately after training impair LTM but not STM (Schafe & LeDoux,

2000; Maren et al., 2003) (see earlier Figure 7.1). These findings are in line with results from a variety of learning paradigms in diverse organisms: mRNA and protein synthesis are universally required for LTM consolidation (Flexner et al., 1965; Davis & Squire, 1984; Goelet et al., 1986; Kandel et al., 1986; McGaugh, 2000a; Kandel, 2001; Schafe et al., 2001; Dudai, 2002; Dubnau, Chiang & Tully, 2003). Consistent with the behavioral results, inhibitors of both transcription and translation impair L-LTP in the amygdala, but not E-LTP (Huang, Martin & Kandel, 2000).

How do gene transcription and protein synthesis account for stable LTM? This question is only beginning to be answered. Two themes seem to be emerging regarding structural synaptic changes important for LTM stabilization. The first is that cytoskeletal rearrangement and the growth of new synapses contribute to long-term plasticity and memory (Bailey & Kandel, 1993; Woolf, 1998; Rampon & Tsien, 2000; Lamprecht & LeDoux, 2004; Sweatt, 2004a). Recent reports suggest that such events may underlie LTM for fear-conditioning. For example, interference with the RhoGAP pathway in the LA, which is necessary for structural plasticity during development, impairs LTM for fear-conditioning whilst leaving STM intact (Lamprecht, Farb & LeDoux, 2002). A recent report also indicates that mutant mice lacking *stathmin* in the LA, a cytosolic regulator of microtubule dynamics, have specific impairments in both STM and LTM for fear-conditioning as well as deficient LTP (Shumyatsky et al., 2005). The second theme involves AMPARs. Modulation of AMPAR function and distribution may be important for both STM and LTM for fear-conditioning. For instance, phosphorylation of AMPARs by αCaMKII may increase their Na^+ conductance and drive them into synapses (Barria, Derkach & Soderling, 1997; Barria et al., 1997; Benke et al., 1998; Hayashi et al., 2000; Krapivinksi et al., 2004). Such a mechanism is believed to underlie STM at a minimum (Malinow, 2003). A recent finding indicates that AMPARs are trafficked to LA synapses following fear-conditioning with implications for both STM and LTM (Rumpel et al., 2005). Indeed, if growth of new synapses is necessary for LTM consolidation it is very likely that new AMPARs will be needed to populate these synapses and make them functional. Further research in this area will be critical for a thorough understanding of LTM consolidation mechanisms.

Extracellular modulation

The experiments discussed in the preceding section have contributed greatly to our understanding of how molecules, within cells of a defined fear-conditioning circuit, participate in LTM consolidation. However, another important area of research concerns how molecules outside this basic circuit act to modulate, or in some cases mediate, important LTM mechanisms. LA principle neurons important for fear-conditioning do not operate in a vacuum, but are embedded in a neural network that is regulated by GABAergic inhibition, modulatory transmitters, neuropeptides, growth factors, diffusible gases, and likely many other known and unknown molecules. Although there has not been a systematic examination of such "extracellular" modulators in fear-conditioning, research in select areas is beginning to shed light on the potential significance of these factors to LTM consolidation. We will highlight recent findings concerning GABAergic inhibition, nitric oxide signaling, brain-derived neurotrophic factor (BDNF) signalling, and catecholamine modulation in fear-conditioning to illustrate this point.

GABAergic inhibition. Approximately 25% of neurons in the LA are GABAergic inhibitory interneurons (McDonald & Augustine, 1993) and both feed forward (Rainnie, Asprodini & Shinnick-Gallagher, 1991; Li, Armony & LeDoux, 1996; Woodson, Farb & LeDoux, 2000) and feedback (Smith, Pare & Pare, 2000; Szinyei et al., 2000; Samson, Dumont & Pare, 2003) inhibition has been reported in this nucleus. Inhibitory neurons in the LA are subject to modulation by serotonin (Rainnie, 1999; Stutzmann & LeDoux, 1999), norepinephrine (Li et al., 2002; Braga et al., 2004), glucocorticoids (Johnson et al., 2005), cannabinoids (Katona et al., 2001; McDonald & Mascagni, 2001), and dopamine (Brinley-Reed & McDonald, 1999). Thus, GABAergic inhibition in the LA may exert a powerful effect on sensory processing related to fear-conditioning and may be a site of influence for a number of modulatory systems.

Several recent reports show that both input and output synapses of LA GABAergic interneurons are plastic. LTP of excitatory inputs to interneurons depends on post-synaptic Ca^{2+} entering through either Ca^{2+}-permeable AMPARs or NMDARs, as blockers of each impair LTP (Mahanty & Sah, 1998; Bauer & LeDoux, 2004). Calcium-dependent LTP has also been demonstrated at synapses between interneurons and pyramidal cells (Bauer & LeDoux, 2004). Thus, inhibitory interneuron transmission in the LA may also be subject to long-term plasticity during fear-conditioning, although further work will be necessary to investigate this possibility.

One recent study directly examined LA GABAergic interneuron contributions to LTP and LTM for fear-conditioning (Shumyatsky et al., 2002). This group used an unbiased screen to identify genes selectively expressed in the LA in response to fear-conditioning. One gene identified was *Grp*, encoding gastrin-releasing peptide (GRP). GRP is a neuropeptide released from axon terminals in conjunction with glutamate that binds the GRP receptor (GRPR). Interestingly, subsequent experiments revealed that the GRPR was expressed on LA GABAergic interneuron dendrites and that GRPR activation excites interneurons leading to greater inhibition of pyramidal cells. Mutant mice lacking GRPR in the LA have enhanced LTM for fear-conditioning and enhanced LTP in the LA. STM was normal in these mice, suggesting further that LA inhibitory transmission may selectively modulate processes important for consolidation of fear-conditioning. Clearly, further work is necessary to more fully characterize the contributions of LA inhibitory transmission to fear-conditioning.

BDNF signaling. Given the hypothesis that memory stabilization involves synaptic growth and rearrangement, a number of laboratories are beginning to investigate molecules known to mediate these actions during development. For instance, neurotrophins such as BDNF are known to play an important role in activity-dependent synaptogenesis and synaptic organization during central nervous system (CNS) development (Alsina & Cohen-Cory, 2001; Aguado et al., 2003; Seil, 2003).

BDNF is soluble and can diffuse through the extracellular space and bind its receptor, a transmembrane tyrosine kinase (TrkB) (Klein et al., 1991; Soppet et al., 1991; Squinto et al., 1991). BDNF is also expressed in adult brain and has been shown to be released from pre-synaptic terminals in response to activity (Kohara et al., 2001; Nawa & Takei, 2001). Largely based on these properties, it was hypothesized that BDNF may be an important molecule for mediating activity-dependent morphological changes related to LTM consolidation.

The past 15 years have seen a fairly intense examination of BDNF in mammalian memory and synaptic plasticity (Lu & Chow, 1999; Chao, 2000; Lu & Gottschalk, 2000; Tyler et al., 2002; Yamada & Nabeshima, 2003). However, the bulk of this research concerns hippocampal plasticity and hippocampus-dependent learning paradigms. Based on this research, potential mechanisms for BDNF in LTP and LTM are emerging. BDNF released from pre-synaptic terminals binds TrkB receptors on both the pre- and post-synaptic membranes. Upon binding BDNF, TrkB receptors dimerise, autophosphorylate, and activate several intracellular kinase pathways important for plasticity and memory, including ERK/MAPK, PKC, and phosphatidylinositol-3 kinase (PI-3k). Results of BDNF signaling in the hippocampus include: (1) enhanced pre-synaptic release of transmitter; (2) phosphorylation of NMDARs and increased channel conductance; (3) facilitation of protein synthesis; and (4) modulation of dendritic growth or complexity (for review see Yamada & Nabeshima, 2003). Thus, BDNF may modulate nearly every phase of LTM and LTP. Indeed, there are now numerous demonstrations of learning or memory and LTP impairments after manipulations that disrupt hippocampal BDNF signaling (Patterson et al., 1996; Linnarsson, Bjorklund & Ernfors, 1997; Minichiello et al., 1999; Mu et al., 1999; Saarelainen et al., 2000; Alonso et al., 2002).

Investigations of BDNF in fear-conditioning are now emerging, and early indications suggest an important role for this neurotrophin. BDNF expression and TrkB phosphorylation in the LA transiently increases in the hours after associative fear-conditioning consistent with a role in LTM consolidation (Rattiner et al., 2004; Ou & Gean, 2006). Studying BDNF signaling in fear-conditioning is difficult because there are no antagonists of TrkB available; however, two recent fear-potentiated startle studies employed clever techniques to disrupt this pathway. The first group used viral infection of LA cells to express a dominant negative TrkB fragment that prevents receptor dimerisation and subsequent kinase activity (Rattiner et al., 2004). This treatment impaired LTM for fear conditioning. The second group infused a scavenger construct into the LA to bind BDNF and thus prevent it from binding endogenous TrkB (Ou & Gean, 2006). This treatment also impaired LTM for fear conditioning. Lastly, this same group infused BDNF into the

LA before fear-conditioning and found enhanced LTM. Unfortunately, no tests of STM or LTP were included in these studies. The combined results provide compelling evidence that BDNF in the LA participates in memory for fear-conditioning; however, further work will be necessary to clarify its specific role in the various phases of memory, and potentially, LTP.

Nitric oxide signaling. Most studies of LA processes in fear-conditioning have focussed on purely post-synaptic mechanisms of synaptic plasticity. However, accumulating evidence is beginning to support the notion that fear-conditioning and LTP involve both pre- and post-synaptic alterations. For instance, paired-pulse facilitation (PPF) in the LA, a form of plasticity believed to be mediated by pre-synaptic changes in transmitter release (Zucker, 1989), is occluded by fear-conditioning and LTP (McKernan & Shinnick-Gallagher, 1997; Huang & Kandel, 1998; Tsvetkov et al., 2002). Since NMDARs and post-synaptic Ca^{2+} are required for LTP induction in the LA, this raises the question of how plasticity could be initiated post-synaptically but expressed on both sides of the synapse. A potential answer to this question is a "retrograde messenger;" a molecule generated post-synaptically during LTP induction that travels back across the synapse to act on a pre-synaptic effector mechanism (Medina & Izquierdo, 1995; Hawkins, Son & Arancio, 1998). A number of candidate molecules have been proposed, including nitric oxide (NO), carbon monoxide, platelet-activating factor, and arachadonic acid (O'Dell et al., 1991; Williams et al., 1993).

NO is an interesting candidate molecule because it has a mechanism consistent with retrograde regulation of pre-synaptic release and because its involvement in other forms of learning and plasticity is well-documented (Chapman et al., 1992; Holscher & Rose, 1992; Bohme et al., 1993; Bernabeu et al., 1995; Bernabeu et al., 1996; Suzuki et al., 1996; Bernabeu et al., 1997a; Zou et al., 1998). NO is synthesized by the enzyme nitric oxide synthase (NOS) in post-synaptic neurons in response to NMDAR-mediated Ca^{2+} influx (Bredt et al., 1990, 1991; Bredt & Snyder, 1992). It is also a soluble gas that may diffuse across the lipid membranes to facilitate pre-synaptic transmitter release (Hawkins, Kandel & Siegalbaum, 1993; Hawkins, 1996).

Recently, Schafe et al. (2005) investigated the possibility that NO serves as a retrograde messenger in the LA important for LTP and fear-conditioning. They found that NOS is present in pyramidal cell dendrites. They next examined the role of NO in fear-conditioning and LTP using two different drugs: 7-Ni, an inhibitor of NOS, and carboxy–PTIO, an NO scavenger that does not cross lipid membranes. Infusion of either drug into the LA before fear-conditioning severely impaired LTM but left STM intact, consistent with a role for NO in consolidation of fear-conditioning.

Importantly, because carboxy–PTIO does not penetrate lipid membranes, and thus does not enter neurons when infused into the extracellular space, the results with this drug are consistent with a role for NO as a retrograde messenger. Other experiments demonstrated that both 7-Ni and carboxy–PTIO impair LA LTP. Together, these results suggest that LTP and LTM for fear-conditioning require NO signaling in the LA and that NO may serve as a retrograde messenger to enhance pre-synaptic neurotransmission (Schafe et al., 2005).

Catecholamine modulation. As noted in the introduction, not all events that lead to learning result in LTM for the experience. So what factors determine whether or not learning and STM are consolidated into LTM? One attractive hypothesis is that emotional events are much more likely to be committed to LTM than non-emotional events (Gold & McGaugh, 1975; Christianson, 1992; LeDoux, 2000). A long history of research implicates neuromodulators, such as stress hormones and catecholamines (McGaugh et al., 1982; McGaugh & Introini-Collison, 1987; McGaugh & Gold, 1989; Ferry, Roozendaal & McGaugh, 1999; Ferry & McGaugh, 2000; McGaugh, 2002a; McGaugh & Roozendaal, 2002), in determining whether learning is stabilized into LTM. More elusive, however, have been demonstrations of how these modulators may interact with synaptic plasticity in well-defined neural circuits to facilitate memory formation. Analyzing the role of catecholamines in synaptic plasticity related to fear-conditioning may be an excellent strategy for answering such questions.

Dopamine (DA) transmission has been extensively studied in relation to learning and memory, and is beginning to be studied in fear-conditioning. LA neurons receive mesencephalic DA-containing afferents (Fallon & Ciofi, 1992; Asan, 1998; Brinley-Reed & McDonald, 1999) and DA is released in the LA in response to stress (Inglis & Moghaddam, 1999). DA can act on either D1 or D2 receptor subclasses, both of which are expressed in the LA (Maltais et al., 2000). D1 receptors are positively coupled to AC and cAMP production whereas D2 receptors may be negatively coupled to AC, but also have other actions (Neve, Seamans & Trantham-Davidson, 2004). Pre-training LA infusions of D1 antagonists impair LTM for fear-conditioning and D1 agonists enhance LTM (Guarraci, Frohardt & Kapp, 1999; Greba & Kokkonidis, 2000). Unfortunately, STM was not assessed in these studies. Pre-training LA infusion of D2 antagonists also impairs LTM for fear-conditioning (Guarraci et al., 2000; Greba, Grifkins & Kokkonidis, 2001) but leaves STM intact (Guarraci et al., 2000). Together, these findings suggest that DA in the LA positively modulates LTM consolidation.

Recent electrophysiological work in the LA is beginning to explain how DA might modulate LTP and consolidation of fear-conditioning. In short, DA enhances pyramidal cell excitability and simultaneously suppresses feedforward inhibition (Rosenkranz & Grace, 2002; Bissiere, Humeau & Luthi, 2003; Kroner et al., 2005). Consistent with these findings, antagonizing DA receptors with haloperidol blocks *in vivo* LA LTP caused by behavioral fear conditioning (Rosenkranz & Grace, 2002). *In vitro*, DA application enables LTP with a weak induction protocol (Bissiere, Humeau & Luthi, 2003). Removing GABAergic inhibition with picrotoxin negates the DA requirement for LTP, supporting the notion that DA suppresses feedforward inhibition and gates LTP induction. Together, these data strongly suggest that DA modulates LTP and LTM consolidation in the LA by enhancing pyramidal neuron excitability, both directly and indirectly.

An emerging theme in neuroscience research is that neuromodulators may be necessary for stabilizing plasticity and STM initiated by Hebbian mechanisms (Bailey et al., 2000). Thus, intracellular cascades triggered by NMDAR activation may be too weak to activate transcription and translation processes. However, neuromodulators converging on the same signaling cascades may provide amplification necessary to recruit LTM processes. For instance, norepinephrine (NE) acting on β-receptors triggers production of cAMP and activation of PKA. When coupled with PKA activation triggered by Ca^{2+} influx through NMDARs and LVGCCs, this signal may be strong enough to translocate to the nucleus and initiate CREB-dependent transcription necessary for L-LTP and LTM.

NE signaling, especially through the β-receptor, has been extensively studied in relation to memory consolidation, mainly in aversive learning paradigms that do not require the LA for LTM storage (Izquierdo & Medina, 1997; Cahill, Pham & Setlow, 2000; Gibbs & Summers, 2002; McGaugh, 2002b; McGaugh & Roozendaal, 2002; Cartford, Gould & Bickford, 2004). The basic finding in these studies is that β-adrenergic antagonists administered after training impair LTM consolidation. Surprisingly, a systematic analysis of NE contributions to consolidation of fear-conditioning has not been done and the available data are inconclusive (Lee et al., 2001b; LaLumiere, Buen & McGaugh, 2003; Debiec & LeDoux, 2004; Grillon et al., 2004). However, a recent report on E-LTP and L-LTP in LA fear-conditioning pathways suggests that this topic warrants further investigation. Huang, Marin and Kandel (2000) found that blockade of β-adrenergic receptors with propranolol disrupted L-LTP but not E-LTP. Additionally, direct stimulation of β-receptors produced L-LTP on its own. These results are consistent with the hypothesis that Hebbian plasticity requires modulatory signaling to transform E-LTP to L-LTP and suggests that a similar profile may be observed with careful study of STM and LTM of fear-conditioning. In any case, the interaction of Hebbian plasticity mechanisms with neuromodulatory mechanisms in fear conditioning is likely to be a fruitful area of consolidation research in coming years.

RECONSOLIDATION OF FEAR CONDITIONING

Memory consolidation theory, for over a century, defined the direction of learning and memory research (McGaugh, 2000a). Treatments administered shortly after training proved to be a successful method for determining the cellular and molecular mechanisms of consolidation processes. A basic assumption of consolidation theory was that stabilized memories were no longer susceptible to disruption, at least not by the same manipulations that prevented the conversion of labile STM to LTM. However, studies in the late

1960s (Misanin, Miller & Lewis, 1968; Schneider & Sherman, 1968) and early 1970s (Lewis, Bregman & Mahan, 1972; Lewis & Bregman, 1973), followed by a few studies in the 1980s (Judge & Quartermain, 1982) and 1990s (Przybyslawski & Sara, 1997; Roullet & Sara, 1998; Przybyslawski, Roullet & Sara, 1999), reported that administration of amnesic treatments shortly after memory retrieval impairs subsequent memory expression. The conclusion drawn from these findings was that memory "reconsolidates" each time it is reactivated (Sara, 2000).

Interest in memory reconsolidation was revived after a publication by Nader et al. (2000a, 2000b). Using auditory fear-conditioning, these workers demonstrated that consolidated memory was disrupted by intra-LA infusions of anisomycin, a protein synthesis inhibitor, administered immediately after reactivating the memory with a CS presentation. In order to convincingly demonstrate reconsolidation, it is important to show that reconsolidation and consolidation share similar patterns of time-dependent susceptibility: post-training (consolidation) or post-retrieval (reconsolidation) amnesic treatments should affect LTM, but leave STM intact (Schafe & LeDoux, 2000). Nader et al. (2000a, 2000b) showed that infusions of anisomycin into the LA had no effect on post-reactivation STM, but profoundly affected performance 24 hours later (post-reactivation LTM was impaired). Drug infusions administered several hours after memory retrieval, or without explicit memory reactivation, had no effect on post-reactivation LTM (Figure 7.6).

The demonstration of reconsolidation within the well-defined auditory fear-conditioning circuit offered the possibility that the cellular basis of this controversial process might be discernable. Although molecular examinations of the reconsolidation process lag far behind consolidation, recent studies using amygdala-specific manipulations are beginning to uncover its mechanism. A recent study reported that activation of the ERK/MAPK cascade is required for the reconsolidation (Duvarci, Nader & LeDoux, 2005). Others demonstrated that blocking the expression of an immediate-early gene, zif-268, impairs reconsolidation (Lee et al., 2005a). Further, post-retrieval intra-LA blockade of noradrenergic signaling by the β-adrenergic receptor antagonist propranolol disrupts reconsolidation of auditory fear learning (Debiec & LeDoux, 2004). Thus, similar to consolidation, reconsolidation appears to rely on specific intracellular signals, such as kinases and gene transcription, and extracellular modulators, such as norepinephrine.

Since the publication of the original study by Nader et al. (2000a), and the renewal of interest in the subject, reconsolidation has been demonstrated in a wide variety of species, including snail (Sangha, Scheibenstock & Lukowiak, 2003b), sea slug (Child et al., 2003), crab (Pedreira, Perez-Cuesta & Maldonado, 2002; Frenkel, Maldonado & Delorenzi, 2005; Merlo et al., 2005), honeybee (Stollhoff, Menzel & Eisenhardt, 2005), rat (Debiec, LeDoux &Nader, 2002; Milekic & Alberini, 2002; Eisenberg et al., 2003; Debiec & LeDoux, 2004; Lee, Everitt & Thomas, 2004; Torras-Garcia et al., 2005), mouse (Bozon, Davis & Laroche, 2003; Suzuki et al., 2004; Boccia et al., 2005; Inda, Delgado-Gsarcia & Carrion, 2005; von Hertzen & Giese, 2005), and human (Walker et al., 2003). Reconsolidation has also been shown in a variety of learning paradigms, such as auditory fear conditioning (Nader, Schafe & LeDoux, 2000a; Debiec & LeDoux, 2004; Duvarci & Nader, 2004; Duvarci, Nader & LeDoux, 2005), contextual fear conditioning (Pedreira, Perez-Cuesta & Maldonado, 2002; Debiec & LeDoux, 2004; Suzuki et al., 2004), inhibitory avoidance learning (Milekic & Alberini, 2002; Tronel, Milekic & Alberini, 2005), Morris water maze task (Suzuki et al., 2004), instrumental learning (Wang et al., 2005), olfactory discrimination learning (Torras-Garcia et al., 2005), conditioned taste aversion (Eisenberg et al., 2003), and others.

As evidence for a reconsolidation process continues to accumulate, the question arises as to whether or not reconsolidation is simply a recapitulation of the cellular and molecular events that underlie original memory consolidation (Nader, Schafe & LeDoux, 2000b; Sara, 2000; Schafe et al., 2001; Dudai, 2004; Alberini, 2005; Nader, Hardt & Wang, 2005). In support of this notion, both processes share important properties. First, both consolidation and reconsolidation are characterized by similar temporal profiles (amnesic treatments affect LTM, but spare STM) (e.g. Nader, Schafe & LeDoux, 2000a; Debiec, LeDoux & Nader, 2002; Lee et al., 2005a). Second, both consolidation and reconsolidation share some of the same biochemical requirements, such as protein synthesis (e.g. Gruest, Richer & Hars, 2004; Inda, Delgado-Garcia & Carrion, 2005; Wang et al., 2005). Furthermore, both consolidation and reconsolidation require the activation of transcription factors: CREB

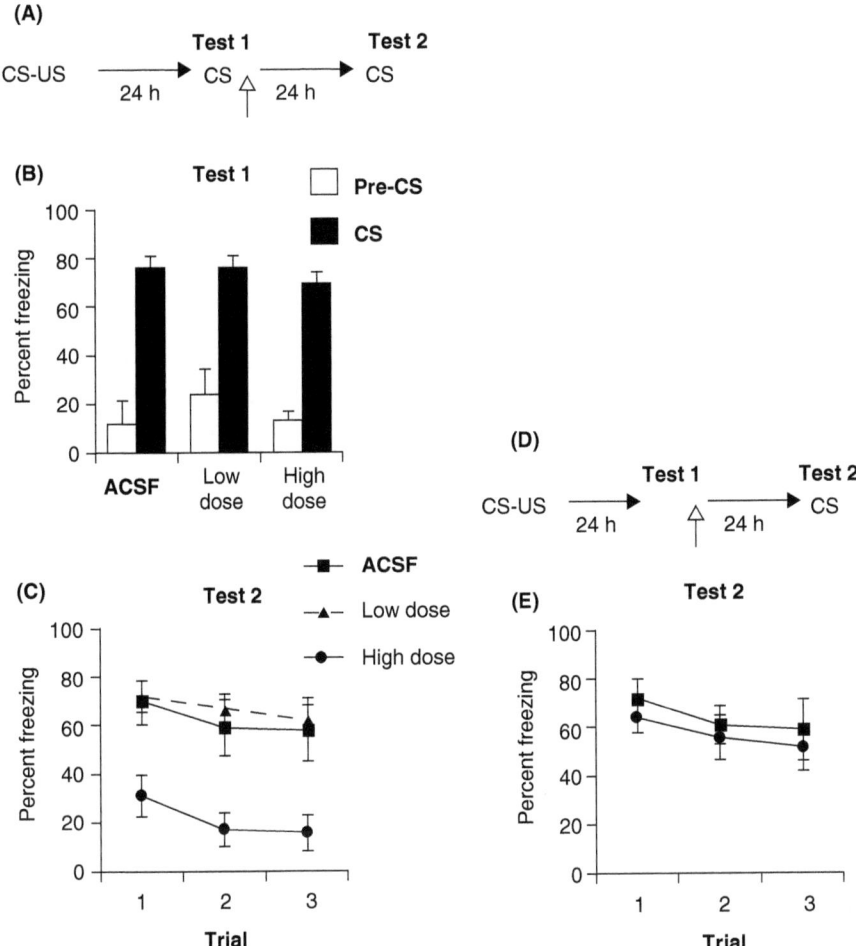

Figure 7.6. Protein synthesis in the LA is required for reconsolidation of auditory fear conditioning. LA infusions of anisomycin after memory reactivation with a CS result in impaired freezing 24 hours later (c). LA infusions of anisomycin have no effect on STM after reactivation (not shown) or LTM with no reactivation (e). Reproduced with permission from Nature (Nader et al., 2000).

(Kida et al., 2002) and zif-268 (Jones et al., 2001; Bozon, Davis & Laroche, 2003). Third, in certain learning tasks, both consolidation and reconsolidation involve the same anatomical sites. For example, protein synthesis in the amygdala is implicated in the consolidation and reconsolidation of both auditory fear conditioning (Nader, Schafe & LeDoux, 2000a; Schafe & LeDoux, 2000) and instrumental incentive learning (Wang et al., 2005).

Despite these similarities, mounting evidence suggests that there are distinctions between consolidation and reconsolidation. First, while consolidation and reconsolidation both have temporal windows, they do not always follow the same time-course of susceptibility to amnesic treatments (Przybyslawski, Roullet & Sara, 1999; Debiec, LeDoux & Nader, 2002). Second, a few studies have demonstrated biomolecular dissociations between consolidation and reconsolidation processes (Lee, Everitt & Thomas, 2004; von Hertzen & Giese, 2005). Using contextual fear conditioning in rats, Lee, Everitt and Thomas (2004) found that consolidation involves BDNF but not the transcription factor zif-268, whereas reconsolidation requires zif-268 but not BDNF. Moreover, using the same behavioral paradigm in mice, von Hertzen and Giese (2005) demonstrated that consolidation upregulates the expression of serum- and glucocorticoid-induced kinase 3 gene (*SGK3*) in the hippocampus, as well as the nerve growth factor-inducible gene B (*NGFI-B*), whereas reconsolidation engages *SGK3* but not *NGFI-B*. Third, some studies report that distinct neural circuits are involved in molecular processes underlying consolidation and reconsolidation (Bahar, Dorfman & Dudai, 2004; Tronel, Milekic & Alberini, 2005). Bahar, Dorfman and Dudai (2004) found that the protein synthesis in the amygdala is necessary for consolidation but not for reconsolidation of conditioned taste aversion. Furthermore, using inhibitory avoidance learning in rats, it has been demonstrated that consolidation requires the transcription factor CCAAT enhancer binding protein β (*C/EBP β*) in the hippocampus but not in the amygdala, whereas reconsolidation engages *C/EBP β* in the amygdala but not in the hippocampus (Taubenfeld et al., 2001; Tronel, Milekic & Alberini, 2005). Taken together, these findings suggest that reconsolidation may only partially recapitulate consolidation processes.

Despite evidence demonstrating the occurrence of reconsolidation processes, some authors report a failure to observe the reconsolidation effect (Biedenkapp & Rudy, 2004; Cammarota et al., 2004; Hernandez & Kelley, 2004). Furthermore, it has also been reported that post-retrieval memory disruption may have a transient character (Lattal & Abel, 2004). In some situations consolidation and extinction compete (Duda & Hart, 1973). It has not yet been determined whether post-retrieval memory susceptibility to disruption involves storage or retrieval alterations. And, as of yet, there have been no demonstrations that disrupting reconsolidation reverses synaptic plasticity related to consolidation of behavioral LTM. Despite these concerns, reconsolidation remains an exciting area of research and examination of this process within fear-conditioning circuits will likely provide important information about the stability of LTM traces.

SUMMARY AND CONCLUSIONS

Pavlovian fear-conditioning in the rodent has proved an invaluable tool for investigating the brain mechanisms of mammalian memory formation and storage. Because a basic neural circuit has been identified, and because learning induces synaptic plasticity in this circuit, fear-conditioning studies may provide unique insight into the cellular and molecular mechanisms of behavioral memory. We have surveyed evidence that associative fear conditioning changes the way sensory information is processed in the amygdala, a region important for emotional expression. Specifically, conditioning enables sensory afferents to gain control over defensive responding through strengthening of synapses in the LA. This synaptic plasticity depends on transmembrane receptors and ion channels that recognize patterned activity and stimulate intracellular kinase pathways. Kinases then modify existing synaptic proteins leading to short-term plasticity and STM, and also initiate transcription and translational processes necessary for the structural changes subserving LTP and LTM formation. We also provided examples of extracellular signaling pathways that exert powerful influence over processes in this basic circuit. The result is a fairly comprehensive view of the brain mechanisms responsible for acquisition and consolidation of Pavlovian fear-conditioning. Although the picture is far from complete, this research strategy has proven successful and future research in this area promises to fill in the gaps.

An emergent theme from fear-conditioning research is that learning and memory are related but distinct processes. Not all learning is stored for future use. A key factor in determining whether learning will be consolidated into LTM appears to be the emotional significance of the experience. And, importantly, consolidation processes may be disrupted, at least within the hours following the learning experience. The revival of reconsolidation studies indicates that even stable, consolidated memories may be returned to a labile state with reminder treatments. These features of emotional memory storage raise the exciting possibility that memory permanence or strength may be modifiable, which could have important implications for treatment strategies related to pathological fear. Indeed, the susceptibility of consolidation to disruption has been exploited to reduce the impact of traumatic experiences in humans (Pitman et al., 2002; Vaiva et al., 2003). However, such preventative interventions are severely limited because they require treatment within a narrow time window after the experience. Reconsolidation, on the other hand, may be a more easily controlled process with equally important therapeutic potential.

8 · Emotional and cognitive reinforcement of rat hippocampal long-term potentiation by different learning paradigms

Volker Korz and Julietta U. Frey

Long-term potentiation (LTP) is a widely studied physiological model for learning and memory formation. Like the finding that long-term memory (LTM) but not short-term memory (STM) is protein-synthesis dependent (Flexner et al., 1965; Grecksch & Matthies, 1980; Izquirdo et al., 2002) at least two phases of LTP may be distinguished: a protein synthesis-independent early-LTP, decaying to baseline within three to four hours, and a protein synthesis-dependent late-LTP, which lasts for at least eight hours (Krug, Lossner & Ott, 1984; Frey et al., 1988). The induction of hippocampal late-LTP requires the associative activation of heterosynaptic inputs, such as the activation of glutamatergic and dopaminergic receptors in area CA1 (Frey, Hartmann & Matthies, 1989; Frey, Schröder & Matthies, 1990; Frey, Huang & Kandel, 1993) or glutamatergic and nordrenergic receptors in the dentate gyrus (DG) (Frey et al., 2001). Under low-stress conditions, such as during novelty detection or water lavage in water-deprived rats, the transformation of early-LTP into late-LTP in the DG depends on the activation of β-adrenergic receptors (Seidenbecher, Reymann & Balschaum, 1997) and on protein synthesis (Straube et al., 2003). A recent study under high-stress conditions showed that a two-minute swim in a water tank 15 minutes after a weak exposure to tetanus also prolongs early-LTP for up to 24 hours in naïve animals. The prolongation is dependent on protein-synthesis but not on β-adrenergic activation (Korz & Frey, 2003a). By use of a high-stress paradigm, LTP is maintained by corticosterone-induced activation of mineralocorticoid receptors (MR). Because the activated MR-complex acts as a transcription factor and modulates the activity of other factors (Diamond et al., 1990; Cato, Ponta & Herrlich, 1992; Malkoski, Handanos & Dorin, 1997), the resulting translational products may include plasticity-related proteins (PRPs) being involved in the maintenance of LTP (Frey & Morris, 1998a; 1998b; Sajikumar & Frey, 2004).

A cellular mechanism for associative consolidation of LTP has been suggested (Frey & Morris, 1997; Frey & Morris, 1998b; Sajikumar & Frey, 2004): a weak input inducing early-LTP sets a "synaptic tag," being active for a distinct time. PRPs whose synthesis is induced by temporally related modulatory heterosynaptic inputs (Sajikumar & Frey, 2004) may be captured and processed by this tag, leading to a reinforcement of the potentiation of the tagged synapse. Early-LTP is thus transformed into late-LTP. Recently it has been demonstrated that long-term depression (LTD) shows the same tagging features as LTP. LTP and LTD may interact dependently by tags, a phenomenon introduced as "cross-tagging" (Sajikumar & Frey, 2004). Thus, at the cellular level, plastic phenomena that are thought to underlie memory formation may be transferred into a long-lasting memory by associative events.

Reinforcement by novelty detection is typically a rapid process that occurs after a single exposure to the arousing situation, and may be blocked by a brief pre-exposure to the stimuli making the novelty familiar (Straube et al., 2003). The term "emotional tagging" has been introduced to characterize this form of LTP-reinforcement related to the rapid consolidation of relevant new information (Richter-Levin & Akirav, 2003). LTP may also be protein synthesis-dependently reinforced by mastering a spatial task in a holeboard (Uzakov, Frey & Korz, 2005). In contrast to emotional reinforcement this cognitive reinforcement appears only after repetitive training in the same environment and is related to the formation of a long-lasting spatial reference memory, suggesting that proteins involved in memory consolidation also contribute to the maintenance of hippocampal LTP.

Different cellular signalling is involved in emotional and cognitive reinforcement of LTP. A preliminary concept of brain and cellular mechanisms which might be involved in emotional and cognitive reinforcement is presented and discussed.

Novelty, acute stress, and LTP-reinforcement

To address the question of whether there are different reinforcement processes occurring during different learning paradigms, the novelty response in the water maze and the holeboard have been tested. As shown in Figure 8.1, statistically significantly increased protein synthesis amplitude is seen only in water maze-naïve animals compared to controls, indicating reinforcement of early-LTP through novelty detection. Water maze-pretrained, as well as holeboard-naïve animals, showed no reinforcement of early-LTP. The lack of reinforcement in water maze-pretrained rats is not an effect of decreased stress by habituation to the procedure. Water maze-pretrained animals show similar corticosterone levels (the main stress-related glucocorticoid in rats) as naïve animals (Figure 8.2). Although a significant difference in serum corticosterone levels may be noted between the different groups, post hoc tests reveal a significant increase in corticosterone in all groups compared to undeprived controls with the exception of the group tested one hour after swim stress. Thus, the reinforcement in naïve animals depends on the novelty of the situation in combination with stress. Elevated corticosterone levels in a familiar environment have no effect on LTP. The peak in corticosterone, however, is only transient and corticosterone levels return back to baseline within one hour. In contrast, in holeboard-naïve animals no reinforcement may be observed, although the corticosterone levels are increased in response to the holeboard exposure. This elevation of the steroid is, as for the water maze, similar irrespective of the pre-experience of the animals, as indicated by enhanced corticosterone levels of holeboard-trained animals

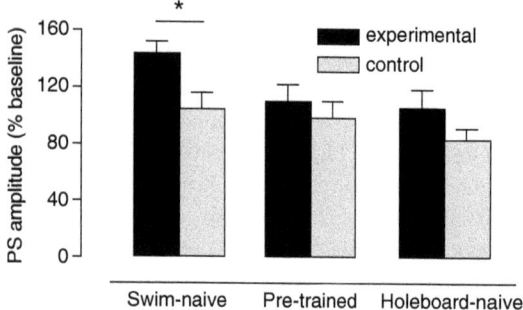

Figure 8.1. Novelty and swim stress reinforces hippocampal long-term potentiation (LTP) but has no effect in water maze-pretrained animals and holeboard-naïve animals. The population-spike amplitudes at a timepoint (24 hours) as percentage from baseline are compared between experimental groups and controls that remained in the recording box throughout the experiment. Asterisks indicate significant differences. Means and standard equivalent of the means (SEM) are given.

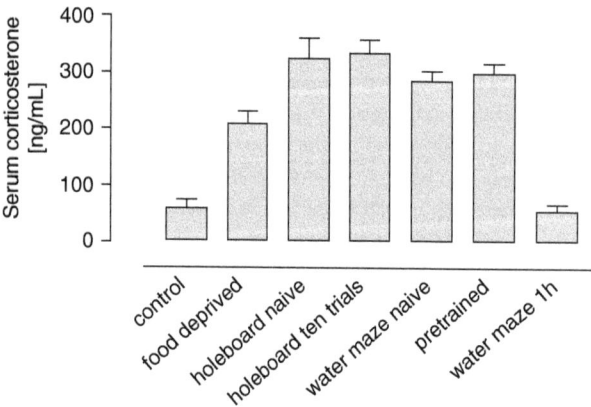

Figure 8.2. Pretraining in the water maze and the number of trials received on the holeboard has no effect on the serum corticosterone response. Food-deprived animals show chronically elevated corticosterone levels compared to non-deprived controls. The increase of corticosterone in response to swim stress is only transient and back to baseline after one hour (water maze, one hour). Means ± standard equivalent of the means (SEM) are given. The dataset was partially adopted from Uzakov, Frey & Korz, 2005) and complemented with additional experiments.

(Figure 8.2) in comparison to controls, which remained in the recording box. The food-deprived control group, however, reveal a substantial difference between the animals trained in the water maze or on the holeboard. The latter show, as an adaptation to food deprivation, chronically elevated corticosterone levels as compared to undeprived animals.

Spatial training

Holeboard

In the spatial training situation in the holeboard, trial-related reinforcement may be observed (Figure 8.3 (A)). Reinforcement may be noted only for animals which experienced a specific amount of trials (i.e. 8–10 trials), which is sufficient to establish a spatial reference memory. Consequently, by plotting individual protein synthesis amplitude at a time-point (24 hours) against the number of reference memory errors made during the last trial, a positive correlation with spatial learning is apparent (Figure 8.3 (B)). Animals that are characterized by only a few or no reference memory errors typically show increased protein synthesis amplitude 24 hours after the last trial. Comparing the amount of working memory errors with individual protein synthesis amplitude does not result in significant correlation (Figure 8.3 (C)). Interestingly, the LTP-reinforcement parallels a sudden learning effect between trials seven

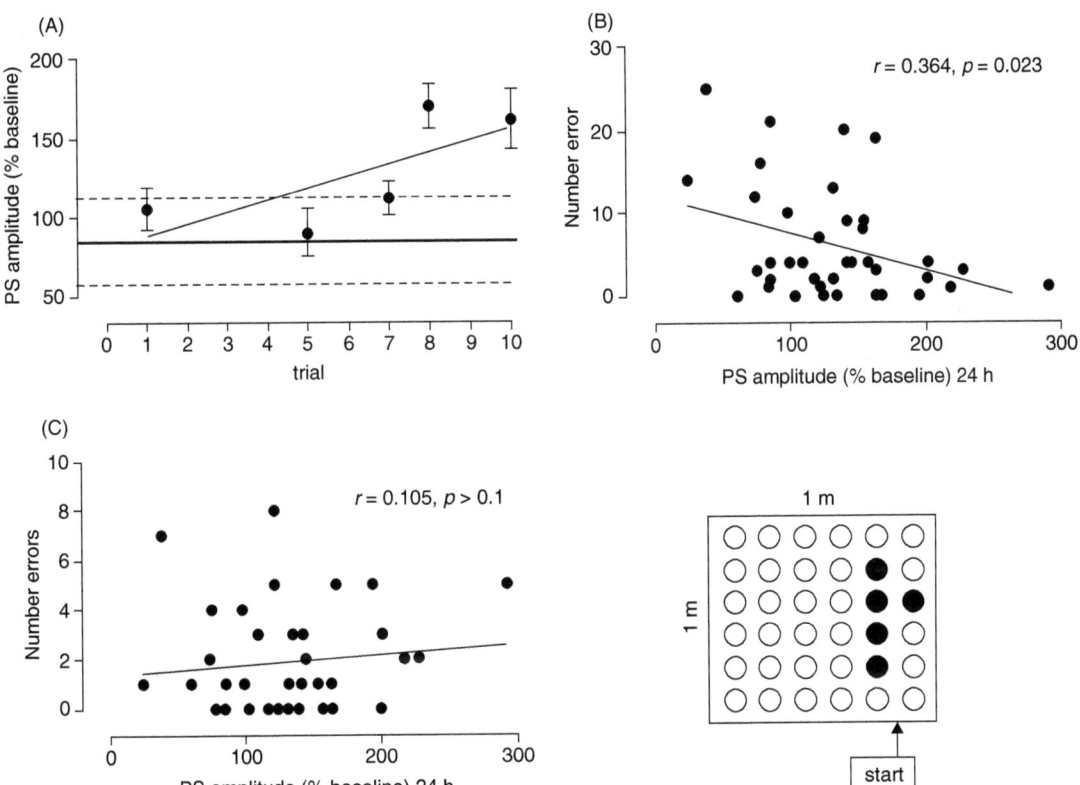

Figure 8.3. Animals trained on a holeboard show a trial-related reinforcement of long-term potentiation (LTP) and a positive correlation of maintenance of LTP with spatial learning. (A) The population-spike amplitudes (PSA) at a timepoint (24 hours) significantly increased after completion of seven trials. The horizontal solid line indicates mean PSA of pseudo-trained animals and the standard deviation (SD) (dashed lines). Means ± standard equivalent of the means (SEM) are given; (B) The 24-hour PSA are negatively correlated with the numbers of reference memory errors made during the last trial. Values of animals from all groups are given; (C) No correlation between working memory errors and PSA could be found. The inset shows the pattern of baited holes (filled circles) from the 36 holes (open circles) used during training.

and eight, with rapid, decreased reference memory errors in the latter as compared to the first. LTP-reinforcement, as well as the consolidation of the reference memory, is protein synthesis-dependent (Uzakov, Frey & Korz, 2005).

Water maze

Animals that are trained in the water maze show more heterogenous results (Figure 8.4): significant differences between groups in protein synthesis amplitude are detected at a timepoint (24 hours) ($F_{3,44} = 9.80$; $p < 0.001$); however, LTP-reinforcement is not trial-related (see Figure 8.4 (A)). The five-trial training protocol results in statistically significant decreased escape latencies as an effect of received trials, but the animals still need longer latencies (60.7 s ± 14.9 s) to find the platform during the last trial as compared to animals trained over 10 trials. Thus, the five-trial protocol results in an intermediate learning effect whereas 10-trial training permitted formation of a robust reference memory, the escape latencies being similar over the last four trials. Surprisingly, in the well-trained 10-trial group no reinforcement is seen compared to controls, in contrast to the weakly-trained five-trial group. Therefore, a positive correlation between protein synthesis amplitude and escape latencies for the individuals of both groups may be noted (Figure 8.4 (B)). However, this reflects less of an inverse correlation as compared to holeboard-trained animals, rather than indicating that reinforcement takes place only during initial stages of reference memory formation. A stable reference memory is established much faster by the water maze than by holeboard training. The highest protein synthesis amplitude may be found in animals that received a probe trial after 10-trial training, thus searching in the right place but without success in finding the platform.

In order to investigate whether heterosynaptic or hormonal events are involved in emotional and cognitive reinforcement, antagonists for the receptors of different neurotransmitters and for corticosterone have been applied (Korz & Frey, 2003a; Uzakov, Frey & Korz, 2005). Both kinds of reinforcement processes depend

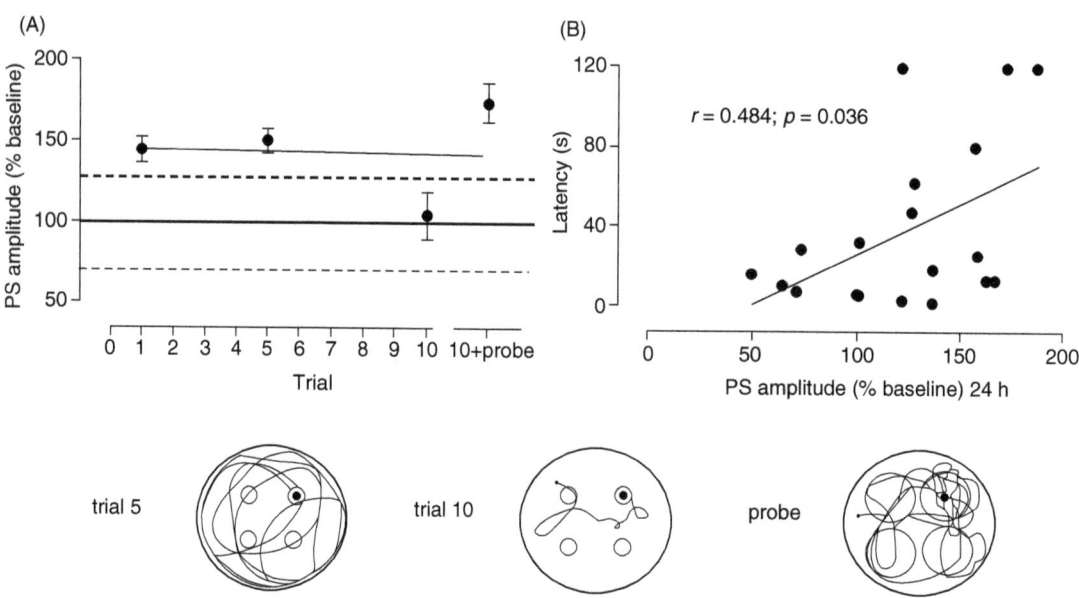

Figure 8.4. Animals trained in a water maze show no trial-related reinforcement of long-term potentiation (LTP) and a positive correlation of population-spike amplitudes (PSA) and escape latencies. (A) The PSA at a timepoint (24 hours) are significantly increased in naïve animals (trial 1), in weakly trained animals, and in well-trained animals with subsequent probe trial. Well-trained animals without probe trial show no reinforcement. The horizontal solid line indicates mean PSA of pretrained animals and standard deviation (SD) (dashed lines). Means ± standard equivalent of the means (SEM) are given; (B) The 24-hour PSA are positively correlated with the escape latencies of the last trial. Values of animals from all groups are given. The inset shows representative swim paths of individual animals.

on protein synthesis. For emotional reinforcement no heterosynaptic activity is required, but it does depend on the activation of the corticosterone-binding MR and on an intact amygdala. In contrast, cognitive reinforcement requires dopaminergic and β-adrenergic activity. Accordingly, application of dopamine and noradrenaline receptor antagonists also blocks the formation of a reference memory (Uzakov, Frey & Korz, 2005).

Specific, functional time windows before or after the weak tetanus during which heterosynaptic activity is required to reinforce LTP have been described earlier (Frey et al., 2001; Straube et al., 2003). These vulnerable intervals are specific for different behavioral tasks (Seidenbecher, Reymann & Balschun, 1997, Straube et al., 2003) and for different brain areas being involved in the modulation of DG-LTP (Frey et al., 2001). Different time intervals tested for the emotional reinforcement paradigm only suggest an effective time window of one hour for swim stress after the weak tetanus.

DISCUSSION

LTP-reinforcement: the role of learning paradigms

An association of maintained LTP with spatial learning in both learning paradigms during early phases of training only has been observed: the positive correlation of protein synthesis amplitude and escape latencies results from the faster consolidation of spatial memory in the water maze. Processes maintaining an already-established reference memory, however, do not reinforce LTP. Because LTP reinforcement and the formation of a lasting reference memory on the holeboard depends on protein synthesis (Uzakov, Frey & Korz, 2005) this result suggests that the proteins involved in memory formation also contribute to the maintenance of LTP. Similar results are obtained in water maze-trained animals; however, the data are less homogenous. There is at least some convergence between the data obtained for the holeboard and the water maze: the effective time window and the finding that reinforcement takes place only at initial stages of memory formation. Since animals learn the task in the water maze faster than they do on the holeboard, the crucial phase in the former is about the fifth trial, and, in the latter, about the eighth to tenth trial.

Well-trained animals show no reinforcement. This may be explained in various ways: (1) in well-trained animals memory is already stored in extrahippocampal structures, thus the cortex and the hippocampus are no longer needed for memory recall; (2) the hippocampus is still involved but the recall or reconsolidation does not require protein synthesis; (3) further potentiation is blocked by molecular mechanisms to avoid interference of an established memory trace with the processing of additional information. This blockade is abrogated when the expectancies of the animals about the location of the platform do not match the actual situation, as during the probe trial. Interference of emotional reinforcement with cognitive processes (Dolan, 2002; Shors, 2004; Sweatt, 2004b) is possible if the frustration of the unsuccessful search for the platform is sufficient to switch into panic and stress. At very least the vain search for the platform requires a complete reorientation and shift of attention of the animal (Mason & Iversen, 1978; Selden et al., 1990; Bouret & Sara, 2004) that may lead to a tonic activation of locus ceruleus (LC) neurons, which is seen to be related to behavioral flexibility and scanning attention (Aston-Jones, Rajkowski & Cohen, 1999). Therefore, reinforcement induced by probe trial experience may depend on noradrenergic activation.

This finding is in line with the cellular tagging hypothesis (Frey & Morris, 1997, 1998a, 1998b): the weak glutamatergic input representing sensory information that is usually transiently stored was transformed into LTM by the temporally related protein synthesis inducing strong noradrenergic input in the DG. During emotional reinforcement the hormonal signal may act as a second strong input, resulting in the synthesis of PRPs.

Emotional and cognitive reinforcement: a current concept

Emotional reinforcement
Figure 8.5 summarizes our current hypotheses of the brain structures and cellular mechanisms involved in emotional and cognitive reinforcement. Sensory stimulation represented by the weak glutamatergic input may be transformed into a long-lasting memory trace by an associative second strong input that induces the synthesis of PRPs. The nature of the second input depends on the relevance and the severity of the environmental stimuli. If the

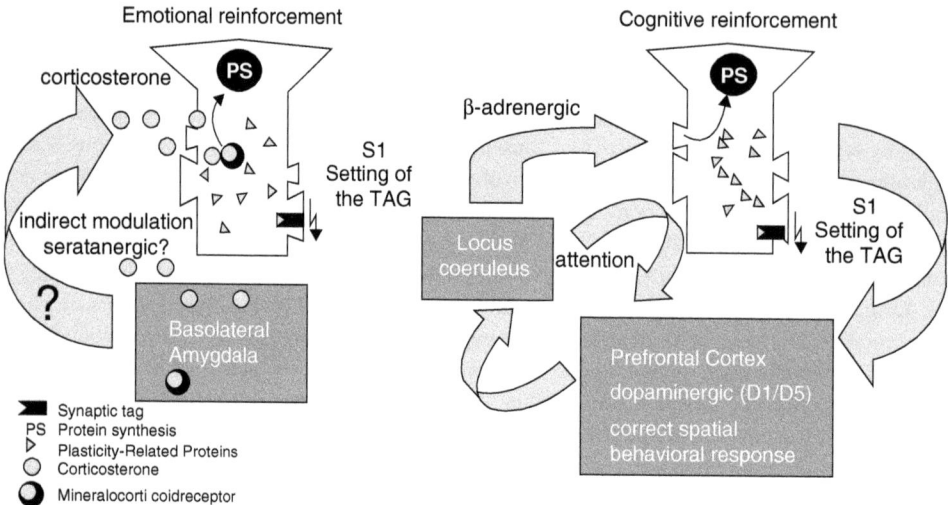

Figure 8.5. Preliminary model of the cellular and brain mechanisms involved in emotional and cognitive reinforcement. For a detailed description, see Discussion.

information is of high, acute, and possibly life-saving relevance that requires immediate storage of its context, corticosterone signals may induce the synthesis of PRPs by activation of intracellular MR. The ligand–receptor complex then translocates into the nucleus and induces gene transcription. The synaptic tag set by the weak tetanus may capture and process PRPs, and reinforcement can take place. By lesion experiments the basolateral amygdala (BLA) has been found to be required for emotional reinforcement (Korz & Frey, 2003b). Owing to the global application of the MR antagonists by intra-cerebroventricular injection, at present it cannot be determined whether the activation of hippocampal MR is not required or is not sufficient for reinforcement (Akirav & Richter-Levin, 1999). Intra-amygdalar activation of MR receptors may stimulate the amygdala resulting in heterosynaptic activity in the hippocampus (Abe, 2001; Frey et al., 2001; McGaugh, 2002, 2004). There is also evidence that the hippocampus modulates amygdalar activity (Maren & Fanselow, 1995; Seidenbecher et al., 2003; Richardson, Strange & Dolan, 2004). Because the amygdala has no direct projection to and from the hippocampus, indirect modulatory activities have to be assumed. In earlier studies a cholinergic or muscarinergic (via the medial septum) and a β-adrenergic (via the LC) (Frey, Bergardo & Frey, 2003; Almaguer-Melian et al., 2003), dependent LTP reinforcement of DG-LTP by electrical stimulation of the BLA has been noted (Frey et al., 2001). The swim-stress experiments fail to show such a dependency so the putative neuromodulatory input remains to be identified. Because of interactions between serotonergic and MR receptors (Semont et al., 1999) a serotonergic activation seems to be a good candidate.

Viewing at the systems level we may speculate that emotional reinforcement supports the very rapid integration of the context of a brief putative life-threatening situation into LTM (Fanselow, 2000; Sapolsky, 2003; Frank, Stanley & Brown, 2004). This allows immediate recall and avoidance of the situation in the future, enhancing the chance of survival on an evolutionary level. Such situations are usually accompanied by activation of the hypothalamo–pituitary–adrenal (HPA) axis, which results in an increased release of glucocorticoids from the adrenal cortices. This has been described as the "fight or flight" syndrome (Selye, 1950). Thus, if the increase in corticosterone is only transient, as in our case, then the brief pulse of corticosterone may function as cellular signal related to a specific situation and can be associated with time-related sensory input. However, even under basal conditions with low levels of corticosterone MR may be almost saturated

Table 8.1. Different neuromodulatory systems involved in emotional and cognitive reinforcement

	Emotional reinforcement	Cognitive reinforcement
Receptors involved		
β-adrenergic (propranolol: 2 µg; timolol: 25 µg)	−	+
dopaminergic (SCH23390: 1 µg)	−	+
muscarinergic (Atropine: 0.677 µg)	−	not tested
Serotonergic	not tested	not tested
Corticosterone:		
glucocorticoid (Mifepristone: 150 ng)	−	−
mineralocorticoid (Spironolactone: 150 ng)	+	not tested
Protein synthesis Dependence (Anisomycin 240 µg)	+	+
Brain structures Involved		
Basolateral amygdala	+	not tested

Note: Receptor systems that have been pharmacologically antagonized and brain regions that have been lesioned are shown. − indicates no dependence; + indicates dependence of reinforcement on the respective receptors or brain structures. Substances used are given in brackets.

because of their high binding affinity (De Kloet, Oitzek & Joels, 1993; van Steensel et al., 1996). But how can MR then function as a cellular signal of increased corticosterone levels? One possibility is that levels of hippocampal MR are transiently upregulated after a brief stressful event, so that the balance in occupied MR and glucocortiocid receptors (GR) is shifted toward the MR, facilitating the neuromodulatory activity of the MR over that of the GR. In fact, upregulation of MR mRNA and proteins has been found after swim stress (Gesing et al., 2001; Korz, Ahmed & Frey, 2004). In contrast, if corticosterone is chronically elevated, as in holeboard-trained animals, any association with specific brief environmental events is no longer possible. In addition, chronic stress causes massive changes on the cellular and brain level, for example on the regulation of MR as well as GR densities (Duffy et al., 2001; Meyer et al., 2001; Sebaai et al., 2001; Karandrea, Kittas & Kitraki, 2002; Hugin-Flores et al., 2004; Shors, 2004). However, whether the upregulation of corticosterone in response to the holeboard transfer may modulate DG-LTP in well-trained animals remains to be investigated.

Cognitive reinforcement

In holeboard-trained animals a significantly higher protein synthesis amplutide in animals which make fewer reference errors may be seen, indicating the establishment of a long-lasting reference memory error. Interestingly, an improvement in the time to find all pellets is not sufficient to reinforce LTP, suggesting that the proteins involved in the formation of a reference memory may also contribute to the maintenance of LTP. As indicated in Table 8.1, cognitive reinforcement depends on β-adrenergic and dopaminergic activation. According to our results the main modulatory heterosynaptic input in the DG is β-adrenergic (Frey et al., 2001) and because of our intra-cerebroventricular application of the antagonists, which does not permit limitation of effects to specific brain regions, we assume that the direct modulatory input in the DG is β-adrenergic. The dopaminergic response is probably located in the prefrontal cortex (PFC) and may be related to rewarding. The reward, however, is not the finding of food but the experience of having learned the correct locations of food pellets. Animals that were food-deprived and transferred repeatedly into the holeboard, where they received food pellets without spatial learning, showed no reinforcement of LTP (Uzakov, Frey & Korz, 2005). In addition, it has been shown that the performance of a correct spatial behavioral response rather than food consumption is correlated with a brief pulse of dopamine (Phillips, Ahn & Floresco, 2004). The PFC and the LC, the main sources of noradrenaline within the brain, project mutually to each other (Tassin *et al.*, 1986; Jodo & Aston-Jones, 1997; Condes-Lara, 1998; Charifi *et al.*, 2001; Bouret & Sara, 2004). Dopaminergic activity within the PFC might therefore result in increased noradrenergic release via the LC to the perforant path input in the DG (Chang *et al.*, 2001; Walling & Harley, 2004). Thus, a learning-related noradrenergic input to the DG granule cells may induce the synthesis of PRPs from which the synaptic tag set by the weak glutamatergic input can benefit, and thus, early-LTP is then transformed into a late-LTP. Increased LC activity has been related to a behavioral state of attention and flexibility (Devauges & Sara, 1990; Harley, 1991; Aston-Jones, Rajkowski & Cohen, 1999; Mansour *et al.*, 2003). Since the holeboard situation is not acutely perilous, rapid consolidation of contextual information may hinder flexible responses to complex environmental stimuli (Nadel & Bohbot, 2001; Summers, 2001), which may support food-gathering behavior.

Further work on the cellular and the systems level is still required for a detailed characterization of the processes underlying memory consolidation related to emotions and cognition (Lamprecht & LeDoux, 2004). The specific constraints of a testing situation should be considered in correlating cellular and systemic processes.

Cognitive and emotional reinforcement: different processes or two sides of the same coin?

At present we cannot decide whether cognitive and emotional reinforcement are essentially different processes or a similar process realized by different mechanisms. One can argue that learning cannot happen anyway without emotions. Emotions, "good" or "bad" ones, function as reinforcers of behavior in response to related experiences (Selye, 1956) and have to be regulated at an optimal level for supporting memory consolidation (Conrad *et al.*, 1999). At least we can conclude from our data that emotional reinforcement may be somewhat "simpler" than cognitive reinforcement in terms of the brain structures and transmitter systems that are involved. Emotions seem to play a role in both processes, however, it has been pointed out that stress is not a response to an environmental situation but a response to endogenous valuations of that situation. Thus, stress is not a quality *per se* but depends on ontogenetical factors such as social pre-experiences (Levine, 1957; 2005). In addition, the effects of stress strongly depend on the organism's ability to cope with the situation (Conner *et al.*, 1971), for example restraint stress induces the development of stomach ulcers in rats which is prevented if the animal has the chance to bite a brush during restraint stress (Vincent *et al.*, 1984). Thus, during emotional and cognitive reinforcement both emotional and cognitive components may be activated but with shifting emphasis on these components. From an evolutionary point of view an increase of the probability of survival may be the pay-off of the rapid and "simple" emotional consolidation. The cost may be a decreased flexibility in response to more complex or changing environments. Response flexibility can be achieved by more complex consolidation and reconsolidation processes involving different brain structures and a complex interplay of signalling systems such as during cognitive reinforcement.

ACKNOWLEDGEMENTS

We thank J. Maiwald for technical assistance. This study was supported by Land Saxony-Anhalt (LSA3475A/1102M) and by a grant of the Volkswagenstiftung AZ: I/77922.

9 · Estrogen and hippocampal synaptic plasticity

Michael Foy, Michael Baudry, and Richard F. Thompson

Long-term potentiation (LTP) of synaptic transmission in the hippocampus and neocortex is considered to be a cellular model of memory trace formation in the brain, at least for certain forms of memory (Landfield & Deadwyler, 1988; Bliss & Collingridge, 1993; Baudry, Davis & Thompson, 2000). The large body of work on the molecular and synaptic mechanisms underlying LTP (Bear & Malenka, 1994; Geinisman, 2000) is now matched with a growing number of studies suggesting the critical role of LTP in behavioral learning and memory (Shors & Matzel, 1997; Morris et al., 2003). Nonetheless, whether LTP is or is not the substrate of the synaptic modifications which occur during learning in forebrain structures of vertebrates, studies of its mechanisms have revealed the existence of a number of processes that undoubtedly play critical roles in memory formation (Bi & Poo, 2001). In area CA1 of the hippocampus, the most widely studied form of LTP involves glutamate N-methyl-D-aspartate (NMDA) receptor activation for its induction, and an increase in α-amino-3-hydroxy-5-methyl-4-isoxazoleproprionate (AMPA) receptor function for its expression and maintenance. In addition, Teyler and associates have demonstrated a second form of tetanus-induced LTP in CA1 that is independent of NMDA receptors, and involves voltage-dependent calcium channels (Grover & Teyler, 1990).

ESTROGEN AND COGNITION

A small, but growing animal literature indicates that 17β-estradiol, the most potent of the biologically relevant estrogens, facilitates some forms of learning and memory, in particular for hippocampal-dependent tasks. A post-training injection of 17β-estradiol was found to facilitate retention in the Morris water maze (Singh et al., 1994), and a cholinergic agonist enhanced this effect (Packard & Teather, 1997). In another series of studies, the effects of 17β-estradiol or raloxifene (a selective estrogen receptor modulator) were evaluated on the acquisition of a delayed matching to position in a T-maze task and on hippocampal acetylcholine release in ovariectomized rats. The results showed that 17β-estradiol, but not raloxifene, enhanced the T-maze task performance, and that 17β-estradiol and a high dose of raloxifene increased potassium-stimulated acetylcholine release in hippocampus (Gibbs et al., 1994). On the other hand, some studies have found little or no effect of estrous cycle (and thereby of endogenous levels of 17β-estradiol) on tasks involving spatial memory (Galea et al., 1995; Berry, McMahan & Gallagher, 1997; Warren & Juraska, 1997; Woolley, 1998).

In humans, depletion of estrogen which occurs after the menopause increases the susceptibility of women to Alzheimer's disease (Paganini-Hill & Henderson, 1996), whereas estrogen replacement in post-menopausal women improves verbal memory (Asthana et al., 2001; Resnick & Maki, 2001). Healthy post-menopausal women with estrogen replacement scored significantly higher on tests of immediate and delayed paragraph recall compared with healthy post-menopausal women who were not taking estrogen replacement (Kampen & Sherwin, 1994). Other evaluations of estrogen replacement therapy in patients with Alzheimer's disease have indicated that estrogen does not seem to alleviate cognitive impairments associated with the disease (Mulnard et al., 2000), although it does seem to have a beneficial effect as a preventive treatment (Tang et al., 1996), most apparent among younger post-menopausal women (Henderson et al., 2005).

ESTROGEN ACTIVATION AND THE BRAIN

The nervous system is a major target of hormones and contains specific receptors for many types of steroid hormones, including estrogen, aldosterone, androgen, and corticosterone. Cell fractionation studies that examined 17β-estradiol binding in cytosol and nuclear fractions have shown that estradiol-receptor complexes are translocated from the cytosol to the nuclear fraction of cells. Once in the nucleus, the estradiol-receptor complex exerts its effect via alterations in gene transcription, followed by changes in protein synthesis. This constitutes the classic genomic mechanism of action for a variety of steroid receptors that interact with, and alter DNA transcription and gene expression. This process is characterized by a long latency and duration of action on the scale of hours to days. Many of the effects of estrogen and other hormones in the brain may be readily explained by this genomic mechanism of action (McEwen & Alves, 1999).

In contrast to the genomic mechanism described above, several steroid hormones have been found to produce rapid short-term effects, mostly on electrophysiological properties of neurons, with latencies and durations on the scale of milliseconds to minutes. Since activation of transcriptional and translational mechanisms by intracellular steroid hormone receptors requires a longer latency, physiological responses occurring with extremely short latencies have been presumed to involve nongenomic, specific membrane actions.

ESTROGEN AND HIPPOCAMPAL ELECTROPHYSIOLOGY

For more than 30 years, electrophysiological reports have shown estrogen to promote changes in synaptic excitability and plasticity within the nervous system. In a pioneering study, decreased thresholds for hippocampal seizures were found in animals primed with estrogen and also during pro-estrus, the time of the estrus cycle when estrogen levels are at their highest levels (Terasawa & Timiras, 1968). In humans, changes in electrical activity of nervous system tissue have been correlated with hormonal factors that appear to play a role in catamenial epilepsy, a form of epilepsy in which the likelihood of seizures varies during the menstrual cycle. Many women with catamenial epilepsy experience a sharp increase in seizure frequency immediately before menstruation, a time when estrogen concentrations relative to those of progesterone are at their highest levels (Backstrom, 1976). Changes in hippocampal responsiveness correlated with estrogen activity have also been reported in a study that found the induction of LTP to be maximal in female rats during the afternoon of pro-estrus, when endogenous estrogen concentrations are at their highest levels (Warren et al., 1995). Furthermore, ovariectomized rats treated with estrogen were found to show a facilitation for the induction of hippocampal LTP (Cordoba Montoya & Carrer, 1997).

ESTROGEN AND HIPPOCAMPAL EXCITABILITY

The development of *in vitro* models to study the mechanisms of neuronal plasticity have allowed researchers the ability to investigate how estrogen regulates synaptic excitability within the nervous system and, in particular, within the hippocampus. However, the binding of tritiated estradiol in hippocampus does not approach that seen in the hypothalamus and related diencephalic structures (McEwen, Gerlach & Micco, 1975; McEwen & Alves, 1999). Nonetheless, studies by Teyler and colleagues using the *in vitro* hippocampal slice preparation have shown that gonadal steroids dramatically affected neuronal excitability within specific pathways of the rodent hippocampus (Teyler et al., 1980; Vardaris & Teyler, 1980). In the initial experiments, extracellular monosynaptic population field responses recorded from area CA1 of hippocampal slices from male and female rats were monitored before and after the application of 17β-estradiol (100 pM) to the slice incubation medium (artificial cerebrospinal fluid; aCSF). In male rats, 17β-estradiol produced a rapid (< 10 minutes) enhancement of population field responses evoked by stimulation of the afferents to CA1 pyramidal cells (Figure 9.1). This was the first published report demonstrating that picomolar concentrations of the gonadal steroid 17β-estradiol directly enhanced glutamatergic synaptic transmission in hippocampus (Teyler et al., 1980).

ESTROGEN AND NONGENOMIC MECHANISM OF ACTION

Although the genomic mechanism of action involving estrogen has been the traditional framework for interpreting the mechanisms of the effect of estrogen on cell function, an increasing number of reports document the effects of acute applications of estrogenic steroids that are too rapid (occurring in fewer than 10 minutes) to be accounted for exclusively by a genomic pathway. In particular, the existence of rapid

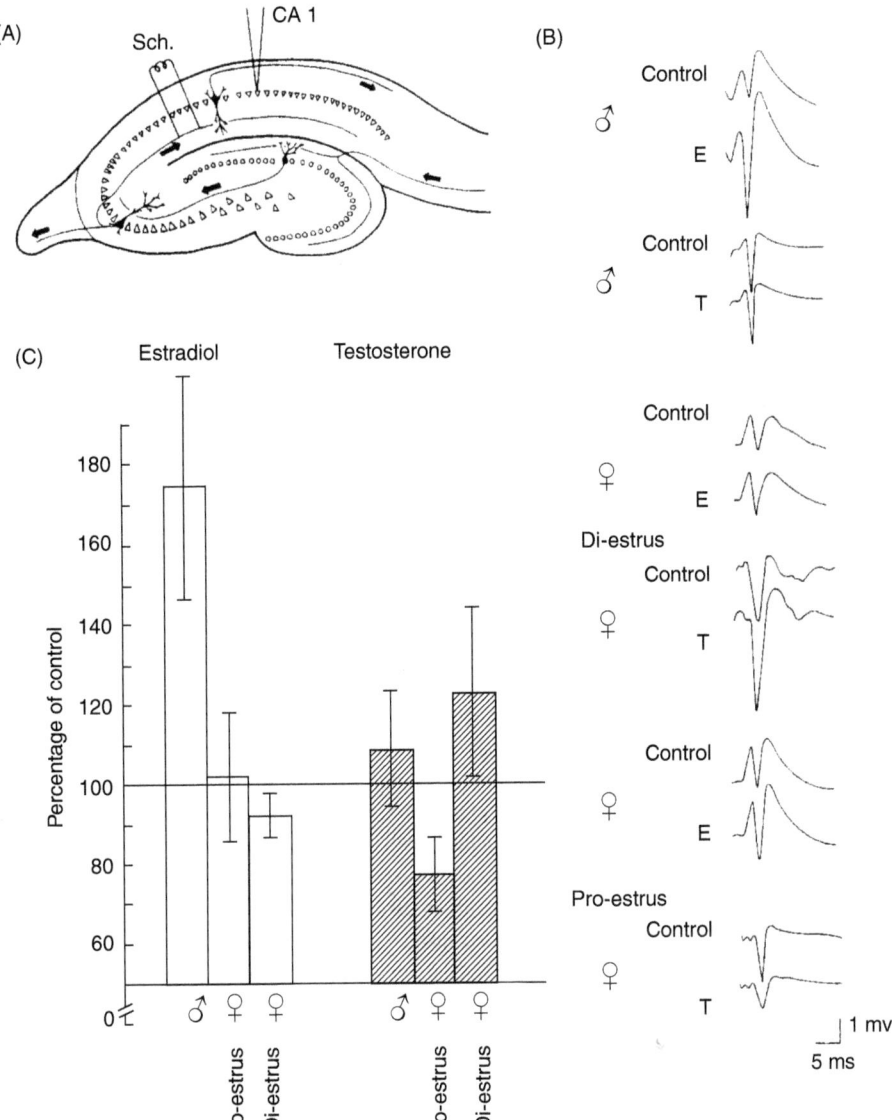

Figure 9.1. Hippocampal slice preparation. (A) Diagram of a transverse hippocampal slice. Stimulating electrodes are located in the afferent pathway, which contains the Schaffer (Sch.) collaterals. Recording micropipettes are situated in the pyramidal cell body layer in CA1. Cells of this subfield receive monosynaptic input from the CA3 pyramids via the Sch. collateral system. (B) Representative field potentials from slice preparations in various experimental conditions. Extracellular population-spike responses to a given stimulus intensity are shown from the control period (before steroid administration) and after the administration of either 10–10 M 17β-estradiol (E) or 10–10 M testosterone (T). Potentials from slices from males and pro-estrus and di-estrus females are shown for comparison. All potentials are single sweeps recorded at the same voltage and timescales. (C) Summary of the major experimental outcomes. Values on the ordinate are mean percentages of spike amplitudes after steroid administration. Data (mean ± SEM) for each condition are from 6–10 animals, each contributing one slice. Reprinted with permission from Teyler et al. (1980).

estrogenic steroid-induced changes in neuronal excitability suggests other, nongenomic mechanisms involving direct interactions with sites on the plasma membrane that alter or regulate a variety of ion channels and neurotransmitter transporters (Pfaff & McEwen, 1983; Wong, Thompson & Moss, 1996). Although the mechanism of action of gonadal steroids in hippocampus is not entirely understood, there is reason to believe that it is receptor-mediated. There was no facilitation of the field responses when the inactive estrogen 17α-estradiol was added to the hippocampal slice medium (Foy & Teyler, 1983), and the further addition of 17β-estradiol no longer resulted in an increased response as observed in the presence of 17β-estradiol alone (Foy & Teyler, 1983; Wong & Moss, 1991, 1992). Similar results were found with the use of the estrogen receptor-blocker tamoxifen when applied to hippocampal slices prior to the addition of 17β-estradiol (Foy, 1983). The ability of 17α-estradiol and tamoxifen to block the effects of 17β-estradiol on hippocampal excitability provides strong evidence that the rapid physiologic modulation of gonadal hormones is most likely caused by a plasma membrane receptor-mediated phenomenon.

ESTROGEN AND AMPA RECEPTOR ACTIVATION

In vitro intracellular recordings of CA1 neurons from adult ovariectomized female rats have shown that the administration of 17β-estradiol increases synaptic excitability in part by enhancing the magnitude of AMPA receptor-mediated responses (Wong & Moss, 1992). The rapid onset of the increased excitability, and its blockade by 6-cyano-7-nitroquinaxaline (CNQX, an AMPA receptor antagonist) but not by D-2-amino-5-phosphonovalerate (D-APV, a competitive NMDA receptor antagonist), supported a post-synaptic membrane site of action most likely mediated by non-NMDA type of glutamate receptors. Later studies using whole-cell recordings found that acute 17β-estradiol application potentiated kainate-induced currents in a subpopulation of CA1 cells (Gu & Moss, 1996), although a direct interaction between 17β-estradiol and the receptor channel was not indicated (Wong & Moss, 1994).

An example of this effect from our work (Figure 9.2) shows intracellular recordings of excitatory post-synaptic potentials (EPSPs) from a CA1 pyramidal neuron in hippocampal slices from an adult male rat in response to brief depolarizing pulses separated by 67 ms in a medium containing 50 μM D-APV and 1.0 mM Mg^{2+} (to block NMDA receptors). There is a rapid and dramatic increase in EPSP amplitude within four minutes of 17β-estradiol infusion (Figure 9.2 (A)). Some cells exhibited depolarization-induced action potentials within eight minutes of 17β-estradiol infusion (Figure 9.2 (B)) (Foy et al., 1999).

ESTROGEN AND NMDA RECEPTOR ACTIVATION

A large body of evidence demonstrates that the 17β-estradiol-mediated regulation of synapse formation is dependent on NMDA receptor activation. Morphological studies during the course of neuronal development conducted in cultured neurons prepared from embryonic day 18 rat embryos have shown that estrogenic steroids exert a growth-promoting, neurotrophic effect on hippocampal and cortical neurons via a mechanism that requires activation of NMDA receptors (Brinton et al., 1997a; Brinton et al., 1997b). *In vivo* studies using adult ovariectomized female rats have also revealed an increased number of dendritic spines in hippocampal CA1 pyramidal cells after 17β-estradiol treatment that could be prevented by blockade of NMDA receptors, but not by AMPA or muscarinic receptor antagonists (Woolley & McEwen, 1994). Other reports using adult ovariectomized female rats provided evidence that chronic 17β-estradiol treatment increased the number of NMDA receptor binding sites and NMDA receptor-mediated responses (Gazzaley et al., 1996; Woolley et al., 1997). These studies indicate that estrogen and NMDA receptors are heavily involved in synapse formation.

ESTROGEN AND NMDA RECEPTOR-MEDIATED EPSPS

The possibility of a direct regulation of NMDA receptor-mediated synaptic transmission by 17β-estradiol may not have been detected previously (e.g. Wong & Moss, 1992) because tests of this hypothesis had not been conducted under optimal conditions. Because of the voltage-dependent blockade of the NMDA receptor channel by Mg^{2+} and the slow kinetics of the channel opening relative to that of the AMPA receptor, there is

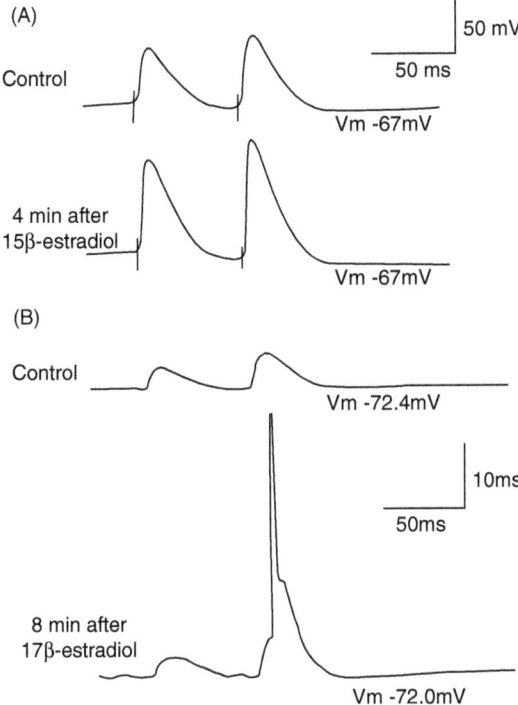

Figure 9.2. Fast potentiation by 17β-estradiol perfusion of pharmacologically isolated, non-NMDA-receptor-mediated EPSPs. The AMPA receptor was isolated pharmacologically by applying the NMDA-receptor antagonist D-APV (50 μM) and the GABA-receptor antagonist bicuculline (5 μM) to the bath. (A) 1 nM 17β-estradiol potentiates non-NMDA-mediated EPSPs in aCSF containing D-APV and 1.0 mM Mg^{2+}. (B) Some cells showed action potentials during continuous perfusion with 17β-estradiol but not in control conditions (before 17β-estradiol perfusion). aCSF, artificial cerebrospinal fluid; AMPA, α-amino-3-hydroxy-5-methyl-4-isoxazoleproprionate; D-APV, D-2-amino-5-phosphonovalerate; EPSP, excitatory post-synaptic potential; GABA, gamma-aminobutyric acid; NMDA, N-methyl-D-aspartate. Reprinted with permission from Foy et al. (1999).

only a minor NMDA receptor-mediated component of the EPSP evoked by low-frequency stimulation of glutamatergic afferents. This NMDA receptor component may be enhanced with low Mg^{2+} concentrations or high-frequency stimulation patterns used to induce the depolarization accompanying the summation of overlapping EPSPs (Xie, Berger & Barrioneuvo, 1992). In experiments using low Mg^{2+} concentrations and in the presence of the AMPA receptor antagonist 6,7-dinitroquinoxaline-2,3-dione (DNQX), acute application of 17β-estradiol in hippocampal slices from adult male rats resulted in a rapid increase in the amplitude of NMDA receptor-mediated EPSPs evoked by stimulation of the Schaffer collaterals (Foy et al., 1999). The effect of 17β-estradiol on pharmacologically isolated NMDA receptor-mediated synaptic responses was such that concentrations of 17β-estradiol >10 nM induced seizure activity in hippocampal neurons, and lower concentrations (1 nM) markedly increased the amplitude of NMDA receptor-mediated EPSPs (Figure 9.3).

ESTROGEN AND HIPPOCAMPAL LTP

To investigate the effect of estrogen on synaptic plasticity associated with learning and memory function, estrogen was applied to hippocampal slices from adult male rats before the slices were exposed to high-frequency stimulation designed to induce LTP. When LTP was assessed after high-frequency stimulation, functional EPSP (fEPSP) values were increased significantly for the 17β-estradiol treated slices compared with control aCSF slices (Figure 9.4). fEPSP mean increase of slope was 192% (experimental) versus 154% (control). Thus, hippocampal slices from adult male rats treated with 17β-estradiol exhibited a pronounced, persisting, and significant increase in LTP as measured by both population fEPSP slope and fEPSP amplitude recordings (Foy et al., 1999).

To further evaluate the effects of 17β-estradiol on the magnitude of hippocampal LTP, the intensity of afferent stimulation to Schaffer collaterals in slices perfused with 17β-estradiol was decreased in order to produce baseline values similar to pre-17β-estradiol levels immediately before the delivery of the high-frequency stimulation train used to elicit LTP (Bi et al., 2000). Under these conditions, 17β-estradiol still produced an increase in the amplitude of LTP from adult male rat hippocampal slices when compared to that obtained in control (aCSF) slices (Figure 9.5). These findings indicate that the estrogen-induced enhancement of hippocampal LTP is not attributed simply to a change in basal EPSP level, but is more likely due to biochemical activation of an intracellular cascade, presumably mediated by activation of a src tyrosine pathway that enhances NMDA receptor function.

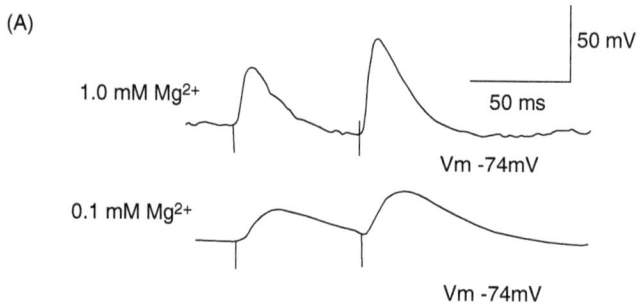

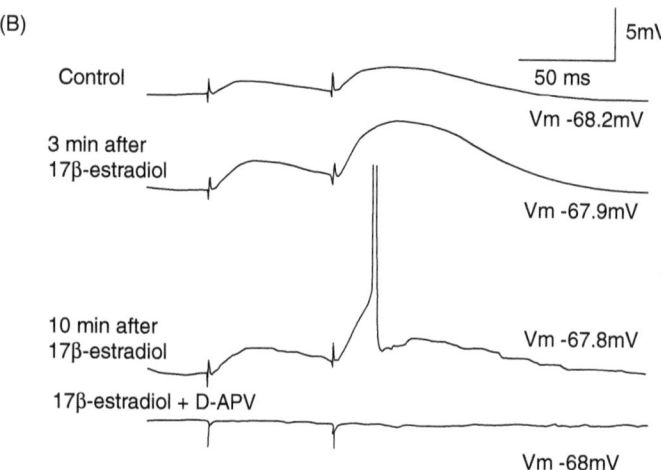

Figure 9.3. EPSPs in CA1 pyramidal cells. The amplitude of NMDA receptor-mediated EPSPs in CA1 pyramidal cells increases shortly after addition of 1 nM 17b-estradiol to the perfusion medium, which contains 5 μM bicuculline. (A) Top: EPSPs evoked when slices are perfused with medium that includes 1.0 mM Mg^{2+} but not the non-NMDA receptor antagonist 6,7-dinitroquinoxaline-2,3-dione (DNQX). Bottom: NMDA-receptor-mediated EPSPs evoked 10 minutes after medium was switched to include 0.1 mM Mg^{2+} and 10 μM DNQX. (B) 1 nM 17b-estradiol potentiates the isolated EPSPs within three minutes. Potentiation, which occurred in nine out of 12 cells, was observed in EPSPs evoked by paired-pulse stimulation and peaked within 10 minutes. In five out of nine cells, the potentiated EPSPs reached threshold and generated action potentials during 17b-estradiol perfusion. The potentiated EPSPs were blocked by the NMDA receptor antagonist D-APV. aCSF, artificial cerebrospinal fluid; AMPA, α-amino-3-hydroxy-5-methyl-4-isoxazoleproprionate; D-APV, D-2-amino-5-phosphonovalerate; EPSP, excitatory post-synaptic potential; NMDA, N-methyl-D-aspartate. Reprinted with permission from Foy et al. (1999).

ESTROGEN AND HIPPOCAMPAL LTP IN FEMALES

In another series of studies, estrous cycle changes in rodents were correlated with changes in synaptic plasticity. Hippocampal slices from cycling female rats in di-estrus (low estrogen concentration) and pro-estrus (high estrogen concentration) were prepared in aCSF, followed by LTP induction via high-frequency stimulation. The difference in LTP values between these groups after high-frequency stimulation was dramatic: slices from rats in the pro-estrus phase exhibited LTP representing about a 50% increase over baseline, whereas slices from rats in the di-estrus phase had LTP values representing about a 25% increase over baseline (Bi et al., 2001) (Figure 9.6). These findings help to support the results from the original work of Teyler et al. (1980) which identified changes in baseline synaptic transmission that were correlated with the phase of the estrus cycle in female rats at the time of hippocampal slice preparation.

Since the electrophysiological study above has shown that female rats in pro-estrus exhibited an

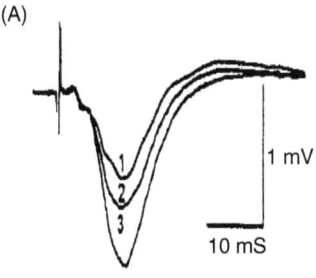

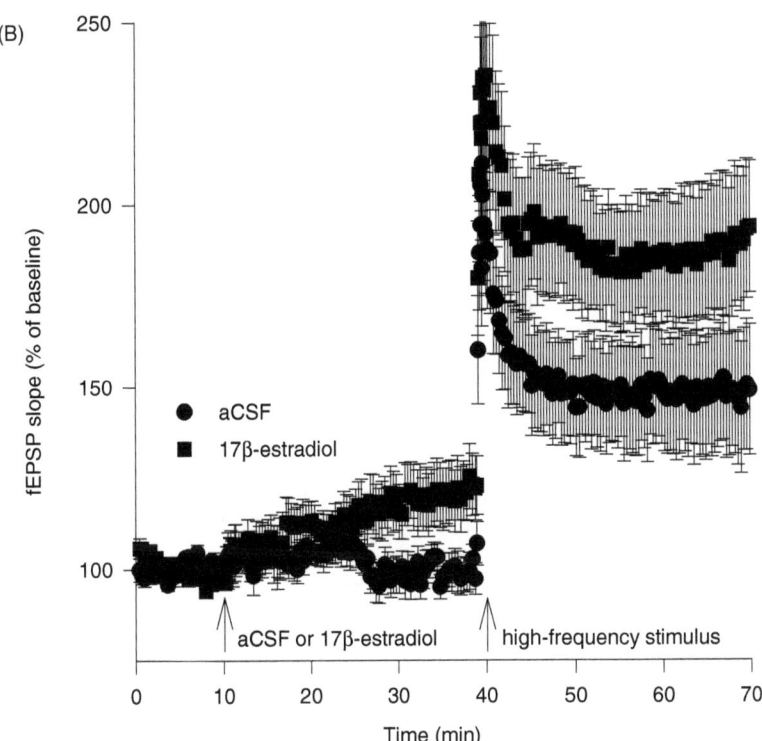

Figure 9.4. Functional EPSP (fEPSP) recordings in area CA1. (A) All hippocampal slices were perfused with aCSF for 10 minutes to obtain fEPSP slope and amplitude percentage baseline data. After 10 minutes of baseline recording, experimental slices were perfused with 100 pM 17β-estradiol. Control slices continued to be perfused with aCSF. Thirty minutes later all slices received high-frequency stimulation, designed to induce LTP. (A1) fEPSP recording at the end of the 10-minute baseline period. (A2) fEPSP recording at the end of the 30-minute perfusion with 17β-estradiol. (A3) fEPSP recorded for 30 minutes after high-frequency stimulation in slices perfused with 17β-estradiol. (B) fEPSP slope responses in area CA1. Data points represent averaged fEPSP slope ± SE (taken at each 20-second sweep) for experimental (17β-estradiol-treated) and control (aCSF) hippocampal slices. aCSF, artificial cerebrospinal fluid; D-APV, D-2-amino-5-phosphonovalerate; EPSP, excitatory post-synaptic potential; fEPSP, functional EPSP. Reprinted with permission, from Foy et al. (1999).

increased magnitude of hippocampal LTP as compared to females in di-estrus, a current study is reported here that examined the effect of 17β-estradiol on hippocampal LTP during the two critical time periods in the rat estrous cycle: pro-estrus and di-estrus. The estrous cycles of adult (aged three to five months) Sprague–Dawley rats were monitored for 10 days before any physiological experiments, and

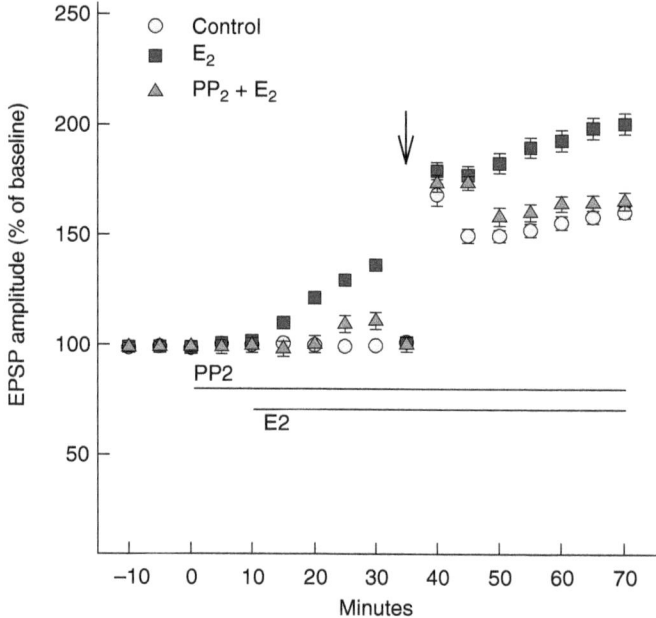

Figure 9.5. Effects of src inhibition on 17β-estradiol (E2)-mediated enhancement of EPSP amplitude and degree of LTP in hippocampal slices. A stimulating electrode was located in CA3 and a recording electrode in the stratum radiatum of CA1. Extracellular EPSPs were evoked by stimulation every 30 seconds, and the EPSP amplitude measured. After recording a stable baseline, an inhibitor of src (PP2, 10 μM) and E2 (1 nM) were added at the times indicated. After resetting the stimulation intensity to obtain EPSPs of the same amplitude as before treatment with either PP2 or E2, high-frequency stimulation (HFS) was delivered, and low-frequency stimulation resumed. PP2 blocked the estrogen-mediated enhancement of LTP but did not affect LTP itself. Data (expressed as percentages of predrug values) are mean ± SEM of six to ten experiments. EPSP, excitatory post-synaptic potential. Reprinted with permission from Bi et al. (2000).

hippocampal slices were prepared from rats that were either in pro-estrus or di-estrus. Recording and stimulating electrodes were positioned in the dendrites of area CA1 and Schaffer collaterals, respectively. Baseline stimulation (0.05 Hz, 100 μs) was adjusted to elicit 50% of the maximum fEPSP amplitude. After 10 minutes of stable baseline stimulation, aCSF or 17β-estradiol at a concentration of 100 pM (experimental group) was perfused to the slices for 30 minutes, and LTP was induced by a brief period of high-frequency stimulation (five trains of 20 pulses at 100 Hz). Subsequent synaptic responses were monitored for 30 minutes post-LTP induction. LTP induced in area CA1 was increased in slices from pro-estrus rats compared with slices from di-estrus rats, as seen previously (Bi et al., 2001). However, 17β-estradiol treatment increased LTP in slices from di-estrus rats while it decreased LTP in slices from pro-estrus rats (Figure 9.7). These observations suggest that 17β-estradiol alters hippocampal LTP in female rats, depending on the state of their estrous cycle (i.e. on the levels of circulating 17β-estradiol). In cycling female rats, when endogenous circulating levels of 17β-estradiol are at their highest levels (i.e. pro-estrus), LTP magnitude is increased, and exogenously applied 17β-estradiol acts to decrease LTP magnitude, possibly through the activation of some type of mechanism associated with an inhibitory or ceiling effect. When endogenous circulating levels of 17β-estradiol are at their lowest levels (i.e. di-estrus), the situation is completely reversed from that observed in the pro-estrus state. Here, LTP magnitude is decreased, and exogenously applied 17β-estradiol acts to increase LTP magnitude.

These results suggest that the cyclic changes in estrogen levels occurring during the estrous cycle in

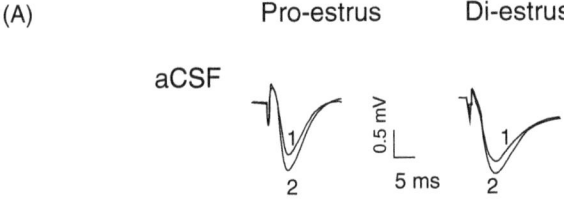

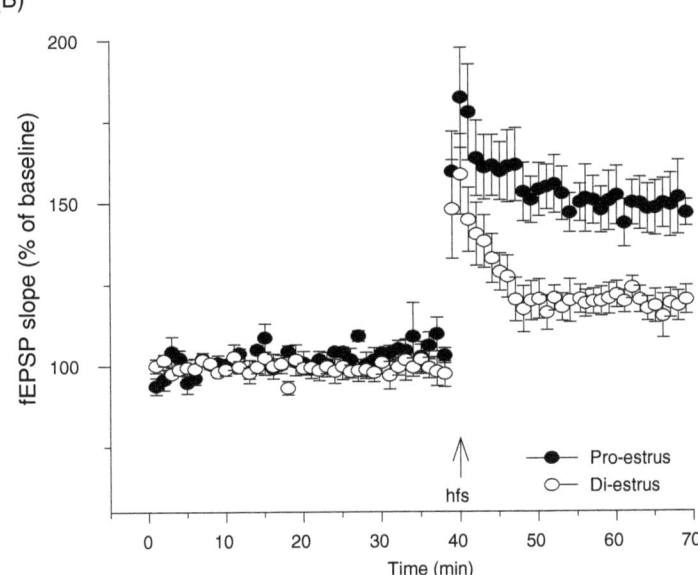

Figure 9.6. LTP in field CA1 of hippocampal slices from female rats in pro-estrus and di-estrus. Hippocampal slices from female rats in either pro-estrus or di-estrus were prepared. fEPSP amplitude and slope values were obtained for each slice and averaged across slices to produce one mean before and after the train of high-frequency stimulation (hfs). fEPSP amplitudes and slopes were normalized for the 10-minute pre-hfs period for each slice. ANOVA and planned two-tailed Student's t tests for pre-hfs and post-hfs periods were used to evaluate the effects of estrous on fEPSP slope and amplitude. (A) Representative waveforms from female rats in pro-estrus and di-estrus for pre-hfs (1) and post-hfs (2) periods. (B) Mean ± SEM of fEPSP slopes recorded in slices from female rats in pro-estrus (filled circles, $n = 6$) and di-estrus (open circles, $n = 5$). Reprinted with permission from Bi et al. (2001).

female rats are associated with changes in the magnitude of LTP recorded from hippocampal CA1 cells, and corroborate work mentioned earlier indicating the facilitation of LTP induction by estrogen in ovariectomized female rats (Cordoba Montoya & Carrer, 1997), and increased LTP in the afternoon of pro-estrus of female rats (Warren et al., 1995).

ESTROGEN, LTP OR LTD AND AGING

It has been reported that in aging, when memory function declines, the processes of synaptic plasticity in the hippocampus are altered. Specifically, LTP is impaired and the opposite process of long-term depression (LTD) is enhanced (Landfield & Lynch, 1977; Barnes, 1979; Landfield, Pitler & Applegate, 1986; Barnes et al., 1992; Barnes, 1994; Geinisman, Detoledo-Morell & Heller, 1995; Norris, Korol & Foster, 1996; Foster & Norris, 1997; Norris, Halpain & Foster, 1998; Foster, 1999). This effect of aging on LTD has been replicated and, further, a profound action of estrogen on this process in aged male rats has been reported (Vouimba, Foy & Thompson, 2000). In these studies, LTD was induced in the CA1 region of

(A)

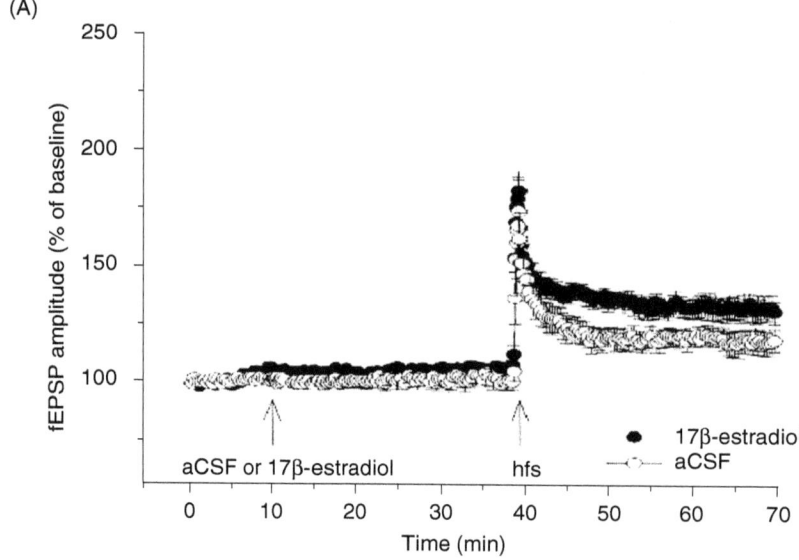

(B)

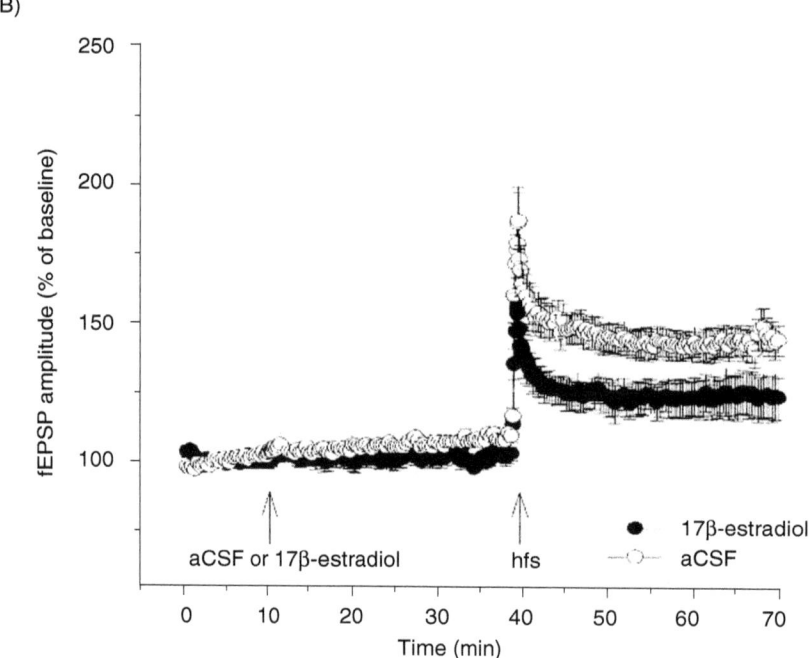

Figure 9.7. LTP in field CA1 of hippocampal slices from female rats in pro-estrus and di-estrus during treatment with 17β-estradiol (experimental) and aCSF (control). (A) Mean ± SEM of fEPSP amplitudes recorded following tetanus in slices from female rats in di-estrus. 17β-estradiol (filled circles) enhances LTP relative to control aCSF (open circles). (B) Mean ± SEM of fEPSP amplitudes recorded following tetanus in slices from female rats in pro-estrus. 17β-estradiol (filled circles) impaired LTP relative to control aCSF (open circles). fEPSP, functional excitatory post-synaptic potential; LTP, long-term potentiation. Reprinted from Foy et al. (2004).

hippocampal slices using standard conditions (potentiation of Schaffer collaterals at 1 Hz for 15 minutes) in adult (aged three to five months) and aged (aged 18–24 months) Sprague–Dawley male rats. In agreement with earlier studies, we find that the standard protocol for inducing LTD resulted in little or no LTD in slices from adult animals, but in marked LTD in slices from aged animals (Figure 9.8 (A)) (Foster & Norris, 1997; Foster, 1999). Infusion of 17β-estradiol in slices causes a slight increase in synaptic transmission (baseline), as in previous studies. It has little effect on LTD in slices from adult animals, but markedly attenuates LTD in slices from aged animals (Figure 9.8 (B)). Thus, the prevention by estrogen of the age-related enhancement of LTD may account, in part, for the protective effects reported in some studies of estrogen on memory functions in aged organisms (see below).

ESTROGEN AND TWO FORMS OF LTP

As noted earlier, Teyler and associates discovered a form of LTP in CA1 pyramidal neurons that is independent of NMDA receptors and involves voltage-dependent calcium channels (Grover & Teyler, 1990). This form of LTP is most strongly induced by very high frequency tetanus of Schaffer collaterals (e.g. 200 Hz for one second) and is blocked by nifedipine but not by the NMDA receptor antagonist D-APV. The NMDA receptor-dependent form of LTP in CA1 also, of course, involves calcium influx via NMDA receptor channels, is best induced by lower frequencies of titanic stimulation (e.g. 25 Hz), and is blocked by D-APV but not by nifedipine. The standard tetanus for LTP induction (i.e. 100 Hz for one second) induces both forms of LTP (Cavus & Teyler, 1996; Morgan, Coussens & Teyler, 2001).

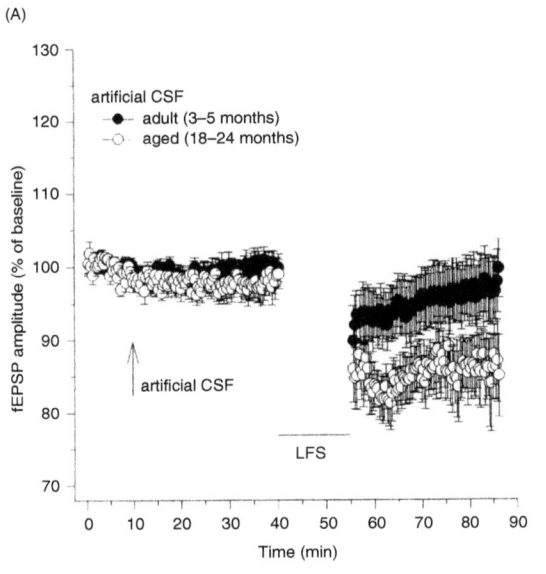

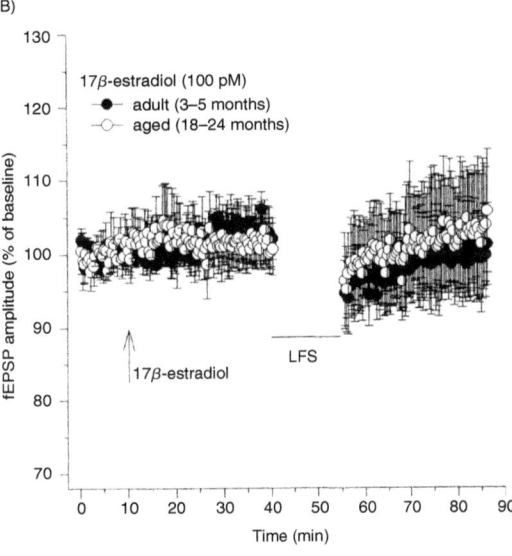

Figure 9.8. Long-term depression. (A) Normalized fEPSP amplitudes (percentage of baseline) obtained in slices from adult (filled circles, $n = 11$) and aged (open circles, $n = 8$) rats bathed in normal perfusion media (aCSF). After an initial baseline period, low-frequency stimulation (LFS) induced robust LTD in slices from aged rats, but not adult rats. The horizontal bar indicates when 900 pulses at 1 Hz were delivered. (B) Effect of 17β-estradiol on LTD in slices from aged and adult rats. Before addition of 17β-estradiol, fEPSP amplitudes recorded during the baseline period were similar in both control and experimental groups. LFS delivered to slices from aged rats perfused with 17β-estradiol (open circles, $n = 6$) did not induce robust LTD. Following LFS, slices from adult rats displayed little change in their synaptic response in the presence of 17β-estradiol (filled circles, $n = 6$) compared with their control (aCSF). aCSF, artificial cerebrospinal fluid; fEPSP, functional excitatory post-synaptic potential; LTD, long-term depression; Reprinted with permission from Vouimba, Foy and Thompson (2000).

The effects of acute application of 17β-estradiol (CA1 slice) on both forms of LTP in hippocampal slices from male rats are now under investigation. By use of 25-Hz tetanus of Schaffer collaterals, 17β-estradiol and nifedipine is infused and extracellular field EPSPs are recorded. 17β-estradiol causes the expected increase in synaptic transmission and pronounced enhancement of LTP; nifedipine has no effect at all on either process, implying that under this condition, LTP is mostly dependent on NMDA receptor activation and estrogen facilitated LTP by increasing NMDA receptor-dependent function (Zeng et al., 2004).

When using 100-Hz tetanus of Schaffer collaterals (in the CA1 region of hippocampal slices from adult male rats), both extracellular field EPSPs and intracellular EPSPs have been recorded from pyramidal neurons while 17β-estradiol and nifedipine were infused. 17β-estradiol alone causes the expected increase in synaptic transmission and pronounced enhancement of LTP, but both effects of 17-β-estradiol are reduced in magnitude by nifedipine. Therefore, under this condition, it would seem that 17β-estradiol is acting by modulating both L-type voltage-gated calcium channels and NMDA receptors. Intracellularly recorded EPSPs in response to paired subthreshold stimuli with a short interstimulus interval (50 ms) in the presence of 17β-estradiol indicate an increase of EPSP amplitude to both stimuli without changes in the paired-pulse ratio, strongly supporting a post-synaptic origin of the effects of 17β-estradiol (Akopian, Foy & Thompson, 2003).

The possibility that 17β-estradiol may modulate calcium influx through L-type calcium channels is consistent with the effects of aging on synaptic transmission and plasticity in hippocampus. Thus, aging is associated with enhanced activity of voltage-gated calcium channels in hippocampal CA1 neurons (Campbell et al., 1996), and blocking calcium influx through L-type calcium channels inhibits LTD induction and enhances LTP in aged animals in the CA1 region of hippocampal slices (Norris, Halpain & Foster, 1998). Blocking L-type calcium channels in hippocampus has also been reported to enhance memory in several paradigms and particularly to enhance learning and memory processes in aged animals (Quevedo et al., 1998; Power et al., 2002; Disterhoft, Wu & Ohno, 2004).

ESTROGEN AND CELLULAR NEUROPROTECTION

In a well established model of estrogen-induced neuroprotection, primary cultures of dissociated hippocampal neurons were prepared in media that contained the excitatory amino acid, glutamic acid. Neuronal injury in the cell cultures resulting from glutamate excitotoxicity was assessed by the quantitative measurement of lactate dehydrogenase (LDH) release in the culture medium. A five-minute treatment with 100 μM glutamate caused significant cell death compared to control conditions (no glutamate exposure), an effect that was significantly decreased after pre-exposure to 17β-estradiol (Nilsen & Brinton, 2002a).

In another *in vitro* study using primary cultures of dissociated hippocampal neurons with microfluorimetry and calcium imaging techniques, 17β-estradiol was found to potentiate glutamate-induced increase in intracellular calcium by about 70% compared to glutamate exposure alone (Nilsen & Brinton, 2002b). The estrogen enhancement of this glutamate response is in agreement with other reports showing enhancement of NMDA receptor activation, LTP, and memory by estrogens (Foy et al., 1999; Rice et al., 2000). Estrogen-induced neuroprotection against excitotoxic glutamate may involve the mitogen-activated protein kinase (MAPK) cascade found in primary cortical neuron cultures (Singer et al., 1999; Nilsen & Brinton, 2003). The impact of estrogen on cellular neuroprotection has been found to be quite specific and dramatic. However, the mechanisms involved in these actions need to be further elucidated to comprehend the complex and indirect ways in which estrogen interacts with cellular signaling pathways.

MOLECULAR MECHANISMS OF ESTROGEN EFFECTS IN THE BRAIN

Recent results from several laboratories have provided a general framework to understand the mechanisms underlying the multiple effects of 17β-estradiol on synaptic structure and function (Lee & McEwen, 2001). 17β-estradiol, at physiological concentrations, interacts with Erα and Erβ receptors, to produce both direct and indirect genomic effects. The direct genomic effects are attributed to the interactions between 17β-estradiol and traditional cytoplasmic receptors followed by the regulation of transcription, through interactions with regulatory estrogen response elements of a variety of genes. In neurons, these genes include anti-apoptotic genes of the bcl-2 family, probably responsible for the neuroprotective effects of 17β-estradiol observed in a number of models of neuronal death. In astrocytes, these genes include GFAP (downregulation) and laminin (upregulation), which might be involved in the sprouting responses observed following lesions, as well as in normal astrocyte activation observed in brains from old animals (Kohama et al., 1995). The indirect, or nongenomic actions of 17β-estradiol are believed to be mediated through plasma membrane-associated estrogen receptors, whose actions have been associated with the activation of various protein-kinase cascades (Losel & Wehling, 2003). Subsequent "nongenomic to genomic signalling" pathways of the effects of 17β-estradiol might be linked to the stimulation of the phosphoinositol-3 (PI3) kinase or Akt system (Datta, Brunet & Greenberg, 1999; Simoncini et al., 2000) and/or of a G protein, and/or of Src tyrosine kinase and ERK/MAP kinase pathways (for review see Bjornstrom & Sjoberg, 2005; Singh et al., 2000).

The MAP kinase pathway occupies a central place in the regulation of synaptic plasticity (Mazzucchelli & Brambilla, 2000; Sweatt, 2001). Pharmacologic manipulations directed at blocking this pathway have consistently produced impairments in synaptic plasticity and learning and memory, and this pathway is activated with LTP-inducing tetanus or in different learning paradigms (Brambilla et al., 1997; Berman et al., 1998; Blum et al., 1999; Selcher et al., 1999). Endogenous estrogen levels in cycling female rats produce a tonic phosphorylation or activation of extracellular signal-regulated kinase 2 (ERK2)/MAP kinase (Bi et al., 2001). This activation of the MAP kinase pathway is also linked to the regulation of glutamate ionotropic receptors and might be involved in the "cognitive enhancing" effects of 17β-estradiol. Indeed, the acute estrogen-mediated enhancement of LTP is mediated by activation of a src tyrosine kinase pathway (Bi et al., 2000). Thus, acute application of the src inhibitor PP2 in the perfusing medium of hippocampal slices from adult male rats abolishes the estrogen enhancement of both synaptic transmission and of LTP, but has no effect on LTP itself (see earlier, Figure 9.5). Similarly, this pathway might also be involved in the neuroprotective effects of 17β-estradiol as MAP kinase inhibitors have consistently been shown to block the neuroprotective effects of 17β-estradiol in a variety of models of neurodegeneration. Moreover, growth factors and other factors providing neuroprotection, such as platelet-derived growth factor (PDGF), also use the MAP kinase pathway for their neuroprotective effects.

Interestingly, it appears that Erα stimulation is critically involved in the neuroprotective effect of estrogen, as Erα knock-out mice are not protected by 17β-estradiol against ischemia-induced neuronal damage (Dubal et al., 2001). Furthermore, recent results obtained from the same knock-out mice suggest the possible existence of novel 17β-estradiol receptors responsible for the activation of the ERK/MAP kinase pathway (Singh et al., 2000). These results indicate that several steps described in Figure 9.9 remain to be elucidated.

SUMMARY OF THE EFFECTS OF ESTROGEN ON SYNAPTIC PLASTICITY

The studies mentioned in this chapter establish several fundamental characteristics of the effects of estrogen on synaptic transmission in the mammalian central nervous system (CNS). Estrogen acts rapidly via presumed membrane mechanisms to enhance both NMDA and AMPA receptor or channel responses elicited by glutamate released from excitatory pre-synaptic terminals.

17β-estradiol may also markedly enhance hippocampal LTP in CA1 neurons of adult male rats. The enhancement of LTP after acute 17β-estradiol application is attributed to an increase in NMDA receptor and AMPA receptor functions. Both possibilities are consistent with intracellular data. Changes in estrogen levels of cycling female rats have also been correlated with changes in synaptic plasticity, as measured by changes in LTP magnitude. This finding suggests a mechanism by which naturally fluctuating endogenous hormone levels may affect a cellular model associated with important aspects of learning or memory storage, or both, in the mammalian CNS.

To the extent that LTP is a mechanism involved in processes of coding and storage of information, that is, in memory formation, estrogen appears to enhance these processes. Indeed, the estrogen enhancement of LTP suggests a possible mechanism by which estrogen can exert its facilitatory effects on memory processes in humans. Clinical evidence indicates that estrogenic

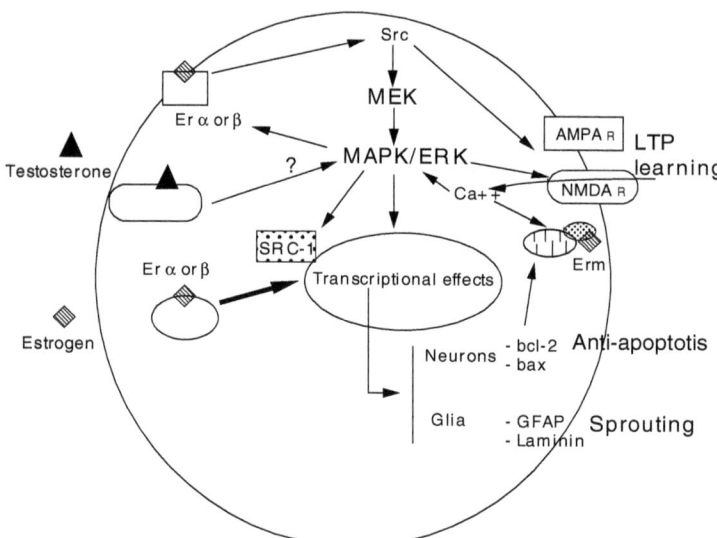

Figure 9.9. General hypotheses that link estrogen or testosterone with the MAPK/ERK pathway, NMDA receptors and synaptic plasticity in brain. NMDA, *N*-methyl-D-aspartate. Reprinted from Foy *et al.* (2004).

steroids may enhance cognitive functions in humans, in particular in post-menopausal women (Henderson, 1997; Kawas *et al.*, 1997; Henderson, 2000); however, prospective observational studies have not found a protective effect of estrogen on either cognition or the incidence of dementia (Barrett-Conner & Kritz-Silverstein, 1993; Matthews *et al.*, 1999). Understanding the mechanisms underlying the changes in some of the effects of estrogen associated with aging will represent a significant advance in understanding the mechanisms involved in decline in cognitive function with aging.

ACKNOWLEDGEMENTS

This work was supported in part by National Institutes of Health Grants, NIA AG-14751, AG-26572, AG-23742, and by funds from Loyola Marymount University and the University of Southern California.

10 · Steroid-induced hippocampal synaptic plasticity: sex differences and similarities

Russell D. Romeo, Elizabeth M. Waters, and Bruce S. McEwen

INTRODUCTION

During perinatal development steroid hormones structurally organize various regions of the central nervous system (CNS) by modulating neuronal events such as cellular proliferation and death, neuronal migration and differentiation, and neurite extension and elaboration (Arnold & Breedlove, 1985). However, these effects of steroid hormones do not stop in early development, but continue to affect the CNS throughout the lifespan of the individual (Arnold & Breedlove, 1985; Romeo, 2003). For instance, in the adult brain, steroid hormones have also been shown to influence factors such as neuronal survival (Nordeen et al., 1985), neurogenesis (Tanapat et al., 1999), neurite outgrowth (Toran-Allerand, 1976), synaptogenesis (Woolley, 1998), receptor expression (Handa et al., 1996), neurotransmitter synthesis (Luine, McEwen & Black, 1977), and neuronal excitability (Mermelstein, Becker & Surmeier, 1996).

The structure and function of the male and female CNS differ on several parameters. Most of the structural differences between the male and female CNS are established early in development by the perinatal hormonal milieu. Examples of these structural differences include sexually dimorphic brain and spinal cord nuclei, in which particular cell groups are larger in either males or females (Simerly, 2002). However, other sex differences in the structure and function of the CNS are more subtly affected by circulating steroid hormone levels in adulthood, and may play important roles in sex differences in physiology and behavior.

This chapter presents a brief overview of basic steroid hormone biochemistry and mechanisms of action. The sex differences and similarities in steroid-induced synaptic plasticity in the adult brain, primarily drawing from studies using *in vivo* models are then considered. Specifically, the effects of estrogen and progesterone, as well as the androgens and glucocorticoids, will be discussed. Particular emphasis will be placed on steroid-induced synaptic plasticity in the hippocampus, a brain region vitally important in learning and memory (Eichenbaum, 1997). This topic is relevant to the growing body of evidence for the actions of sex hormones outside of the reproductive neuroendocrine axis. It also tells an important and emerging story about the non-genomic as well as genomic actions of steroids at the cellular and molecular level.

STEROID HORMONE BIOCHEMISTRY

In mammals, all steroid hormones (e.g. corticoids, androgens, progestins, estrogens) are synthesized from cholesterol via enzymatic cleavage (for a review see Payne & Hales, 2004). For example, testosterone is produced from a series of enzymatic reactions that convert cholesterol to progesterone, then to testosterone acetate, and finally to testosterone. Testosterone is the precursor of estrogen, in that estrogen is formed by the aromatization of testosterone by the aromatase enzyme. These enzymatic reactions that produce steroids may take place in the endocrine glands or the target tissues on which they act, such as the brain. In females, the ovaries are the major endocrine gland that produce and secrete estrogen and progesterone, whereas the testes are the major source of testosterone in males. Testosterone is commonly considered the "male sex steroid" and estrogen the "female sex steroid," but it is important to note that testosterone, estrogen, and progesterone are found in both sexes. Indeed, estrogen formed locally in the CNS of males by the aromatization of testosterone is in large part responsible for the masculinization of the male brain and plays a fundamental role in mediating male reproductive behavior.

The corticoids are also derived from cholesterol. Similar to testosterone, progesterone is also the steroid precursor to the glucocorticoids, such as cortisol (the primary glucocorticoid in primates) and corticosterone (the primary glucocorticoid in most rodent species). Specifically, 11-deoxycortisol and 11-deoxylcorticosterone act on progesterone to form cortisol and corticosterone, respectively, in the adrenal cortex.

STEROID HORMONE MECHANISMS OF ACTION

Steroid hormones act on a variety of cell types through many different mechanisms. This chapter concentrates on how steroid hormones act on hippocampal neurons, and specifically on how steroids affect synaptic function in males and females.

The classic mechanism of steroid hormone action may be broken down into four basic steps (Tsai & O'Malley, 1994). First, the hormone travels through the circulatory system and crosses the blood–brain barrier (or is formed locally) to act on neuronal or glial target cells. Second, the steroid diffuses through the plasma membrane of the cell and binds to its intracellular receptor located in either the cytoplasm or the nucleus. Third, the hormone–receptor complex then undergoes a conformational change allowing the hormone–receptor complex to migrate, dimerize, and bind to particular areas of the DNA known as hormone response elements (HREs). Finally, once the hormone–receptor complex binds to its HRE, transcription of a particular gene is initiated allowing the transcription, or repression, of genes that ultimately produce proteins altering the functioning of the neuron.

Although many actions of the steroid hormones in the nervous system are achieved through the above-mentioned process, recent advances in our understanding of hormone action have revealed many indirect genomic as well as non-genomic actions of steroids in neuronal and non-neuronal elements (McEwen, 2001). For instance, estrogen has been shown to bind to receptors located in the cellular membrane of neurons, which activate second-messenger cascades to initiate gene transcription (Lee & McEwen, 2001). These indirect genomic effects of steroids are not mediated through HREs, but instead are mediated by DNA regulatory sites such as activator protein 1 (AP-1) and the cAMP response element (CRE). In addition to these indirect genomic effects, steroids may act through non-genomic mechanisms such as directly affecting ion channel permeability and acting as antioxidants (Lee & McEwen, 2001). Interestingly, steroids may also act at specific local sites in the neuron, which do not involve direct interactions with the DNA. Recent studies in male and female rats have determined that extranuclear androgen and estrogen receptors are located in axon terminals and spines of hippocampal CA1 pyramidal cells and glia (Milner et al., 2001; Towart et al., 2003; Tabori et al., 2005). Various mRNAs for synaptic proteins and translation machinery have been localized in dendritic spines suggesting that translation of such proteins can occur in the dendrite itself (Steward & Schuman, 2001). Thus, steroid hormones may act directly on the synaptic apparatus through these extranuclear receptors and local mRNAs to affect synaptic function and synaptogenesis. These genomic, indirect-genomic and non-genomic mechanisms of steroid hormone action are presented in figure 10.1.

SEX DIFFERENCES IN THE STRUCTURE AND FUNCTION OF THE HIPPOCAMPUS

Historically, the first brain region observed to differ structurally between males and females was the hypothalamus (Arnold & Gorski, 1984). For example, in the late 1970s a nucleus within the anterior hypothalamus was discovered to be two and a half to five times larger in males compared to females (Gorski et al., 1978). This nucleus is known as the sexually dimorphic nucleus of the medial pre-optic area (SDN-MPOA), and it plays an important role in mediating the display of male sexual behavior (Christensen, Nance & Gorski, 1977). However, other structural differences between the sexes have been noted in extra-hypothalamic areas not involved in sex differences in reproductive behavior or physiology.

One such extra-hypothalamic brain area that displays sexual dimorphisms is the hippocampus, a brain region that is critical to learning and memory and spatial navigation (Eichenbaum, 1997; see also Figure 10.2). In the pyramidal cell layers of the hippocampus, males have larger CA1 and CA3 pyramidal cell field volumes and cell body sizes than females (Isgor & Sengelaub, 1998). There are also sex differences in number of glial cells in the CA1 and CA3 region so that males have a greater number of astrocytes in the CA3 region, whereas females have more astrocytes in the CA1 region (Conejo et al.,

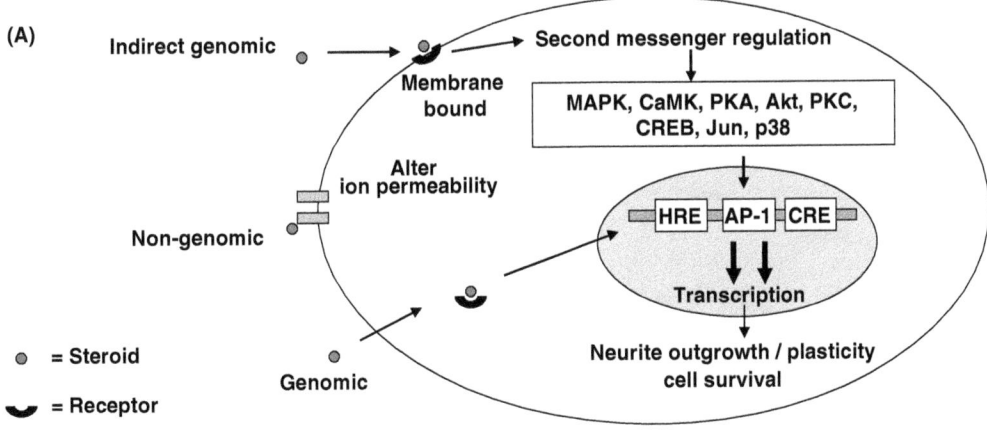

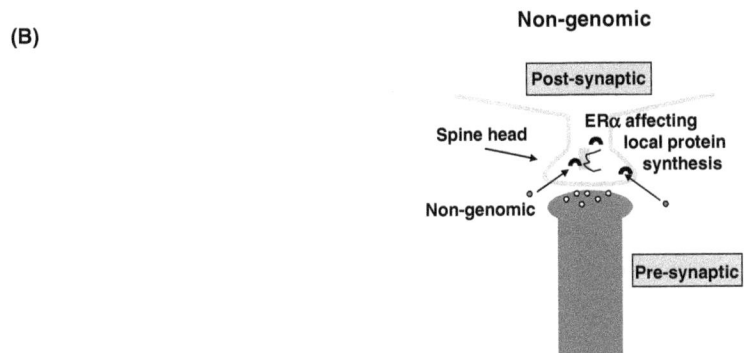

Figure 10.1. The various mechanisms of steroid hormone action. (A) The indirect genomic, non-genomic, and genomic actions of steroid hormones on neurons. Indirect genomic mechanisms include the activation of membrane-bound receptors linked to second-messenger cascades that activate gene transcription through DNA binding domains such as AP-1 or CRE, which ultimately alters neuronal function. A non-genomic mechanism of action includes the ability of estrogen to influence ion permeability. In the classic genomic mechanism, estrogen binds to the cytoplasmic or nuclear form of the steroid receptor which permits the steroid–receptor complex to translocate to the nucleus. This steroid–receptor complex then binds to the hormone response elements (HRE), which activates transcription of the genome, and in turn alters neuronal function. (B) A putative non-genomic effect of estrogen acting through ERα on local protein synthesis in the spine head. This mechanism would presumably allow for the rapid regulation of spine-specific mRNAs and proteins by estrogen. Reprinted with permission from Romeo, Waters and McEwen (2004).

2003). Male rats and humans also have greater dendritic branching compared to females in the CA1 region of the stratum radiatum, the dendritic field of the apical dendrites of the CA1 pyramidal cells (Barrera et al., 2001; Markham et al., 2004). Yet, in the CA3 region, males are reported to have more apical dendritic excrescences in stratum lucidum than females, but females have a greater number of primary dendrites than males (Gould et al., 1990a).

In contrast to the sex difference in dendritic branching, female rats in the pro-estrous stage of their reproductive cycle (i.e. when estrogen levels are at their peak) have higher levels of dendritic spines in the CA1 region of the hippocampus compared to males (Shors, Chua & Falduto, 2001). The density of dendritic spines is important to synaptic plasticity as neurons receive the vast majority of excitatory inputs via these membranes protuberances. These spines also

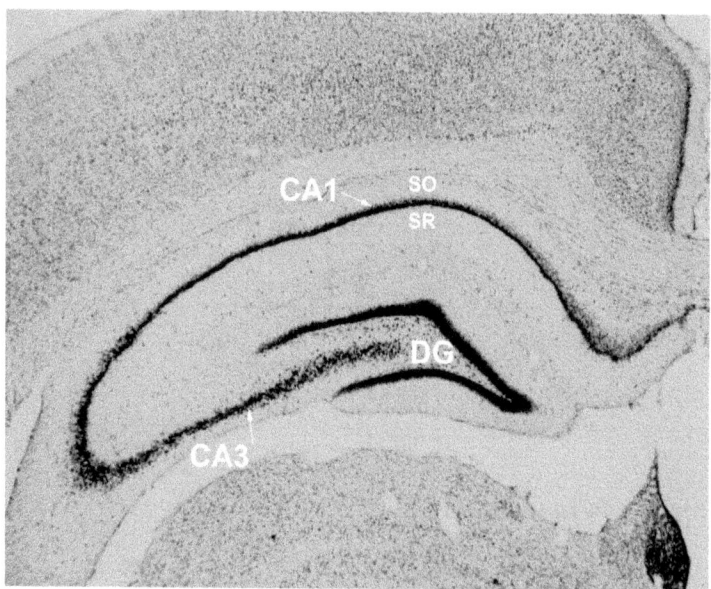

Figure 10.2. Nissl-stained coronal section of the hippocampus. Abbreviations: CA1 and CA3 pyramidal cell layers; DG, dentate gyrus; SO, stratum oriens; SR, stratum radiatum. Reprinted with permission from Romeo, Waters and McEwen (2004).

contain α-amino-3-hydroxy-5-methylisoxazole-4-proprionic acid- (AMPA) and N-methyl-D-asparate- (NMDA) type ionotropic glutamate receptors. Thus, changes in the morphology or number of spines, and the amount and trafficking of the AMPA and NMDA receptors in the spine, have been proposed as substrates for hippocampal synaptic plasticity at excitatory synapses (Hering & Sheng, 2001; Collingridge, Isaac & Wang, 2004).

In addition to sex differences in the pyramidal cell layers, there are also sexual dimorphisms in the dentate gyrus (DG) of the hippocampus. Specifically, males have a larger DG than females (Roof, 1993; Tabibnia, Cooke & Breedlove, 1999) along with greater synaptic connectivity in this region (Parducz & Garcia-Segura, 1993). Furthermore, the subiculum, the source of major hippocampal efferents, is larger in males compared to females (Andrade, Madeira & Paula-Barbosa, 2000). Thus, at the morphological level, the hippocampal formation demonstrates a number of sexual dimorphisms.

There have also been reports of sex differences in hippocampal neuronal excitability and performance on various hippocampal-dependent memory tasks. For instance, a number of studies have demonstrated that long-term potentiation (LTP), a putative electrophysiological correlate of learning and memory, is easier to induce in male compared to female rats (Maren, De Oca & Fanselow, 1994; Smith, Jones & Wilson, 2002; Yang et al., 2004). Furthermore, males demonstrate superior performance on certain hippocampal-dependent behavioral tests compared to females (Perrot-Sinal et al., 1996; Isgor & Sengelaub, 2003) or use different strategies to solve similar spatial tasks (Roof & Stein, 1999; Kanit et al., 2000). It should be noted however, not all studies have shown a male-biased superiority on all hippocampal-dependent tasks (Bucci, Chiba & Gallagher, 1995; Perrot-Sinal et al., 1996; Healy, Braham & Braithwaite, 1999). It has also been suggested that females may do worse than males in the Morris water maze because they experience more stress and have higher glucocorticoid levels (Beiko et al., 2004).

The above mentioned studies use intact males and females, and do not necessarily control for the stage of the estrous cycle of the female, which may dramatically influence hippocampal morphology (Woolley et al., 1990), physiology (Warren et al., 1995), and performance on learning and memory tasks (Healy, Braham & Braithwaite, 1999). Thus, when investigating sex differences in the structure and function of the hippocampus it is imperative to consider the hormonal milieu experienced by the animals being studied. Indeed, the next sections will highlight some of the profound differences exhibited by the male and female hippocampus in response to different types and levels of steroid hormones.

STEROID-INDUCED SYNAPTIC PLASTICITY

Similar to gross neuroanatomical sex differences, the first brain region in which steroid hormones were shown to influence morphological plasticity in a sex-dependent fashion was the hypothalamus (Raisman & Field, 1973). For example, dendritic spines in the ventromedial (Frankfurt et al., 1990) and arcuate nuclei (Matsumoto & Arai, 1980) of the hypothalamus were shown to exhibit sex differences in spine density so that estrogen-treated females had a greater number of dendritic spines compared to similarly treated males. Research conducted over the last 15 years, however, has revealed that other brain regions not typically associated with reproductive behavior or physiology are also significantly affected by gonadal and adrenal steroids in adulthood, namely the hippocampus (Gould, Woolley & McEwen, 1991).

ESTROGEN-INDUCED HIPPOCAMPAL SYNAPTIC PLASTICITY

In several species, estrogen has been shown to increase dendritic spine density, pre- and post-synaptic proteins, and synaptic connectivity in the CA1 region of the female hippocampus (Gould et al., 1990a; Woolley et al., 1990; Woolley & McEwen, 1992; Brake et al., 2001; Leranth, Shanabrough & Redmond, 2002; Choi et al., 2003; Lee et al., 2004; Li et al., 2004). For instance, female rats that lacked estrogenic stimulation, due to removal of their ovaries, had significantly fewer dendritic spines on the apical dendrites of the CA1 pyramidal cells of the hippocampus compared to females that received estradiol injections after ovariectomy (Gould et al., 1990b; Woolley & McEwen, 1993) (Figure 10.3). These findings were further confirmed by exploiting the natural fluctuations in estradiol that the female rat undergoes during her four- to five-day estrous cycle. Specifically, females that were sacrificed on the day of pro-estrus, when estrogen levels are at their highest, had a significantly greater number of dendritic spines on their CA1 cells compared to females that were sacrificed on the day of estrus, when estrogen levels are relatively low (Woolley et al., 1990; Woolley & McEwen, 1992). These studies were supplemented by experiments that demonstrated that the increase in spine density was accompanied by an increase in synaptic density as well (Woolley, Wenzel & Schwartzkroin, 1996). It is important to note that there are species differences in the effects of estrogen on spine formation.

For instance, in female rats and monkeys, estrogen increases CA1 spine density (Woolley, 1998; Leranth, Shanabrough & Redmond, 2002), whereas in mice estrogen changes the morphology of the spine, but not spine number (Li et al., 2004).

In parallel with these estrogen-induced structural changes, females experiencing estrogenic stimulation, either by experimental manipulation or because of their estrous cycle, exhibit superior performance on certain hippocampal-dependent learning and memory tasks (Luine, 1997; Daniel et al., 1998; Luine et al., 1998; Bimonte & Denenberg, 1999; Gibbs, 1999; Sandstrom & Williams, 2001; Heikkinen et al., 2002; Luine, Jacome & MacLusky, 2003; Gibbs et al., 2004; Li et al., 2004; Sandstrom & Williams, 2004) and enhanced LTP (Warren et al., 1995; Cordoba Montoya & Carrer, 1997; Good, Day & Muir, 1999).

The mechanisms through which estrogen affects CA1 spine density in females are still being elucidated. Estrogen increases the expression of NMDA receptors in the dendritic fields of the CA1 pyramidal cells, namely the stratum oriens and stratum radiatum (Weiland, 1992; Daniel & Dohanich, 2001). This effect of estrogen on NMDA receptors is necessary for estrogen-induced spine formation, so that females simultaneously treated with NMDA receptor antagonists and estradiol fail to show estrogen-induced increases in hippocampal spine density (Woolley & McEwen, 1994).

Estrogen-induced hippocampal synapse formation and NMDA receptors involve multiple sites of estrogen action through both genomic and non-genomic mechanisms. It does appear that the estrogen-induced increase in spines is mediated by nuclear estrogen receptors (ERs) (McEwen, Tanapat & Weiland, 1999). However, CA1 pyramidal cells are devoid of appreciable levels of nuclear ERs (Weiland et al., 1997), but

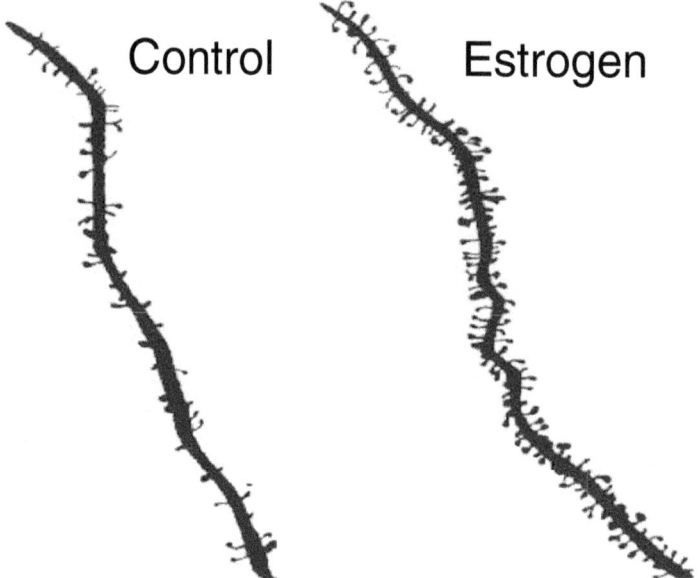

Figure 10.3. Camera lucida drawings of a dendrite from a female not experiencing estrogenic stimulation (left, control) compared to a female exposed to estrogen (right, estrogen). Note the marked increase in spine density in the estrogen-treated female. Adapted from Woolley and McEwen, 1992.

do contain extra-nuclear ERs (Milner et al., 2001; Towart et al., 2003). Instead, the effects of estrogen on spine density appear to be mediated trans-synaptically through multiple systems that do possess nuclear ERs. First, estrogen increases acetylcholine (ACh) activity in the forebrain of female rats (Luine & McEwen, 1983; Luine, Renner & McEwen, 1986; Gibbs et al., 1994; Gibbs, 1996, 1997, 2000) and ACh release into the hippocampus (Gabor et al., 2003). This estrogen-mediated increase in cholinergic activity to the hippocampus has been proposed to contribute to the estrogen-induced increase in hippocampal NMDA receptors (Daniel & Dohanich, 2001). A second trans-synaptic mechanism involves the ability of estrogen to disinhibit CA1 pyramidal cells via reduction of inhibitory inputs from the hippocampal GABAergic interneurons (Murphy et al., 1998; Rudick & Woolley, 2001), which contain nuclear ERs (Weiland, 1992b). This disinhibition likely contributes to the greater excitability demonstrated by estrogen-treated CA1 pyramidal cells.

In addition to these trans-synaptic effects of estrogen, rapid, non-genomic effects of estradiol on signaling cascades implicated in hippocampal spinogenesis have been revealed (Akama & McEwen, 2003; Lee et al., 2004). Interestingly, ERα has been localized in dendritic spines and the axonal terminals of CA1 pyramidal cells in the female (Milner et al., 2001; Towart et al., 2003), suggesting that estrogen may act on these extra-nuclear receptors to initiate rapid, non-genomic effects directly on the synaptic apparatus and local mRNAs in the spines and terminals. It is interesting to note that a reduction in dendritically localized ERα has been associated with the failure of estrogen to induce spine synapses in aging female rats (Adams et al., 2002). Thus, both trans-synaptic effects of estrogen operating through classic genomic mechanisms and local, non-genomic effects of estrogen on the synaptic apparatus may underlie estrogen-induced spine formation in the female hippocampus.

Although estrogen increases CA1 spine density in females, this effect does not occur in males (Lewis, McEwen & Frankfurt, 1995; Leranth, Petnehazy & MacLusky, 2003; Lee et al., 2004). That is, castration reduces CA1 spine density in males, but estrogen treatment does not reverse this decrease (Leranth, Petnehazy & MacLusky, 2003). The mechanisms through which estrogen fails to induce spinogenesis in males are not completely understood. A recent study, however, indicates that estrogen treatment does not increase hippocampal NMDA binding in males, a necessary effect for estrogen-mediated spinogenesis in females (Woolley & McEwen, 1994). Furthermore, estrogen fails to increase cholinergic activity in the male forebrain (Luine & McEwen, 1983; Luine, 1985; Luine, Renner & McEwen, 1986), possibly mediating the lack of estrogen-induced NMDA receptors. It is also interesting to note that males have fewer extranuclear ERs located in their dendritic spines, suggesting a lack of local estrogenic action on the spines of male pyramidal cells.

The sex difference in the ability of estrogen to induce hippocampal spine formation in adulthood appears to be dependent upon the perinatal hormonal milieu experienced by the developing individual. For instance, males treated with an aromatase inhibitor early in development, which blocks the normally occurring conversion of testosterone to estrogen neonatally, show estrogen-induced spinogenesis in adulthood (Lewis, McEwen & Frankfurt, 1995). Interestingly, similarly treated males also show increases in forebrain cholinergic activity upon estrogen treatment in adulthood (Luine, Renner & McEwen, 1986). These data indicate that the sexually dimorphic response of the hippocampus to estrogen in adulthood may be modulated by neonatal hormonal manipulations. Whether females treated neonatally with androgens fail to show estrogen-induced spinogenesis in adulthood has not been established.

The role of estrogen on hippocampal GABAergic interneurons in males and the implications of having fewer dendritically localized ERs in the spines compared to females are presently unknown. Furthermore, the exact role played by the neonatal hormonal milieu experienced by the developing female remains to be described. Thus, future studies will need to address the contributions of trans-synaptic and local effects of estrogen on this sex difference, in addition to understanding the early organizational effects of the neonatal hormonal environment.

PROGESTERONE-INDUCED HIPPOCAMPAL SYNAPTIC PLASTICITY

In rats, the four- to five-day estrous cycle leads to dynamic changes in both estrogen and progesterone levels experienced by the female (Shaikh & Shaikh, 1975). However, few studies have addressed the role of progesterone on hippocampal synaptic plasticity. The studies that have investigated the effects of progesterone on spine density are usually conducted in females simultaneously experiencing estrogenic stimulation. For instance, it has been shown that progesterone modulates dendritic spine density and synaptic number after estrogen stimulation in the CA1 region of the female hippocampus (Woolley et al., 1990b; Woolley & McEwen, 1993). Specifically, in females that lack ovaries, steroid replacement revealed that progesterone initially enhances estrogen upregulation of dendritic spines then rapidly decreases spine number (Woolley & McEwen, 1993). The progesterone-induced decrease in spine number could be blocked with progesterone receptor antagonists (Woolley & McEwen, 1993). In hippocampal cell cultures, progesterone prevents the estrogen-induced increase in spines (Murphy & Segal, 2000) and progesterone metabolites inhibit neurite outgrowth (Brinton, 1994). Similar to the rat, female rhesus macaques that are treated with estrogen and progesterone show a decrease in both pre- and post-synaptic proteins in the CA1 region of the hippocampus (Choi et al., 2003). However, the females that received progesterone treatment alone had an increase in the pre-synaptic protein synaptophysin (Choi et al., 2003). Future studies will need to address whether progesterone in the absence of estrogen has spine-promoting effects, in addition to its synaptic pruning effect observed when in the presence of estrogen.

Progesterone originates principally in the ovaries in females (Smith, 1975); in contrast, males lack this gonadal source but can synthesize progesterone in the adrenal gland (Kalra & Kalra, 1977) and locally in the hippocampus (Ibanez et al., 2003). Although the actions of progesterone on dendritic spine number have not been determined in males, administration of the progesterone receptor antagonist RU486 to males after birth reduces sexual behavior and measures of fear (Lonstein, Quadros & Wagner, 2001), suggesting that progesterone plays a role in organization of the male brain. Future studies will need to address whether locally synthesized or adrenally produced progesterone may affect spine density in males.

The mechanisms for the effect of progesterone on the hippocampus are currently under investigation.

Progesterone receptors are present in dendritic spines and the cell body of CA1 neurons in females, suggesting that progesterone may have a local, non-genomic action in addition to a classical genomic action. Progesterone treatment of hippocampal cell cultures may increase miogen-activated protein kinase (MAPK) activation and induce nuclear translocation of phosphoERK (Nilsen & Brinton, 2003), as well as block phosphorylation of cAMP response-element binding protein (CREB) induced by estrogen (Murphy & Segal, 2000). Progesterone may also reduce the overall excitability of neurons. For instance, progesterone attenuates glutamate-induced increases in intracellular calcium concentrations in females (Nilsen & Brinton, 2002b), whereas progesterone metabolites have been shown to modulate GABAergic inhibition in males (Gulinello, Gong & Smith, 2002).

Knowledge of sex differences in the action of progesterone is limited by the rarity of female–male comparisons in the same study. In addition, hippocampal progesterone receptor localization needs to be clarified in females and males. Further research is needed to characterize the actions of progesterone versus those of its metabolites, and also to elucidate the mechanisms responsible for the structural and functional outcomes associated with progesterone exposure.

ANDROGEN-INDUCED HIPPOCAMPAL SYNAPTIC PLASTICITY

Although estrogen does not induce spine formation in males, a recent study has shown testosterone, acting through its androgenic metabolite 5α-dihydrotestosterone (DHT), increases hippocampal CA1 spine density in male rats (Leranth, Petnehazy & MacLusky, 2003). The mechanisms through which androgens promote hippocampal spine formation are not completely understood. DHT does increase hippocampal NMDA receptors in males (Romeo et al., 2005), but, interestingly, does not affect cholinergic forebrain activity or input to the hippocampus (Romeo et al., 2005). In contrast to the paucity of nuclear ERs in the female pyramidal cells, males have nuclear and extra-nuclear androgen receptors (AR) in the pyramidal cells of the CA1 hippocampal region (Sar et al., 1990; Kerr et al., 1995; Xiao & Jordan, 2002; Tabori et al., 2005). Thus, the ability of DHT to increase NMDA receptors in the CA1 region may be directly mediated at the level of the pyramidal cells, obviating trans-synaptic effects of the hormone. This would indicate that androgen-induced increases in spine synapses may be more dependent on the direct actions of the steroid on the pyramidal neurons than the trans-synaptic mechanisms required for estrogen-induced increases in hippocampal spinogenesis in females (Figure 10.4).

Unlike the sex difference in the inability of estrogen to induce spines in males, androgen-treated females do evince CA1 spine induction (Leranth, Hajszan & MacLusky, 2004). However, the magnitude of the induction is less in females than males (Leranth, Hajszan & MacLusky, 2004). Females do process nuclear androgen receptors in their CA1 pyramidal cells, albeit with lower expression than the male (Xiao & Jordan, 2002). Thus, similar to androgen-induced spinogenesis in males, androgens may act directly on the CA1 pyramidal cells in females. It is currently unknown whether androgens increase hippocampal NMDA binding or affect cholinergic inputs to, or GABAergic interneurons in, the female hippocampus.

In males, DHT has been shown to increase the duration of action potentials in the CA1 region and lower after-hyperpolarization, suggesting that DHT may increase hippocampal excitability (Pouliot, Handa & Beck, 1996). However, DHT and testosterone have also been shown to decrease LTP in the CA1 region of males (Harley et al., 2000). It should be noted, however, that these animals were implanted chronically with DHT or testosterone and may have been experiencing supraphysiological levels of these steroids. The role of androgens and, specifically, DHT in hippocampal-dependent learning and memory is also unclear. For example, DHT has been shown to increase cognitive performance on hippocampal-dependent tasks in male rodents and humans (Frye & Lacey, 2001; Cherrier, Craft & Matsumoto, 2003; Frye et al., 2004), while another study reported that DHT had no effect on hippocampal-dependent working memory in aged male rats (Bimonte-Nelson et al., 2003). Again, dose and treatment regimens may have significantly affected the outcome of these behavioral studies.

The role of neonatal hormonal exposure on the later response of the hippocampus to androgens in adulthood is presently unknown. This response does not appear to be completely sexually differentiated neonatally as females do show some androgen-induced hippocampal

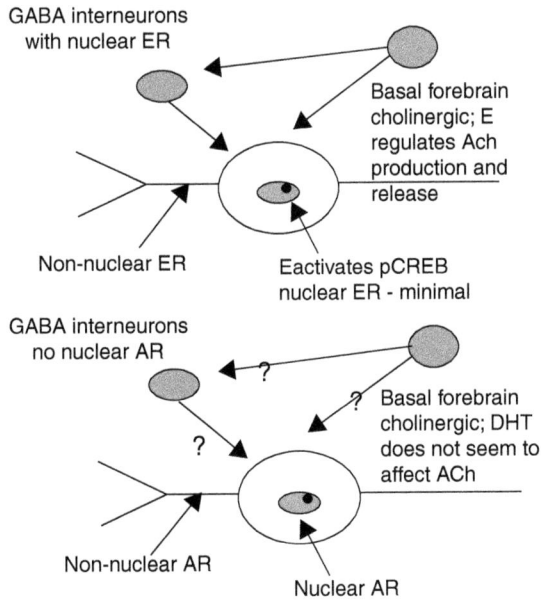

Female CA1 neurons
1. Cholinergic input regulated by E; plays a strong role in E induction of NMDA receptors.
2. NMDA receptor blockade prevents E induction of spines.
3. Fimbria-fornix lesions block E effect on spines.
4. Cholinergic input may operate via GABA interneurons as well as directly on CA1.

CONCLUSION: Spine synapse induction by E is heavily dependent on trans-synaptic influences and non-nuclear E.

Male CA1 neurons
1. Cholinergic input does not appear to be regulated by DHT.
2. DHT induces NMDA receptor binding; not known if cholinergic mediates induction.
3. Not known if NMDA receptor blockade prevents DHT-induction of spines.
4. Fimbria-fornix lesions have lesser effect in male on DHT induction of spines.
5. GABA interneurons lack nuclear AR

CONCLUSION: Spine synapse induction by DHT may be less dependent on trans-synaptic and more influenced by nuclear AR.

Figure 10.4. Sex differences in CA1 synapse formation. The mechanisms through which estrogen and DHT affect spine synapse formation differentially in females and males. Reprinted with permission from Romeo, Waters and McEwen (2004).

spinogenesis in adulthood (Leranth, Hajszan & MacLusky, 2004). However, whether females treated neonatally with androgens would show a greater hippocampal response to testosterone or DHT in adulthood remains to be demonstrated. Furthermore, whether testosterone or DHT have any effect on learning and memory in females also needs to be addressed.

GLUCOCORTICOID-INDUCED HIPPOCAMPAL SYNAPTIC PLASTICITY

In contrast to the spine-promoting effects of estrogen and testosterone on the male and female hippocampus, high levels of corticosterone, induced by repeated stressors or through pharmacologic manipulation, have been shown to decrease the branching of hippocampal CA3 pyramidal cells and cause dendritic atrophy in male rats (Magarinos & McEwen, 1995a, 1995b). In addition to decreasing dendritic branching, stress has been shown to increase spine density on the CA3 pyramidal dendrites in the male hippocampus (Sunanda, Rao & Raju, 1995).

Similar to chronic stress, males exposed to an acute stressor also demonstrate an increase in hippocampal spine density, but females show the opposite pattern so that acute stress decreases spine number (Shors, Chua & Falduto, 2001). Behaviorally, acute stress impairs spatial memory in males, but enhances spatial memory in females, regardless of the stage of their estrous cycle (Conrad et al., 2004). Interestingly, chronic stress reduces spatial memory in both males and females, but females show faster improvement in their spatial memory compared to males upon recovery from chronic stress (Conrad et al., 2003). The mechanisms through which stress affects hippocampal dendritic branching, spinogenesis, and learning and memory are not well understood, but NMDA receptors appear to play a role. For instance, the corticosterone-induced decrease in CA3 pyramidal cells may be mimicked by glutamate treatment (Magarinos & McEwen, 1995b), whereas blockade of the NMDA receptor abrogates the stress-induced increase and decrease in spines in males and females, respectively (Shors, Falduto & Leuner, 2004).

Future research will need to elucidate whether these effects of stress are mediated solely by stress-induced increases in corticosterone or whether other steroids or chemical messengers are involved. Furthermore, future studies will need to investigate whether the perinatal hormonal milieu has any influence on how the hippocampus responds differently to stress steroids in adult males and females.

SUMMARY

It is clear from the research reviewed above that the influence of steroid hormones on hippocampal structure and function may differ vastly between the sexes. For instance, it appears that while estrogen only influences spinogenesis in females, the androgens primarily affect spine density in males. Moreover, corticosterone affects spine density in both males and females, but in different directions. There are certain similarities, however, through which steroids affect synaptic plasticity in males and females. For example, these effects all appear to involve NMDA receptor activation and regulation.

The sex differences exhibited by the hippocampal formation in response to different steroids may have important clinical implications. For example, increased attention has been focussed on the ability of estrogen to ameliorate neurodegenerative diseases such as Alzheimer's disease (Henderson, Watt & Buckwalter, 1996). The mechanism by which estrogen mediates these beneficial effects is not well understood. However, one hypothesis is that estrogen mediates these effects on memory and cognition by acting on the hippocampus, a brain region impaired in patients with Alzheimer's disease (Kasa, Rakonczay & Gulya, 1997). Thus, if the beneficial effect of estrogen on Alzheimer's disease is, in part, through the effect of estrogen on hippocampal spine growth and synaptogenesis, then males would benefit far less than females from any therapeutic effect of estrogen.

Conversely, progesterone has been shown to positively affect the outcome after ischemic injury in both females (Chen, Chopp & Li, 1999; Alkayed et al., 2000; Murphy, Littleton-Kearney & Hurn, 2002) and males (Gibson & Murphy, 2004). In addition, testosterone and DHT have been reported to protect against excitotoxic damage in the male hippocampus (Ramsden, Shin & Pike, 2003), while DHT has been reported to reverse the accumulation of β-amyloid protein upon castration (Ramsden et al., 2003). These results are intriguing as men suffering from Alzheimer's disease have lower serum and brain androgen levels compared to aged men without Alzheimer's disease (Hogervorst et al., 2001; Rosario et al., 2004). It is unknown whether androgens affect these parameters in the female brain, but suggests that androgens may provide protection against age-related decrements in hippocampal function and plasticity in males and females.

Although future studies will undoubtedly lead us to a greater understanding of these phenomena, the data reviewed above indicate that when studying synaptic plasticity the sex and hormonal milieu of the individual may significantly influence the outcome and interpretation of the research. Furthermore, given the possible clinical implications of steroid-induced hippocampal plasticity, it is imperative to further our understanding of how steroids affect synaptic function in males and females and what are the similarities and differences regarding these effects.

ACKNOWLEDGEMENTS

Research in this review was supported by grants MH065749 (R.D.R.) and NS07080 and AG16765 (B.S.M.)

Part III
Cell–cell signaling molecules in synaptic plasticity

11 · MHC class I in activity-dependent structural and functional plasticity

Lisa M. Boulanger

INTRODUCTION

The major histocompatibility complex (MHC) class I is a large family of vertebrate genes first discovered as the molecular basis for the rejection of grafted tissue. Subsequently, it was found that MHC class I plays a much broader role in discriminating from non-self during bacterial and viral infections and eliminating some cancers. MHC class I is expressed on the surface of most nucleated cells in the body where it presents antigens derived from cytosolic proteins. Stabilizing these antigens at the cell surface enables interactions with antigen-specific cytotoxic T cells, which can then kill infected cells and cancers. This ability to quickly and accurately identify and dispatch cellular sources of foreign or aberrant proteins is one of the central features of the adaptive immune response.

Surprisingly, growing evidence suggests that these immune functions are only the tip of the iceberg, and that MHC class I proteins also perform critical roles outside the immune system, specifically in the activity-dependent development and plasticity of the mammalian brain. Recent studies have implicated MHC class I in the establishment of precise neuronal circuitry, and in changes in synaptic strength thought to underlie certain forms of learning and memory. In this chapter, I summarize recent studies characterizing the expression and function of MHC class I in the nervous system, and discuss what the results of these studies suggest about the relationship between MHC class I and activity-dependent structural and functional plasticity.

EXPRESSION OF MHC CLASS I IN NEURONS

It has long been known that neurons and glia may be induced to express high levels of MHC class I after injury, infection, or treatment with cytokines (Lampson & Fisher, 1984; Wong et al., 1984; Maehlen et al., 1988; Streit, Graeber & Kreutzberg, 1989; Pereira, Tscharke & Simmons, 1994; Neumann et al., 1995; Fujimaki et al., 1996; Neumann et al., 1997; Redwine, Buchmeier & Evans, 2001; Foster et al., 2002). However, neurons were generally not thought to express MHC class I in the basal state (Lampson & Fisher, 1984; Lampson, Whelan & Seigal, 1988; Joly, Mucke & Oldstone, 1991; Joly & Oldstone, 1992; Drew et al., 1993; White, Keane & Whittemore, 1994; Fujimaki et al., 1996) (reviewed in Lampson, 1995). This idea is supported by the fact that some viruses persist in the central nervous system (CNS), as compared with other tissues (Joly, Mucke & Oldstone, 1991; Rall, Mucke & Oldstone, 1995), and that tissue grafts into the brain are not immediately rejected, as they are elsewhere (Head & Griffin, 1985). These functional properties earned the CNS the label of "immune privileged."

Several studies have since shown, however, that MHC class I is expressed in characteristic patterns by normal, uninjured neurons in the developing and adult brain (Corriveau, Huh & Shatz, 1998; Lidmn, Olsson & Piehl, 1999; Linda et al., 1999; Huh et al., 2000) (Figure 11.1). MHC class I mRNA or protein, or both, have been detected in subsets of neurons of the developing lateral geniculate nucleus (LGN), developing and adult hippocampus, primary somatosensory cortex, layer IV of primary visual cortex (Corriveau, Huh & Shatz, 1998; Huh et al., 2000), substantia nigra pars compacta (Lidman, Olsson & Piehl, 1999), dorsal root ganglion, adult brainstem and spinal cord (Linda et al., 1999), and vomeronasal organ (Ishii, Hirota & Mombaerts, 2003; Ishii & Mombaerts, 2008; Loconto et al., 2003). In parallel, it has been shown that the immune privilege of the brain is in part the result of actively suppressed immune responses to MHC class I–peptide complexes in the CNS, rather than a lack of MHC class I protein on neurons (Streilein, 1993).

Although MHC class I is detectable in neurons, its expression in the brain differs in many ways from that seen in other nucleated cells. First, levels of MHC class I in neurons are lower than in many other cell types (Corriveau, Huh & Shatz, 1998; Huh et al., 2000). Lower MHC class I levels may reduce the risk of autoimmunity (Singer, 1997) in the brain, where postmitotic neurons may be relatively difficult to replace. Second, MHC class I is expressed in most

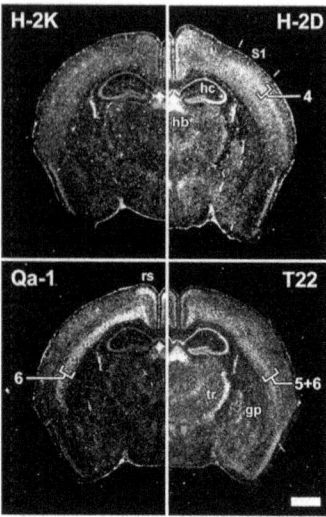

Figure 11.1. Expression of mRNAs for four different MHC class I genes in coronal sections of adult mouse brain, detected by in situ hybridization. Hybridization of probes specific for two classical MHC class I genes (K and D) and two nonclassical MHC class I genes (Qa-1 and T22) are shown. S1, primary somatosensory cortex; hc, hippocampus; hb, habenula; rs, retrosplenial cortex; gp, globus pallidus; tr, thalamic reticular nucleus. Numbers indicate strongly labelled cortical layers. Scale, 1mm. With permission from Huh *et al.*, 2000.

tissues at relatively constant levels, in the absence of infection or other challenge (Janeway *et al.*, 2001). In the brain, in contrast, MHC class I levels are highly dynamic, even in the absence of infection or injury, and the levels and localization of MHC class I mRNAs change dramatically over the course of normal development (Corriveau, Huh & Shatz, 1998; Huh *et al.*, 2000). Third, some so-called "non-classical" MHC class I proteins have not yet been detected in other tissues, but are strongly (and in some cases exclusively) expressed in specific regions of the CNS. For example, members of the M10 family of non-classical MHC class I's are only detected in the vomeronasal organ (VNO), a specialized pheromone-sensing olfactory structure (Ishii, Hirota & Mombaerts, 2003; Loconto *et al.*, 2003; Ishii & Mombaerts 2008). Finally, most antigen-presenting cells co-dominantly express all of the classical MHC class I alleles in an individual's genome (Janeway *et al.*, 2001), but it is as yet unknown if a given CNS neuron expresses the full array of classical MHC class I proteins. In situ hybridization studies indicate that the mouse classical MHC class I genes H2-D and H-2K are expressed in distinct, partially overlapping subsets of neurons in the adult CNS (Huh *et al.*, 2000), and that the expression of these genes differs from that of the nonclassical MHC class I genes T22 and Qa-1 (Huh *et al.*, 2000 (Figure 11.1). In addition, members of the non-classical MHC class I M10 family are expressed in subpopulations of sensory neurons in the VNO, such that each neuron may express none, one or a few of these non-classical MHC class I molecules (Ishii, Hirota & Mombaets, 2003; Loconto *et al.*, 2003; Ishii & Mombaerts 2008). Together, these results indicate that different neurons will express different complements of MHC class I proteins. It will be of interest to examine the expression patterns of the many members of the classical and non-classical MHC class I families in the brain at the single-cell level to ascertain if there is a novel molecular logic to their neuronal expression.

The subcellular localization of neuronal MHC class I is also unusual, compared with non-neuronal cell types, most of which express MHC class I over the entire cell surface. In several brain regions, neuronal MHC class I expression is found selectively in the somato-dendritic compartment. For example, in the retinogeniculate and geniculocortical systems, MHC class I is high in the soma and dendrites of neurons post-synaptic to actively remodeling axons (Corriveau, Huh & Shatz, 1998). In acutely dissociated hippocampal neurons *in vitro*, MHC class I protein is detectable throughout dendrites (Goddard *et al.*, 2007). Similarly, in acutely dissociated cortical neurons *in vitro*, MHC class I protein is initially detected in clusters in both axons and dendrites, but becomes selectively localized to dendrites as the cultures mature (Wampler & McAllister, 2004). In these cells, MHC class I is detected at both excitatory and inhibitory synapses (Wampler & McAllister, 2004). In sensory neurons of the VNO, in contrast, a fluorescently tagged MHC class I fusion protein is targeted to both axonal and somato-dendritic compartments (Ishii and Mombaerts, 2008).

In the somatodendritic compartment, some of the MHC class I is found at synapses. Immunostaining of hippocampal neurons *in vitro* shows MHC class I immunostaining throughout the dendrites, including some that is localized to spines and apposed to pre-synaptic markers (Goddard & Shatz, 2003). Furthermore, MHC class I immunoreactivity is enriched in synaptosomal preparations of whole adult rat brain (Huh *et al.*, 2000).

MHC class I expression in the brain is high in sites of ongoing activity-dependent plasticity, including the developing visual system and the adult hippocampus and cerebellum (Corriveau, Huh & Shatz, 1998; Huh *et al.*, 2000). Protein or mRNA encoding several MHC class I receptors or receptor components, including CD3ζ (Corriveau, Huh & Shatz, 1998; Huh et al., 2000), T cell receptor beta subunit (TCR β) (Syken & Shatz, 2003a), PIRB (Syken et al., 2006), Digr1 and Digr2 (Syken & Shatz, 2003b), Ly49 (Zohar et al., 2008), and killer immunoglobulin receptor (KIR) (Bryceson et al., 2005) have also been found in the brain, and most are detected in highly plastic regions of the CNS. Thus the expression patterns of MHC class I as well as candidate MHC class I receptors in the CNS are consistent with a role for MHC class I signaling in activity-dependent plasticity (Corriveau, Huh & Shatz, 1998; Huh et al., 2000; Boulanger, Huh & Shatz, 2001).

ELECTRICAL ACTIVITY REGULATES NEURONAL EXPRESSION OF MHC CLASS I

In many parts of the developing brain, spontaneous electrical activity drives the establishment of the final, adult pattern of connections. In the developing mammalian visual system, where this process is relatively well-understood, each eye sends axons to a proximal target in the thalamus, the LGN. Initially, these inputs are overlapping, but later in development they segregate into eye-specific layers through a process that requires new gene expression as well as electrical activity arising in the retinas. One expectation of genes that are required for such activity-dependent plasticity is that at least a subset of them will be regulated by the relevant electrical activity (Corriveau,Huh & Shatz, 1998). To identify genes involved in activity-dependent remodeling, Corriveau, Huh and Shatz (1998) performed an unbiased genetic screen for LGN-expressed genes that are regulated by endogenous electrical activity during developmental remodeling of the retino-geniculate projection. Unexpectedly, this screen showed that a member of the MHC class I family is strongly down-regulated in the absence of normally occurring spontaneous electrical activity that drives remodeling of retinal afferents in the developing LGN (Corriveau, Huh & Shatz, 1998). This initial developmental remodeling of the retino-geniculate projection finding has led to further characterization of the expression patterns, regulation (above), and effects (below) of MHC class I in the CNS.

In addition to early spontaneous activity, MHC class I is upregulated by later, visually driven activity, since tetrodotoxin (TTX) in the eye at this stage reduces MHC class I expression specifically in the LGN layer subserving that eye. MHC class I can also be regulated by exogenous activity: MHC class I levels increase in the adult hippocampus and neocortex after kainic acid-induced seizures, which dramatically increase activity (Corriveau, Huh & Shatz, 1998).

Thus, in many brain regions *in vivo*, MHC class I is upregulated by electrical activity and downregulated when activity is reduced (Corriveau, Huh & Shatz, 1998). In contrast, some studies have found that, dissociated hippocampal neurons *in vitro* upregulate MHC class I in response to interferon gamma (IFN-γ) only under electrically silenced conditions (Neumann et al., 1995; 1997). There are many possible explanations for the contrast in these results. One is that the experiments are not comparable in the cell type or system used; the experiments that found MHC class I was upregulated in silent neurons were performed on dissociated hippocampal neurons *in vitro* (Neumann et al., 1995; 1997), whereas the experiments that found MHC class I was upregulated by activity examined a variety of neuronal populations *in vivo* (Corriveau, Huh & Shatz, 1998; Huh et al., 2000). Furthermore, recent preliminary studies suggest that one crucial feature may be whether the activity blockade is local or global. In dissociated cortical neurons *in vitro*, global depression of activity via bath application of TTX increases the density of MHC class I immunoreactivity (Wampler & McAllister, 2004). This is reminiscent of previous results showing that MHC class I is upregulated in dissociated hippocampal neurons *in vitro* in response to IFN-γ treatment, but only if activity is globally blocked with TTX (Neumann et al., 1995; 1997). However, when activity is blocked in a single cell (through overexpression of a gene encoding a channel that clamps the voltage of the cell at a hyperpolarized resting potential), activity blockade instead decreases the expression of MHC class I (Wampler & McAllister, 2004), mimicking earlier results *in vivo* (Corriveau, Huh & Shatz, 1998). Thus MHC class I expression may be differentially regulated by activity depending on the cell type, local network properties, activity levels, and relative level of activity in neighboring cells. Given this possibility, it would be interesting to know if more subtle modulations of the level or pattern of activity, such as patterns that trigger synaptic plasticity, are also a relevant signal for the regulation of MHC class I expression.

Together, these results suggest that electrical activity can bi-directionally regulate the expression of MHC class I in a given neuron or population of neurons. This regulation can occur pre-natally or post-natally, and may occur in response to spontaneous or sensory-evoked endogenous activity as well as exogenous activity. In addition, the circuit activity and relative levels of activity in neighboring neurons may contribute to the extent or even sign of the change in MHC class I levels in a given neuron.

MHC CLASS I IN REMODELING OF DEVELOPING AND ADULT CONNECTIONS

MHC class I was identified in a blind screen for genes involved in developmental activity-dependent remodeling, and is highly expressed in regions of ongoing remodelling in a manner that is regulated by activity (Corriveau, Huh & Shatz, 1998; Huh et al., 2000). These results together suggest that MHC class I may be involved in activity-dependent structural plasticity (Corriveau, Huh & Shatz, 1998; Huh et al., 2000; Boulanger, Huh & Shatz, 2001; Boulanger & Shatz, 2004). To test this hypothesis directly, Huh et al. (2000) examined activity-dependent structural plasticity in mice genetically deficient for cell surface MHC class I. These mice are genetically deficient for β2-microglobulin (β2m), the obligatory light chain that associates with many MHC class I heavy chains, and TAP1, a transporter required to load intracellular peptides onto many MHC class I molecules. In the absence of both of these products, cell surface levels of MHC class I are greatly reduced (Zijlstra et al., 1990; Jackson & Peterson, 1993; Dorfman et al., 1997; Neumann et al., 1997). In addition, Huh et al. (2000) examined mice deficient for CD3ζ, a component of several known receptors for MHC class I in the immune system (Love et al., 1993). These mice are all immune-compromised, but are outwardly normal if kept in a clean facility, their gross neuroanatomy is normal (Huh et al., 2000). All experiments on these mice were performed blind to genotype.

Since MHC class I was identified in a screen for genes involved in the remodeling of the developing visual system, and MHC class I is highly expressed in the developing LGN during the period when eye-specific layers are forming, Huh et al. (2000) examined retinogeniculate remodeling in MHC class I-deficient mice. They found that in mice deficient for MHC class I signaling, the final, activity-dependent steps in the remodeling of retinal ganglion cell projections to the LGN are selectively disrupted. Although the inputs from the two eyes arrive in the LGN, they do not undergo normal activity-dependent remodeling to form the tightly localized, eye-specific regions seen in wild types. As a result, the projections to the ipsilateral LGN are significantly larger in MHC class I-deficient mice, and aberrant, ectopic projections occur. These results are consistent with a role for endogenous MHC class I in the removal of inappropriate connections during normal development (Huh et al., 2000; Boulanger, Huh & Shatz, 2001; Boulanger & Shatz, 2004). Furthermore, that CD3ζ-deficient mice were indistinguishable from MHC class I-deficient mice with regard to retinogeniculate remodeling defects suggests that MHC class I may modify remodeling through a CD3ζ-containing receptor in the developing brain, perhaps in a manner analogous to the T-cell-mediated elimination of infected or cancerous cells as in the immune system (Huh et al., 2000). In addition, recent studies have identified a role for another neuronally-expressed putative MHC class I receptor, PIRB, in activity-dependent remodeling of adult thalamocortical projections following sensory deprivation (Syken et al., 2006). Of note, retino-geniculate development is normal in PIRB-deficient mice (Syken et al., 2006), indicating that multiple MHC class I receptors are expressed by neurons and subserve distinct functions in brain development and plasticity.

In addition to the visual system, activity drives the remodeling of many other developing connections. For instance, during motor development, there is selective elimination of innervation of autonomic post-ganglionic neurons, skeletal muscle fibers, and somatic motoneurons (Edstrom et al., 2004). This remodeling of motor connections, like remodeling of the developing visual system, is driven by electrical activity, suggesting they may share significant mechanistic features. However, recent studies of activity-dependent remodeling of climbing fiber innervation of cerebellar Purkinje cells indicate that remodeling proceeds normally in mice deficient for b2m and the classical MHC class I genes H-2D and H-2K (Letellier, 2008). Thus MHC class I is likely required for some, but not all, activity-dependent remodeling in the developing brain. Given the widespread use of activity-dependent remodeling to achieve mature connectivity throughout the central and peripheral nervous system, the fact that MHC class I-deficient animals are outwardly normal also supports spatial selectivity in MHC class I regulation of activity-dependent circuit development. One intriguing possibility is that this selectivity reflects functional specialization of the differentially expressed members of the large MHC class I gene family.

In addition to its role in normal developmental activity-dependent remodeling of developing visual circuits, MHC class I has also been implicated in pathologic remodeling, both as a consequence of normal aging and in response to acute injury. MHC class I is upregulated in normal, aged motoneurons (Edstrom et al., 2004), at a time when there is a pronounced, progressive loss of afferent boutons. This correlation raises the possibility that the pathologic retraction of aged projections, like the normal developmental regression of inappropriate visual projections, is mediated by MHC class I.

Neurons also upregulate MHC class I in response to axotomy (Olsson et al., 1989; Linda et al., 1998; Sabha et al., 2008) or infection (Foster et al., 2002), both of which can induce selective withdrawal of affected pre-synaptic terminals from cell bodies and dendrites. In the facial nucleus, peripheral nerve transection leads to pruning of synapses from the affected neuronal somas, a process known as "synaptic stripping". Similarly, transection of spinal motor nerves leads to a reduction in the number of synapses made onto the somas of these cells in the spinal cord. Mice lacking MHC class I exhibit greater synapse reduction following axotomy, and this aberrant over-pruning affects inhibitory synapses more severely (Oliveira et al., 2004; Thams et al., 2008). Thus endogenous MHC class I acts at these synapses to reduce degeneration or enhance regeneration of inhibitory terminals. Subsequent studies have shown that more subtle natural variation in MHC class I levels in different mouse strains correlates with regenerative capability and suppression of synaptic stripping after nerve lesion, with higher levels associated with higher numbers of remaining connections (Sabha et al., 2008). These results are in contrast to those seen in the developing retino-geniculate, where endogenous MHC class I increases pruning of inappropriate connections. It is likely that this apparent contradiction simplys reflect mechanistic differences between developmental processes and responses to injury, as well as the diverse roles MHC class I may play, depending on cell type, specific MHC class I genes expressed, and developmental age. Together, these observations indicate that endogenous MHC class I participates in the remodeling of diverse synaptic connections, at a variety of ages, under both normal and pathologic conditions, in the central and peripheral nervous systems. In addition, they suggest that MHC class I can differentially modify synapse stability in developing central synapses versus injured peripheral synapses.

MHC CLASS I IN ACTIVITY-DEPENDENT SYNAPTIC PLASTICITY

How is activity-dependent remodeling accomplished, on a cellular level? The prevailing hypothesis is that changes in the strength of individual synapses, such as long-term potentiation (LTP) and long-term depression (LTD), lead to the stabilization or withdrawal, respectively, of the affected connections. If this model is correct, it is possible that in MHC-deficient mice, the lack of normal activity-dependent remodeling might arise from defects

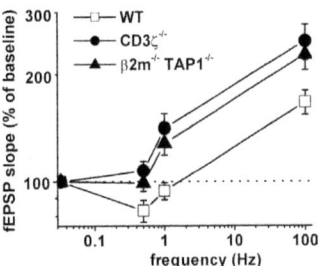

Figure 11.2. MHC class I signaling is required for normal bi-directional synaptic plasticity in the adult hippocampus. In wild-type adult mice, low-frequency stimulation leads to longlasting synaptic depression, whereas higher frequencies result in potentiation. In mice deficient for either MHC class I ($\beta 2m^{-/-} TAP1^{-/-}$) or CD3$\zeta$, the frequency–response curve is systematically shifted in favor of potentiation. With permission from Huh et al. (2000).

in activity-dependent modification of synaptic strength. To test this directly, Huh et al. (2000) examined synaptic plasticity at CA3–CA1 synapses in the adult hippocampus, where the cellular mechanisms of synaptic plasticity are relatively well understood, and where MHC class I is strongly expressed (Corriveau, Huh & Shatz, 1998; Huh et al., 2000). These authors found long-term potentiation (LTP) to be significantly enhanced in MHC class I-deficient mice, and synaptic plasticity is systematically shifted in favor of potentiation across a wide range of stimulation frequencies (Figure 11.2).

The enhancement of potentiation may be due to an amplification of existing plasticity mechanisms, rather than the unmasking of mechanistically distinct LTP since LTP is fully blocked by N-methyl-D-aspartate receptor (NMDAR) antagonists in both the wild-type and MHC-deficient mice. As in the developing retinogeniculate projection, CD3ζ-deficient mice phenocopy the synaptic plasticity shift of MHC class I-deficient mice (Figure 11.2), suggesting a direct or indirect role for a CD3ζ-containing receptor in activity-dependent synaptic plasticity in the adult hippocampus (Huh et al., 2000).

These changes in potentiation and depression cannot be attributed to altered inhibition in the absence of MHC class I, since they persisted in the presence of GABA$_A$-blockers. Importantly, these changes in synaptic plasticity are also not likely an indirect effect of immune compromise, since more severely immunocompromised RAG1-deficient mice have normal LTP at these synapses (Huh et al., 2000). Whilst many knockout mice have been found to exhibit a loss of LTP, which may arise from nonspecific impairment of basic cellular functions, only a handful of other mutant mice exhibit enhanced LTP and loss of LTD at these synapses. These include mice mutant for PSD-95 (Migaud et al., 1998) and cadherin-11 (Manabe et al., 2000). Thus MHC class I is one of a small group of synaptically localized proteins required for normal bi-directional synaptic plasticity in the adult hippocampus (Huh et al., 2000; Boulanger, Huh & Shatz, 2001; Boulanger & Shatz, 2004). In future studies it will be important to determine the cellular and molecular mechanisms whereby MHC class I affects hippocampal plasticity. It will also be of interest to characterize the rules of normal synaptic plasticity at the developing retinogeniculate synapse, and to subsequently examine if MHC-deficient mice have functional alterations in plasticity at these synapses which precede the failure of retinogeniculate remodeling.

MECHANISMS OF ACTIVITY-DEPENDENCE

It is as yet unclear why MHC class I seems to be selectively required for activity-dependent but not activity-independent events in brain development, function, and response to injury. One possibility is that the relatively low basal levels of MHC class I in neurons, combined with the fact that MHC class I is upregulated by activity, effectively restrict its action to periods of high electrical activity. For example, the electrical activity that drives the final refinement of retinal projections upregulates the expression of MHC class I in the LGN, and MHC acts to promote removal of synaptic connections in the LGN selectively at this time (Corriveau, Huh & Shatz, 1998). In this model, MHC class I levels function as a molecular readout of activity levels. Interestingly, a component of many MHC class I receptors in the immune system, CD3ζ, is also expressed in neurons in the developing and adult brain, including the developing LGN and adult hippocampus (Corriveau, Huh & Shatz, 1998; Huh et al., 2000), but its levels are not regulated by spontaneous activity in the developing LGN (Corriveau, Huh & Shatz, 1998). Similarly, the light chain of many MHC class I, β2m, is not upregulated in response to increased activity (Corriveau, Huh & Shatz, 1998). Thus it is possible that MHC class I co-subunits and receptors are constitutively expressed, and that the abundance of MHC class I heavy chain abundance is limiting and confers activity-dependence to its signaling.

However, some of the abnormal responses to activity in MHC class I-deficient mice occur rapidly within minutes (e.g. hippocampal synaptic plasticity), before activity-driven transcription and translation of MHC class I protein are likely to be significant. One possibility, as yet untested, is that physiological activity in the adult hippocampus induces rapid relocalization of existing MHC class I protein. Another possibility is that the lack of MHC class I has affected earlier events in brain development and changed the starting state of the synapse. The question of whether MHC class I acts early in development, throughout life, or acutely, for instance only in the presence of plasticity-inducing patterns of activity, is a pressing topic for future study.

One implication of the above results is that MHC class I could potentially participate in a negative feedback loop involving the regulation of neuronal activity. For instance, elevated levels of activity can increase expression of MHC class I in the adult hippocampus (Corriveau, Huh & Shatz, 1998). In turn, endogenous MHC class I at these synapses limits the magnitude of LTP and permits LTD (Huh et al., 2000). Thus neurons with high activity levels will express more MHC class I, and their plasticity may shift in favor of synaptic weakening, thereby reducing their activity and maintaining them within a functional dynamic range. Conversely, less-active neurons may downregulate MHC class I, which in the hippocampus shifts the balance of plasticity in favor of potentiation. In this way, MHC class I may be important in the adaptive regulation of synaptic strength at these and other synapses. Such homeostatic adjustments may occur early in development, permanently changing the "set-point" of adult synapses, or may occur acutely, as the levels of activity at adult synapses change in response to experience. Recent studies in hippocampal neurons in vitro indicate that MHC class I is necessary for homeostatic upregulation of postsynaptic PSD-95 expression in response to chronic activity blockade (Goddard et al., 2007).

MHC CLASS I SIGNALING IN NEURONS

MHC class I is expressed in the somatodendritic compartment in the developing retinogeniculate and geniculocortical projections (Corriveau, Huh & Shatz, 1998; Huh et al., 2000), and in cultured hippocampal neurons, where it is detected in dendritic spines apposed to sites of pre-synaptic markers (Goddard et al., 2007). However, mice deficient in MHC class I show pre-synaptic abnormalities at these same synapses, including aberrant remodeling of developing retinal ganglion cell arbors (Huh et al., 2000), and a modest increase in pre-synaptic ultrastructure in the adult hippocampus (Goddard et al., 2007). These data together suggest that post-synaptically expressed MHC class I can induce show retrograde signaling to affect pre-synaptic structures (Figure 11.3).

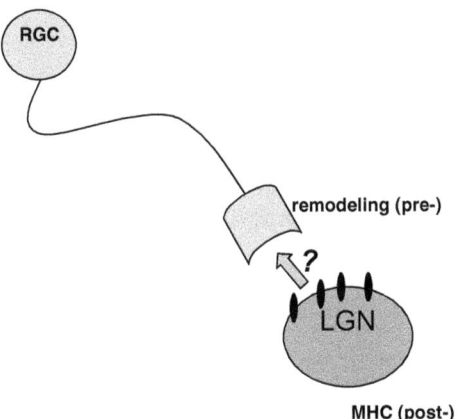

Figure 11.3. MHC class I signaling may involve retrograde signaling. In the developing visual system, and in other regions of the developing and adult brain, MHC class I is expressed in the somatodendritic compartment. In the absence of normal MHC class I signaling, remodeling of retinal ganglion cell (RGC) axons, which are pre-synaptic to MHC-expressing LGN neurons, is abnormal. This suggests that MHC class I may signal through an unknown retrograde messenger at these and other synapses. With permission from Boulanger and Shatz (2004).

How could such trans-synaptic signaling be mediated? One possibility is that MHC class I engages signaling through traditional neuronal retrograde messengers, such as nitric oxide (NO) or endocannabinoids. Alternatively, TCRs or other classical immunoreceptors for MHC class I may be expressed by neurons and located on the pre-synaptic cell (Syken & Shatz, 2003a, 2003b; Boulanger, Huh & Shatz, 2001; Syken et al., 2006). Indeed, several recent studies have found evidence for mRNA and/or protein encoding components of numerous known immunoreceptors in neurons (see above). Of note, although retinogeniculate remodeling is impaired in MHC class I-deficient mice, it is normal in mice lacking either TCRbeta (Syken and Shatz, 2003) or PIRB (Syken et al., 2006), suggesting that retinogeniculate remodeling involves as-yet unidentified MHC class I signaling pathways.

Downstream signaling molecules for known MHC class I immunoreceptors are also found in neurons, some of which (including Grb-2 and fyn) are localized to the pre-synaptic terminal. Interestingly, many of these downstream immune signaling molecules already have known roles in activity-dependent plasticity in the CNS (reviewed in Boulanger, Huh & Shatz, 2001).

There is also evidence for postsynaptic functions of postsynaptically expressed MHC class I, suggesting MHC class I may interact with neuronal proteins in cis. In hippocampal neurons in vitro, chronic activity blockade with TTX causes upregulation of the postsynaptic scaffolding protein PSD-95. In MHC class I-deficient neurons, however, PSD-95 immunoreactive puncta do not increase in size following TTX treatment (Goddard et al., 2007). Although synapsin puncta also fail to scale up in response to TTX in MHC class I deficient neurons, this is likely due to the fact that presynaptic parameters are already scaled up in the basal state (Goddard et al., 2007). Thus in hippocampal neurons in vitro, MHC class I can have both pre-and post-synaptic effects; the known presynaptic effects appear to be developmental in origin, while the known postsynaptic effects arise upon induction of homeostatic plasticity. Together, the evidence to date supports both presynaptic (in trans) and postsynaptic (in cis) binding partners for MHC class I in neurons, and suggests these binding partners may mediate distinct synaptic functions. This is perhaps not surprising, since numerous cis and trans binding partners for MHC class I have been identified in non-neuronal cell types, each with specialized functions.

CONCLUSIONS

MHC class I is expressed by normal neurons throughout life. During development, expression of MHC class I is regulated by the naturally occurring electrical activity that sculpts projections. In the adult, MHC class I expression is primarily dendritic and remains sensitive to both natural and pathological changes in activity. Independent studies have determined that MHC class I is required for normal developmental synapse remodeling, activity-dependent synaptic plasticity, and responses to injury at various sites in the central and peripheral nervous systems. In particular, MHC class I appears to be required for activity-dependent weakening and withdrawal of synaptic connections in the uninjured nervous system, although it may have neuroprotective effects in the face of pathological remodeling following nerve injury. Challenges for the future include identifying and characterizing the multiple MHC class I binding proteins in neurons, and determining if different MHC class I proteins are functionally specialized in the brain. These studies will permit a molecular-level understanding of how MHC class I, which is best known for its role in adaptive immunity, translates activity into changes in synaptic strength and neuronal connectivity in vivo. Thus MHC class I may be critically involved in translating activity into changes in synaptic strength and neuronal connectivity *in vivo*.

ACKNOWLEDGMENTS

The author is grateful to Carla Shatz, in whose laboratory many of these experiments were performed. The author's work is supported by NIH grant 1F32EY07016 and a Junior Fellowship from the Harvard Society of Fellows to LMB.

12 · Cytokine induction of neuronal receptor trafficking: relevance to synaptic function and excitotoxicity

Dmitri Leonoudakis, Steven P. Braithwaite, Michael S. Beattie, and Eric C. Beattie

INTRODUCTION

Injury or disease in the central nervous system (CNS) increases the amount of tumor necrosis factor alpha (TNFα) to which neurons may be exposed. This cytokine is central to the inflammatory response that occurs after injury and during prolonged CNS disease, contributing to the process of neuronal cell death. Previous studies have addressed how long-term apoptotic signaling pathways initiated by TNFα might influence these processes, but the effects of inflammation on neurons or synaptic function in the short timescale of minutes after exposure are largely unexplored. Our published studies examining the effect of TNFα on trafficking of α-amino-3-hydroxy-5-methyl-4-isoxazolepropionate- (AMPA-) type glutamate receptors (AMPARs) in hippocampal neurons demonstrated that glial-derived TNFα causes a rapid (fewer than 15 minutes) increase in the number of neuronal, surface-localized, synaptic AMPARs leading to an increase in synaptic strength. This suggests TNFα signal transduction acts to facilitate increased surface localization of AMPARs from internal post-synaptic stores. Importantly, an excess of surface localized AMPARs may predispose the neuron to glutamatergic excitotoxicity and excessive levels of intracellular calcium levels leading to cell death. This may suggest a new mechanism for excitotoxic TNFα-induced neuron death initiated minutes after neurons are exposed to the products of the inflammatory response.

In this chapter we review the importance of AMPAR trafficking in normal neuronal function and how abnormalities, mediated by glial-derived cytokines such as TNFα, may be central in causing neuronal disorders. We have further investigated the effects of TNFα on different neuronal cell types and present new data here using cortical and hippocampal neuron cultures. Lastly, we have expanded our investigation of the temporal profile of the action of this cytokine relevant to neuronal damage. We conclude that TNFα-mediated effects on AMPAR trafficking are common in diverse neuronal cell types and are very rapid in their onset. The abnormal AMPAR trafficking elicited by TNFα may present a novel target that may aid in the development of new neuroprotective drugs.

THE BASIC MECHANISMS OF EXCITATORY POST-SYNAPTIC FUNCTION

Glutamate is the major excitatory neurotransmitter in the mammalian CNS and exerts its effects via a number of both pre- and post-synaptic receptors with differing physiologic properties. The ionotropic receptors (iGluRs) are ligand-gated ion channels which, upon binding of glutamate, allow entry of cations into neurons and are thus the major mediators of synaptic transmission. These receptors have been further subdivided based upon their biophysical and pharmacological properties into the N-methyl-D-aspartate (NMDA), AMPA, and kainate receptors (Hollmann & Heinemann, 1994; Chittajallu et al., 1999). Each of these classes of receptor has been demonstrated to have differential localizations and contributions to the mechanisms of normal synaptic function. Proper control of synaptic strength is essential for the normal functioning of the CNS, including the formation and maintenance of memory. Thus, the mechanisms underlying both basic synaptic transmission and plasticity have been of keen interest to the neuroscience field. Synaptic transmission is an inherently regulated process determined by changes in release of glutamate from pre-synaptic terminals and the localization and activity of receptors on the post-synaptic membrane.

Recently, post-synaptic iGluRs have become a particular focus of numerous laboratories as it has become apparent that receptor expression at the post-synaptic membrane is highly dynamic (Figure 12.1); for reviews see Carroll *et al.*, 2001; Malinow & Malenka, 2002; Song & Huganir, 2002; Bredt & Nicoll, 2003).

Dominating this field are the AMPARs, which mediate the vast majority of fast excitatory synaptic transmission and are exclusively expressed post-synaptically. Recent evidence has also demonstrated that NMDA receptors (NMDARs) (reviewed in Carroll & Zukin, 2002) and kainate receptors (Hirbec *et al.*, 2003) are also dynamically trafficked to affect synaptic function. Changes in post-synaptic responsiveness to pre-synaptically released glutamate may be mediated by alterations in the conductance of pre-existing surface-expressed receptors or by a change in their number at the post-synaptic membrane. Although changes in channel conductance as a result of direct iGluR subunit phosphorylation are important (Benke *et al.*, 1998), a wealth of data has led to the additional findings that the surface expression of AMPARs is highly regulatable, and the amount of this surface expression is directly relevant to synaptic efficacy (reviewed in Malinow & Malenka, 2002; Collingridge, Isaac & Wang, 2004). The rapid mobility of AMPARs occurs also in a constitutive fashion, with a continual turnover of AMPARs at the synaptic membrane mediated by exocytosis and endocytosis (Lin *et al.*, 2000). However, there is also a significant regulation of AMPAR trafficking mediated by synaptic activity and subsequent receptor activation (Carroll *et al.*, 1999; Lissin *et al.*, 1999; Beattie *et al.*,

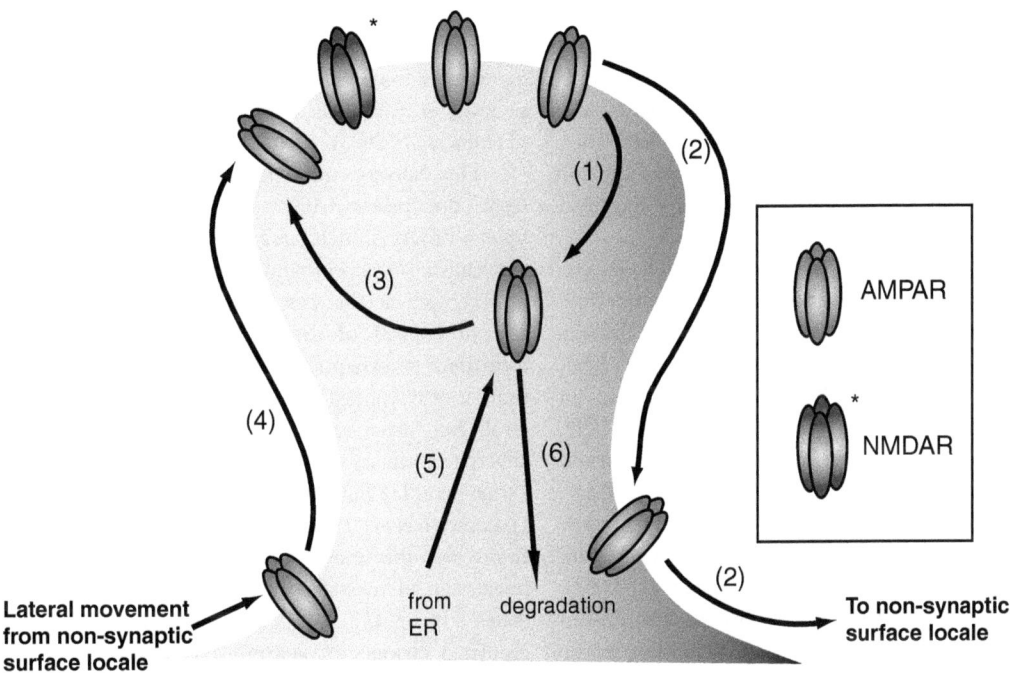

Figure 12.1. Paths of glutamate receptor trafficking within the dendritic spine during normal and abnormal synaptic function. α-amino-3-hydroxy-5-methyl-4-isoxazolepropionate- (AMPA-) type glutamate receptors (AMPARs) may be rapidly removed from the synapse on dendritic spines by endocytosis (1) or lateral movement to extra-synaptic surface locales (2). Endocytosed AMPARs may be trafficked to lysosomes for degradation (6) or recycled back to the synapse (3). AMPARs may also be transported into the synapse after synthesis in somatic or local dendritic endoplasmic reticulum (5) and then exocytosed to the synapse (3). Lateral movement of extra-synaptic receptors to the synapse (4) may also replenish receptors at this location. (*) recent work has shown that NMDARs also are trafficked into and out of the synapse though possibly at a different timecourse than AMPARs (reviewed in Carroll & Zukin, 2002). The increase in surface receptors attributed to increased delivery via pathways (3) and (4) or due to a reduction in removal via pathways (1, 6) and (2) may increase chances for excitotoxicity during acute trauma and neuronal overactivation during epilepsy. Reprinted with permission from Leonoudakis *et al.* (2004).

2000; Ehlers, 2000; Lin et al., 2000). These activity-dependent changes in AMPAR trafficking have been linked to the modulation of synaptic strength, as occurs during certain forms of long-term potentiation (LTP) (Malenka & Nicoll, 1999) and long-term depression (LTD) (Carroll et al., 2001). The mechanisms underlying this regulation, both in the delivery of receptors to the post-synaptic membrane and in their removal, have been fruitful areas of research.

The dynamics of AMPAR expression at the synaptic membrane are highly complex. In normal non-injury conditions, only AMPARs present at the post-synaptic surface which are immediately apposed to the pre-synaptic bouton contribute to the synaptic response. The complement of receptors at the active post-synaptic site may be modulated by a number of mechanisms (see earlier Figure 12.1): (1) internalization to intracellular stores; (2) lateral diffusion to extra-synaptic sites on the plasma membrane where they are not activated by glutamate; (3) delivery of new AMPARs from intracellular stores; (4) lateral diffusion from extra-synaptic sites on the neuronal surface to the post-synaptic active site; (5) new synthesis and delivery of receptors from the endoplasmic reticulum; and (6) degradation of receptors via lysosomal pathways.

Increases in AMPAR surface expression have been correlated with the development of LTP through both the unsilencing of "silent synapses" (synapses lacking AMPARs) (Isaac, Nicoll & Malenka, 1995; Liao, Hessler & Malinow, 1995) and the increase in AMPARs at pre-existing synapses (Shi et al., 1999; Hayashi et al., 2000). Phosphorylation is an important process involved in this regulation of AMPAR-mediated delivery to the post-synaptic membrane through kinases, including CaMKII and PKA (Hayashi et al., 2000; Lee et al., 2000; Esteban et al., 2003). Interaction with intracellular proteins has also proved important. The disruption of AMPAR binding to the PDZ domain containing proteins or overexpression of such interacting proteins perturbs receptor surface expression (Hayashi et al., 2000; Rumbaugh et al., 2003). Another mechanism for the delivery of AMPARs to synaptic membranes is the requirement for transmembrane proteins, such as stargazin or related transmembrane AMPAR regulatory proteins (TARPs) (Chen et al., 2000; Tomita et al., 2003).

Endocytosis of receptors from the synaptic membrane is a dynamin-dependent process (Carroll et al., 1999), its regulation being dependent on phosphorylation and interaction with intracellular proteins. An increase in post-synaptic calcium levels has been linked with the physiologic process of LTD (Mulkey & Malenka, 1992) and the regulated internalization of AMPARs (Beattie et al., 2000). The calcium-sensitive phosphatase calcineurin has also been linked to both of these processes (Mulkey & Malenka, 1992; Beattie et al., 2000) indicating the role of AMPAR phosporylation in the processes of receptor internalization and LTD. A strong body of literature has also demonstrated that AMPAR interacting proteins including N-ethylmaleimide-sensitive fusion protein (NSF), glutamate receptor interacting protein (GRIP), and protein interacting with C-kinase (PICK) are linked with the regulated internalization of AMPARs (reviewed in Braithwaite, Meyer & Henley, 2000; Passafaro, Piech & Sheng, 2001; Malinow & Malenka, 2002)). TARPs may also regulate the internalization of AMPARs as their dissociation from each other appears to be necessary for removal of AMPARs from the synaptic membrane (Tomita et al., 2003).

The delivery and removal of receptors to and from the post-synaptic membrane is dependent upon a variety of highly regulated mechanisms. These mechanisms have profound effects on the complement of receptors at the post-synaptic membrane giving rise to control of synaptic transmission and the formation of synaptic plasticity. Transcription and translational control over AMPAR subunit protein levels have been shown to change the amount and subtype specificity of AMPAR surface expression over a time period of hours (Aronica et al., 1997; Pellegrini-Giampietro et al., 1997; Tanaka et al., 2000). New data show that this may be accomplished through the translation of messages encoding AMPARs in dendrites (Ju et al., 2004). However, the rapidity of regulated changes in synaptic transmission and the requirement for a proper contingent of synaptic substructure proteins make it likely that trafficking mechanisms (along with degradation) are major contributors to the dynamic regulation of post-synaptic functioning (O'Brien et al., 1998; Ehlers, 2000; Man et al., 2000; Stein et al., 2003; Tao et al., 2003; Ehrlich & Malinow, 2004). Clearly, abnormalities in any of these trafficking systems will lead to dysfunction in synaptic transmission and the development of neuronal disorders.

ABNORMAL TRAFFICKING OF AMPARs AS A MECHANISM OF NEURONAL DYSFUNCTION

The precise control of synaptic efficacy is certainly important for the normal functioning of the CNS, and thus abnormalities in synaptic function may lead to neurological disorders or toxicity. Much research has indicated that overactivation of iGluRs can lead to neurotoxicity, primarily through excessive calcium influx which leads to the induction of apoptotic processes (Choi, 1992; Lee, Zipfel & Choi, 1999; Mattson et al., 2000; Weiss & Sensi, 2000; Mattson & Chan, 2003). This iGluR-mediated neurotoxicity has been implicated in the neuronal loss associated with acute traumas to the CNS and disorders such as Alzheimer's disease and Parkinsonism. In addition, hyperactivity of iGluRs, without leading to neurotoxicity, may lead to seizures such as occur in epilepsy. On the other hand, suboptimal iGluR activity may also lead to neurological deficits, as in schizophrenia (Tsai & Coyle, 2002), and may also underlie the cognitive deficits in disorders such as Alzheimer's disease (Kamenetz et al., 2003). Addictive behavior has been demonstrated to involve inappropriate plasticity of glutamatergic transmission (Tzschentke & Schmidt, 2003), again indicating that imbalances in iGluRs at synapses may have serious neurobiological consequences. As the trafficking of AMPARs to the post-synaptic membrane is a major mechanism for normal neuronal function it is clear that abnormal trafficking might lead to these conditions of neuronal dysfunction. AMPARs are a major regulatory contributor to fast excitatory synaptic transmission. Although the majority of AMPARs are not permeable to Ca^{2+}, the major intracellular mediator of cell death, there is a subset of receptors at synaptic sites which contain no GluR2 subunits (GluR2-lacking AMPARs) making them Ca^{2+} permeable (Ogoshi & Weiss, 2003). In addition, activation of AMPARs is required to relieve the voltage-dependent block of NMDARs, these being the major iGluRs which permit Ca^{2+} entry into neurons (Herron et al., 1986). Therefore abnormal trafficking of AMPARs may both directly and indirectly lead to excitotoxicity.

THE TRIGGER OF ABNORMAL AMPAR TRAFFICKING IN NEURONAL DYSFUNCTION

The trafficking of AMPA receptors in the normal state of neurons is a constitutive process regulated by activity. In neuronal disorders there must be an additional trigger that disrupts the trafficking process. Research has indicated that a major contributory factor to AMPAR trafficking are cytokines released from glial cells, including the inflammatory cytokine tumor necrosis factor (TNFα) (Beattie et al., 2002a). TNFα is intimately involved in inflammation, immune activation, differentiation, and cell death, but its complex actions are increasingly implicated in diseases of the CNS (Perry et al., 1995; Allan & Rothwell, 2001). TNFα and other cytokines are rapidly induced in response to tissue injury or infection and in the CNS (Perry et al., 1995; Allan & Rothwell, 2001). TNFα administered to cultured primary neurons causes a dose-dependent cytotoxicity (Zhao et al., 2001) as well as neuronal apoptosis (Reimann-Philipp et al., 2001). However, some in vitro studies have reported neuroprotective effects of TNFα (Cheng, Christakos & Mattson, 1994; Bruce et al., 1996). Whether TNFα is damaging or protective appears to depend on many factors. TNFα present during acute injury appears to be damaging, but its long-term presence may be protective (Wilde et al., 2000). The effect of TNFα on distinct cell types is greatly influenced by the presence of specific receptors on target cells. TNFα binds specifically to two distinct and co-expressed receptors, TNFR1 and TNFR2, which are both found on neurons (Vitkovic, Bockaert & Jacque, 2000). The activation of TNFR1 appears to be damaging to neurons, whereas activation of TNFR2 is protective (Peschon et al., 1998; Fontaine et al., 2002; Yang et al., 2002). Furthermore, the presence or absence of compounds that modify TNFα action also greatly influences possible neuroprotective or neurotoxic effects (Schubert et al., 1997; Carlson et al., 1998). Therefore there is a clear linkage between the actions of TNFα released by glia and neuronal AMPAR trafficking events. TNFα is involved in regulation of neuronal function and AMPAR localization, but the complex actions and temporal profile of a neuron's response to this cytokine determines its ultimate effects on neuronal pathology.

INTERPLAY BETWEEN TNFα AND AMPAR TRAFFICKING AFTER ACUTE NEURAL TRAUMA

A number of studies have demonstrated interplay between TNFα and AMPAR activity in neuronal disorders. Best characterized are the conditions of acute neural trauma or ischemic insults where there are both major inflammatory responses and neuronal damage. Acute neural trauma or ischemic insults release glutamate from intracellular vesicles, resulting in high extracellular concentrations that are thought to induce both necrotic and apoptotic neuronal cell death by activating iGluRs and increasing intracellular calcium levels (Choi, 1994). Neurotrauma also produces a complex ongoing cascade of inflammatory reactions that includes the liberation of cytokines and chemokines, and the recruitment of immune cells from the periphery as well as activation of resident microglia (Bethea et al., 1999; Popovich et al., 2002). Considerable evidence suggests that this inflammatory response may contribute to secondary injury in trauma or stroke (Allan & Rothwell, 2001), although there is also evidence for a neuroprotective role for some aspects of the inflammatory cascade (Bruce et al., 1996; Kim et al., 2001; Shaked et al., 2004).

It is likely that inflammation and excitotoxicity together contribute to neuronal damage beyond the initial insult of acute trauma. For example, microglial activation can release glutamate (Barger & Basile, 2001), glutamate receptors on microglia and neurons can stimulate the release of cytokines, including TNF-α (Matute et al., 2001), and TNFα can rapidly and dramatically increase the surface localization and trafficking of AMPA-type iGluRs (Beattie et al., 2002a). Indeed, it has been shown that the co-injection of TNFα with AMPAR agonists enhances necrotic cell death in the spinal cord, and this enhancement was blocked by the specific AMPAR antagonist 6-cyano-7-nitroquinoxaline-2,3-dione (CNQX) (Hermann et al., 2001). Low extracellular potassium levels induce increased surface expression of AMPARs as well as increases in AMPA-mediated currents in cerebellar granule cells (Ha et al., 2002). These cells also show increased susceptibility to excitotoxicty, indicating the importance of AMPAR trafficking in a different class of neurons. Together, these observations suggest that rapid changes in AMPAR plasma membrane localization and/or sensitivity may be mediated by cytokines in the inflammatory cascade after injury, and that this interaction may contribute to the cascade of neuronal cell death after injury or ischemia.

AMPAR antagonists have been shown to be neuroprotective in models of spinal cord injury; 2,3-dihydroxy-6-nitro-7-sulphamoylbenzo(f)-quinoxaline; 6-nitro-7-suplphamoylbenzo(f)-quinoxaline-2,3-dione (NBQX) was effective in limiting both gray and white matter damage after spinal cord contusions in rats (Wrathall, Teng & Marriott, 1997). Contusion injury to the cord rapidly increases levels of TNF-α (Bethea et al., 1999; Pan et al., 1999) and other inflammatory cytokines (McTigue et al., 1998; McTigue et al., 2000) as well as increasing extracellular glutamate. Astrocytic damage may result in deficits in glutamate uptake and clearance away from the extracellular space. Thus, initial damage may initiate secondary injury resulting from the interplay between all of the major cellular players in the gray matter: neurons, microglia, and astrocytes, leading to a cascade of increasing extracellular glutamate. This increase in glutamate release alongside a rapid increase in surface AMPARs could dramatically raise the potential for neurotoxicity. The spread of this cascade to the white matter also puts myelinating oligodendrocytes and axon tracts at risk.

Oligodendrocytes also have AMPARs (McDonald et al., 1998; Park, Lui & Fehlings, 2003) and are susceptible to cell death after spinal cord injury, ischemia, or other insults. It has been proposed that TNFα and glutamate cooperate to kill oligodendrocytes by apoptosis in models of multiple sclerosis (MS) (see Matute et al., 2001), although direct effects on AMPAR trafficking, have not been tested. It has been shown that all of the features of oligodendrocyte apoptotis observed *in vivo* (in spinal cord injury, e.g. Crowe et al., 1997; Springer, Azbill & Knapp, 1999; Beattie et al., 2002b) and in models of MS may be shown to be mediated *in vitro* via AMPA and kainate receptors (Sanchez-Gomez et al., 2003). Takahashi, Svoboda and Malinow (2003) support the idea that cytokines (IL-1B and TNFα) may potentiate AMPAR-mediated oligodendrocyte cell death *in vitro*. Although a cascade of inflammation and excessive surface AMPAR localization is attractive as a simple hypothetical pathway to neuron death, the situation is likely to be more complicated. First, TNFα and TNFα receptor knock-out mice show both increased and decreased damage after CNS injury, depending upon the circumstances (Bruce et al., 1996; Bethea et al., 1999; Kim et al., 2001). Second, whereas injections of glutamate or TNFα may cause inflammatory-like changes in the spinal cord (Hermann et al., 2001), others have reported that TNFα and its co-inflammatory cytokine, IL1β, may have opposite effects on AMPAR-mediated cell death after injections into the brain (Allan, 2002).

It seems likely that there are at least two somewhat distinct stages of cell death after spinal cord injury and other CNS injuries: an acute, necrotic cell death, prominent in neurons, which is exacerbated by the effects of cytokines on AMPAR trafficking, and a longer-term, apoptotic wave of cell death, especially prevalent in oligodendrocytes that are away from the primary lesion (Beattie et al., 2002b). These changes are more likely to be mediated by long-term transcriptional or translational events, but initial changes in AMPAR surface localization soon after injury may influence these long-term events even if this AMPAR trafficking alteration is only transient. In addition, De Krueger and Simasko (2003) showed that TNFα increased AMPARs and susceptibility to neuronal cell death, dependent upon protein synthesis. In contrast, decreased expression of iGluR subunits in both gray and white matter has been observed in more chronic (e.g. one week after trauma) injuries (Grossman et al., 1999; Park, Lui & Fehlings, 2003). This could, of course, reflect the death of cells that expressed AMPARs early after injury, and which may have been affected by alterations in AMPAR trafficking.

GLIAL-DERIVED TNFα RAPIDLY INCREASES AMPAR SURFACE LOCALIZATION IN HIPPOCAMPAL NEURONS

The established intimate physical apposition of glia and neurons has profound implications for the mechanisms underlying both synaptic plasticity and the neural damage caused by a variety of brain insults. For example, astrocytes often respond to activity with increases in intracellular calcium levels, and may release a variety of chemical substances which may transiently modulate neural processes (Bezzi et al., 2001). Recently, it has been shown in cultured retinal ganglion neuron cultures that astrocytes are required for normal synaptogenesis and synaptic stability, and that the required unidentified factor is a diffusible extra-cellular signal (Pfrieger & Barres, 1997; Ullian et al., 2001). These experiments looked at the effects of astrocytes on synapse number and function over the course of days and did not address whether glia are required for the more rapid modulation of synaptic strength.

The possibility that TNFα might influence the surface expression of AMPARs, and thereby synaptic strength, was suggested by the observations that TNFα enhanced the responses of brainstem neurons to excitatory afferent inputs (Nägler, Mauch & Pfrieger, 2001) and greatly potentiated the cell death induced by injection of the AMPAR agonists into the spinal cord, an effect that was blocked by an AMPAR antagonist (Mauch et al., 2001). Since neurons are exposed to a large increase in TNFα after injury (Perry et al., 1995; Allan & Rothwell, 2001), we sought to determine how this might change synaptic function during and immediately after TNFα exposure. To address this, our group utilized the model system of primary dissociated hippocampal neuron cultures. We demonstrated that bath application of exogenous TNFα causes a rapid (fewer than 15 minutes) and significant increase in the number surface-localized AMPARs containing the GluR1 subunit (Beattie et al., 2002a). The increase in surface expressed GluR1 was visualized and measured by immunofluorescence microscopy and was confirmed using electrophysiologic recordings. There was no change in the amount of surface localization of the NMDA receptor subunit NR1 at this timepoint (Beattie et al., 2002a). It was also determined by co-localization of GluR1 with the synaptic marker synaptophysin that the increase in the delivery of surface AMPARs was both to synaptic and extra-synaptic sites. The rapid timeframe of AMPAR delivery suggests a trafficking effect and not a change in protein translation. However, new data showing the close apposition of the endoplasmic reticulum and mRNA with the base of dendritic spines suggest that protein production for the synapse may be faster than previously believed (Ju et al., 2004; Zukin et al., 2004; reviewed in Glanzer & Eberwine, 2003). In support of a trafficking mechanism for this action it was recently demonstrated that there is a pool of AMPARs within recycling endosomes that supply the synapse with receptors in as few as 20 minutes after

NMDAR-mediated stimuli that induce LTP (Lu et al., 2001; Park et al., 2004). These data also suggest that these newly delivered AMPARs likely arise from pre-existing intracellular pools (Eshhar et al., 1993).

Since TNFα may be produced by both neurons and glia (Allan & Rothwell, 2001), especially after injury to the CNS, we asked which of these cells might be potential sources of TNFα in our hippocampal cultures. We first looked microscopically to examine the staining pattern of TNFα in our hippocampal neurons and astrocytes. Both showed staining but astrocytes displayed dense perinuclear staining that suggested TNFα in biosynthetic pathways of these cells. Neurons displayed diffuse staining (Figure 12.2).

To test the hypothesis that TNFα can be supplied by astrocytes in our cultures more definitively, we prepared astrocyte-conditioned media from neuron-depleted cultures and applied it to our mixed cultures. An increase in both surfaced-localized GluR1 subunit-containing AMPARs and synaptic strength was observed on neurons after exposure of our mixed cultures to the conditioned media for 15 minutes (Figure 12.3). When soluble TNFα was specifically removed from this astrocyte conditioned media using either an anti-TNFα antibody or a soluble form of the TNFR1 receptor, this increase in surface AMPARs and synaptic strength was blocked. Furthermore, treatments of astrocyte cultures with GM 6001, a metalloproteinase inhibitor that inhibits release of soluble TNFα from the cell surface, completely prevented the conditioned media from having any effect on hippocampal neurons. In complementary hippocampal acute slice electrophyisiology experiments, we found that blocking TNFα signaling with soluble TNFR1 blocked the enhancing effect of TNFα on synaptic strength (Beattie et al., 2002a).

These data showing the rapid response of increased AMPAR surface localization through the bath application of exogenous and astrocyte-derived TNFα raise further questions. How long does this response last with continuous TNF application? Are both TNFα receptors, TNFR1 and TNFR2, responsible for the signaling that causes this activity? Do neuron classes other than those from the hippocampus respond similarly to TNFα? These questions underscore the importance of future, more detailed, studies of the effects of the actions of TNFα. Clearly, TNFα affects the trafficking of AMPARs and is a major contributory

(A)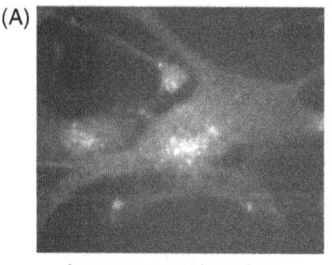

Astrocytes stained for TNFα

(B)

Neurons stained for TNFα

Figure 12.2. Cultured astrocytes exhibit perinuclear TNFα staining: (A) Astrocytes stained for TNFα. This staining was often punctate and perinuclear in appearance. Astrocyte cultures lacking neurons were grown as described in Beattie et al. (2002a) and stained as described below. Co-staining with a GFAP-specific antibody (R&D Systems; Cat. No. AF-510-NA; applied at 1 μg/mL) also confirmed the identity of these cells as astrocytes (data not shown); (B) Neurons also frequently exhibited TNFα immunoreactivity that was more evenly distributed and less punctate than observed in astrocytes. An absence of co-staining with a GFAP-specific antbody (R&D Systems; Cat. No. AF-510-NA; applied at 1 μg/mL) confirmed the identity of these cells as neurons (data not shown). Staining for TNFα was performed using a goat anti-TNFα antibody (R&D Systems; Cat No. AF-510-NA, applied at 1 μg/mL), which was applied to fixed and permeablized cells for one hour at room temperature. Cells were then washed in the blocking solution (PBS, 1% BSA, 0.8 mg/mL of saponin) three times. The signal was visualized by the application of a donkey antigoat antibody conjugated to Cy3 (Jackson Immunoresearch), at a dilution of 1:800 for 45 minutes at room temperature. The cells were then washed three times in PBS and mounted. Reprinted with permission from Leonoudakis et al. (2004).

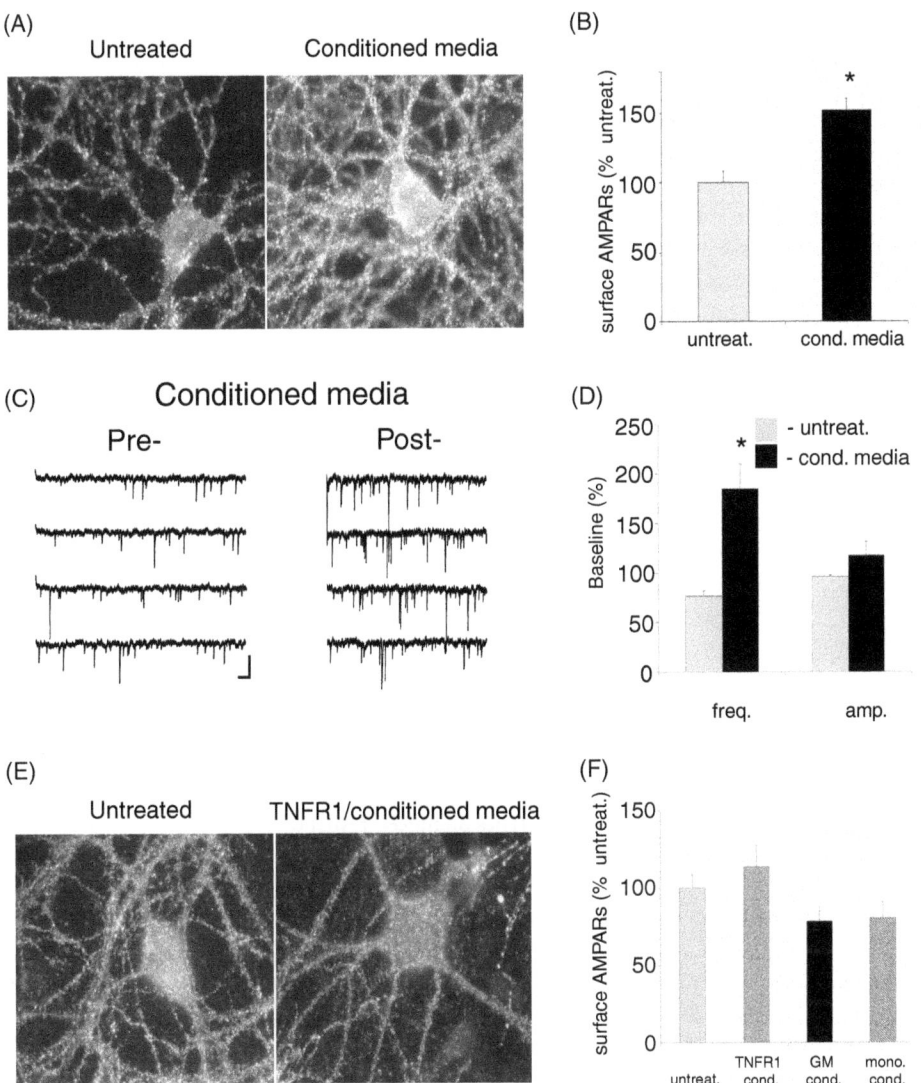

Figure 12.3. Astrocyte-conditioned media increases surface expression of α-amino-3-hydroxy-5-methyl-4-isoxazolepropionate- (AMPA-) type glutamate receptors (AMPARs) and synaptic strength via TNFα. (A) Examples of surface AMPAR staining in untreated and conditioned media-treated neurons. (B) Quantification of effects of conditioned media on surface AMPAR staining ($*p < 0.01$; untreated 100% ($\pm 9\%$), $n = 45$; conditioned media 152% ($\pm 9\%$), $n = 37$). (C) Examples of miniature excitatory post-synaptic potential currents (EPSCs) before and after application of conditioned media (calibration bars: 20 pA, 500 ms). (D) Mean percent change in miniature EPSC frequency and amplitude in cells treated with control or conditioned media ($n = 7, 8$) for untreated and conditioned media-treated cells; ($*p < 0.01$; percent initial mEPSC frequency: conditioned media 185% ($\pm 25\%$); normal media 76% ($\pm 5\%$); percent initial mEPSC amplitude: conditioned media 117% ($\pm 14\%$); normal media 96% ($\pm 2\%$)). (E) Examples of surface AMPAR staining in an untreated cell and a cell treated with conditioned media containing TNFR1. (F) Quantification of effects of conditioned media containing TNFR1, anti-TNFα antibody or the matrix metalloproteinase inhibitor GM 6001 ($n = 31$–45 for each group; untreated 100% ($\pm 9\%$)); TNFR1 and conditioned media 113% ($\pm 13\%$); anti-TNFα antibody and conditioned media 80% ($\pm 10\%$); GM 6001 and conditioned media 78% ($\pm 9\%$). Figure taken from Beattie et al., 2002a, with permission.

factor to damage in a number of neuronal disorders, especially where inflammatory responses occur. There is, however, much evidence that TNFα may have differing roles, either to exacerbate neuronal damage or to protect neurons, depending upon the situation. With this complex scenario of TNFα as potential friend or foe to the injured neuron, it is vital that the role of TNFα in AMPAR trafficking is studied in different neuron classes and the temporal aspects of its action are understood. We have therefore begun more detailed studies of the action of TNFα on AMPAR trafficking, and will compare hippocampal and cortical neurons with regard to the TNFα-induced AMPAR trafficking response.

TNFα EXPOSURE TO NEURONS CAUSES AN INCREASE IN SURFACE LOCALIZED GLUR1 AMPAR THAT CONTINUES THROUGH 30 MINUTES BUT PEAKS AT 15 MINUTES

Is the increase in surface localized AMPARs induced by TNFα treatment maintained throughout the timecourse of TNFα application? To address this question, we have treated cultured hippocampal neurons with or without TNFα for 15 minutes. To observe and quantify surface localized AMPARs, we immunofluorescently labeled non-permeabilized neurons with an antibody that recognizes an extra-cellular epitope of GluR1 and analyzed surface GluR1 fluorescence images captured by digital microscopy. Like our previous data which showed ∼200% increase in surface AMPARs (Beattie et al., 2002a), we observed a dramatic increase in surface-localized GluR1 of 195% (± 10%) in neurons treated with TNFα for 15 minutes longer than untreated control neurons (Figure 12.4 (A, B, and D)).

Treatment of neurons with TNFα for 30 minutes still showed an increase in surface-localized GluR1 (149% (± 10%)) over that of untreated control neurons (Figure 12.4 (A, C, and D)). However, these neurons showed a decrease in surface-localized GluR1 from the peak of response at 15 minutes with TNFα (Figure 12.4 (D)). These data indicate that after TNFα exposure, surface-localized GluR1 reaches a maximum around 15 minutes, and though surface GluR1 levels fall by the 30-minutes' exposure, AMPAR surface localization is still abnormally high. This implies that perhaps neurons have mechanisms in place to sense and compensate for excess surface AMPARs. The molecular pathways which underlie these mechanisms may be important in protecting against excitotoxicity.

TNFα CAUSES RAPID, INCREASED SURFACE EXPRESSION OF AMPARs IN CULTURED PRIMARY CORTICAL NEURONS

So far, our studies have been confined to the well-established culture system of primary hippocampal neurons. This begs the question: is the phenomenon of induction of surface-localized AMPARs by TNFα restricted to hippocampal neurons or a general phenomenon among other neuronal subtypes? In order to begin to address this question, we have cultured primary neurons derived from the cerebral cortex and exposed mature neurons to TNFα for 15 minutes. We observed and quantified surface GluR1 by immunofluorescent microscopy as described above. Our data showed an increase in surface-localized GluR1 of 147% (± 11%) in neurons treated with TNFα for 15 minutes longer than untreated control neurons (Figure 12.5). These data indicate that other neuron types do indeed respond to TNFα in a similar manner to hippocampal neurons. However, we did observe ∼50% lower level of surface-localized AMPARs in response to TNFα at 15 minutes compared to hippocampal cultures. This may reflect a difference in TNFα sensitivity between neuron types (perhaps due to different expression and type of TNFα receptors) or simply a lower available pool of GluR1. Since presently we have only tested the 15-minute TNFα treatment time point for cortical neurons we cannot comment on a possible peak of response in this class of neurons. Planned future timecourse experiments will address this.

These data demonstrate that two diverse classes of neurons are sensitive to exposure to TNFα and suggest that this cytokine may predispose other neuron types to increased sensitivity to excitotoxicity. In support of these data, neurons of the spinal cord are also sensitive to TNFα and show increased excitotoxicity when exposed to TNFα, and this toxicity may be blocked by use of the AMPAR antagonist CNQX (Hermann et al., 2001). Understanding the mechanism by which TNFα induces increased AMPAR surface

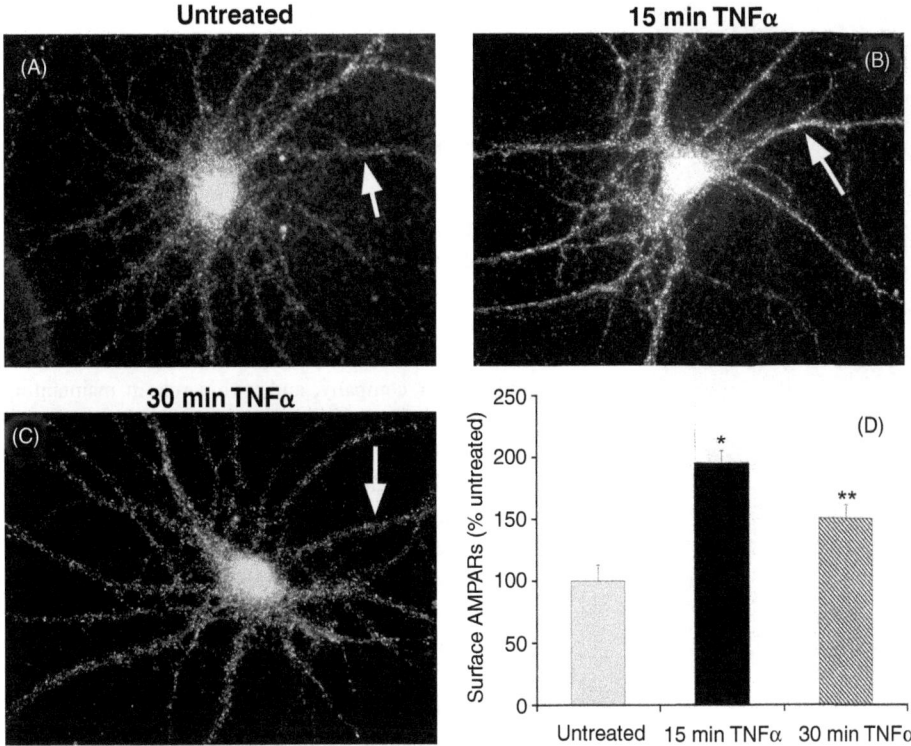

Figure 12.4. TNFα causes increased surface expression of α-amino-3-hydroxy-5-methyl-4-isoxazolepropionate- (AMPA-) type glutamate receptors (AMPARs) over 15 and 30 minutes in cultured primary hippocampal neurons. Examples of surface AMPAR staining in (A) untreated and (B) neurons treated with TNFα for 15 minutes or (C) 30 minutes. (D) Quantitation of effects of TNFα on surface AMPAR staining ($n = 50$–70 for each group pooled from at least three different experiments; $*p$, 0.08; $**p$, 0.4; untreated, 100% ($\pm 14\%$); 15 minutes of TNFα, 195% ($\pm 10\%$); 30 minutes of TNFα, 149% ($\pm 10\%$)). Arrows point at processes where very defined punctuate staining appears after TNF application (B, C). Experimental methods: after treatment of 17–25-day-old neuron cultures (see Beattie et al., 2002b for culture methods) with 6 nM TNFα at 37 °C for the indicated times, neurons were chilled on ice, washed with cold PBS, and surface AMPARs were visualized by indirect immunofluorescence using a rabbit antibody to the extracellular N-terminus of GluR1 (Ab-1, Oncogene Research Products) at a dilution of 1:20 for one hour at 4°C. Neurons were then washed with cold PBS and fixed with 4% paraformaldehyde/4% sucrose in PBS. The non-permeablized cells were then blocked with 3% BSA, 1% goat serum in PBS and a donkey antirabbit secondary Ab conjugated to Alexafluor 568 (Molecular Probes) was applied at a dilution of 1:800 for 45 minutes at room temperature, followed by thorough washing with PBS and mounting on slides with Fluoromount G (Electron Microscopy Services). Neurons were visualized and images were captured with immunofluorescence microscopy as described in Beattie et al. (2002b). For individual experiments, images for all conditions were analyzed using identical acquisition parameters, and untreated and treated cells from the same culture preparation were compared with one another. Images from each experiment were obtained using a threshold equal to the average background fluorescence in untreated control cells. The total area of fluorescently labeled, surface AMPARs was measured automatically by the Metamorph software and divided by the total cell area, which was determined by setting a lower threshold level to measure background fluorescence produced by the fixed cells. For each experiment, the fluorescence of all cells was normalized by dividing by the average fluorescence of the untreated control cells. Each experimental manipulation was repeated a minimum of three times using different culture preparations. N values represent the number of microscope fields examined. Statistical significance between individual experimental groups and the control group was determined by use of Student's t-test. Error bars in figures represent SEM. Reprinted with permission from Leonoudakis et al. (2004).

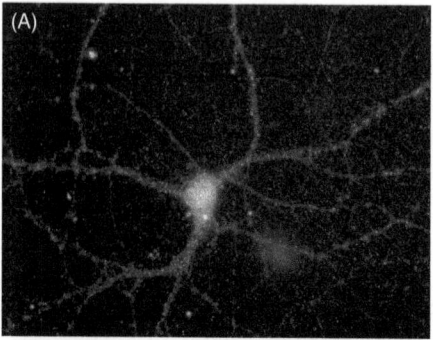
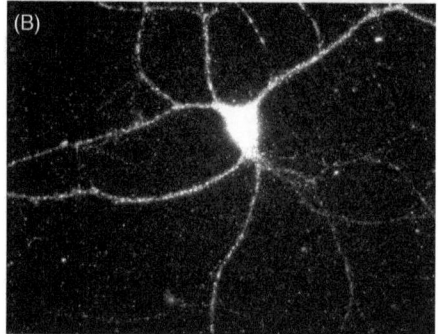
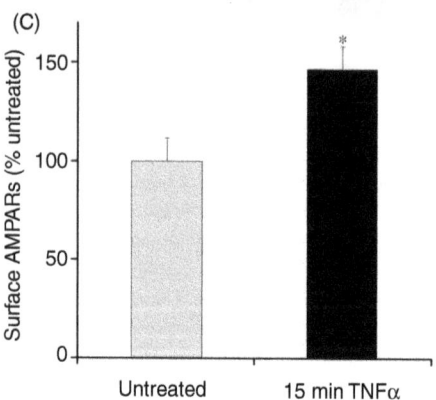

Figure 12.5. TNFα causes rapid, increased surface expression of α-amino-3-hydroxy-5-methyl-4-isoxazolepropionate- (AMPA-) type glutamate receptors (AMPARs) in cultured primary cortical neurons. Examples of surface AMPAR staining in (A) untreated and (B) cortical neurons treated with TNFα for 15 minutes. (C) Quantification of effects of TNFα on surface AMPAR staining ($n = 70$ for each group pooled from four different experiments; $*p$, 0.01; untreated, 100% ($\pm 11\%$); 15 minutes of TNFα, 147% ($\pm 11\%$)). Cortical neurons were prepared, treated, and visualized as described in Figure 12.4. Reprinted with permission from Leonoudakis et al. (2004).

localization will lead to a better understanding of this phenomenon and may lead to preventative therapies to treat the earliest stages of excitotoxicity.

FUTURE DIRECTIONS

Our research has brought us to the fascinating and complex intersection of the nervous and immune systems. In this rapidly blurring border zone, a cytokine whose home has historically been outside the CNS currently finds itself at home. It finds itself seated securely in the brain, with specific receptors on neurons and glia for company, and with a role in maintaining normal synaptic function. But this "new resident" of the brain likely becomes an unwelcome tenant in the CNS during acute injury when it becomes abnormally abundant and destroys the delicate chemical and electrical balance of neurons and glia. Past research has focussed on the long term apoptotic signaling characteristics of TNFα to attempt to understand the mechanisms of this carnage. We have begun to investigate the rapid effects of TNFα on the surface localization of an ionotropic glutamate receptor whose precise regulation is central to normal synaptic function as well as to neuron viability.

We have tested the response of AMPAR trafficking to TNF exposure. This does not in itself directly address changes in the neuron's excitotoxic vulnerability. However, as discussed above, a large body of literature links excessive iGluR activation and increased calcium-induced excitotoxicity. Indeed, co-stimulation of AMPARs and TNFα receptors in the spinal cord (which alone are non-toxic) induces dramatic excitotoxicity (Hermann et al., 2001). Thus, we predict future studies examining the excitotoxic sensitization of various neuron cultures with TNFα will show a correlation between abnormal AMPAR trafficking and neuron death. An important test to go beyond any such correlation toward a better understanding of a mechanism underlying excitotoxicity will be to test the effect of TNFα on neurons during a blockage of AMPAR cell-surface delivery. If this type of experiment reduces excitotoxic death *in vivo* or *in vitro* during extended exposure of TNFα, an exciting new drug target for neuroprotection will have been uncovered.

The surprising and exciting observation that TNFα signaling feeds into the control of AMPAR trafficking leads to more questions, some of which we list below:

(1) Which TNFα receptor (TNFR1 or TNFR2) is activating the AMPAR trafficking?
(2) What other receptors or surface proteins are trafficked differentially with TNFα signaling and are there particular subtypes of AMPARs that are preferentially affected? The existence of GluR2-lacking, calcium-permeable AMPARs is gaining acceptance. Indeed, evidence exists for a decreased expression of GluR2 subunits in various neurological disorders, including epilepsy and amyotrophic lateral sclerosis (ALS). If GluR2-lacking AMPARs are included in the TNFα-induced surface-localized receptor population, this would have clear implications for excitotoxicity. In acute neurotrauma and in diseases such as epilepsy, glutamate is found in high concentrations in synaptic as well as extra-synaptic locations. Calcium-permeable AMPARs in extra-synaptic sites would provide a dangerous source of calcium influx even in the absence of depolarization, thus short-circuiting the need for NMDARs to induce long term changes dependent upon calcium signaling. In addition, calcium influx through calcium permeable AMPARs may contribute to calcium-induced apoptosis.
(3) How long does the trafficking change last in culture and does it happen in *in vivo* models?
(4) Is TNFα and abnormal AMPAR-trafficking part of longer-term, more gradual neuron death such as seen in neurodegerative diseases? Studies in epilepsy and ALS discussed above give us correlative data which suggest this may be the case. What long-term changes might occur as a result of the rapid rise in calcium concentration which we believe is maintained for 30 minutes in hippocampal neurons?
(5) What are the main sources of TNF after injury or during neurological disease? Here we show the potential for astrocytes to be the source but what of microglial activation and neurons themselves along with blood supply source of TNF? As discussed above, TNF does cross the blood–brain barrier after acute trauma.
(6) Is there any contribution in the short or long term from TNFα signaling to an increase in AMPAR protein production? Even though our observations of this very rapid AMPAR trafficking suggest no contribution from new protein translation, new data discussed above suggest that the synapse may have a very rapid protein translation machinery nearby. Thus more careful studies controlling for protein translation are planned. Regardless of the method by which more AMPARs get to the surface, it is clear at least that over 30 minutes of exposure to TNFα an increase in potentially active surface AMPARs occurs. Thus, questions of how this might set in motion long-term changes in neuron health are still of great interest.

We believe that there is enough precedent to believe that these questions will lead to a uniquely useful view of the underlying mechanisms of excitotoxicity. We hope that our future studies based on these questions will provide the neuroscience community with new drug targets for the reduction of neuron damage and death that occurs after acute trauma and during neurodegenerative diseases.

13 · Neurotrophin signaling among neurons and glia during formation of synapses

Sarina B. Elmariah, Ethan G. Hughes, Eun Joo Oh, and Rita J. Balice-Gordon

The assembly of central nervous system (CNS) synapses is a complex and highly dynamic process, requiring the coordinated exchange of anterograde and retrograde signals between pre- and post-synaptic neurons and surrounding glia (reviewed in Garner et al., 2002). Whilst synapses may be formed in the absence of glia, recent work has demonstrated that activity-dependent and independent interactions between neurons and glia modulate synapse formation and function (Pfrieger & Barres, 1997; Haydon, 2001; Ullian et al., 2001; Cohen-Cory, 2002; Zhang et al., 2003; Ullian et al., 2004).

Several lines of evidence suggest that astrocytes coordinately modulate pre- and post-synaptic development of excitatory synapses by soluble as well as contact-mediated factors. Glial-derived soluble factors have been shown to enhance neurite outgrowth, synapse formation, and function (Aoyagi et al., 1994; Pfrieger & Barres, 1997; Ullian et al., 2001; Ullian et al., 2004). In purified retinal ganglion cell and spinal motor neuron cultures, glia-conditioned medium dramatically increased the number of pre-synaptic contacts made between neurons, quantal size, and the efficacy of transmitter release (Pfrieger & Barres, 1997; Nagler, Mauch & Pfrieger, 2001; Ullian et al., 2001; Ullian et al., 2004). Mauch and colleagues (2001) showed that astrocyte-derived cholesterol complexed to apolipoprotein E (ApoE) was both necessary and sufficient to increase the number of functional pre-synaptic terminals. Treatment with glia-conditioned media also increased the number of post-synaptic α-amino-3-hydroxy-5-methyl-4-isoxazolepropionate (AMPA) receptor clusters and responsiveness to glutamate. Once mature synapses have formed, astrocytes continue to provide soluble factors that regulate synaptic transmission. In hippocampal slices, peri-synaptic astrocytes release glutamate, in response to γ-aminobutyric acid ionotropic family B (GABA$_B$) receptor activation, which potentiates interneuron GABA release and inhibitory transmission (Kang et al., 1998). Recent work has demonstrated that thrombospondin-1 and -2, expressed by immature astrocytes, may induce the formation of immature synapses between retinal ganglion cells *in vitro* and *in vivo*. These immature synapses are pre-synaptically active but post-synaptically silent, suggesting that multiple additional, but as yet unknown, factors coordinate pre- and post-synaptic maturation (Christopherson et al., 2005). Glial contact with neurons has also been shown to enhance synapse formation. In hippocampal neurons grown *in vitro*, integrin-mediated contact between astrocytes and pyramidal neurons induced neuron-wide activation of protein kinase C (PKC) signaling. PKC activation, in turn, promoted the maturation of excitatory pre-synaptic terminals, but had no effect on post-synaptic AMPA-type glutamate receptor (AMPAR) cluster number or localization (Hama et al., 2004). As synapses mature, astrocytes continue to modulate synaptic function by potentiating or suppressing activity at pre-synaptic glutamatergic terminals via the release of glutamate or adenosine triphosphate (ATP), respectively (Zhang et al., 2003; Fiacco & McCarthy, 2004). Ephrin-Eph signaling between neurons and glial cells has also been suggested to modulate post-synaptic spine morphology. EphrinA3, localized to astrocytic processes that envelop spines, interacts with EphA4 receptors in dendritic spines of pyramidal neurons, providing repulsive cues that refine spine morphology (Murai et al., 2003). Moreover, pre-synaptic ephrinB1 ligands and post-synaptic EphB2 receptors have been shown to modulate the clustering of N-methyl-D-aspartate (NMDA) receptors in cortical and hippocampal neurons *in vitro* (Dalva et al., 2000). These studies underscore the complexity of glial regulation of synapse formation and function by mechanisms that include soluble as well as contact-mediated factors.

Based on these and other studies, it is clear that astrocytes actively modulate synapse formation by multiple mechanisms. Neurotrophins are potent morphogenetic modulators that are sensitive to activity-dependent regulation and, as discussed below, induce the formation of new synaptic contacts and modify the structure of existing synapses by pre- as well as post-synaptic mechanisms. Astrocytes express a subset of tropomyosin-related kinase (Trk) receptors and have been shown to release as well as recycle neurotrophins (Alderson et al., 2000). In this chapter we discuss how neurotrophins and Trks modulate synapse formation and function, and discuss the evidence which suggests that neurotrophin signaling is one of several mechanisms by which astrocytes or other glial cells influence synaptic circuitry in the developing and mature nervous system.

NEUROTROPHINS, TRKS, AND P75 RECEPTORS

The neurotrophins are a family of neuronal growth factors that include nerve growth factor (NGF), brain-derived neurotrophic factor (BDNF), neurotrophin-3 (NT3), and neurotrophin-4/5 (NT4/5). Neurotrophins bind with high-affinity ($K_d \sim 10^{-9}$ M) to members of the Trk receptors. NGF binds preferentially to TrkA and BDNF, and NT4/5 bind to TrkB, and NT3 binds to TrkC. Neurotrophin binding results in dimerization and autophosphorylation of Trk receptors, triggering tyrosine kinase activation and initiation of one or more signal transduction cascades, including mitogen-activated protein kinase (MAPK), PI3K, or PLC-γ signaling pathways (see Huang & Reichardt, 2001; Patapoutian & Reichardt, 2001; Chao, 2003). These signals induce rapid changes in the structure and function of local proteins, or are propagated to the nucleus where they activate transcription factors that alter gene expression.

In addition to full-length (FL) receptors, TrkB and TrkC have differentially spliced variants known as truncated (t) Trk isoforms, which bind neurotrophin with the same affinity as FL receptors but lack the intracellular kinase domain. Truncated receptors are expressed by neurons as well as by glial cells, and they have traditionally been thought to function only through their ability to sequester neurotrophins or to dominant-negatively inhibit activation of Trk kinases by forming non-signaling heterodimers (see Huang & Reichardt, 2001). Recent work has shown that TrkB.t1 mediates BDNF-induced calcium signaling in astrocytes (Rose et al., 2003). Thus, the pleiotropic effects of neurotrophin signaling are attributed, in part, to the activation of diverse Trk receptors.

All neurotrophins also bind with low affinity to a structurally distinct receptor from the tumor necrosis factor (TNF) receptor superfamily, the p75 neurotrophin receptor (p75NR). The p75 receptor serves numerous roles in the developing nervous system, including increasing the affinity and specificity of Trk–neurotrophin interactions when binding neurotrophins as a co-receptor and inducing apoptosis in neurons and other cell types in the absence of Trk activation. p75 may also act as a high-affinity receptor for uncleaved peptide precursors of the neurotrophins, known as proneurotrophins (Lee F. S. et al., 2001). Pro-neurotrophins are secreted during neural injury and may stimulate apoptosis in oligodendrocytes and vascular smooth muscle cells by selectively binding to and activating p75 receptors (Lee R. et al., 2001; Beattie et al., 2002). Neurotrophin activation of p75-mediated apoptosis has also been associated with the elimination of neurons that do not make appropriate connections during development (Lee R. et al., 2001). In addition, recent work has suggested that p75 plays a role in modulating peripheral nerve myelination via interactions with neurotrophins as well as with myelin-associated glycoprotein and the Nogo receptor complex (Cosgaya, Chan & Shooter, 2002; Wang et al., 2002). Although p75-mediated signaling appears to play several roles in the developing and mature nervous system, this signaling has not yet been implicated in synapse formation or function and remains to be carefully addressed.

ACTIVITY-DEPENDENT NEUROTROPHIN SYNTHESIS, TRANSPORT, AND RELEASE

At developing and mature synapses, neurotrophins are exchanged between pre- and post-synaptic neurons via anterograde, retrograde, and trans-synaptic transport. Neurotrophins are synthesized and packaged into vesicles in neuronal cell bodies, and are subsequently transported to pre-synaptic axon terminals or post-synaptic dendrites for local secretion (see Poo, 2001). The synthesis, transport, and release of neurotrophins are tightly regulated by neural activity. In cultured neurons, neuronal depolarization produced by application of glutamate, elevated extra-cellular K^+, or high-frequency stimulation increased levels of BDNF messenger ribonucleic acid (mRNA) and subsequent BDNF secretion (Zafra et al., 1991; Lindholm et al., 1994). In hippocampal slices, normal physiologic activity capable of potentiating glutamatergic synapses is sufficient to increase BDNF mRNA (Patterson et al., 1992). In contrast, GABAergic inhibition of neuronal activity decreases BDNF mRNA in hippocampal neurons (Berninger et al., 1995). In studies conducted in vivo, blockade of visual input induces a downregulation in BDNF mRNA in rat visual cortex (Castren et al., 1992). Consistent with activity-dependent changes in expression, BDNF and TrkB mRNA transcripts are targeted to dendrites in an activity-dependent fashion where they have been proposed to undergo local translation (Tongiorgi, Righi & Cattaneo, 1997). Trans-synaptic neurotrophin transport is also modulated by activity. Kohara and colleagues (2001) showed that green fluorescent protein (GFP)-tagged BDNF is transferred from pre-synaptic terminals to post-synaptic sites in hippocampal neurons in culture and that BDNF transfer declines as neuronal activity decreases.

TrkB expression, membrane incorporation, and accumulation at synapses are also modulated by synaptic activity (Tongiorgi, Righi & Cattaneo, 1997; Meyer-Franke et al., 1998; Du et al., 2000). Depolarization and elevated cyclic adenosine monophosphate (cAMP) levels induce the rapid recruitment of TrkB from internal stores to the plasma membrane of hippocampal neurons (Meyer-Franke et al., 1998; Du et al., 2000). Upon binding ligand, neurotrophin–Trk complexes are internalized in endocytic vesicles which may continue transducing signal and/or may eventually be re-released at nerve terminals or post-synaptic dendrites to synaptic partners in an activity-dependent manner (see Poo, 2001).

NEUROTROPHINS AS SYNAPTIC MODULATOR: PRE-SYNAPTIC TERMINAL FORMATION AND FUNCTION

Neurotrophins, in particular BDNF, profoundly modulate the growth, remodeling, and stability of dendrites and axons in hippocampal, cortical, and cerebellar neurons (Cohen-Cory & Fraser, 1995; McAllister, Katz & Lo, 1996, 1997; Alsina, Vu & Cohen-Cory, 2001). By increasing the complexity of axonal and dendritic arbors, neurotrophins increase the number of potential contact sites between pre- and post-synaptic neurons and thereby modulate the density of synaptic innervation. Following the establishment of these nascent connections, BDNF and TrkB signaling actively promote the maturation and stabilization of CNS synapses via pre- as well as post-synaptic mechanisms.

During synaptogenesis, retrograde Trk signaling modulates the development of pre-synaptic terminals, in part by regulating the expression and/or distribution of synaptic vesicle proteins (see Vicario-Abejon et al., 2002). BDNF and NT3 treatment in E16-dissociated and P7-slice hippocampal cultures resulted in an increase in the number of synaptic vesicles docked at active zones (Collin et al., 2001; Tyler & Pozzo-Miller, 2001). Consistent with these observations, several groups report that BDNF treatment of hippocampal and cortical neuronal cultures resulted in increased synaptobrevin expression (Takei et al., 1997; Tartaglia et al., 2001; Yamada et al., 2002). In accordance with the effects of exogenous neurotrophins, mice deficient in BDNF or TrkB exhibit marked reductions in the total number of docked vesicles at hippocampal synapses, and a redistribution of docked vesicles to areas far from active zone in the cerebellar synapses (Martinez et al., 1998; Pozzo-Miller et al., 1999; Carter et al., 2002). Moreover, hippocampal sections from TrkB-deficient mice exhibit an overall reduction in syntaxin 1 and SNAP25 immunoreactivity (Martinez et al., 1998). Consistent with reduced exocytotic machinery, TrkB-deficient mice exhibit impaired Ca^{2+}-evoked glutamate and GABA release in cortical synaptosomal preparations from post-natal mice (Carmona et al., 2003). Despite these reports, the mechanisms by which neurotrophins modulate synaptic vesicle machinery remain somewhat controversial, as several groups report that expression of synaptobrevin, syntaxin 1, or SNAP25 is not altered following BDNF treatment in culture (Vicario-Abejon et al., 1998; Vicario-Abejon et al., 2002) or upon examination of BDNF-deficient mice (Pozzo-Miller et al., 1999). Thus, whereas neurotrophins modulate the formation of pre-synaptic axon terminals, the precise cellular and molecular mechanisms remain to be determined.

Once nascent synaptic contacts have been formed, BDNF signaling promotes the maturation of pre-synaptic terminals and modulates ongoing neural transmission at excitatory and inhibitory synapses. In dissociated and slice cultures from the hippocampus and cortex, BDNF treatment resulted in increased mean excitatory post-synaptic current (mEPSC) frequency and FM 1–43 turnover at excitatory terminals (Rutherford, Nelson & Turrigiano, 1998; Collin et al., 2001; Tyler & Pozzo-Miller, 2001). Reducing BDNF signaling by scavenging BDNF with TrkB–IgG fusion proteins resulted in a decrease in mEPSC frequency in hippocampal neurons (Cabelli et al., 1997). Consistent with these findings, BDNF enhances glutamatergic transmission at mature hippocampal synapses, in part attributed to increases in pre-synaptic glutamate release (Kang & Schuman, 1996; Carmignoto et al., 1997; Li et al., 1998a). At hippocampal and cortical synapses in vivo and in vitro, BDNF facilitates long-term potentiation (LTP), attenuates long-term depression (LTD), and promotes homeostatic and competitive interactions that refine neural circuitry (Korte et al., 1995; Akaneya, Tsumoto & Hatanaka, 1996; Cabelli et al., 1996; Huang et al., 1999; Turrigiano & Nelson, 2000; 2004). At hippocampal and cerebellar synapses in BDNF-deficient mice, high-frequency stimulation and paired-pulse facilitation are impaired, consistent with compromised pre-synaptic release (Pozzo-Miller et al., 1999; Carter & Regehr, 2002). These studies demonstrate that BDNF acts pre-synaptically to modulate the dynamics of glutamatergic release at developing and mature synapses.

In addition to modulating excitatory pre-synaptic terminals, BDNF and TrkB signaling also influence the construction and function of inhibitory synapses. BDNF promotion of inhibitory axon outgrowth and synapse formation has been well characterized in hippocampus, cortex, and, in particular, in cerebellum in vitro and in vivo (Huang et al., 1999; Marty, Wehrle & Sotelo, 2000; Seil & Drake-Baumann, 2000; Yamada et al., 2002). Accordingly, fewer GABAergic synapses are observed in cerebellar slices following scavenging of endogenous BDNF (Seil & Drake-Baumann, 2000) and in conditional TrkB deletion mutants (Rico, Xu & Reichardt, 2002). Far less is known about how TrkB signaling modulates the formation of pre-synaptic inhibitory terminals. Pre-synaptic TrkB activation increases glutamic acid decarboxylase (GAD) expression in interneurons and enhances pre-synaptic GABA release and uptake at inhibitory terminals (Vicario-Abejon et al., 1998; Marty, Wehrle & Sotelo, 2000; Yamada et al., 2002). Consistent with these findings, increased mean inhibitory post-synaptic current (mIPSC) frequency and enhanced quantal content of GABA-containing vesicles have been observed in hippocampal and cerebellar neurons after treatment with BDNF (Bao et al., 1999; Bolton, Lo & Sherwood, 2000). These studies underscore the role of the retrograde actions of BDNF and TrkB in regulating the formation and function of inhibitory CNS synapses.

Astrocytes modulate inhibitory pre-synaptic terminal formation and function via soluble as well as contact-mediated factors that are not neurotrophins. An increase in inhibitory axon outgrowth and in the number of inhibitory synaptic contacts are observed when neuronal cultures are treated with astrocyte-conditioned medium (ACM), and this astrocyte effect is activity- and neurotrophin-independent. Thus, astrocytes release soluble factors that modulate interneuron maturation and the formation of inhibitory contacts in the absence of neuronal activity or neurotrophin signaling. ACM as well as contact with astrocytes increased the number of pre-synaptic terminals immunopositive for synaptophysin and the vesicular GABA transporter (VGAT). However, an increase in the number of functional inhibitory synaptic terminals, assayed by AM 1–43 labeling, was only observed when neurons were plated in direct contact with astrocytes. Taken together, these results suggest that astrocytes modulate the inhibitory pre-synaptic terminal formation and function, at least in vitro, by both secreted and contact-mediated factors.

The mechanisms by which neurotrophins enhance the formation and function of pre-synaptic terminals remain unclear. Neurotrophins may modulate the assembly or stabilization of synaptic vesicles through post-translation modifications of soluble N-ethylmaleimide-sensitive fusion protein attachment protein receptor (SNARE) complex proteins (Vicario-Abejon et al., 1998; Pozzo-Miller et al., 1999; Tartaglia et al., 2001). Consistent with this hypothesis, BDNF-deficient mice exhibit decreased synaptosomal levels of the vesicular proteins, synaptobrevin and synaptophysin, although levels from whole hippocampal extracts appear similar to controls (Pozzo-Miller et al., 1999). Thus, by altering SNARE protein associations with other vesicular proteins, neurotrophins may regulate vesicle distribution and properties of vesicular release. Recent work demonstrates that Synapsin 1 is phosphorylated after BDNF activation of MAPK, promoting synaptic vesicle mobilization and enhancing neurotransmitter release (Jovanovic et al., 2004). Long-term and acute neurotrophin signaling may also influence, and be influenced by, Ca^{2+} levels in pre-synaptic terminals. BDNF and NT3 potentiate neurotransmitter release, owing to increased pre-synaptic Ca^{2+} at developing *Xenopus* neuromuscular synapses (Lohof, Ip & Poo, 1993; Stoop & Poo, 1995; 1996). BDNF-induced MAPK activation has been shown to increase the number of P/Q-type Ca^{2+} channels expressed in hippocampal neurons (Levine et al., 1995; Baldelli, Forni & Carbone, 2000). These studies show that BDNF and TrkB signaling modulate the formation, maturation, and function of pre-synaptic terminals at CNS synapses, although the mechanisms by which such modulation occurs remain to be determined.

NEUROTROPHINS AS SYNAPTIC MODULATORS: POST-SYNAPTIC NEUROTRANSMITTER RECEPTOR CLUSTERS

In addition to their roles in establishing pre-synaptic terminals and regulating neurotransmitter release, neurotrophins are emerging as key modulators of post-synaptic specializations. Prolonged neurotrophin signaling promotes quantal scaling of excitatory and inhibitory synapses in postnatal neural networks by modulating post-synaptic receptors (Rutherford et al., 1997; Rutherford, Nelson & Turrigiano, 1998). At glutamatergic synapses, BDNF has been shown to promote maturation of silent synapses by increasing the translocation of the AMPAR subunit GluR2 to the neuronal surface in neocortical neurons (Narisawa-Saito et al., 1999; Narisawa-Saito et al., 2002). In hippocampal neurons, BDNF rapidly increases the phosphorylation and potentiates the conductance of NMDA receptors (Levine et al., 1996; Levine et al., 1998). At inhibitory synapses in hippocampal cultures, BDNF treatment has been shown to increase $GABA_A$ receptor subunit expression, differentially modulate post-synaptic $GABA_A$ receptor cluster number, regulate $GABA_A$ receptor recruitment to the neuronal surface, and modulate $GABA_A$ channel conductance (Brunig et al., 2001; Kilman, van Rossum & Turrigiano, 2002; Yamada et al., 2002; Jovanovic et al., 2004). Examination of current density in hippocampal cultures at times before many synapses have formed suggest that astrocytes contribute to the maintenance of $GABA_A$ receptors at the neuronal cell surface (Liu et al., 1996; Liu et al., 1997). Where the upregulation of GABA current density required Ca^{2+} elevation in astrocytes, this effect was not dependent on direct contact with astrocytes, as ACM treatment mimicked the effects observed in co-cultures (Liu et al., 1996; Lui et al., 1997). Thus, soluble factors released by astrocytes modulate the distribution of $GABA_A$ receptors prior to synapse formation. These studies demonstrate that neurotrophins may modulate synaptic integrity and strength by regulating the expression, distribution, and kinetics of post-synaptic neurotransmitter receptors at excitatory and inhibitory synapses.

We recently showed that FL and truncated TrkB are diffusely distributed throughout the dendrites and soma of rat hippocampal neurons grown *in vitro*, and that some TrkB is localized to some, but not all, post-synaptic specializations (Figure 13.1 (A)) (Elmariah et al., 2004). Manipulation of TrkB-mediated signaling resulted in dramatic changes in the number and synaptic localization of post-synaptic NMDA receptor (NMDAR) and $GABA_A$ receptor ($GABA_A$R) clusters (Figure 13.1 (B, C)). BDNF treatment resulted in an increase in the number of NMDAR and $GABA_A$R clusters and increased the proportion of clusters apposed to pre-synaptic terminals. Downregulation of TrkB signaling resulted in a decrease in receptor cluster number and synaptic localization (Figure 13.1 (B)). Examination of the time course of the effects of BDNF on receptor clusters showed that the increase in $GABA_A$R clusters preceded the increase in NMDAR clusters by at least 12 hours (Figure 13.1 (C)). Moreover, the TrkB-mediated effects on NMDAR clusters were dependent on $GABA_A$R activation. Whilst tetrodotoxin (TTX), amprenavir (APV), and 6-cyano-7-nitroquinoxaline-2,3-dione (CNQX) treatment had no effect, blockade of $GABA_A$Rs with bicuculline abolished the BDNF-mediated increase in NMDAR cluster number and synaptic localization (Figure 13.1 (D)). On the other hand, application of exogenous GABA prevented the decrease in NMDAR clusters induced by BDNF scavenging (Figure 13.1 (E)). Taken together, these results suggest that TrkB-mediated signaling modulates the clustering of post-synaptic $GABA_A$Rs whose activity is required for a subsequent upregulation of NMDAR clusters (Elmariah et al., 2004).

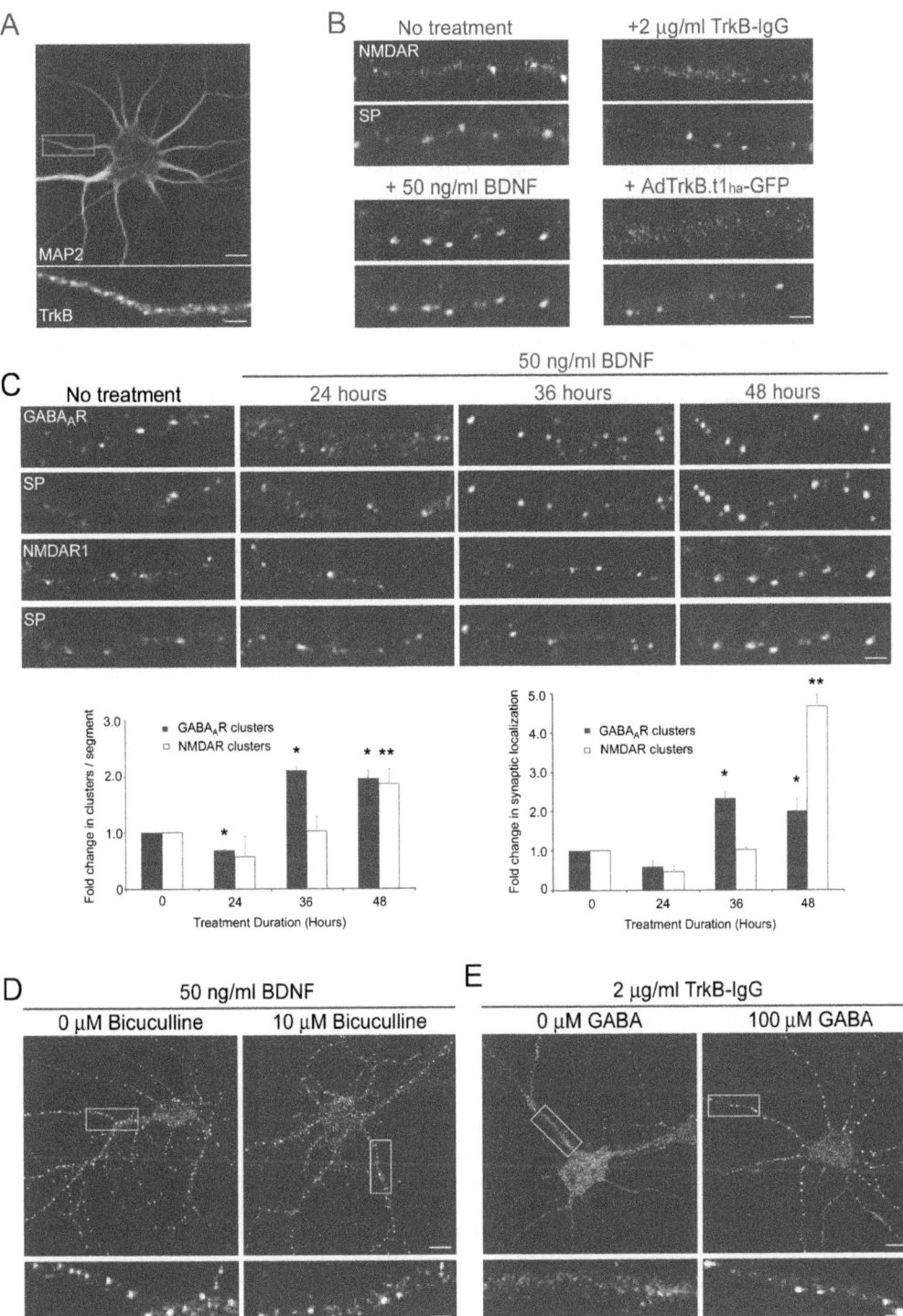

Figure 13.1. TrkB signaling modulates neurotransmitter receptor cluster formation and synaptic localization in hippocampal neurons *in vitro*. Rat hippocampal neurons at 10 days *in vitro* (*div*) were immunostained and analyzed with confocal microscopy. (A) Immunostaining in permeabilized neurons with an antibody against all isoforms of TrkB (bottom) and MAP2

These results support a model in which TrkB-mediated signaling directly increases inhibitory interneuron synapses on to pyramidal neurons, leading to a homeostatic increase in NMDAR clusters and their synaptic localization (Figure 13.2). One important question raised by this simple model is what mechanism limits homeostatic increases in excitatory and inhibitory synapses, preventing synapse number from constantly increasing. One possibility is that activity-dependent mechanisms underlying synaptic competition and loss may counterbalance these effects as circuits develop. Post-synaptic effects of neurotrophin signaling are also evident at neuromuscular synapses, where bi-directional neurotrophin signaling influences post-synaptic development via direct and indirect mechanisms. Retrograde signaling by BDNF, NT3, and NT4 released by muscle cells regulates neuregulin expression in pre-synaptic nerve terminals, which in turn stimulates acetylcholine receptor (AChR) synthesis in post-synaptic muscle fibers (Loeb & Fischbach, 1997; Loeb et al., 2002). Anterograde neurotrophin signaling at neuromuscular synapses is revealed in experiments involving dominant-negative disruption of post-synaptic TrkB signaling. This results in the disassembly of existing AChR clusters in muscle fibers *in vivo* and *in vitro* (Gonzalez et al., 1999), suggesting a role for TrkB signaling in receptor cluster maintenance. Taken together, this work at neuromuscular synapses *in vivo* and at CNS synapses *in vitro* suggests that TrkB-dependent signaling plays an important role in modulating post-synaptic neurotransmitter clusters. In the CNS as least, this signaling may be part of a mechanism that balances inhibitory and excitatory synaptic transmission in developing neural circuits.

Neurotrophin signaling at tripartite synapses

A role for neurotrophin signaling in glial modulation of synapse formation and function, and vice versa, has

[Caption for Figure 13.1. (cont.)] top shows that TrkB is distributed diffusely throughout dendrites. Scale bar = 10 μm. Areas within white boxes are shown below at higher magnification. Scale bar = 2 μm. (B) Compared with no-treatment controls (top left), treatment with 50 ng/mL brain-deprived neurotrophic factor (BDNF) (bottom left) for 48 hours resulted in an increase in the number and synaptic localization of *N*-methyl-D-aspartate (NMDA) receptor (NMDAR) clusters in pyramidal neurons. In contrast, treatment with 2.0 μg/mL TrkB–IgG (top right) to scavenge endogenous BDNF or infection with an adenovirus encoding truncated TrkB (AdTrkB.t1$_{ha}$; bottom right) to dominant-negatively downregulate TrkB signaling resulted in a decrease in NMDAR cluster number and synaptic localization compared with no-treatment controls. Hippocampal neurons were immunostained with an antibody against NMDAR1 (top) and synaptophysin (SP; bottom) to visualize pre-synaptic terminals. Scale bar = 2 μm. (C) Time course of the effects of BDNF on GABA$_A$R and NMDAR clusters. *Top panels* Treatment with 50 ng/mL BDNF for 36 (middle right) and 48 (right) hours increases the number of GABA$_A$R clusters compared with untreated neurons (left). GABA$_A$R clusters visualized with antibody against GABA$_A$ β2/3 subunits (top) and SP is shown below. Scale bar = 2 μm. *Bottom panels* An increase in NMDAR clusters is observed only after 48 hours (bottom right). NMDAR clusters visualized with antibody against NMDAR1 (top); overlay of NMDAR (red) and SP (green) is shown below. Scale bar = 2 μm. *Graph left*, Quantification of the fold change in GABA$_A$R (black bars) or NMDAR (white bars) clusters per 20 μm dendrite segment in BDNF-treated compared with untreated neurons at 24, 36, and 48 hours. Each bar represents the ratio of the number of clusters per segment in BDNF-treated neurons compared with untreated controls at the timepoint indicated. * (GABA$_A$Rs); ** (NMDARs) indicates significant change from a ratio of 1 observed at 0 hours of treatment. *Graph righ*, Quantification of the fold change in the proportion of synaptically localized GABA$_A$R (black bars) and NMDAR (white bars) clusters after BDNF for 24, 36, and 48 hours. Each point represents the ratio of the percentage of synaptically localized clusters in BDNF-treated neurons compared with untreated controls at the timepoint indicated. * (GABA$_A$Rs), ** (NMDARs) indicates significant change from a ratio of 1 observed at 0 hours of treatment ($p<0.005$; χ^2 and other tests). (D) Hippocampal neurons treated with 50 ng/mL BDNF and 10 μM bicuculline, a GABA$_A$ receptor antagonist, for 48 hours were analyzed after immunostaining at 10 *div* with an antibody against NMDAR1. Treatment with BDNF + bicuculline led to a decrease in NMDAR cluster number compared with neurons treated with BDNF alone. Scale bar = 10 μm. Areas within white boxes are shown below at higher magnification. Scale bar = 2 μm. (E) Hippocampal neurons treated with 2 μg/mL TrkB–IgG and 100 μM GABA for 48 hours were analyzed after immunostaining at 10 *div* with an antibody against NMDAR1. Treatment with TrkB-IgG + 100 μM GABA led to an increase in NMDAR cluster number compared with neurons treated with TrkB–IgG alone. Scale bar = 10 μm. Areas within white boxes are shown below at higher magnification. Scale bar = 2 μm. Reprinted with permission from Elmariah et al. (2004).

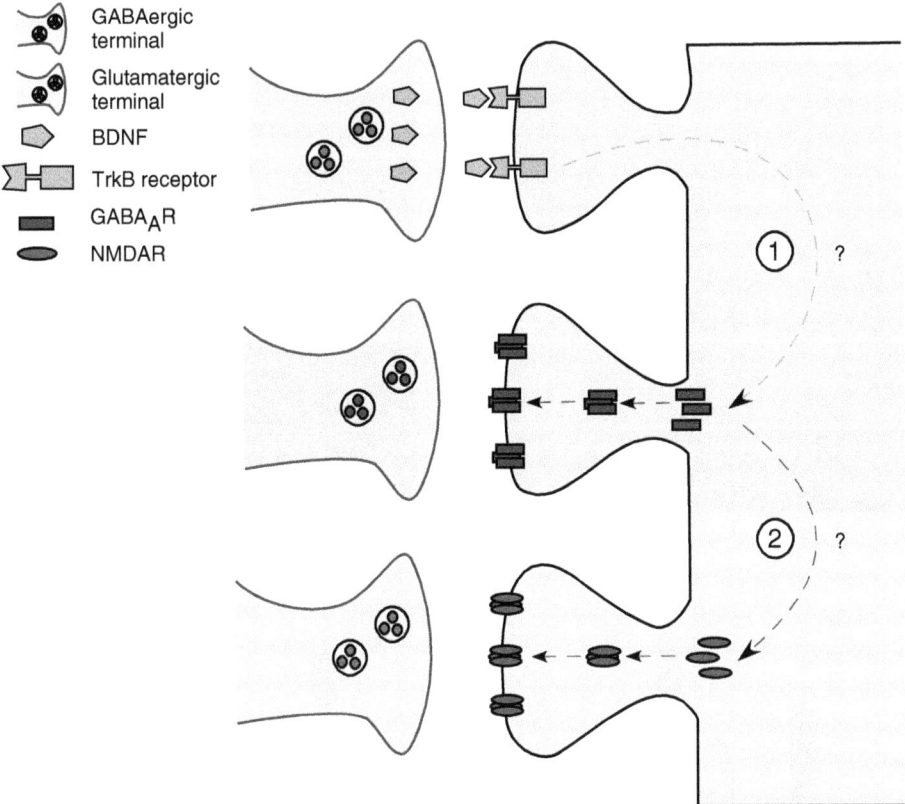

Figure 13.2. Neurotrophin signaling modulates $GABA_A$ and NMDA receptor clustering at hippocampal synapses. TrkB-mediated signaling may be part of a homeostatic mechanism that balances excitatory and inhibitory synaptic activity in developing neural circuits. (1) Activation of post-synaptic TrkB signaling by BDNF (top) increases the number and synaptic localization of $GABA_A$ receptor clusters (middle) in pyramidal neurons. Upregulation of functional GABAergic synapse results in a relative imbalance in inhibition and excitation, thereby reducing excitability in target pyramidal neurons. (2) Reduced neuronal excitability in turn leads to a compensatory increase in NMDA receptor cluster (bottom) number and synaptic localization in pyramidal neurons. Adapted with permission from Elmariah et al. (2004).

begun to emerge. Several groups have reported that primary astrocytes in culture produce NGF, BDNF, and NT3, and that the expression of these neurotrophins and their receptors in astrocytes is modulated by cAMP signaling (Furukawa et al., 1987; Houlgatte et al., 1989; Rudge et al., 1992; Condorelli et al., 1994). Thus, astrocytes may release neurotrophins that affect neurons which, in turn, enhance synapse formation and function. Similarly, neurons release neurotrophins, and these may affect astrocytes which, in turn, signal back to neurons. Furthermore, glia may regulate neurotrophin release from neurons. For example, in hippocampal cell cultures, stimulated astrocytes release activity-dependent neurotrophic factor (ADNF), which induces NT3 release from post-synaptic neurons. Retrograde NT3 then acts on the pre-synaptic pyramidal neurons to increase glutamatergic synapse formation (Blondel et al., 2000). Finally, astrocytes may modulate the expression of neuronal Trk receptors indirectly by influencing neuronal activity. As the expression and distribution of neurotrophin and Trk mRNA and protein are highly sensitive to neuronal activity levels, glial modulation of neuronal excitation, for example by the release of glutamate or ATP, may regulate Trk signaling between neurons. These and other cellular and molecular mechanisms by which neurotrophin and

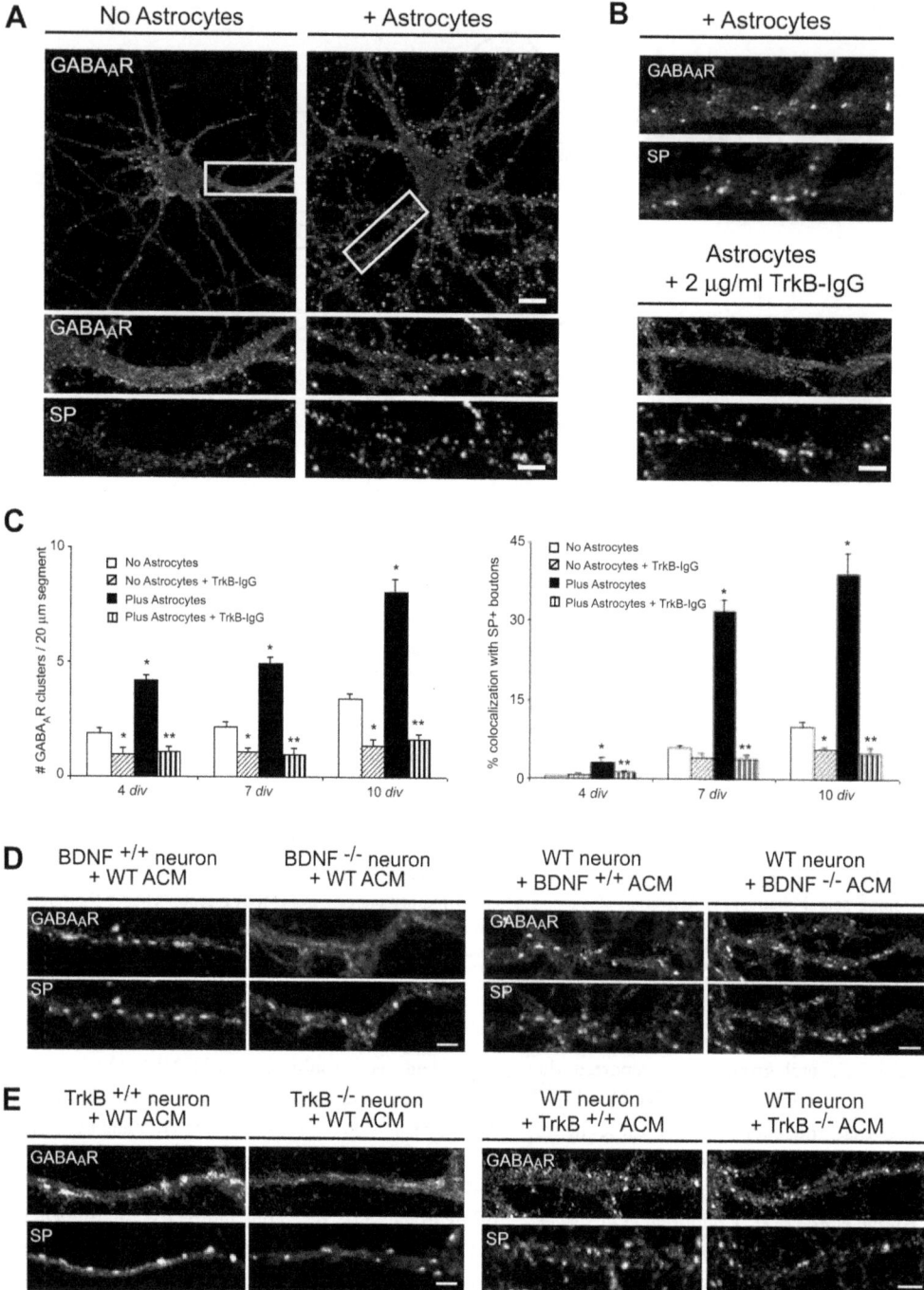

Figure 13.3. Astrocytes promote inhibitory synapse formation in hippocampal neurons via TrkB-mediated signaling. Hippocampal neurons were cultured in the presence or absence of astrocytes or astrocyte-conditioned medium (ACM) and were immunostained with antibodies against $GABA_AR$-$\beta 2/3$ (top) and SP (bottom) to visualize inhibitory presynaptic terminals. (A) The number of SP+ and VGAT+ pre-synaptic terminals and the number of $GABA_AR$ clusters increased in neuron–astrocyte

other signaling affect, and are affected by, glia remain to be explored.

The development of inhibitory synapses in dissociated cultures of embryonic rat hippocampal neurons grown in the presence and absence of astrocytes or astrocyte-conditioned medium indicates that astrocytes enhance inhibitory synaptogenesis by promoting the formation and post-synaptic localization of $GABA_AR$ clusters via neurotrophin and Trk-mediated signaling (Elmariah et al., 2005). Astrocytes increase the number and synaptic localization of $GABA_AR$ clusters during the first week in vitro (Figure 13.3 (A)). Scavenging endogenously released BDNF prevents the astrocyte-induced increase in post-synaptic $GABA_AR$ clusters, but this has no effect on pre-synaptic GABAergic terminal number (Figure 13.3 (B, C)). Moreover, TrkB-deficient astrocytes increased post-synaptic $GABA_AR$ clusters in wild-type neurons. However, wild-type astrocytes fail to induce an increase in the synaptic localization of $GABA_AR$ clusters in TrkB- or BDNF-deficient neurons (Figure 13.3 (D, E)). NT3-mediated signaling decreases the synaptic localization of $GABA_AR$ clusters in the presence of astrocytes. Taken together, these results suggest that neurotrophin or Trk signaling is not required in astrocytes, but is required in neurons, to increase post-synaptic $GABA_AR$ clusters (Elmariah et al., 2005). One hypothesis is that astrocytes upregulate activity-independent release of BDNF from pre- or post-synaptic neurons which acts in a paracrine or autocrine fashion to upregulate $GABA_AR$ clusters (Figure 13.4). Once mature networks have formed, activity-dependent BDNF and TrkB signaling then provide ongoing modulation of post-synaptic $GABA_AR$ clusters. Taken together, this work suggests that astrocytes regulate the formation of inhibitory synapses by modulating the number of post-synaptic $GABA_AR$ clusters, and these effects are mediated, in part, by neurotrophin and Trk signaling in neurons that are enhanced by astrocytes.

CONCLUSIONS AND FUTURE DIRECTIONS

There are several areas in which our understanding of how neuronal and glial signals coordinate and modulate synapse assembly, maturation, and function is currently lacking. First, our understanding of the cellular and molecular mechanisms underlying neurotrophins and

[Caption for figure 13.3. (cont.)] co-cultures (right) compared with pure neuronal cultures (left) at 10 div. The proportion of $GABA_AR$ clusters apposed to pre-synaptic terminals also increased in neuron–astrocyte co-cultures. Scale bar = 10 μm. Areas within white boxes are shown below at higher magnification. Scale bar = 2 μm. (B) Treatment of neuron–astrocyte co-cultures with 2.0 μg/mL of TrkB–IgG (bottom) to scavenge endogenous BDNF resulted in fewer post-synaptic $GABA_AR$ clusters compared with no-treatment controls (top) at 10 div. No change was observed in the number of SP+ boutons after BDNF scavenging. $GABA_AR$ clusters visualized with antibody against $GABA_A$ β2/3 subunits (top); SP (bottom) is shown below. Scale bar = 2 μm. (C) Graph left Quantification of TrkB–IgG effects on the number of $GABA_AR$ clusters per 20 μm dendrite segment at different ages in vitro. * Indicates significant difference compared with no-treatment in pure neuronal cultures ($p<0.001$); **indicates significant decrease compared with no-treatment in neuron–astrocyte co-cultures ($p<0.001$). Graph right Quantification of TrkB–IgG effects on the synaptic localization of $GABA_AR$ clusters. * Indicates significant difference compared with no-treatment in pure neuronal cultures ($p<0.001$); ** indicates significant decrease compared with no-treatment in neuron–astrocyte co-cultures ($p<0.001$). (D) Left pair of panels Hippocampal neurons were cultured from $BDNF^{-/-}$ mice and wild-type (WT) littermates in the presence of WT ACM. ACM treatment increased the number and synaptic localization of $GABA_AR$ clusters in $BDNF^{+/+}$ neurons (left) but had no effect on $GABA_AR$ clusters in $BDNF^{-/-}$ neurons (right). Right pair of panels ACM collected from $BDNF^{+/+}$ (left) and $BDNF^{-/-}$ (right) astrocytes increased $GABA_AR$ cluster number and synaptic localization by a similar magnitude in WT neurons. $GABA_AR$ clusters visualized with antibody against $GABA_A$ β2/3 subunits (top); SP (bottom) is shown below. Scale bar = 2 μm. (E) Left panels Hippocampal neurons were cultured from $TrkB^{-/-}$ mice and WT littermates in the presence of WT ACM. ACM treatment increased the number and synaptic localization of $GABA_AR$ clusters in $TrkB^{+/+}$ neurons (left), but had no effect on $GABA_AR$ clusters in $TrkB^{-/-}$ neurons (right). Right panels ACM collected from $TrkB^{+/+}$ (left) and $TrkB^{-/-}$ (right) astrocytes increased $GABA_AR$ cluster number and synaptic localization by a similar magnitude in WT neurons. $GABA_AR$ clusters visualized with antibody against $GABA_A$ β2/3 subunits (top); overlay of $GABA_AR$ (top) and SP (bottom) is shown below. Scale bar = 2 μm. Reprinted with permission from Elmariah et al. (2004).

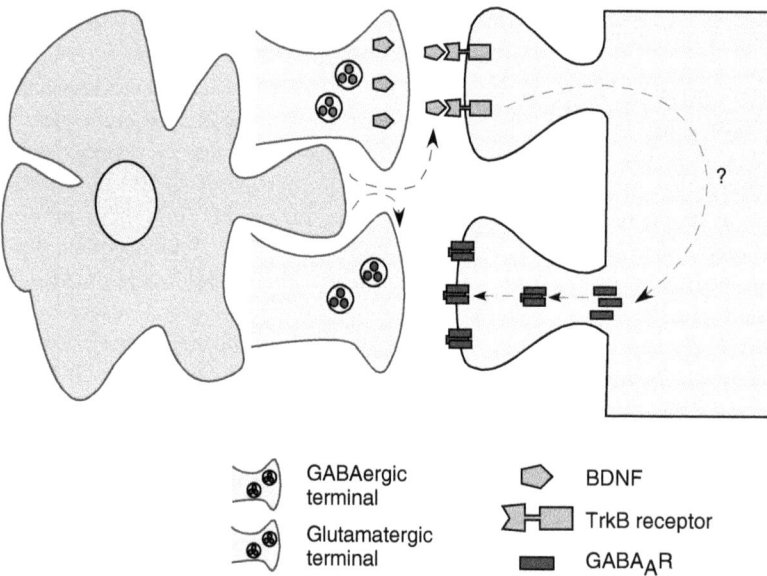

Figure 13.4. Astrocytes regulate BDNF and TrkB-mediated modulation of inhibitory synaptogenesis. Astrocytes coordinately modulate the pre- and post-synaptic development of inhibitory connections. Soluble factors released by astrocytes promote the formation of inhibitory pre-synaptic terminals via neutrotrophin and TrkB-independent signaling. In contrast, astrocytes enhance BDNF and TrkB signaling between neurons and thereby promote the formation and post-synaptic localization of GABA$_A$R clusters. Our data suggest that neurons, not astrocytes, are the relevant source of BDNF and site of TrkB activation for post-synaptic GABA$_A$R modulation. Astrocytes may increase TrkB-mediated modulation of post-synaptic GABA$_A$R clusters, either by enhancing neuronal BDNF release and/or TrkB activation in neurons, via signals that are, as yet, unknown. Reprinted with permission from Elmariah et al. (2004).

other cell–cell signals that modulate synaptogenesis is in its infancy. Second, while neurotrophin and glial modulation of synaptic morphology and activity has been extensively studied in dissociated cell and slice cultures, comparatively less is known about how neuron–glia signaling contributes to the construction and modulation of CNS synapses *in vivo*. The combination of imaging approaches to observe development *in vivo* and molecular approaches to manipulate neuronal and glial signaling using targeted genetic approaches will be critical to our understanding of how different cells and different molecules, including the neurotrophins, interact to shape synaptic connectivity. Addressing these issues will be critical to understanding how synaptic connections are formed and function in the developing and mature brain.

ACKNOWLEDGEMENTS

We thank M. Maronski, M.O. Scott, and H.Y. Zhou for technical assistance; and Drs. S. Gibbs, J. Panzer, S. Potluri, E. Vernon Pitts and D. Hess, X. Peng, Y. Song, S. Zevallos, and R. Wyatt for helpful discussions. This work was supported by grants from the NIH (NS/AR40763) and NSF (0130822) to R.B.-G. and by an NIH grant NRSA (NS43821) to S.B.E.

14 · Regulation of neurogenesis by neurotrophins: implications in hippocampus-dependent memory

Bai Lu and Jay H. Chang

The remarkable plasticity of the brain is exemplified by its ability to generate new neurons throughout life. This process, also known as neurogenesis, was first observed in the songbird brain (Nottebohm, 1989), but later demonstrated in almost all mammalian species examined, including human (Gage, 2000). Although neurogenesis occurs in many brain areas during development, in the adult active generation of new neurons is restricted to two main areas: the subgranular zone (SGZ) of the hippocampus and the subventricular zone (SVZ) of the lateral ventricles. The whole process of neurogenesis involves multiple, sequential steps (Kempermann et al., 2004a). First, neural stem cells proliferate to give rise to rapidly amplifying neural progenitor cells (NPCs). These multipotential NPCs then undergo cell fate determination, in which some commit themselves to the neuronal lineage. These cells then differentiate into immature neurons and migrate to their final locations. During this process some of the differentiated and undifferentiated cells die. A subset of surviving new neurons, however, gradually develops functional neuronal properties, including axons and dendrites, and the capacity to fire action potentials and receive synaptic inputs. Ultimately, these cells mature into fully functional neurons, and may be integrated into the existing neural network (Song, Stevens & Gage, 2002; van Praag et al., 2002). This complex process is tightly regulated at all steps by a variety of intrinsic and extrinsic factors. Defining diffusible factors which regulate this process is one of the most active areas in neural stem cell biology (Cameron, Hazel & McKay, 1998; Gould & Gross, 2002).

In this chapter, the role of neurotrophins in neurogenesis during development and in the adult is presented. Particular emphasis is given to the functional significance of neurogenesis in learning and memory, and how neurotrophins regulate this process. Recent experiments have linked adult neurogenesis to long-term synaptic plasticity in the hippocampal dentate gyrus (DG), and have shown a relationship between neurotrophins, adult neurogenesis, and dentate synaptic plasticity. These findings provide new insights into the mechanisms by which gene–environment interaction controls cognition and brain plasticity.

Several lines of evidence suggest that neurotrophins, a family of secretory neurotrophic proteins (Lewin & Barde, 1996), may serve as excellent candidates as diffusible factors that regulate neurogenesis. For example, brain-derived neurotrophic factor (BDNF) and neurotrophin-3 (NT-3) are expressed in highly neurogenic areas such as the olfactory bulb and hippocampus (Maisonpierre et al., 1990; Friedman, Ernfors & Persson, 1991; Guthrie & Gall, 1991; Lauterborn, Isackson & Gall, 1994). Many stimuli which facilitate neurogenesis, such as enriched environment and physical exercise, also enhance the expression of neurotrophins (Falkenberg et al., 1992; Torasdotter et al., 1998; Russo-Neustadt, Beard & Cotman, 1999). Moreover, a number of in vitro experiments have demonstrated that neurotrophins inhibit the proliferation and stimulate the differentiation of NPCs (Ghosh & Greenberg, 1995; Vicario-Abejon et al., 1995; Barnabe-Heider & Miller, 2003). Changes in BDNF levels, either by infusion of BDNF into the brain or by gene knockout, also alters neurogenesis in the adult brain (Pencea et al., 2001; Lee, Duan & Mattson, 2002).

The functional significance of adult neurogenesis remains unclear. It has been hypothesized that adult neurogenesis in the hippocampus is involved in learning and memory (Gould et al., 1999a; Kempermann, 2002). There is also good evidence that neurotrophins are important factors that regulate learning and memory (Tyler et al., 2002; Lu, 2003). An attractive hypothesis is that at least some of the functional effects of neurogenesis are mediated by neurotrophins. In this chapter, we will try to put the studies of the effects of neurotrophins on neurogenesis into the conceptual framework of learning and memory. We first summarize experiments examining the effects of neurotrophins on various aspects of neurogenesis. Next, we discuss the functional significance of neurogenesis in the adult hippocampus. We review the evidence accumulated thus far that links adult neurogenesis with learning and memory. We further discuss the possibility that adult neurogenesis contributes to learning and memory by regulating long-term potentiation (LTP) at dentate synapses, and whether neurotrophins could play a role in this process. Readers are referred to some recent reviews for a more general discussion of neurogenesis (Cameron, Hazel & McKay, 1998; Gould et al., 1999a; Gage, 2000; Temple, 2001; Alvarez-Buylla & Lim, 2004; Kempermanm, Wiskott & Gage, 2004b; Lie et al., 2004).

REGULATION OF NEUROGENESIS BY EXTRINSIC FACTORS

Studies more than 30 years ago showed that enriched environment promotes biochemical and morphological changes in discrete regions of the brain, including the hippocampus (Cummins et al., 1973; Greenough, 1975; Rosenzweig et al., 1978). More recent studies have demonstrated that animals in enriched environments exhibit enhanced neurogenesis in the adult hippocampal dentate gyrus (Kempermann, Kuhn & Gage, 1997, 1998; Nilsson et al., 1999; van Praag, Kempermann & Gage, 2000; Gould & Gross, 2002). This was first observed in songbirds living in the wild (Barnea & Nottebohm, 1994). Voluntary exercise, such as running, has similarly been shown to enhance neurogenesis in the hippocampus, but not in the olfactory bulb (van Praag et al., 1999a; van Praag, Kempermann & Gage, 1999b). The enhanced neurogenesis observed in enriched and exercised animals has been associated with improved spatial memory performance (Kempermann, Kuhn & Gage, 1998; van Praag et al., 1999a; Brown et al., 2003). Moreover, enriched environment has been shown to attenuate memory deficits in animals with hippocampal lesions (Dalrymple-Alford & Benton, 1984) or brain injury (Will & Kelche, 1992). Together these studies strongly argue for a functional role for adult neurogenesis. However, direct evidence proving that such experience-induced neurogenesis is required for memory tasks is lacking.

In contrast to enrichment and exercise, stressful experiences suppress dentate neurogenesis during development and in adulthood (Gould & Tanapat, 1999). For instance, a single exposure of rat pups to the odor of a natural predator during the first post-natal week, a period when neurogenesis is maximal, results in a decrease in proliferation of NPCs in the developing DG (Tanapat, Galea & Gould, 1998). This reduction in dentate neurogenesis is associated with elevated plasma corticosterone levels. Acute corticosterone treatment was previously shown to decrease the number of new neurons in the adult dentate (Cameron & Gould, 1994). In contrast, removal of endogenous corticosterone by adrenalectomy results in a significant increase in the number of new granule neurons (Cameron & McKay, 1998). Together these studies strongly suggest that the stress-induced decrease in neurogenesis in the dentate is modulated by glucocorticoid levels. A similar reduction in cell proliferation has been observed after acute exposure to stressful stimuli in the adult dentate of a number of different species, including rat (Tanapat et al., 2001), tree shrew (Gould et al., 1997), and marmoset (Gould et al., 1998). Moreover, repeated stress has been shown to produce prolonged suppression of cell proliferation in the dentate of adult tree shrews (Fuchs & Flugge, 1998). Hippocampal neurogenesis continues throughout adulthood but declines with old age (Kuhn, Dickinson-Anson & Gage, 1996; Cameron & McKay, 1999). This decline in neurogenesis is also thought to be caused, at least in part, by increased levels of corticosteroids (Nichols, Zieba & Bye, 2001).

REGULATION OF NEUROGENESIS BY NEUROTROPHINS

Several extrinsic factors such as enriched environment and physical activity have been shown to enhance the expression of neurotrophins (Neeper et al., 1996; Oliff et al., 1998; Russo-Neustadt, Beard & Cotman, 1999; Young et al., 1999; Cirelli & Tononi, 2000; Ickes et al., 2000; Berchtold et al., 2001). Conversely, extrinsic factors such as stress decrease levels of neurotrophins (D'Sa & Duman, 2002; Pham et al., 2002). Although there is no direct evidence demonstrating that neurotrophins mediate the effects that such extrinsic factors have on neurogenesis, there is increasing evidence indicating that neurotrophins do play an important role in neurogenesis.

In principle, neurotrophins could regulate neurogenesis by promoting NPC proliferation, by influencing cell fate decisions, or by supporting differentiation and/or survival of new neurons. Whilst the specific stages at which neurotrophins act have not been fully dissected, *in vitro* studies using cultured NPCs isolated from embryonic and post-natal brain have provided valuable insights. In the presence of mitogens such as fibroblast growth factor-2 (FGF-2) or epidermal growth factor (EGF), NPCs proliferate and form neurosphere clusters (Reynolds & Weiss, 1992; Reynolds, Tetzlaff & Weiss, 1992). Application of NT-3 to cultured FGF-2-responsive NPCs derived from embryonic cortex or hippocampus has been shown to antagonize the proliferative effects of FGF-2 and enhance the differentiation of the NPCs into neurons (Ghosh & Greenberg, 1995; Vicario-Abejon *et al.*, 1995). Both BDNF and nerve growth factor (NGF) have similarly been shown to channel the differentiation of EGF-responsive NPCs derived from fetal forebrain into neurons but not oligodendrocytes (Lachyankar *et al.*, 1997; Benoit *et al.*, 2001). Using a slightly different approach, Barnabe-Heider and Miller (2003) employed function-blocking antibodies to show that both endogenous BDNF and NT-3 regulate survival and neuronal differentiation of cultured cortical progenitor cells. Together, these studies using cultured NPCs from developing brain strongly argue that neurotrophins play an important role in the survival and differentiation, but not proliferation, of NPCs.

In the adult mammalian brain, neurogenesis only occurs in the SVZ of the lateral ventricle and the SGZ of the hippocampal dentate. Progenitors from the SVZ enter the rostral migratory stream (RMS) and eventually differentiate into neurons in the olfactory bulb (Luskin, Parnavelas & Barfield, 1993; Goldman, 1995). In the SGZ, hippocampal progenitors proliferate and migrate into the granule layer where they differentiate into hippocampal granule neurons (Bayer, Yackel & Puri, 1982; Kaplan & Bell, 1984; Stanfield & Trice, 1988; Kuhn, Dickinson-Anson & Gage, 1996). NPCs derived from adult brain have the capacity to self-renew and to generate both glia and neurons *in vitro*, much like those derived from embryonic brain (Gage *et al.*, 1995; Palmer, Ray & Gage, 1995). For instance, cultured NPCs from adult hippocampus cease dividing upon removal of FGF-2 from the culture medium, and some begin to differentiate into either glial or immature neuronal cells (Palmer, Takahashi & Gage, 1997). Application of BDNF or NT-3 to these cultures has been shown to significantly increase the number of mature neurons expressing γ-aminobutyric acid (GABA), acetylcholinesterase, tyrosine hydroxylase, or calbindin without having an effect on the number of new neurons (Takahashi, Palmer & Gage, 1999). However, this only occurred once immature neurons were observed following retinoic acid treatment, which suggests that neurotrophins may act at a later maturation stage of neurogenesis.

Several approaches have been used in conjunction with thymidine incorporation or bromodeoxyuridine (BrdU) labeling to study the role of neurotrophins in adult neurogenesis *in vivo*. One approach has been to supply exogenous neurotrophins to adult brain, either by infusion or viral infection. Intraventricular infusion of BDNF into the lateral ventricle of adult rat brain has been shown to increase the number of BrdU-labeled cells in the olfactory bulb, as well as other brain areas (Zigova *et al.*, 1998; Pencea *et al.*, 2001). These cells were co-labeled with neuronal markers, indicating that exogenous neurotrophin could have a direct impact on the generation or survival of new neurons (Zigova *et al.*, 1998). Adenoviral expression of BDNF in the adult ventricular zone has been shown to similarly increase in the number of BrdU-positive neurons in the olfactory bulb and this resulted in neuronal recruitment to the neostriatum (Benraiss *et al.*, 2001). Another strategy to study neurogenesis *in vivo* has been to alter the expression of endogenous neurotrophins by small molecules. For instance, antidepressants effectively stimulate the expression of BDNF and this has been shown to enhance neurogenesis in the hippocampal dentate (Castren, 2004). Lastly, the most convincing evidence that demonstrates that endogenous neurotrophins are required for adult neurogenesis *in vivo* has been obtained by use of neurotrophin knockout mice. For example, BDNF heterozygous mutants exhibit impairment in both basal and dietary restriction-induced neurogenesis (Lee, Duan & Mattson, 2002). Moreover, we have recently demonstrated a severe reduction in the survival or differentiation, but not the proliferation, of NPCs in the adult dentate of NT-3 conditional knockout mice, suggesting a role of NT-3 in adult neurogenesis (Shimazu et al., 2006).

FUNCTIONAL SIGNIFICANCE OF ADULT NEUROGENESIS IN LEARNING AND MEMORY

A longstanding issue in neural stem cell biology is the functional significance of adult neurogenesis. Although direct evidence is still lacking, substantial experimental data support the view that adult neurogenesis in the DG participates in some forms of hippocampus-dependent learning or memory (Gould et al., 1999a; Kempermann, 2002). First, a major area for adult neurogenesis is the hippocampal DG, an area critical for spatial learning and memory (Jones, 1993; Richter-Levin et al., 1995; Lisman et al., 1999). Thousands of new neurons are born in the adult hippocampus each day (Gould et al., 1999b). Second, elevated and synchronized neuronal activity facilitates neurogenesis in the dentate. For example, the number of new neurons, detected by BrdU and neuronal markers, in the adult DG is markedly increased during seizures (Bengzon et al., 1997; Parent et al., 1997; Scharfman, Goodman & Sollas, 2000). As described above, both exercise and enriched environment have been shown to enhance learning and memory (Rampon et al., 2000; van Praag, Kempermann & Gage, 2000; Cotman & Berchtold, 2002), and promote hippocampal neurogenesis. More strikingly, training on hippocampus-dependent learning, such as spatial navigation learning using the Morris water maze and fear-conditioning using a trace protocol, enhances the survival of newly generated neurons in the dentate (Gould et al., 1999b). Interestingly, hippocampus-independent learning, such as cue training in the Morris water maze and fear-conditioning using a delay protocol, had no effect (Gould et al., 1999b).

The studies described thus far have not been sufficient to prove that neurogenesis is required for learning and memory. One useful strategy to establish a causal relationship between adult neurogenesis and learning and memory is to inhibit proliferation of hippocampal NPCs, and then examine its consequences on learning and memory. Two approaches have been taken. In the first approach, inhibition of NPC proliferation by systemic injection of the mitotic inhibitor methylazoxymethanol (MAM) resulted in a significant reduction in the number of newly generated neurons in the DG. This was accompanied by a marked impairment in some forms of hippocampal-dependent memory (trace conditioning), but not hippocampal-independent memory (delay conditioning) (Shors et al., 2001). However, it is difficult to exclude the possibility that MAM may cause non-specific effects, such as interference of protein synthesis or signaling mechanisms required for learning and memory, or even cytotoxic effects. The second approach utilized focal radiation. Brain irradiation significantly inhibited the proliferation of NPCs, leading to a decrease in neurogenesis in the dentate. As a consequence, a hippocampal-dependent place-recognition task, but not hippocampal-independent object-recognition task, was impaired (Madsen et al., 2003; Rola et al., 2004). One concern is that radiation may also damage the stem cell niche in the subgranular zone (Monje et al., 2002). Thus, it is inconclusive whether the memory deficits were due to a direct effect on NPC proliferation or an indirect effect of damaging the stem cell niche. Another limitation of both approaches is that they focus only on proliferation, without addressing a number of other aspects of neurogenesis.

Given the drawbacks, it is necessary to develop new technologies. Genetic manipulation represents a powerful alternative approach to inhibit neurogenesis, with several distinct advantages. First, the combined use of region-specific expression of toxin and inducible systems should allow ablation of new neurons in a spatially and temporally specific manner. Second, by deleting genes specifically expressed during different stages of neurogenesis, it should be possible to block not only NPC proliferation, but also cell fate determination and neuronal differentiation. A number of recent studies have used this approach to study the functional role of neurogenesis. In one study, deletion of methyl-CpG binding protein 1 (MBD1), a gene important for DNA methylation and therefore gene transcription during early development, resulted in a decrease in NPC differentiation, impaired spatial learning and memory, and a reduction in LTP in the dentate (Zhao et al., 2003). Santarelli et al. (2003) studied the role of hippocampal neurogenesis in behavioral responses to antidepressants. Treatment with antidepressant is known to enhance adult neurogenesis, but whether neurogenesis contributes to the antidepressant effect is not known. In serotonin

1A receptor knockout mice, antidepressant-induced neurogenesis was blocked, leading to a significant attenuation of the antidepressant effect of fluoxetine (Prozac®) as measured by a novelty-suppressed feeding test. These results suggest that adult neurogenesis is required for the behavioral effects of antidepressants. Mice lacking endothelial nitric oxide synthase also exhibited reduced neurogenesis, through inhibition of proliferation, not survival, of NPCs. These mice, however, exhibited better performance in a learned helplessness paradigm (Reif et al., 2004). Further experiments are necessary to sort out the functional role of neurogenesis in antidepressant behavior. In a remarkable set of experiments Feng et al. (2001) used forebrain-specific knockout mice for presenilin, a gene whose mutation is associated with early-onset Alzheimer's disease, to study the role of neurogenesis in learning and memory. These mice exhibited a pronounced deficiency in enrichment-induced neurogenesis in the dentate (Feng et al., 2001). However, this deficiency did not affect learning and the ability to form new memories. Rather, contextual memory was significantly reduced in these knockout mice when animals were exposed to enriched environment between training and testing. The authors proposed a provocative hypothesis: adult neurogenesis may help clear old memory traces from the hippocampus, leaving room for new memory processing. However, Wang et al. (2004) reported that the deficiency in neurogenesis observed in a presenilin knockin mutant line was accompanied by defects in the formation of new contextual memory (Wang et al., 2004).

Region-specific knockout mice reveal the contribution of specific aspects of neurogenesis to learning and memory. *In vitro* experiments suggest that FGF-2 stimulates the proliferation but not differentiation of NPCs, while NT-3 enhances the differentiation without affecting proliferation (Ghosh & Greenberg, 1995; Vicario-Abejon et al., 1995). In one study, a line of conditional knockout mice that lacked FGF receptor 1 (FGFR1), a major receptor for FGF-2, in the brain (Zhao et al., 2007) was tested. BrdU-labeling experiments demonstrate that FGFR1 is required for proliferation of NPCs. Generation of new neurons, but not glia, was impaired in the mutant hippocampus. Water maze experiments demonstrated that the FGFR1 mutant mice exhibit significant deficits in memory consolidation, but not in the formation of new memory. Spatial learning was also normal. Thus, proliferative neurogenesis in the adult dentate is important for memory consolidation. Another showed that the survival and differentiation, rather than proliferation, of NPCs in the dentate were significantly impaired in mice lacking NT-3 in the brain (Shimazu et al., 2001; Shimazu et al., 2002). Triple labeling for BrdU, the neuronal marker NeuN, and the glial marker GFAP indicated that NT-3 affects the number of newly differentiated neurons, but not glia, in the dentate. The NT-3 mutant mice also exhibit deficits in hippocampal-dependent, spatial memory. Taken together, these studies indicate that distinct soluble factors, which act at specific stages of hippocampal neurogenesis, may contribute differently to learning and memory.

REGULATION OF HIPPOCAMPAL PLASTICITY: A MECHANISTIC LINK BETWEEN ADULT NEUROGENESIS AND MEMORY?

If we accept that neurogenesis is required for some forms of hippocampal-dependent memories, an immediate question is how newly generated neurons contribute to memory formation or consolidation. An attractive hypothesis is that new neurons have special properties that allow them to integrate into the existing neural network, leading to an alteration of synaptic plasticity (Deisseroth et al., 2004). Young neurons have been shown to exhibit a higher input resistance and a lower threshold for firing action potentials relative to mature neurons (Snyder, Kee & Wojtowicz, 2001; Schmidt-Hieber, Jonas & Bischofberger, 2004). In addition, LTP is more easily induced, although the magnitude of LTP is much smaller. Since LTP in the DG is known to be important for hippocampal-dependent memory, it is conceivable that changes in neurogenesis could alter learning and memory by regulating LTP in the hippocampal dentate. Indeed, running, which has been shown to promote neurogenesis in the DG, not only improved water maze learning, but also selectively enhanced LTP in the medial perforant path (MPP)-granule cell synapses (van Praag et al., 1999a). Low-dose radiation, which effectively reduced the number of BrdU-labeled cells, resulted in an unequivocal reduction in LTP at the MPP synapses (Snyder, Kee & Wojtowicz, 2001). Mice lacking the MBD1 gene also exhibited a parallel reduction in neurogenesis and a selective impairment in LTP at the MPP synapses, but not CA1 synapses (Zhao et al., 2003). The only exception was observed in presenilin mutant mice, which exhibited decreased neurogenesis and impaired memory without any change in dentate LTP (Feng et al., 2001; Wang et al., 2004).

Using field-recording techniques, the role of adult neurogenesis in dentate hippocampal synaptic plasticity has been investigated. In FGFR1 knockout mice, deficits in NPC proliferation were accompanied by impaired LTP in the existing synaptic circuit at MPP-granule neuron synapses (Zhao et al., 2005). Basal synaptic transmission and other forms of short-term synaptic plasticity at these synapses are essentially normal. In contrast, deletion of the NT-3 gene in the brain affects the survival and differentiation, but not proliferation, of NPCs (Shimazu et al., 2001; Shimazu et al., 2002). The NT-3 mutant mice also exhibited selective impairment in LTP in the existing synaptic circuit at the lateral perforant path (LPP) synapses, but not MPP synapses. Again basal synaptic transmission and short-term plasticity at the LPP synapses are normal, except for a slight decrease in paired-pulse facilitation (PPF). In short, blockade of FGF-2 signaling inhibits proliferation of NPCs and impairs MPP LTP, whereas blockade of NT-3 signaling inhibits survival and differentiation of NPCs and impairs LPP LTP. Although specific mechanisms regarding LPP and MPP synapses remain to be worked out, it is possible that adult neurogenesis is necessary for the formation of the synaptic network which forms the structural basis for LTP. Taken together, these studies provide a critical link between neurogenesis and dentate LTP, and suggest that adult neurogenesis contributes to hippocampal-dependent memory by controlling the formation of the synaptic network for LTP at dentate synapses.

Why is adult neurogenesis needed for memory? Since it takes two to four weeks for NPCs to mature into neurons and integrate into the existing synaptic network, it is unlikely that learning-induced neurogenesis mediates the acute formation of new memory. Two simple models for the functional role of adult neurogenesis in memory processes may be proposed. In model A, generation of new neurons and the formation of new synaptic circuits would result in a disruption of information flow in the existing neural network (Figure 14.1 (A)). The original circuit 1→2→3→1 would be changed to, for example, 1→4→2→3. This could provide a structural basis for the erasure of existing memory, and may also lead to better performance in future memory tasks. In model B (Figure 14.1 (B)), ongoing adult neurogenesis, as a consequence of learning or running, may be necessary to maintain the integrity of the neural network underlying long-term memory. A critical element of this model is that new neurons have to be added precisely to replace existing neurons. Without neurogenesis, memory could still form, but it will steadily decay and therefore not be consolidated. A related question is: how could neurogenesis alter dentate LTP? Since changes in LTP have only been reported in MPP and LPP synapses, we consider here only the dendritic integration of new neurons. Imagine that the growing dendrites of a newly generated granule neuron form new synapses with the perforant path axons (Figure 14.1 (C)). According to Schmidt-Hieber, Jonas and Bischofberg (2004), LTP by whole-cell recording of the young neuron should easily be induced. The magnitude of LTP by field recording should also be bigger because of the increase in the number of synapses at the MPP. Another possibility is that new neurons may suppress the GABAergic inputs to the MPP synapses (Figure 14.1 C)), leading to enhanced LTP. Similar mechanisms may apply to the LPP synapses. Based on these ideas, we could speculate that a reduction in the number of new neurons at the dentate would reduce the magnitude of field LTP recorded at MPP synapses. This is precisely what was observed in MBD1 and FGFR1 mutant mice (Zhao et al., 2003; Zhao et al., 2005). On the other hand, inhibition of dendritic growth of the new neurons could contribute to impaired LTP at LPP- but not MPP-synapses, as exemplified in the NT-3 mutant mice (Shimazu et al., 2001; Shimazu et al., 2002).

CONCLUSION AND FUTURE PERSPECTIVES

The concept of adult neurogenesis has now been widely accepted. There are two significant challenges in the field of neurogenesis. The first is to identify specific factors that control different aspects of neurogenesis. Neurotrophins are certainly good candidate factors. Efforts should be made to determine

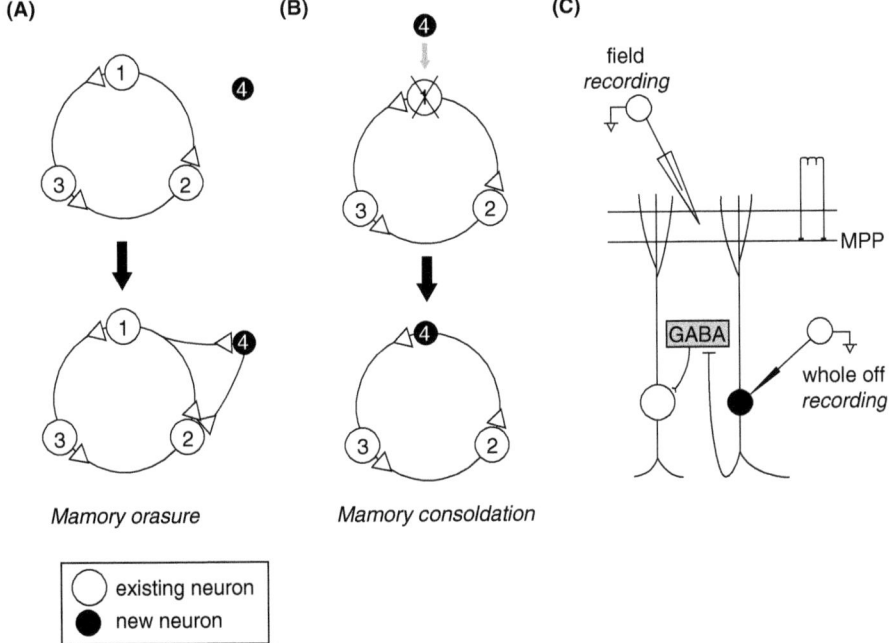

Figure 14.1. Models of functional roles of adult neurogenesis in memory processes. (A) New neurons may disrupt existing neural networks to form new synaptic circuits. This may underlie the erasure of existing memory. (B) Integration of new neurons into existing neural networks to replace dying neurons may serve as a mechanism for memory consolidation. (C) New neurons at the dentate gyrus (DG) may enhance LTP either by increasing the number of synapses at the perforant path or by suppressing GABAergic inputs to the perforant path. Reprinted with permission from Lu and Chang, 2004.

the precise step(s) in which specific neurotrophins regulate neurogenesis as well as the specific brain regions or developmental stages, or both. In particular, it will be important to determine how neurotrophins regulate neurogenesis *in vivo* under physiologically relevant conditions, such as during exercise, learning, and in an enriched environment. The second challenge is to determine the functional significance of adult neurogenesis. Although there is good correlation between adult neurogenesis in the hippocampal DG and learning and memory, a causal link between the two remains to be established. A crucial piece of experimental evidence would be to demonstrate that inhibition of specific aspects of neurogenesis could attenuate learning or memory. A genetic approach appears far superior to the use of inhibitors or radiation. Future efforts should be directed toward ablating new neurons in a spatially and temporally specific manner, using strategies such as region-specific expression of toxin and inducible systems. Moreover, by deleting genes involved in NPC proliferation, fate determination, or neuronal differentiation, we should be able pinpoint the contributions of specific stages of neurogenesis for learning and memory. An attractive hypothesis is that adult neurogenesis contributes to learning or memory by regulating LTP at the dentate synapses. How newly generated neurons integrate into existing synaptic circuits, and how this integration affects dentate LTP remain as difficult issues that will require innovative approaches to be resolved. The relationship between neurotrophins, adult neurogenesis, and dentate synaptic plasticity represents an exciting new area of research that is likely to generate important new insights into our understanding of how the gene–environment interaction controls cognition and brain plasticity.

15 · Focal adhesion-like processes underlie induction of long-term potentiation in the Schaffer collateral–CA1 region of the hippocampus

Richard G. LeBaron, Ruben V. Hernandez, Mary M. Navarro, James E. Orfila, Lisa R. Curry, and Joe L. Martinez Jr

Today, most researchers agree that memories are made of networks of neurons that are connected functionally by either increased efficacy of synaptic transmission or by long-term potentiation (LTP) (Martinez & Derrick, 1996), and are connected physically by changes in the shape of synaptic contacts (Trachtenberg *et al.*, 2002) among other processes (Federmeier, Kleim & Greenough, 2002). In this chapter, we focus on two classes of molecules, integrins and proteoglycans, which connect synapses to extracellular molecules and are ideally placed to alter synapse shape and signaling. In other cell systems these two molecules work with the actin cytoskeleton and intracellular signaling molecules to allow focal adhesions to develop. In addition, they anchor cell membranes to the extracellular matrix, allow changes in cell shape, and function in cell motility (Woods *et al.*, 1986; LeBaron *et al.*, 1988; Mitra, Hanson & Schlaepfer, 2005). In this chapter, we explore the evidence that a focal adhesion-like process is necessary for LTP in the Schaffer collateral–CA1 (Sch–CA1) pathway of the hippocampus.

In the hippocampus and neuronal membranes, α and β subunits of integrins form various heterodimers. Integrins include domains that bind divalent cations which are important for receptor–ligand interactions linking extracellular molecules, the actin cytoskeleton, and cytoplasmic signaling proteins. Integrins are believed to be involved in the process of CA1 LTP because the addition of the integrin-binding ligand Gly-Arg-Gly-Asp-Ser-Pro (GRGDSP; hereafter referred to as RGD) or function-inhibiting integrin antibodies either before or shortly after induction significantly reduces LTP (Staubli, Chun & Lynch, 1998; LeBaron *et al.*, 2003).

Proteoglycans are complex molecules defined as a protein to which one or more linear (typically sulfated) carbohydrate chains (glycosaminoglycans) are attached. Based on the types of monosaccharide that comprise the carbohydrate chains, proteoglycans are grouped into heparan, chondroitin, and keratan sulfates. The core protein of some proteoglycans spans the plasma membrane, which makes available a cytoplasmic tail that is involved in transmembrane signaling. LTP in CA1 is altered significantly when proteoglycan function is antagonized, which implicates these complex molecules in processes underlying synaptic plasticity. Blocking the action of heparan sulfate proteoglycan (HSPG) by a heparin-binding peptide that disrupts focal adhesions impaired the maintenance of CA1 LTP (Navarro *et al.*, 2000). Similarly, applying heparan binding-growth associated molecule (HBGAM), a HSPG-binding protein, to area CA1 of the rat hippocampus antagonizes LTP (Lauri *et al.*, 1998). Thus, two legs of a process that form focal adhesions are involved in the induction of LTP, which provides evidence that a focal adhesion-like process might exist in CA1 synapses.

In our laboratory we have confirmed that RGD impairs Sch–CA1 LTP in hippocampal slices *in vitro*. We also find that the heparin-binding peptide that blocks the formation of focal adhesions also reduces LTP. Although either RGD or heparin-binding peptide can reduce LTP significantly, in our hands it is not inhibited completely by either peptide alone. However, administering the two peptides simultaneously completely eliminates LTP one hour after induction. We interpret these data to indicate that the two molecules function synergistically and support the idea of the involvement of a focal adhesion-like process in LTP. Additionally, we have extended previous findings of a time-dependent effect of RGD on LTP (Staubli, Chun & Lynch, 1998) with evidence that RGD significantly reduces LTP within a few minutes of LTP induction: binding of RGD in CA1 *stratum radiatum* increases within minutes of LTP induction and this effect is Ca^{2+} dependent (Figures 15.1 and 15.2).

Application of three concentrations of RGD (5 μM, 50 μM, and 250 μM) has no detectable effect on baseline responses (Figure 15.2 (A, B, C)). The peptide GRGDSP has a significant effect on LTP at 50 minutes after theta-burst stimulation (TBS), with 250 μM producing a statistically significant decrease in the amount of LTP compared with artificial cerebrospinal fluid (aCSF) controls (Figure 15.2 (C)). The inactive peptide GRADSP, bath applied at 250 μM, does not affect the Sch–CA1 field exitatory postsynaptic potential (fEPSP) either at baseline or at 60 minutes post-TBS.

Our studies, and those of others, show that application of RGD to CA1 during TBS and for up to 40 minutes post-TBS reduces LTP. When RGD is applied 30–45 minutes post-TBS, no reduction is evident (Staubli, Chun & Lynch, 1998). Similarly, LTP is unaffected when 250 μM RGD is applied 30 minutes post-TBS (Figure 15.3 (A)). Unexpectedly, however, application of 250 μM RGD at five minutes post-TBS has no effect on LTP compared with aCSF controls (Figure 15.3 (B)).

These time-dependent results suggest that the effects of RGD on LTP occur in the first few minutes after TBS. If so, a brief application of the peptide should reduce LTP similarly to the extended application. This hypothesis is supported by experiments showing that application of RGD 10 minutes before TBS and removal at five minutes post-TBS decreases LTP to the same degree as the extended application of peptide, which was bath applied for 40 minutes after induction of LTP (Hernandez *et al.*, 2001) (Figure 15.4).

In our studies, RGD has no detectable effect on baseline fEPSPs in the 10 minutes before LTP induction. However, others have shown that when RGD is applied for extended periods of time, increases in the fEPSP slopes and amplitudes are evident, starting 20–30 minutes post-infusion (Kramar *et al.*, 2003). Our studies

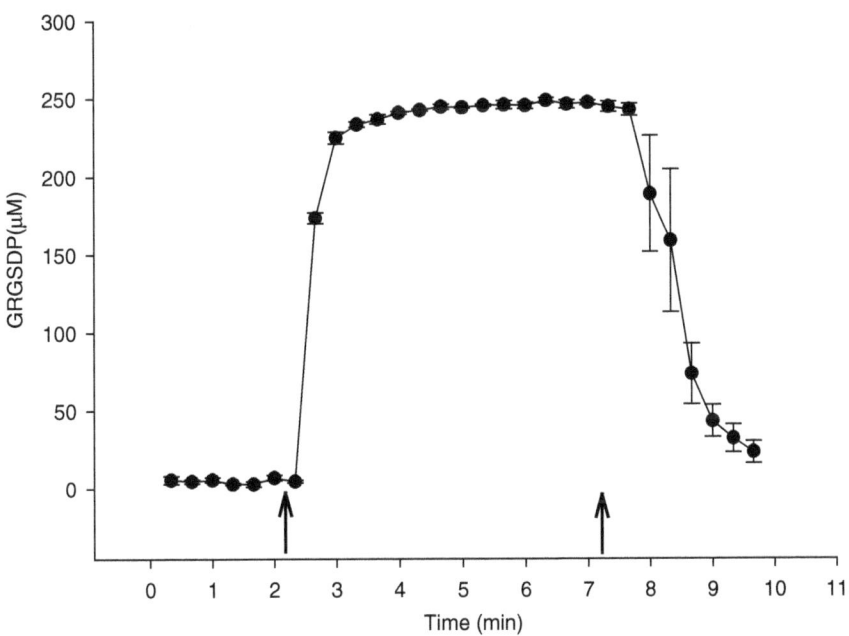

Figure 15.1. The integrin-binding ligand RGD is delivered to and cleared from an interface chamber in minutes. The aCSF exiting the chamber was collected every 20 seconds and the peptide concentration quantified by recording the absorbance at 215 nm. The first arrow indicates entry of the peptide into the chamber and the second arrow indicates the start of the aCSF washout (aCSF without RGD). Data are mean (± SEM) ($n = 5$). Reprinted from LeBaron *et al.* (2004), with permission.

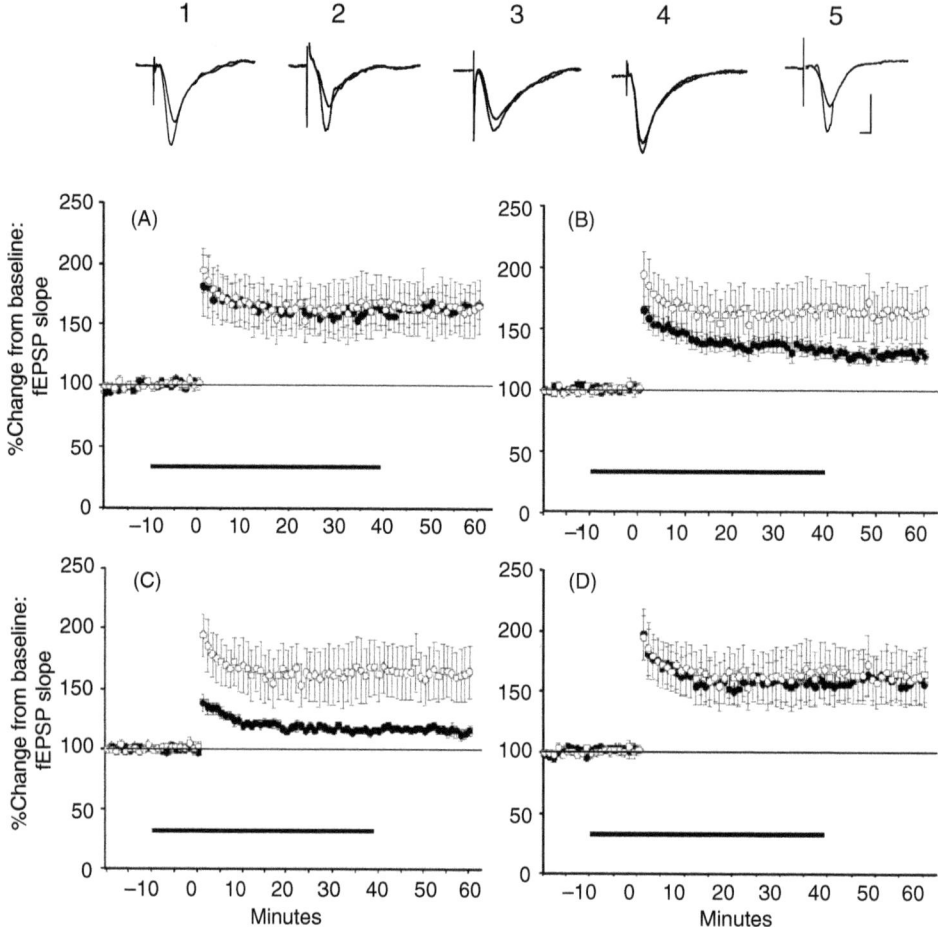

Figure 15.2. The integrin-binding ligand RGD reduces LTP. *Above* Sample traces of fEPSPs before and after TBS. (1) aCSF; (2) 5 μM RGD; (3) 50 μM RGD; (4) 250 μM RGD; and (5) 250 μM GRADSP. Scale bars: 5 ms (horizontal); 0.5 mV (vertical). *Below* Open circles, aCSF; filled circles, peptide. Horizontal bars show when peptide was applied. (A) 5 μM RGD, $n = 7$; (B) 50 μM RGD, $n = 10$; (C) 250 μM RGD $n = 11$; and (D) 250 μM GRADSP, $n = 8$. The 250 μM RGD significantly reduces LTP compared with aCSF only (Dunn's pairwise comparison, $p < 0.05$). TBS was introduced at time $= 0$. Data are mean ($\pm$ SEM). Reprinted from LeBaron et al. (2004), with permission.

induced LTP after a 10-minute application of RGD, a time when little, if any, effect on baseline responses occurred in the study by Kramar and colleagues. Even in our experiments, in which RGD was applied for 40 minutes post-LTP induction, fEPSPs remained decreased significantly compared with controls and did not increase during the one-hour post-LTP recording. The extended application experiments were in the timeframe that Kramar et al. (2003) reported increases in fEPSP in response to RGD. Regarding the effects of RGD on potentials evoked by low-frequency stimulation (LFS), a significant difference between these two studies is that LTP is induced only during our experiments. This difference in the effect of RGD on fEPSPs might indicate that some mechanisms of the α-amino-3-hydroxy-5-methyl-4-isoxazolepropionate (AMPA) receptor (AMPAR) mediated response are activated differentially depending on whether LTP is induced in the presence of RGD. Future studies should reconcile these findings.

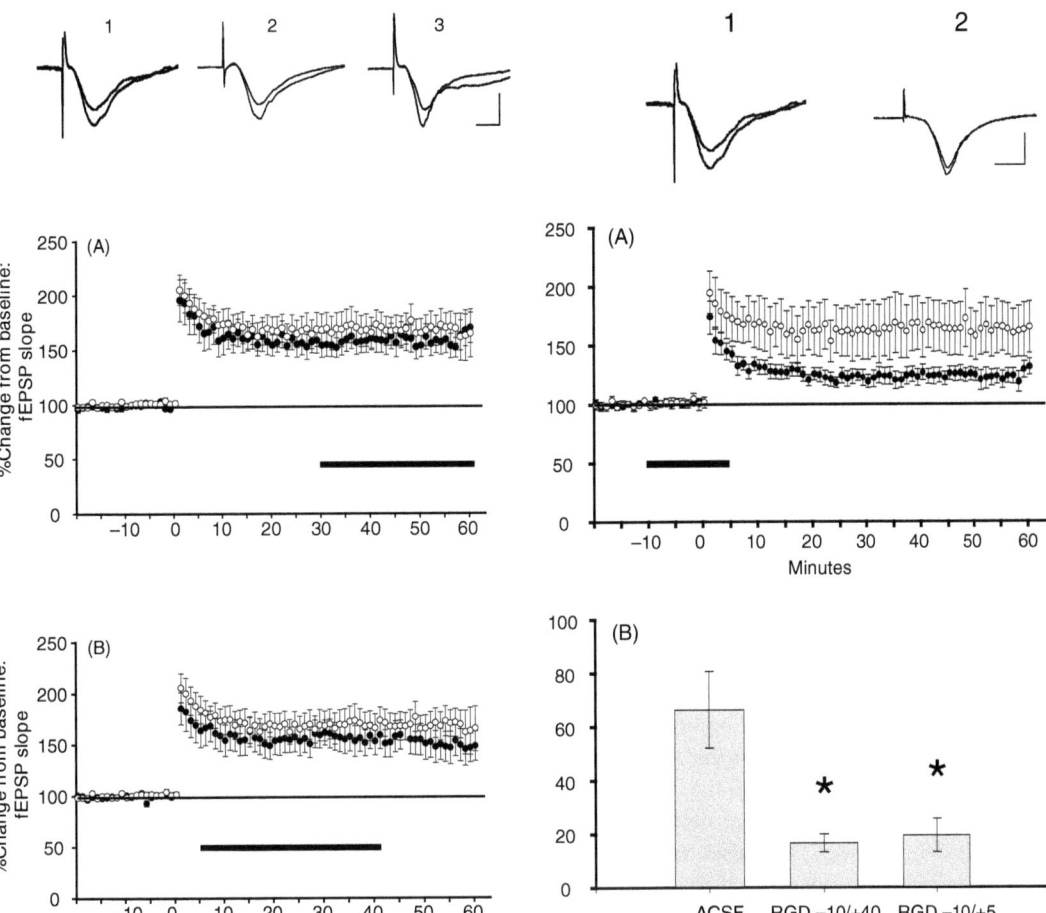

Figure 15.3. The peptide RGD (250 μM) does not antagonize LTP at two time points post-TBS. *Above* Sample traces of fEPSPs before and after TBS. (1) aCSF; (2) 250 μM RGD applied 30 minutes post-TBS; and (3) 250 μM RGD applied five minutes after TBS. Scale bars: 5 ms (horizontal); 0.5 mV (vertical). *Below* Open circles, aCSF only; filled circles show the following experimental conditions: (A) 40-minute application of RGD beginning 30 minutes post-TBS (55.9% (± 11.2%), $n=9$); (B) 35-minute application of RGD beginning five minutes post-TBS (56.3% (± 14.3%), $n=7$). Horizontal bars indicate when peptide was applied. Data are mean (± SEM). Reprinted from LeBaron *et al.* (2004), with permission.

Figure 15.4. Brief application of RGD at the time of TBS reduces LTP. Bath application of 250 μM RGD 10 minutes before TBS, and peptide clearance five minutes post-TBS (15 minutes; brief application, $n=9$) significantly decreased LTP (19.4% (± 6.2%), Student's *t*-test; $p < 0.05$), similar to the extended application (50 minutes). (A) Open circles, aCSF only ($n=12$); filled circles, brief application of RGD groups ($n=9$). (B) Comparison of aCSF, 50-minute application of 250 μM RGD (− 10 to + 40) and 15-minute application of 250 μM RGD (− 10 to + 5). Data are mean (± SEM). *Significant difference from aCSF group (Student's *t*-test; $p < 0.05$). *Above* Sample traces of fEPSPs before and after TBS. (1) aCSF and (2) 250 μM RGD applied 10 minutes before and five minutes after TBS. Scale bars: 5 ms (horizontal), 0.5 mV (vertical). Reprinted from LeBaron *et al.* (2004), with permission.

Another set of experiments examines the effects of RGD on LTP within the 15-minute application period. In these experiments we quickly reduced the peptide concentration in the chamber. Although the peptide was not expected to be cleared completely (see Figure 15.1), the concentration was reduced efficiently with brief washout periods. First, a 10-minute application of 250 μM RGD, with a 30–60-second washout immediately before TBS, does not reduce LTP (Figure 15.5 (A), solid bar) compared with either aCSF (Figure 15.5 (A), open bar) or 250 μM inactive peptide, GRADSP (Figure 15.5 (A), shaded bar). This indicates that RGD might not engage integrins that affect LTP before tetanus. To determine whether a decrease in LTP by post-TBS application of RGD is concentration-dependent within the ranges tested previously (Staubli, Chun & Lynch, 1998; LeBaron et al., 1999), a 40-minute bath application of 500 μM RGD, beginning five minutes post-TBS, has been tested. This does not decrease CA1 LTP (Figure 15.5 (B), solid bar) compared with aCSF controls (Figure 15.5 (B), open bar) (Hernandez et al., 2001).

Therefore, one principal finding of RGD action on LTP is that it alters mechanisms that are crucial for the maintenance of LTP during the first minutes following induction, but it does not necessarily affect early baseline responses. This raises the possibility that integrin-binding activity itself is enhanced by induction of LTP. A fluorophore (Alexa Fluor 488) may be attached to the RGD peptide (RGD-488) to obtain spatial and temporal representations of integrin binding. The fluorescence signal increases at CA1 by use of the TBS 200 stimulation paradigm. Hippocampal slices incubated with RGD-488 exhibit an increased number of stained CA1 processes (Figure 15.6 (A,D)) compared with LFS CA1 (Figure 15.6 (B, E)). Furthermore, integrin-binding activity peaks in two minutes following TBS and returns to levels similar to those in LFS CA1. The binding of RGD-488 depends on the presence of cations that are essential for integrin binding. Including ethylenediamine tetraacetic acid (EDTA) in the incubation solution stops the increase in RGD-488 binding compared with identical experiments without the calcium ion chelator (Figure 15.6 (C,F)). The fluorescence signals from stained processes are compared in Table 15.1. The data indicate a significance difference in the fluorescence signal from RGD-488 staining in TBS CA1 compared with the signal from LFS CA1 and CA1 incubated in EDTA.

A body of evidence indicates that integrins have roles in synaptic transmission at post-synaptic membranes (Staubli, Chun & Lynch, 1998; Chun et al., 2001; Gall et al., 2003; LeBaron et al., 2003; Lin et al., 2003). However, electrophysiological and genetic evidence indicate that integrins in pre-synaptic neurons are important for synaptic plasticity (Orfila et al., 2000; Chavis & Westbrook, 2001; Chan et al., 2003; LeBaron et al., 2003). Currently, the most definitive evidence of integrin location in CA1 cell

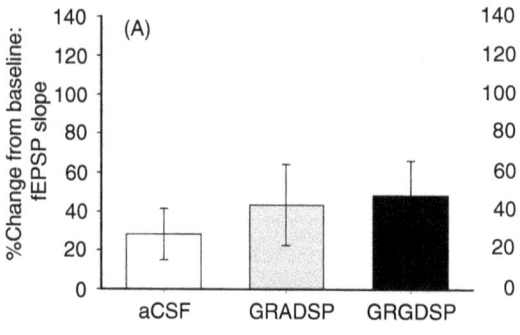
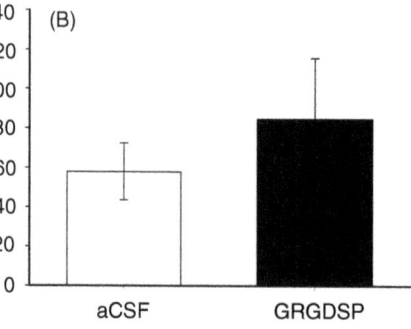
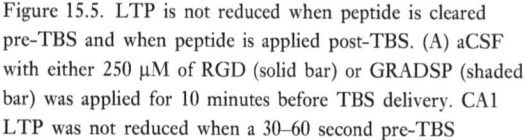

Figure 15.5. LTP is not reduced when peptide is cleared pre-TBS and when peptide is applied post-TBS. (A) aCSF with either 250 μM of RGD (solid bar) or GRADSP (shaded bar) was applied for 10 minutes before TBS delivery. CA1 LTP was not reduced when a 30–60 second pre-TBS washout was used to reduce the concentration of RGD. (B) After induction of robust CA1 LTP, a 40-minute application of 500 μM RGD, starting five minutes after TBS does not reduce LTP. Reprinted from LeBaron et al. (2004), with permission.

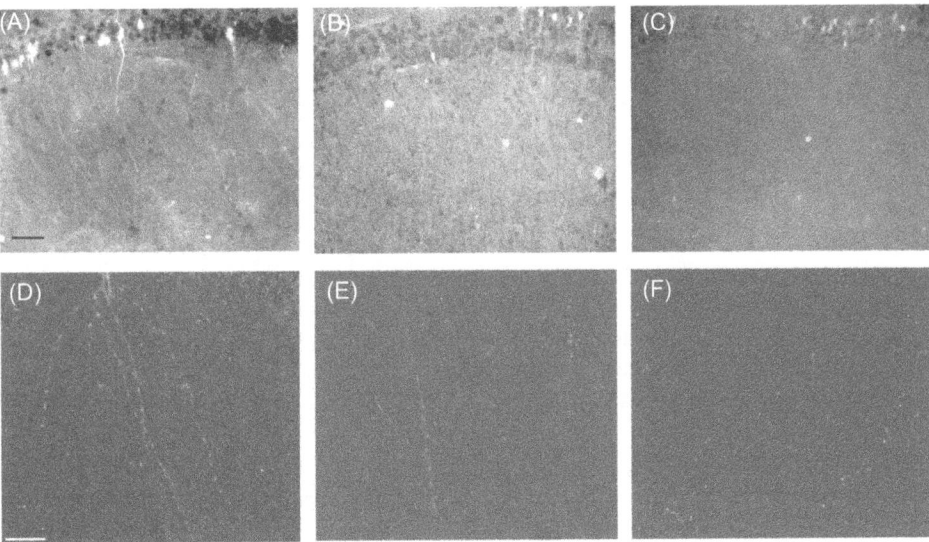

Figure 15.6. Hippocampal tissue labeled with RGD-488. (A,D) Hippocampal tissue labeled with 250 μM of RGD RGD-488 at two minutes post-TBS. (B,E) LFS tissue labeled with 250 μM RGD-488. (C,F) At two minutes post-TBS, hippocampal tissue was labeled with 250 μM RGD-488 in aCSF containing EDTA. Scale bars, 50 μm (A–C); 30 μm (D–F). Reprinted from LeBaron et al. (2004), with permission.

Table 15.1. Quantified fluorescence signal of RGD-488 binding to integrins in TBS and LFS CA1

CA1	Intensity sum (10^9)	Standard deviation (10^9)
2-min post-TBS	8.0*	1.5
LFS	4.8	0.6
EDTA-treated	4.6	0.4

*Significant difference in the staining detected in TBS CA1 compared with LFS and EDTA-treated CA1. Student's t-test, $p < 0.05$, $n = 3$.

types comes from ultrastructural analyses. Our first studies detect integrin α5 in pre-synaptic and post-synaptic neurons. Antibodies to both ectodomains and cytodomains of integrin α5 occur at the CA1 synapse (Figure 15.7). Immunogold particles in pre-synaptic neurons are detected either at or near vesicles (Table 15.2). Previous studies have indicated that integrins and proteoglycans colocalize on the plasma membrane and function together in the development of focal adhesions. Because proteoglycans have been detected with integrins on growth cones of neurons in culture, we reasoned that proteoglycans might have a function at the synapse and underlie the induction of LTP in area CA1.

This hypothesis has been tested using the peptide WQPPRARITGY, which antagonizes the formation and maintenance of focal adhesions (Woods et al., 1993). When this peptide is bath-applied to tissue 10 minutes before TBS until five minutes post-TBS, no detectable effect on baseline responses during the 20-minute pre-tetanus period are seen using a wide range of concentrations: 0.14 μM to 14 μM WQPPRARITGY. Whether the heparin-binding peptide has an effect on baseline responses when application times are extended during LFS must be examined in future studies by testing the mechanisms of proteoglycan-mediated effects on LTP. However, initial trials reveal that 14 μM WQPPRARITGY significantly reduces LTP by 80% (Figure 15.8 (A)). As a control, 14 μM of the scrambled sequence RYPQGIPWTAR has no significant affect on either LTP or baseline responses compared with aCSF controls (Figure 15.8 (B)). Interestingly, combined application of the integrin-binding peptide (250 μM)

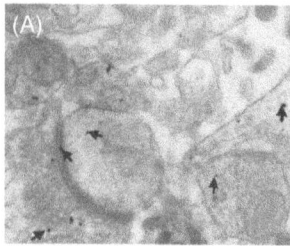 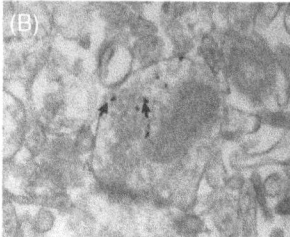 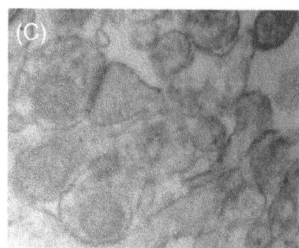

Figure 15.7. Integrin α5 antibodies label pre- and post-synaptic elements in hippocampal tissue. (A) Antibody to an ectodomain of the integrin α5 subunit. (B) Antibody raised against the cytoplasmic tail of the integrin α5 subunit. (C) Normal rabbit IgG. Arrows indicate silver-enhanced gold particles. Magnification, 28 000×. Reprinted from LeBaron et al. (2004), with permission.

Table 15.2. Anti-α5 antibody staining of CA1 synapses

α5 antibody	Pre-synaptic	Post-synaptic	Number of synapses
Ectodomain	113	21	12
Cytodomain	46	3	6
Control	10	10	13

and heparin-binding peptide (14 μM) to the bath containing hippocampal slices blocks LTP completely (Figure 15.8 (C)).

DISCUSSION

Our results reveal that 250 μM RGD is the lowest concentration tested that reduces LTP significantly. Although this concentration does not alter baseline responses recorded 10 minutes before TBS, evidence from studies that did not induce LTP found that extended periods of RGD application (30–60 minutes) increases the fEPSP slopes and amplitudes compared with controls. These studies indicate a possible relationship between the effects of RGD on baseline responses with changes in AMPAR kinetics (Kramar et al., 2003). Further evidence indicates that cytoplasmic signaling molecules function in these integrin-mediated effects on CA1 fEPSPs through binding of peptides and larger ligands (Kramar et al., 2003; Bernard-Trifilo et al., 2005). Therefore, the AMPAR fEPSP response to RGD might depend, in part, on whether LTP is induced.

RGD-containing peptides target integrins and perturb their binding to their natural ligands. The specificity of the tripeptide is significant; a conservative amino acid substitution of GRGDSP to yield GRADSP inactivates the peptide and greatly reduces its ability to perturb cell adhesion (Pierschbacher & Ruoslahti, 1984; Torimoto et al., 1990). Thus GRADSP is a control peptide, and LTP is not reduced by 250 μM GRADSP in our experiments, demonstrating that the LTP-blocking effects of GRGDSP are caused by actions of the ligand peptide. Furthermore, the effect of GRGDSP on LTP is concentration-dependent and, thus, consistent with a ligand–receptor interaction. Therefore, the molecular mechanism of RGD is likely to involve the binding of RGD to cell-surface receptors, such as members of the integrin superfamily.

Additionally, we find that LTP is reduced significantly when GRGDSP is applied during the delivery of a conditioning train, utilizing either a −10- to +5-minute or a −10 to +40-minute application paradigm. In general, the impairing action of RGD in hippocampal CA1 LTP agrees with the results of others (Staubli, Chun & Lynch, 1998). However, there are two unexpected outcomes of our experiments. First, the effective time period for RGD to decrease LTP is shorter by 15 minutes. When RGD is applied five minutes after induction of LTP (+5 to

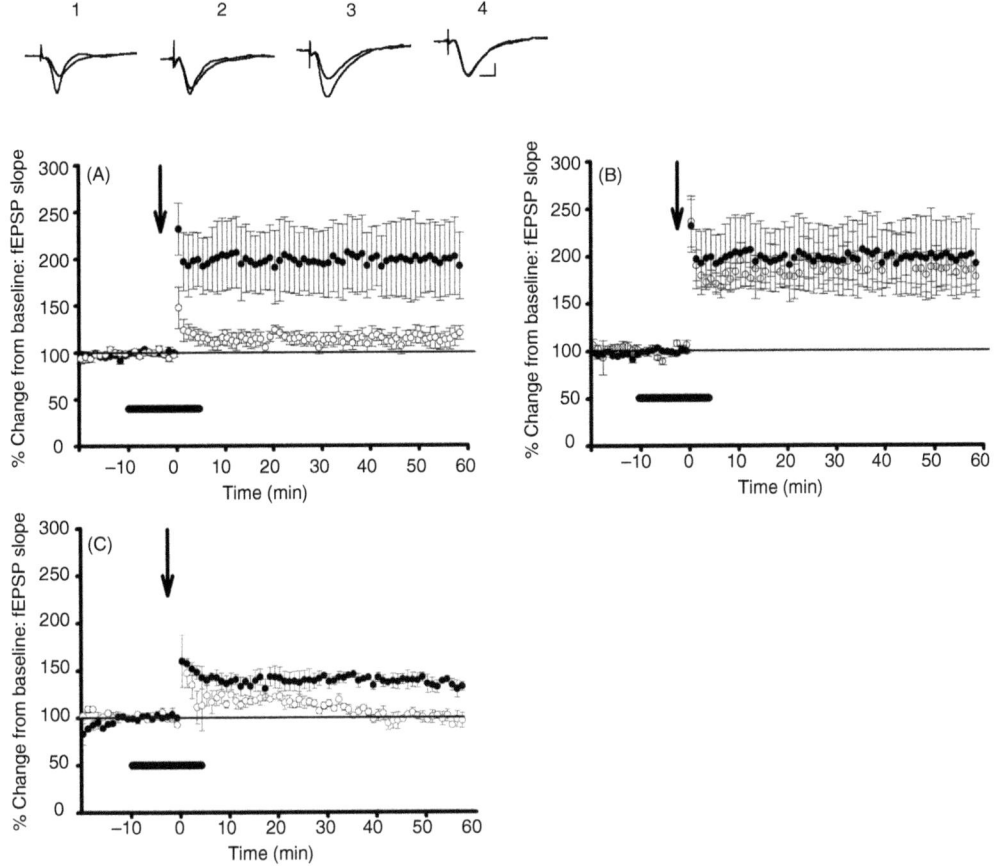

Figure 15.8. LTP is perturbed by peptides that antagonize the formation of focal adhesions. Peptides were bath applied 10 minutes before tetanization and cleared five minutes post-tetanization. (A) Effect of 14 μM of a HSPG binding peptide WQPPRARITGY on Sch–CA1 fEPSP induced by TBS. Application of 14 μM WQPPRARITGY (open circles; $n = 7$) significantly reduced Sch–CA1 fEPSP compared with aCSF controls (filled circles; $n = 7$). (B) Effect of 14 μM of the scrambled peptide RYPQGIPWTAR on Sch–CA1 fEPSP LTP induced by TBS. Application of 14 μM RYPQGIPWTAR, a scrambled peptide, (open circles; $n = 6$) had no effect on Sch–CA1 LTP compared with aCSF controls (filled circles; $n = 7$). (C) Combination of 14 μM WQPPRARITGY and 250 μM RGD significantly reduced Sch–CA1 LTP (2.0% (± 2.0%), open circles; $n = 3$) compared with aCSF controls (37.3% (± 4.8%), filled circles; $n = 3$). Arrows indicate delivery of the conditioning train. *Above* Sample fEPSPs before and after TBS. (1) aCSF; (2) 14 μM WQPPRARITGY; (3) 14 μM RYPQGIPWTAR; and (4) 14 μM WQPPRARITGY plus 250 μM RGD. Scale bars: 5 ms (horizontal) and 0.5 mV (vertical). Reprinted from LeBaron et al. (2004), with permission.

+40 paradigm), no antagonistic effects on LTP are detected, unlike the significant reduction in LTP when RGD is applied during TBS (−10 to +5 and −10 to +40 paradigms). This indicates that the peptide affects LTP only in the first few minutes after induction. Also, concentrations of RGD greater than the effective concentration (250 μM) do not impair LTP when applied five minutes post-TBS (Hernandez et al., 2001), which further indicates that the need for RGD to be present in the first minutes after tetanization is a temporal requirement rather than a concentration effect.

Another unexpected result regarding a possible RGD-mediated effect on early events of potentiation was found in experiments to test for changes in binding of labeled RGD peptide immediately following induction of LTP. First, the labeled peptide itself reduces LTP, indicating that integrins that bind labeled RGD are crucial for LTP. In addition, the binding of labeled-RGD is higher on TBS 200-stimulated CA1 during the first two minutes after LTP induction compared with LFS CA1. This indicates that integrins are either activated as a consequence of LTP induction or are translocated from intracellular vesicles to the cell surface following TBS. The RGD 488 staining is at neuronal processes, as expected of integrins that are involved in Sch–CA1 LTP.

Additionally, RGD binding is Ca^{2+} dependent because it is reduced by the addition of EDTA. To locate integrins, we have recently begun to examine the ultrastructural location of integrin subunits by use of electron microscopy. Preliminary experiments detect anti- $\alpha 5$ integrin antibodies in both pre-synaptic and post-synaptic neurons. If substantiated, these early findings support the conclusions of several investigations that show that integrins function in the synapse (Staubli, Chun & Lynch, 1998; Chavis & Westbrook, 2001; Chan et al., 2003) and also at both pre-synaptic and post-synaptic terminals. However, the contribution of pre-synaptic and post-synaptic integrins to LTP still needs to be elucidated.

The full array of mechanisms by which integrins support LTP is unknown. Current hypotheses suggest that integrins might act as either signaling proteins (Bernard-Trifilo et al., 2005) or be involved in the structural rearrangement of synapses through arrangements of actin (Fischer et al., 1998). These ideas are not mutually exclusive and are both components of our focal adhesion hypothesis. It is clear that integrin binding activates cytoplasmic signaling proteins (Giancotti & Ruoslahti, 1999) including focal adhesion kinase (FAK), which mediates transmembrane signaling and changes in the shape of plasma membranes to facilitate cell motility (Mitra et al., 2005). FAK is recruited to focal adhesions, which occur in most cell types (Schaller et al., 1995). The formation of focal adhesions appears to require integrins and proteoglycans at the cell surface (Woods et al., 1986; LeBaron et al., 1988; Woods et al., 1993; Woods & Couchman, 1994), and occur either at or near synapses. We reason that if focal adhesions are important structures for synaptic plasticity, then proteoglycans should play a role in the induction of CA1 LTP. We have tested this hypothesis, using a heparin-binding peptide derived from the sequence of fibronectin. This peptide (WQPPRARITGY) blocks the formation of focal adhesions and antagonizes their maintenance (Woods et al., 1993). Application of this peptide to the bath reduces CA1 LTP in experiments that are, essentially, identical to the RGD–peptide paradigm (Hernandez et al., 2001; LeBaron et al., 2003). The effect of this peptide is concentration-dependent, with LTP significantly reduced by 14 μM WQPPRARITGY but not affected by the scrambled peptide (which maintains essentially identical charges on the amino acid side-chains). These data agree with previous studies that test HSPG effects in synaptic plasticity (Lauri et al., 1998; Lauri et al., 1999). Finally, mixing both integrin-binding and heparin-binding peptides (250 μM and 14 μM, respectively), blocks LTP completely, which indicates that these two classes of molecules act together, as with focal adhesions, to affect either the signaling or the structural events required for synaptic plasticity.

16 · Signaling to the nucleus in long-term memory

Olena Bukalo and R. Douglas Fields

Persistent changes in synaptic strength engage cellular processes extending to the nucleus. How the messages are generated and transmitted to the nucleus are important areas for current research controversy. Similarly, how the appropriate genes are transcribed and their protein products delivered to the specific synapse to be strengthened (or weakened) are crucial questions of active research. In this chapter we consider how signals reach the nucleus to control long-term memory (LTM), and we examine the relative contribution of synaptic input and post-synaptic action potential firing.

Despite the importance of neuronal firing in theoretical learning rules, most experimental research on how genes are regulated to consolidate changes in synaptic strength and memory concerns synaptic potentials and synapse-to-nucleus signaling molecules. However, the identity of the synapse-to-nucleus signaling molecules and how the nucleus integrates inputs from thousands of synapses to produce an appropriate genomic response are unclear. For these reasons, we have focussed our experimental studies on how action potential firing regulates gene expression necessary for converting early- to late long-term potentiation (LTP), a cellular model of LTM.

The importance of neuronal action potential firing is central to the cellular rules for learning. Environmental input conveyed through action potential firing regulates development of neurons and glia, and initiates the adaptive responses of the nervous system in post-natal life. In 1949, Donald Hebb predicted a form of synaptic plasticity driven by temporal contiguity of pre- and post-synaptic activity. Specific function for the Hebbian synapse would be the conversion of short-term into long-term memory by stabilization of reverberatory activity patterns. According to Hebb's postulate coordinated action potentials firing between two neurons strengthens connections, which would be the basis for learning and memory (Hebb, 1949). The molecular mechanisms for the induction and stabilization of synapse strengthening during learning and memory has become a major focus of cellular neuroscience over the past decades, yet the focus of experimental research and theory has been on synaptic potentials and synapse-generated signals.

It has been known for at least 40 years that gene transcription and protein synthesis are necessary for consolidating STM into LTM, but only recently has the paradox of how a single nucleus could control the strength of tens of thousands of synapses on the same neuron been resolved through elegant experiments supporting the concept of "synaptic tagging" (Frey & Morris, 1997). Through the mechanism of synaptic tagging, a synapse that is strengthened temporarily enters into a state that can selectively utilize the newly synthesized gene product which will consolidate the increased synaptic efficacy. Three aspects of this theory must be elucidated at a molecular level: (1) the mechanism for the synaptic tag to "mark" potentiated synapses; (2) the signaling mechanism that reaches the nucleus to activate gene transcription; (3) the gene products that consolidate the temporary synaptic strengthening into a persistent increase. This chapter will consider each of these, with an emphasis on how synaptic excitatory potentials and somatic action potentials initiate signaling to the nucleus. Research from our laboratory suggests that a synapse-to-nucleus signaling molecule is not necessary for consolidating LTP, and we provide evidence supporting an alternative hypothesis that somatic action potentials are essential in activating gene transcription (Dudek & Fields, 2002).

GENE TRANSCRIPTION AND PROTEIN SYNTHESIS IN SYNAPTIC STRENGTHENING AND LEARNING

Long-term enhancement in synaptic strength (LTP) and the formation of memories share, at least in part, common mechanistic properties at the neuronal level. Both LTP and LTM could be characterized in terms of their temporal storages and depend on new proteins and messenger ribonucleic acid (mRNA) synthesis for consolidation. The formation of STM and the cellular analog an "early phase" of LTP (E-LTP) are unaffected by transcriptional and translational inhibition. In contrast, LTM and "late phase" of LTP (L-LTP) have a greater amplitude and longer duration (more than three hours) and, critically, depend on new mRNA and protein synthesis (Krug, Lossner & Ott, 1984; Otani & Abraham, 1989; Nguyen, Abel & Kandel, 1994; Frey & Morris 1997; Kandel, 2001). In addition, long-lasting forms of long-term depression (LTD), sustained reduction of synaptic responses, which depend on protein synthesis, have recently been described (Huber, Roder & Bear, 2001; Sajikumar & Frey, 2003).

Gene transcription in the nucleus is essential for L-LTP. Although mRNAs and translational machineries are observed in dentrites, and synaptic activity may regulate the local translation of some mRNA (reviewed by Steward & Schuman, 2001), it has been demonstrated that dentrites that have been severed from their cell bodies fail to maintain LTP (Frey et al., 1989; but see also Cracco et al., 2005; Vickers, Dickson & Wyllie, 2005). Thus, an understanding of how signals initiated at the cell membrane reach the nucleus to induce this transient upregulation (and perhaps also repression) of essential gene transcription and protein translation is crucial.

SIGNALING TO THE NUCLEUS

Activated synapses may send signals to the nucleus in a number of ways. The prevailing theory is that local generation of signaling intermediates at the synapse diffuse or are transported into the nucleus (synapse-to-nucleus signaling). This view arose from the observation that pharmacologic inhibition of some synaptic receptors, in particular N-methyl-D-aspartate receptors (NMDARs), blocks expression of many activity-dependent genes (see Platenik, Kuramoto & Yoneda, 2000). Evidence for synapse-to-nucleus signaling has been observed in *Aplysia californica* motor neurons. In this system, the cyclic adenosine monophosphate (cAMP), effector molecule of protein kinase A, and the Ras pathway intermediate mitogen-activated protein kinase (MAPK) are translocated into the nucleus after neural activity (Bacskai et al., 1993; Martin et al., 1997). Other observations supporting this view are that activated transcription factors and calmodulin may transmit information from activated synapses into the nucleus (Meberg et al., 1996; Deisseroth, Heist & Tsien, 1998; Mermelstein et al., 2001; Meffert et al., 2003).

A self-sustaining feedback cycle between synapses and the nucleus is a possibility for LTP; however, the maintenance of two-way pathways between individual synapses and the nucleus appears to be cumbersome. One problem with such mechanisms is that the transport of molecules from synapses is slow and the messengers are likely to be corrupted en route to the nucleus, especially in synapses located on the more distal dendrites. It is unclear how the nucleus would evaluate and integrate thousands or tens of thousands of messages from each synapse on the dendritic tree to determine the appropriate transcriptional response. Shuttling molecular signals into and out of the nucleus in response to synaptic activity undoubtedly occurs in association with many physiological processes, but may not be effective or necessary for LTP.

An alternative mechanism proposes that somatic or back-propagated spikes activate gene transcription essential for L-LTP (Dudek & Fields, 2002; Fields, Lee & Cohen, 2005). The somatic action potential firing is the consequence of integration between synaptic input, and neuronal firing is the logical condition for triggering gene expression to strengthen synapses between two neurons in accordance with learning theory. Back-propagating dendritic action potentials may generate large calcium transients that propagate to the cell soma (Magee & Johnston, 1997; Markram et al., 1997; Nakamura et al., 1999) and, as the nuclear pore does not appear to be a diffusion barrier for calcium ions (Lipp et al., 1997), are also likely to invade the nucleus. Therefore, in contrast to the uncertain identity of the synapse-to-nucleus signaling molecule involved in L-LTP, the well-established role of calcium in intracellular signaling from action potentials (discussed later) could activate gene expression in response to action potential firing.

The concept that synapse-to-nucleus signaling is the prevailing mechanism to activate gene expression in L-LTP was mainly supported on the basis of experiments suggesting that activation of a transcription factor involved in L-LTP and LTM, cAMP response element-binding protein (CREB), could be phosphorylated by calcium influx through neurotransmitter receptors, but not required calcium influx through voltage-sensitive calcium channels (VSCCs) (Bading, Ginty & Greenberg, 1993; Deisseroth, Heist & Tsien, 1996). The demonstration that, in cell culture systems, neurons discriminate between action potentials and excitatory post-synaptic potentials (EPSPs), generated by synaptic stimulation, producing genetic response to synaptic potentials rather than to action potentials (Luckman, Dyball & Leng, 1994; Deisseroth, Bito & Tsien, 1996; Mermelstein et al., 2000), is another example arguing for the synapse-to-nucleus signaling.

To test the hypothesis that CREB and gene expression essential for conversion of E-LTP to L-LTP could be activated in CA1 hippocampal neurons by action potentials in the absence of a synapse-to-nucleus signaling molecule, experiments were performed in rat hippocampal slice in the presence of pharmacological blockers of excitatory synaptic transmission (Dudek & Fields, 2002). Action potentials were induced in the alveus (the white matter that coats the hippocampus and contains axons from the area CA1 pyramidal cells) through a theta-burst stimulation (TBS) delivered antidromicaly. This stimulation resulted in activation of extracellular signal-regulated kinase1/2, robust phosphorylation of CREB, and activation of *zif268*, an immediate early gene associated with LTP and learning (Dudek & Fields, 2002). Antidromic stimulation resulted in sustained L-LTP after a weak synaptic stimulus that otherwise would have resulted only in transient E-LTP. Thus, our experiments support the hypothesis that somatic action potential firing converts E-LTP to L-LTP (and by extrapolation STM to LTM) by activating the gene expression. Since genetic responses occur in the absence of glutamatergic synaptic activity, a synapse-to-nucleus signaling is not necessary for L-LTP.

SPECIFICITY OF SIGNALING – SYNAPTIC TAGGING AND CAPTURE

LTP has among its features synapse specificity. How synapse specificity of L-LTP is achieved, and how are gene products from the nucleus targeted to the few activated synapses in a vast dendritic tree? What are the factors that facilitate potentiated synapses to sequester factors necessary for LTP maintenance? To answer these questions a "synaptic tag" hypothesis has been proposed. It suggests that synaptic specificity is conferred by a synaptic "tag," whose molecular nature is as yet unknown but is conceptualized to function as spatially restricted and persistent markers of synaptic activity and should fulfil some criteria. First, it should be spatially restricted; second, it should be time-limited and reversible; third, it should be able to interact with cell-wide molecular events that occur after strong stimulation (most commonly thought to involve changes in gene expression) to produce long-term, synapse-specific strengthening (see Martin & Kosik, 2002; Kelleher, Govindarajan & Tonegawa, 2004; Reymann & Frey, 2007). The slate of "synaptic tag" candidates is wide-ranging, including persistent protein kinases, changes in adhesion molecules at the synapse, alterations in cytoskeletal elements, activation or trafficking of channels, protein synthesis or degradation, and translation. At present, the challenge is to understand whether and how each may interact with products of transcription to produce enduring changes in synaptic efficacy.

The tag function is to sequester or "capture" proteins newly synthesized as a result of L-LTP induction. Evidence supporting tagging hypothesis was first obtained by Frey and Morris (1997), who reported a novel long-term heterosynaptic facilitation of L-LTP when examining two independent synaptic inputs in the Schaffer collateral pathway. Repeated tetanization of the first input resulted in the establishment of homosynaptic protein synthesis-dependent L-LTP, which was inhibited by pretreatment with a protein synthesis inhibitor (either anisomycin or emetine). One hour later, repeated tetanization was delivered to the second input in the presence of a protein synthesis inhibitor, and normal L-LTP was paradoxically observed, suggesting that the proteins synthesized in response to L-LTP induction in the first input also enabled the establishment of L-LTP in the second. Thus, providing of the newly synthesized proteins produced in response to L-LTP induction is sufficient to convert of E-LTP to L-LTP. The synaptic tag model has been supported further by a number of studies in the rodent hippocampus (see Frey & Morris, 1998; Martin & Kosik, 2002; Kelleher, Govindarajan & Tonegawa, 2004; Pang & Lu, 2004; Reymann & Frey, 2007): most of them have been limited to examining tagging and capture in a dendritic compartment, induced by synaptic stimulation.

In an experiment similar to the work by Frey and Morris (1998), we demonstrated that action potentials initiated antidromically in axons from CA1 neurons in the absence of synaptic activity are sufficient to protect E-LTP from decay (Dudek & Fields, 2002). The results of our study demonstrated that somatic action potentials may "prime" synaptic tagging, and indicated that action potentials were sufficient for inducing the synthesis of mRNAs, which were then available for use by the "tagged" synapses. Our data were confirmed recently by experiments showing that L-LTP induction is facilitated in basilar and apical dendritic compartments of CA1 pyramidal cells by action potentials (Alarcon, Barco & Kandel, 2006), indicating that the mechanism underlying synaptic tagging and capture in different functional compartments is similar. The nature of initial stimulation (synaptic or non-synaptic) could serve as a detection mechanism to activate neuron-wide tags (initiated by action potentials' back-propagation) versus restricted to specific compartment tag (initiated by synaptic potentials).

SELECTIVITY OF CALCIUM SIGNALING TO THE NUCLEUS

Information in the nervous system is coded in the pattern of action potential firing. How the transcription of specific genes may be regulated by appropriate patterns of impulse firing through intracellular calcium fluxes is an important question. Calcium is a ubiquitous second-messenger, which penetrates into cells through gated channels to transmit signals from membrane to cytoplasm and nucleus. How does intracellular calcium achieve the signal specificity associated with electrical stimulation? In neurons there are two general mechanisms that regulate calcium-dependent cellular processes, including gene transcription, in a selective manner: spatial and temporal regulation.

Spatial regulation

Spatial compartmentalization of calcium signals could link different calcium sources to the downstream effector mechanisms responsible for the maintenance of LTP. Calcium signals in different cellular compartments, and consequently different strengths of electrical stimuli, modulate distinct although overlapping sets of genes (reviewed by Bading, 2000; West *et al.*, 2001). Numerous studies on visualization of calcium signal in response to synaptic activity have demonstrated that the calcium transients directly derived from NMDARs are limited to the parts of dendritic tree receiving direct synaptic output, such as dendritic spines (Bliss & Colingridge, 1993). Some types of electrical stimulation are effective to trigger calcium influx through NMDARs and activate the signaling pathways leading to a transcriptional response (Bading, Ginty & Greenberg, 1993; Hardingham, Arnold & Bading, 2001; Hardingham, Fukunaga & Bading, 2002). Since synaptic NMDARs are a considerable distance from the parent dendrite and soma, it is possible that extra-synaptic NMDARs (Lozovaya *et al.*, 2004) would be important to generate the nuclear signaling (Hardingham, Fukunaga & Bading, 2002).

Activation of VDCCs, on the other hand, produces calcium influx mainly in the cell body or proximal dendrites, where the majority of these channels are found (Weick, Kuo & Mermelstein, 2005). Recent experiments have shown that different forms of LTP, which depend on different calcium sources, coexist at CA3–CA1 synapses and, more importantly, that the calcium signals emanating from these discrete sources are spatially segregated (Raymond & Redman, 2006). In these experiments, synaptic stimulation results in elevated levels of cytosolic and nuclear calcium via L-VSCCs and consequently in generation of L-LTP in the Schaffer collateral pathway. Activating nifedipine-sensitive L-VSCCs channels in the absence of synaptic stimulation by action potentials delivered to CA1 axons in theta pattern leads to CREB phosphorylation and expression of IEG *zif268* (Dudek & Fields, 2002). Using a developmentally regulated transgenic mouse model in which calcium or calmodulin signaling was selectively inhibited in the nucleus, has been shown to directly impair cyclic adenosine monophosphate-response element- (CRE-) mediated gene expression and L-LTP, while leaving E-LTP unaffected (Limback-Stokin *et al.*, 2004). In addition, mice lacking the L-VSCCs displayed a reduction in tetanus-induced L-LTP (Moosmang *et al.*, 2005). These studies suggest that transcription-dependent forms of synaptic plasticity require somatic (cytosolic and nuclear) calcium elevations and activation of L-VSCCs is essential in activation of transcription.

Although influx of calcium through NMDARs and through VSCCs is of major significance in synaptic plasticity (Bliss & Collingridge, 1993), the internal stores, functioning as the sink or additional sources of calcium, may contribute to LTP. Action potentials may trigger a release of calcium from internal stores in hippocampal neurons (Alford et al., 1993; Jacobs & Meyer, 1997). By soaking up and storing the brief pulses of calcium associated with each action potential, the intracellular stores may track neuronal activity and be able to signal this information to the nucleus through periodic bursts of calcium (Berridge, 1998). The nuclear calcium release from intracellular stores has been proposed as mechanism mediating gene expression by calcium-mediated synapse-to-nucleus communication (Hardingham, Arnold & Bading, 2001). Induced by tetanic stimulation L-LTP and CREB phosphorylation may be reduced by blocking calcium release from intracellular stores (Lu & Hawkins, 2002).

Temporal regulation

The timing of action potentials may play a critical role in determining the synaptic changes in both sign and magnitude. In particular, LTP can be observed if the pre-synaptic action potential is followed by a post-synaptic one, whereas LTD occurs if the temporal order of the action potentials is reversed. This type of plasticity, named spike timing-dependent plasticity, is believed to be important in shaping the various synaptic plasticities required for learning and adaptation (Dan & Poo, 2004). Temporal regulation of intracellular calcium signaling is particularly important in decoding action potential firing patterns (reviewed by Fields, Lee & Cohen, 2005). On the molecular level, what are the mechanisms that transduce the temporal features of membrane depolarization into appropriate intracellular signaling pathways to stimulate or inhibit transcription of specific genes? Frequencies in the range of 0.1–1 Hz were shown to activate calcium or calmodulin-dependent protein kinase II (CaMKII) (Eshete & Fields, 2001). Action potentials delivered at higher frequencies (5 Hz, 100 Hz) result in activation of MAPK signaling pathways (Dudek & Fields, 2001, 2002) that lead to the phosphorylation of CREB (Dudek & Fields, 2002). Cytosolic calcium oscillation frequency regulates activation of nuclear factor of activated T cells (NFATc) (Dolmetsch et al., 1998; Li et al., 1998b; Tomida et al., 2003). Increasing the frequency of action potentials firing in primary hippocampal neurons resulted in an increase of the nuclear calcium load accompanied by an enhanced, dose-dependent c-fos expression (Hardingham, Arnold & Bading, 2001). In addition, the duration of different pharmacologically induced nuclear calcium transients has been shown to determine the magnitude of gene transcription in cultured hippocampal neurons (Chawla & Bading, 2001).

A threshold number of action potentials should prevent the spurious activation of genes not relevant to plasticity; this threshold could be cell-type specific, allowing it to adjust to the functions and needs of the cells, accordingly. It has been demonstrated that in acute slice preparation, nuclear calcium signals during LTP induction play a role as a digital on/off switch for activating transcription (Johenning & Holthoff, 2007). In that model, nuclear transcription provides a basal pattern of mRNAs in an "all or none" fashion. This plasticity-related basal mRNA pattern might have a mainly static maintenance function, precluding the necessity of graded transcription.

SIGNALING TO REGULATE TRANSCRIPTION IN THE NUCLEUS

What is the target of somatic calcium transients during induction of L-LTP? Through the extensive network of interactions, calcium entering at the plasma membrane or in the nucleus may induce a number of signaling pathways known to be essential for transcription factor activation (see Berridge, 1998; Bading, 2000; West et al., 2001; West, Griffith & Greenberg, 2002; Deisseroth et al., 2003). In general, transcription factors are proteins that bind to a specific DNA sequence in a promoter or enhancer region and, via interaction of their trans-activation domains with the transcription complex of RNA polymerase, activate or suppress transcription of a particular gene.

In the nucleus

A typical representative of the constitutive transcription factor is CREB, the most studied calcium-regulated transcription factor. In fact, CREB remains constitutively bound to CRE (calcium/cAMP responsive element) regulatory sites controlling target genes, and becomes activated by phosphorylation at Serine-133 (Ser-133) (with modulatory influences exerted by other amino acids) (Deisseroth & Tsien, 2002). A fast calmodulin-dependent kinase cascade and a slow MAPK cascade converge on Ser-133 after surface membrane depolarization and calcium influx, and work together to promote CREB-dependent gene expression (Bito, Deisseroth & Tsien, 1996; Deisseroth, Heist & Tsien, 1998; Dolmetsch et al., 2001; Hardingham, Arnold & Bading, 2001). After phosphorylation at Ser-133, CREB recruits the CREB binding protein (CBP), which acts as a transcriptional coactivator (Chrivia et al., 1993; Kwok et al., 1994). CBP promotes transcription through its recruitment of components of the ribonucleic acid (RNA) polymerase transcription machinery and through its function to remodel chromatin structure into a form that is accessible to active transcription. CBP-mediated chromatin remodeling is the critical component of memory consolidation, as has been demonstrated by study with transgenic mice that express CBP with eliminated histone acetyltransferase activity (Korzus, Rosenfeld & Mayford, 2004). Other CREB coactivators in mediating CRE-driven gene transcription have been described. Among them modulators called transducers of regulated CREB activity (TORCs) (Conkright et al., 2003; Iourgenko et al., 2003). Nuclear translocation of TORCs enhances CREB-dependent gene transcription (Bittinger et al., 2004; Screaton et al., 2004), and TORC1 is essential for CRE-driven gene transcription and maintenance of L-LTP in the hippocampus (Zhou et al., 2006).

A novel calcium-binding protein acting as transcriptional repressor, named DREAM (downstream regulatory element antagonistic modulator) has been recently identified (Carrion et al., 1999). DREAM is a calcium-binding protein with specific roles in different cell compartments (Carrion et al., 1999; An et al., 2000). In the nucleus, DREAM acts as a calcium-dependent transcriptional repressor (Carrion et al., 1999; Osawa et al., 2001). It has been proposed that nuclear DREAM remains bound to a downstream regulatory element (DRE) that acts as a gene silencer when nuclear calcium is low, but dissociates upon elevation of calcium causing DRE derepression, activation of downstream genes such as that which encodes prodynorphin (Carrion et al., 1999) and attenuation of pain signaling in vivo (Cheng et al., 2002). In the absence of calcium DREAM may also bind to CREB. As a result, DREAM impairs recruitment of CBP by phosphorylated CREB and blocks CBP-mediated transactivation at CRE sites in a calcium-dependent manner (Ledo et al., 2002).

Outside the nucleus, DREAM, also named KChIP3 (voltage-gated potassium channel-interacting protein) or calsenilin (An et al., 2000), interacts with potassium channels of the Kv4 class directing their trafficking to and inside the plasma membrane (Takimoto, Yang & Conforti, 2002) and regulating in a calcium-dependent manner the gating properties of the channel (An et al., 2000). Phosphorylation of DREAM is regulated by G-protein coupled protein kinases and calcineurin is important for regulation by DREAM Kv4 channel cell surface expression (Ruiz-Gomez et al., 2007). Interestingly, deletion of Kv4.2 gene eliminates dendritic A-type potassium current and enhances induction of LTP in CA1 pyramidal neurons (Chen et al., 2006). The location on apical dendries and properties of the A-type potassium currents make them suitable as key regulators of action potential backpropagation (Hoffman & Johnston, 1998; Chen & Johnston, 2004). In the cytosol DREAM binds to the C-terminal region of the presenilins, transmembrane proteins that function as a part of the γ-secretase protease complex (Buxbaum et al., 1998), and blocks the release of calcium from the endoplasmic reticulum (Lilliehook et al., 2002). Mice deficient for DREAM exhibit lower Aβ formation, associated to Alzheimer's disease and enhanced LTP (Lilliehook et al., 2003).

From cytoplasm to nucleus

Somatic calcium transients are known to activate a number of constitutive transcription factors via calcium-dependent translocation from the cytosol to the nucleus (reviewed by Cruzalegui & Bading, 2000; West et al., 2002; Deisseroth et al., 2003). Compared with CREB, relatively little is known about the mechanisms which support signaling to activate these transcription factors during LTP consolidation. The classic example is a member of the nuclear factor of activated T cells (NFAT) family of transcription factors. Recently identified in hippocampus, NFATc initiates gene expression after a period of heightened synaptic activity and is hypothesized to play a critical role in shaping long-term changes in cell excitability (Graef et al., 1999; Graef et al., 2003; Groth & Mermelstein, 2003; Benedito et al., 2004). NFATc undergoes a striking translocation from the cytosol to the nucleus upon the opening of L-VSCCs in response to electrical activity (Graef et al., 1999). The translocation step is dependent upon calcineurin-mediated dephosphorylation of NFATc. Once in the nucleus NFATc requires the cooperative binding of a phosphorylated nuclear partner to initiate transcription (Crabtree & Olson, 2002; Hogan et al., 2003).

NF-κB is another constitutive transcriptional factor normally found in the cytoplasm. In an inactive form NF-κB is bound to its inhibitor, IκB. The rise of intracellular calcium through opening of L-VSCCs at the plasma membrane and calcium release from intracellular stores activate NF-κB (Lilienbaum & Israel, 2003). Activation of NF-κB proceeds through the site-specific phosphorylation, polyubiquitylation, and proteasome-mediated degradation of the IκB (West, Griffith & Greenberg, 2002). The newly liberated NF-κB complex rapidly translocates into the nucleus, where it engages cognate κB enhancer elements in a variety of cellular target genes (Ghosh & Karin, 2002). As transcriptional regulators, NF-κB proteins may potentially either positively or negatively regulate the expression of genes governing changes in synaptic plasticity and cognitive functions (West, Griffith & Greenberg, 2002). Several reports support a positive link between the activation of NF-κB factors and the induction of LTP or LTD (Meberg et al., 1996; Albensi & Mattson, 2000; Freudenthal, Romano & Routtenberg, 2004) and memory (Yeh et al., 2002; Meffert et al., 2003). In contrast, other studies suggest a negative correlation between NF-κB action and synaptic functions (Furukawa & Mattson, 1998; O'Mahony et al., 2006). Increased NF-κB action is also associated with the accelerated onset of cognitive deficits in an experimental model of Alzheimer's disease (Arancio et al., 2004).

From membrane to nucleus: regulated intramembrane proteolysis

Recent studies have revealed a role for regulated intramembrane proteolysis (RIP) in the generation of nuclear signals from transmembrane proteins. As a result of gamma-secretase-dependent intramembrane proteolytic cleavage, intracellular domains become soluble cytoplasmic messengers that can signal back to the nucleus, in some cases acting directly as transcription factors (Ebinu & Yankner, 2002). Thus, RIP results in quick and direct activation of target genes, bypassing adaptor proteins and kinase cascades (Ebinu & Yankner, 2002). Several of these transmembrane proteins play important roles in neuronal development, including Notch (Redmond & Ghosh, 2001), neuregulins and their ErbB receptors (Buonanno & Fischbach, 2001), or in neuronal degeneration, including amyloid precursor protein (Wilquet & De Strooper, 2004).

One of the RIP examples is C-terminal fragment of the $Ca_v1.2$ subunit of L-VSCCs named CCAT (the calcium channel-associated transcriptional regulator), has been described recently (Gomez-Ospina et al., 2006). The nuclear localization of CCAT is negatively regulated by calcium influx via L-VSCCs and NMDARs. In the nucleus, CCAT binds to a protein associated with an endogenous promoter and regulates the expression of genes important for neuronal signaling and excitability, including gap junction, NMDARs subunits, sodium-calcium exchanger, and potassium channels. By analogy with DREAM it has been proposed that CCAT both regulates transcription and reduces calcium influx trough L-VDCCs (Gomez-Ospina et al., 2006).

GENES ACTIVATED BY CALCIUM SIGNALING TO THE NUCLEUS

Identifying the genes that are regulated by LTP-induced stimuli is essential for understanding the molecular mechanisms underlying L-LTP and LTM formation. Immediate early genes (IEGs) are the first group to be expressed following synaptic activation. IEGs are operationally defined as those mRNAs expressed in the presence of protein synthesis inhibitors, and thereby do not require *de novo* protein synthesis for expression (Guzowski, 2002). Initial evidence into the relationship between intense synaptic activity and rapid transient gene expression came from studies involving the IEG c-*fos* (Morgan et al., 1987) and the role of signaling through CREB (for review of the early IEGs literature, see Sheng & Greenberg, 1990).

IEG proteins may be placed into two broad categories: inducible transcription factors (c-*fos*, c-*jun*, and *zif268*) and direct effector proteins (e.g. activity-regulated cytoskeletal-associated protein (Arc), Homer, brain-derived neurotrophic factor (BDNF), tissue plasminogen activator (tPA)). The expression of late effector genes may require IEGs (Morgan & Curran, 1991; Clayton, 2000) or a second wave of constitutive transcription factors' activation (Bernabeu et al., 1997b; Viola et al., 2000). Late effectors may complete synaptic modifications initiated by IEGs to influence synapse growth, morphology, or function (for reviews see Clayton, 2000; Guzowski, 2002). The proteins of effector

IEGs may influence memory consolidation processes by participating in late structural events which involve synthesis or posttranslational modifications of neural cell adhesion molecules (Fields & Itoh, 1996; Dityatev & Schachner, 2003; Kleene & Schachner, 2004). A direct role for the IEGs c-*fos* (Johnson, Spiegelman & Papaioannou, 1992; Paylor *et al.*, 1994), *zif 268* (Abraham *et al.*, 1993; Worley *et al.*, 1993; Jones *et al.*, 2001), tPA (Pang & Lu, 2004), BDNF (Lu, 2003; Bramham & Messaoudi, 2005), and Arc/Arg3.1 (Tzingounis & Nicoll, 2006) in neuroplasticity and LTM consolidation has been demonstrated in mice, by use of transgenic and knockout strategies, and in rats, by use of antisense methods.

An alternative possibility to the direct involvement of IEGs in synapse strengthening would be an orchestration of the expression of genes that encode proteins involved in synaptic plasticity and memory formation (Tzingounis & Nicoll, 2006). IEGs may act to increase memory storage for later events. Therefore, instead of directly functioning to facilitate consolidation of the experience which induced its expression, an IEG may facilitate consolidation of experiences yet to come. Thus, IEGs could function to enable a metaplastic state, where periods of earlier activity influence the storage of later events (Clayton, 2000).

As the expression of several IEGs is regulated through the action of CREB, the memory impairments caused by CREB antisense treatment (Guzowski & McGaugh, 1997) or in CREB knockout mice (Bourtchuladze *et al.*, 1994) might be attributed to disruption of IEG expression. To explore the role played in hippocampal synaptic plasticity by CRE-driven genes, mice with regulated expression of a constitutively active form of the transcriptional activator CREB, called VP16-CREB, were generated and characterized (Barco, Alarcon & Kandel, 2002). Expression of VP16-CREB lowers the threshold for L-LTP induction and results in L-LTP being independent of new transcription (Barco, Alarcon & Kandel, 2002; Barco *et al.*, 2005; Alarcon, Barco & Kandel, 2006). By analyzing altered gene expression in the hippocampus of VP16–CREB mice, these authors demonstrate that CREB activation turns on the downstream gene, producing mRNA transcripts, such as *c-fos*, *jun B*, and *Arc*, and propose that the products of these genes are captured by the activated synapses implement L-LTP (Barco *et al.*, 2005). Interestingly, another microarray study performed on the mouse hippocampus uncovered that activity-regulated genes regulating various cellular process are clustered together on chromosomes (Park *et al.*, 2006). Although genes in the same cluster have apparently different molecular properties, they are functionally correlated by regulation of LTP. In addition, in some chromosomal clusters activity-induced genes have been shown to be co-regulated by CREB.

A key component of enhanced L-LTP driven by activated CREB is increased expression of BDNF (Barco *et al.*, 2005). The synaptic actions of BDNF depend on the specific site (McAllister, 2002), the precise time window, the extent of BDNF secretion (Lessmann, Gottmann & Malcangio, 2003), and the localization of its cognate receptors (Poo, 2001). BDNF may modulate synaptic transmission within seconds (Lohof, Ip & Poo, 1993; Levine *et al.*, 1995; Kang & Schuman, 1995; Kovalchuk *et al.*, 2002). Of particular interest is that the availability of BDNF is regulated in an activity- and calcium-dependent manner at three different levels, namely by increasing transcription (Ernfors *et al.*, 1991; Shieh *et al.*, 1998; Tao *et al.*, 1998; Chen *et al.*, 2003), translation (Schratt *et al.*, 2004), and secretion (Balkowiec & Katz, 2000; Hartmann, Heumann & Lessmann, 2001; Gartner & Staiger, 2002; Aicardi *et al.*, 2004). Recent studies have established that all these regulatory events are critical steps in the L-LTP. Exogenous BDNF administered to mouse hippocampal slices rescued LTP impaired by protein synthesis inhibition (Pang & Lu, 2004). The cleavage of precursor BDNF to a mature form of BDNF is regulated by tPA, and that regulation is essential to convert E-LTP to L-LTP (Pang & Lu, 2004). BDNF levels increase after learning-related events, presumably as a result of CREB binding to the CRE sequence in its promoter (Patterson *et al.*, 1992; Korte *et al.*, 1995; Tao *et al.*, 1998).

We have demonstrated that signaling pathways associated with L-LTP may be activated by action potentials in hippocampal CA1 neurons in the absence of excitatory synaptic activity, thus eliminating the requirement for a synapse-to-nucleus signaling molecule to regulate transcription of genes necessary for L-LTP (Dudek & Fields, 2002). To determine those genes activated by action potentials firing in the presence or absence of excitatory synaptic activity, we performed custom microarray analysis to investigate the signaling pathways, genes, and proteins involved in hippocampal synaptic plasticity. Our research has identified sets of transcription factors, structural genes, and signaling pathways that are regulated by activity patterns which lead selectively to different types of synaptic changes in the CA1 subfield of hippocampus (Lee et al., 2005b). One of the genes differentially regulated by action potential firing in CA1 neurons with or without excitatory synaptic transmission was mRNA for exon-1 of BDNF. We observed that synaptic activation by TBS increases abundance of BDNF exon-1, but antidromic stimulation in the same pattern during blockade of glutamatergic synaptic transmission decreased the level of mRNA containing exon-1 of the BDNF gene. Therefore, we propose that the pattern of BDNF transcript regulation after firing of the post-synaptic neuron, correlated or uncorrelated with synaptic firing, may participate in the molecular mechanism strengthening and weakening involved in L-LTP consolidation.

TRANSLATIONAL REGULATORY MECHANISMS

What is the cellular mechanism for targeting newly synthesized mRNAs to synaptic sites? L-LTP consolidation requires the selective delivery of new molecular components to the synapses. Neurons do possess a mechanism through which newly synthesized mRNA transcripts are targeted to activated synapses where they mediate the local synthesis of proteins that become part of synapses (reviewed by Steward & Schuman, 2001; Steward & Worley, 2001; Kelleher et al., 2004). Supportive of this hypothesis is the finding that transport of CaMKII mRNA to the dendrites is required for L-LTP and behavioral memory (Miller et al., 2002). Similarly, BDNF and Arc mRNAs underwent a rapid dendritic translocation in response to synaptic stimulation (Steward & Worley, 2001; Tongiorgi, Domenici & Simonato, 2006). These mRNAs are present in RNA granules, in which they are maintained in a translationally silent state (Krichevsky & Kosik, 2001). When neurons are depolarized, the granules unravel (Krichevsky & Kosik, 2001), releasing mRNA. This is accompanied by an increase in the size of the spine, which is attributed to protein synthesis and is correlated with enhanced synaptic transmission (Ostroff et al., 2002). This mechanism would lead to an increase in the translation of a subset of mRNAs, rather than a global increase in protein synthesis. The activation of translation itself or the reorganization of the mRNA in the region of activity may serve as a "tag" that can be recognized by transcription-dependent products (Blitzer et al., 2005). The capture and mobilization of mRNA by stimulated synapses would logically precede synaptic protein synthesis because the mRNAs in granules cannot be translated (Krichevsky & Kosik, 2001), which fits with the observation that tagging in the hippocampus does not require protein synthesis (Martin & Kosik, 2002).

Another level of translation regulation is the regulation the fragile X mental retardation protein (*FMR1*) gene product, which is directly associated with fragile X syndrome (Garber et al., 2006). Fragile site mental retardation protein (FMRP), the protein encoded by *FMR1*, is an RNA binding protein abundant in the brain and involved in the control of local protein synthesis. FMRP is associated with polysomes in the cell body and dendrites, shuttles from the nucleus to cytoplasm and suppresses translation of a selective group of mRNAs, including αCaMKII, microtubule-associated protein and Arc, to which it binds (Brown et al., 2001; Laggerbauer et al., 2001; Li et al., 2001). Current theories suggest that FMRP controls the local translation in response to synaptic activity: L-LTD is enhanced in FMRP null mice (Bear, Huber & Warren, 2004). It has been proposed that translation suppression induced by FMRP occurs through microRNA (miRNA), a class of non-coding RNAs (Jin, Alisch & Warren, 2004; Garber et al., 2006). These miRNAs are not translated into protein, but might modulate translation by binding to mRNAs, thereby regulating their translation and controlling stability of mRNA transcripts (Ambros, 2001; Bartel, 2004). Little is known about signals that induce miRNAs expression or mechanism of their transcriptional regulation in general, and, in particular, how miRNAs may be involved in regulation of synaptic functions in mammals. About 20% to 40% of miRNAs in brain is developmentally regulated and associated with synapse formation (Presutti et al., 2006). Recently it has been demonstrated that BDNF regulates the local translation of some miRNAs (Vo et al., 2005; Schratt et al., 2006). One of them, miR-132, identified as the target of the CREB, in response to BDNF stimulation regulates neurite outgrowth by decreasing levels of

GTPase-activating protein, p250GAP (Vo *et al.*, 2005). Another miRNA, miR-134, inhibits translation of an mRNA encoding a protein kinase, Limk1, that controls spine development, and BDNF treatment relives this repression (Schratt *et al.*, 2006). miRNAs could therefore finely tune the translational response to synaptic activation by affecting the efficiency of translation or the stability of mRNA, thus regulating synaptic strength and growth of spines (Klein, Impey & Goodman, 2005; Presutti *et al.*, 2006). It has been suggested that miRNAs have several properties that make them plausible candidates for recognizing a tag (Martin & Kosik, 2002). Transcriptional activation might disperse miRNAs widely throughout the soma and dendritic tree, where they recognize mRNAs released locally from RNA granules that are positioned at activated synapses.

CONCLUSIONS AND FUTURE DIRECTIONS

The formation of a stable memory trace is thought to require a molecular consolidation cascade. A central concept in most neurobiologic models of memory is that this process involves activity-dependent changes in gene and protein expression, which result in long-lasting alterations in the strength of synaptic connectivity and remodeling of neural networks activated during the encoding of experience. Despite considerable progress having been made in the identification of signaling pathways and transcription factors regulating gene expression in response to neuronal activity, the mechanisms that convey an initial message from the neuronal surface to a nuclear transcription factor, regulating gene expression remains largely unknown.

Action potentials represent powerful integrators of synaptic activity, making them ideally suited to regulate gene expression according to a cell's pattern of firing. We found that signaling pathways associated with L-LTP may be activated by action potentials in hippocampal CA1 neurons in the absence of excitatory synaptic activity, thus eliminating the requirement for a synapse-to-nucleus signaling molecule to regulate transcription of genes necessary for L-LTP (Dudek & Fields, 2002). Action potentials are sufficient to initiate calcium transients which propagate to the soma and nucleus to activate transcription factors (CREB) and IEGs (*zif 268*, BDNF). Temporal regulation of action potential pattern, and whether action potential firing occurs together or without synaptic transmission, may determine which transcription factors are activated and therefore which genes. The temporal dynamics of intracellular calcium, the frequency of action-potential firing, the duration of the burst, and the interval of time between repeated bursts of action potentials may have different effects on different intracellular signaling networks to act upon appropriate transcription factors and other proteins influencing gene expression (Fields, Lee & Cohen, 2005). Together with synaptic tagging mechanisms, an action potential-dependent nuclear signalling mechanism might have a crucial role in the consolidation of specific synaptic changes.

As multiple molecular mechanisms have been found to underlie LTP induction and memory formation, there seem to be multiple molecular synaptic tags which might be differentially recruited by various stimuli and mediate plasticity over different time and structural domains. Action potentials are sufficient to induce tag formation and capture to convert E-LTP to L-LTP (Dudek & Fields, 2002; Alarcon, Barco & Kandel, 2006).

Numerous questions remain. Perhaps the most significant outstanding questions are the nature of the synaptic tag(s) and the identities of the minimal set of proteins required for synaptic capture and the expression of L-LTP. How is mRNA or these proteins transported? Which proteins and mRNAs are packed to proteosomes and how is this process controlled and regulated? Can action potential firing control miRNA regulation of neuronal mRNA translation? Which proteins derived from newly synthesized mRNAs may contribute to the maintenance phase of L-LTP? Additional studies applying technologies that allow an integrative analysis of multiple signaling pathways and transcription factors within single cells and neuronal networks, together with data obtained from gene expression-profiling experiments, will provide new insights into the molecular mechanisms of memory consolidation.

Part IV
Non-traditional transmitters and glia

17 · Diffusible hydrogen peroxide generated by synaptic activity inhibits axonal dopamine release in striatum

Marat V. Avshalumov, Jyoti C. Patel, Li Bao, Duncan G. MacGregor, Zsuzsanna Sidló and Margaret E. Rice

INTRODUCTION

Reactive oxygen species (ROS) are often considered to be potentially damaging by-products of cellular respiration. Increased ROS production and oxidative stress contribute to cell death after acute brain injury, including cerebral ischemia-reperfusion (Cao et al., 1988; Hyslop et al., 1995; Chan, 2004; Starkov, Chinopoulos & Fiskum, 2004), as well as in slowly progressing neurodegenerative disorders such as Parkinson's disease (Cohen, 1994; Ebadi, Srinivasan & Baxi, 1996; Sonsalla et al., 1997; Olanow & Tatton, 1999; Zhang, Dawson & Dawson, 2000; Xu et al., 2002; Wood-Kaczmar, Gandhi & Wood, 2006; Fukae, Mizuno & Hattori, 2007; Dodson & Guo, 2007). This view of ROS is evolving rapidly, however, in light of increasing evidence that ROS also act as cellular messengers that modulate processes from short-term ion-channel activation (Seutin et al., 1995; Krippeit-Drews et al., 1999; Avshalumov & Rice, 2003; Tang et al., 2004; Avshalumov et al., 2005; Bao, Avshalumov & Rice, 2005; Hidalgo, Donoso & Carrasco, 2005) to gene transcription and cell proliferation (Suzuki, Forman & Sevanian, 1997; Sauer, Wartenberg & Hescheler, 2001; Haddad, 2002; Esposito et al., 2004; Rhee et al., 2005; Rhee, 2006; Infanger et al., 2006; Bedard & Krause, 2007). Of the ROS generated by various cellular processes, hydrogen peroxide (H_2O_2) is a particularly strong candidate as a signaling agent because it is membrane-permeable, with possible membrane transport via aquaporins (Bienert, Schjoerring & Jahn, 2006; Bienert et al., 2007). H_2O_2 is also relatively inert; in contrast to other ROS that are reactive free radicals, including superoxide anion ($•O_2^-$) and the hydroxyl radical ($•OH$), H_2O_2 is a mild oxidant (Cohen, 1994; Rhee, 2006).

Generation of ROS occurs from a variety of cellular processes, the most important of which is mitochondrial respiration, which produces $•O_2^-$ from a single-electron reduction of molecular oxygen (Boveris & Chance, 1973; Peuchen et al., 1997; Liu, Fiskum & Schubert, 2002). Additional sources of ROS include enzymes such as reduced nicotinamide adenine dinucleotide phosphate (NADPH) oxidase, which produce $•O_2^-$ that can participate in intracellular signaling cascades (Sauer, Wartenberg & Hescheler, 2001; Lambeth, 2004; Infanger, Sharma & Davisson, 2006). Levels of $•O_2^-$ are managed by mitochondrial and cytosolic forms of superoxide dismutase, which in turn produce H_2O_2 that is regulated primarily by glutathione (GSH) peroxidase, which is free in the cytosol, and catalase, which is confined to subcellular peroxisomes (Cohen, 1994; Peuchen et al., 1997; Dringen, Pawlowski & Hirrlinger, 2005). Additional regulation of H_2O_2 comes from peroxiredoxins, which are abundant, but have lower catalytic efficacy than GSH peroxidase or catalase (Rhee et al., 2001; Hofmann, Hecht & Flohe, 2002). Interaction of either $•O_2^-$ or H_2O_2 with trace metal ions, including iron and copper, may produce the aggressive radical, $•OH$, which is neutralized primarily by the low molecular weight antioxidants GSH and ascorbate (Cohen, 1994). Thus, enzymes and low molecular weight antioxidants work together to regulate ROS and prevent oxidative damage. A role for H_2O_2 and other ROS as neuromodulators requires that oxidant regulation must be more subtle than previously thought, however.

The first evidence that H_2O_2 might regulate neurotransmission was the finding that exogenous H_2O_2 could suppress the amplitude of evoked population spikes (PS) in hippocampal slices, possibly by inhibiting transmitter release (Pellmar, 1986; Pellmar, 1987). Our laboratory confirmed that transmitter release can be inhibited by exogenous H_2O_2 by demonstrating reversible suppression of evoked dopamine (DA) release during H_2O_2 exposure in striatal slices (Chen, Avshalumov & Rice, 2001). More significantly, we found that *endogenous* H_2O_2 inhibits axonal DA release in the striatum (Chen, Avshalumov & Rice, 2001, 2002; Avshalumov & Rice, 2003; Avshalumov et al., 2003) and suppresses the activity of dopaminergic neurons in the substantia nigra pars compacta (SNc) (Avshalumov et al., 2005). These data implicate that activity-dependent H_2O_2 as an endogenous neuromodulator. Regulation of the nigrostriatal DA system is important because of the central role this pathway plays in the control of movement by the basal ganglia. In dopaminergic neurons, intracellular H_2O_2 modulates cell excitability via ATP-sensitive K^+ (K_{ATP}) channels (Avshalumov et al., 2005). By contrast, modulatory H_2O_2 in the striatum is generated downstream from glutamatergic α-amino-3-hydroxy-5-methyl-4-isoxazolepropionate (AMPA) receptors (AMPARs), which are not expressed by dopaminergic axons, indicating that H_2O_2 is generated post-synaptically and then diffuses beyond a synapse to modulate pre-synaptic DA release (Avshalumov et al., 2003).

Neuromodulation by H_2O_2 is not limited to the nigrostriatal system. Other studies have shown that H_2O_2 influences characteristics of LTP in the hippocampus (Auerbach & Segal, 1997; Kamsler & Segal, 2003; Kamsler & Segal, 2004), which has implications for memory formation. Diffusible H_2O_2 also plays a role in neuron–glia signaling in the hippocampus, in which neuronal activation leads to H_2O_2-dependent phosphorylation of myelin basic protein in oligodendrocytes (Atkins & Sweatt, 1999). Thus, H_2O_2 in the brain may act as an intracellular signaling agent and as a diffusible messenger. For H_2O_2 to act at both intracellular and potentially distant targets requires that the brain antioxidant network allows functional levels of H_2O_2 and other ROS, while still preventing oxidative stress, which adds a previously unrecognized dimension to oxidant management. We have proposed (Avshalumov et al., 2004) that key features of such a permissive environment are the predominance of cytosolic GSH peroxidase activity in glia, with sub-compartmentalization of catalase in peroxisomes in both glia and neurons, and the predominance of ascorbate in neurons (Cohen, 1994; Rice & Russo-Menna, 1998; Rice, 2000; Dringen, Pawlowski & Hirrlinger, 2005).

In this chapter, we review evidence that modulatory H_2O_2 in striatum is generated downstream from dendritic AMPARs on medium spiny neurons and that inhibition of DA release by diffusible H_2O_2 is mediated by K_{ATP} channels on DA axons. Additionally, we will discuss characteristics of the neuronal and glial antioxidant networks required to prevent oxidative damage under normal conditions, yet at the same time permit intracellular, as well as cell–cell signaling by H_2O_2.

GLUTAMATE-DEPENDENT MODULATION OF DA RELEASE IN DORSAL STRIATUM REQUIRES AMPAR-DEPENDENT H_2O_2 GENERATION

The question of how, or even *if*, glutamate regulates axonal DA release in the striatum has been a longstanding source of controversy (Cheremy et al., 1986; Leviel, Gobert & Guibert, 1990; Moghaddam & Gruen, 1991; Westerink, Santiago & De Vries, 1992; Keefe, Zigmond & Abercrombie, 1993;Wu et al., 2000). The apparent absence of ionotropic glutamate receptors on DA terminals (Bernard & Bolam, 1998; Chen et al., 1998b) has suggested that any glutamatergic influence must be indirect. Proof of this has been elusive, however, and the effects of glutamate receptor blockade have often been paradoxical. Indeed, in our own studies of the effect of blocking ionotropic glutamate receptors on axonal DA release elicited by local electrical stimulation in striatal slices, we obtained the surprising result that AMPAR-blockade by the selective antagonist GYKI-52466 causes a roughly 100% increase in evoked DA release (Avshalumov et al., 2003) (Figure 17.1 (A)). Blockade of N-methyl-D-aspartate (NMDA) receptors in dorsal striatum has no effect on evoked DA release under the same conditions. These data suggested that glutamate acting at AMPARs normally *inhibits* striatal DA release. Our finding that H_2O_2 may act as an inhibitory intermediate downstream from AMPARs solves the conundrum of how glutamate regulates DA release in the dorsal striatum (Avshalumov et al., 2003).

We initially tested a possible role for H_2O_2 generation in glutamatergic regulation of striatal DA release for two reasons. First, activation of glutamate-receptors enhances mitochondrial H_2O_2 generation in

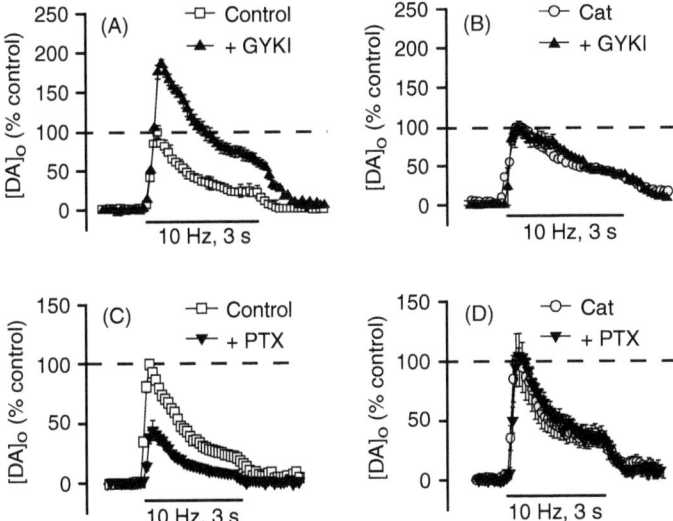

Figure 17.1. Regulation of striatal dopamine (DA) release by glutamate and γ-aminobutyric acid (GABA) requires H_2O_2. (A) α-amino-3-hydroxy-5-methyl-4-isoxazoleproprionate (AMPA) - receptor (AMPAR) blockade by GYKI-52466 (GYKI) (50 μM) causes a ~100% increase in evoked $[DA]_o$ in striatum ($p<0.001$, $n=6$). (B) The effect of AMPAR blockade is prevented by catalase (Cat). (C) $GABA_A$ receptor blockade by picrotoxin (PTX) (100 μM) causes a ~50% decrease in evoked $[DA]_o$ ($p<0.001$, $n=6$). (D) Catalase abolishes the effect of PTX. Responses in the presence of heat-inactivated catalase were the same as control. Data are means (± SEM), shown as percentage of same-site control (modified from Avshalumov et al., 2003).

isolated neurons in culture (Dugan et al., 1995; Reynolds & Hastings, 1995; Bindokas et al., 1996; Carriedo et al., 2000), suggesting that this might also be a source of H_2O_2 generation in striatal slices with more nearly intact microcircuitry. Second, our previous studies had shown that endogenously generated H_2O_2 could suppress DA release: evoked extracellular DA concentration ($[DA]_o$) in both dorsolateral striatum and the shell of the nucleus accumbens falls by 30–40% during inhibition of the H_2O_2-metabolizing enzyme GSH peroxidase by mercaptosuccinate (MCS) (Chen, Avshalumov & Rice, 2002). Neither exogenous H_2O_2 nor MCS application causes a change in tissue DA levels under the conditions tested, indicating that the decrease in evoked $[DA]_o$ is not due to oxidative loss of the releasable pool of DA. Additionally, the decrease in evoked $[DA]_o$ in MCS persists in the presence of a DA uptake inhibitor, indicating that the effect of increased H_2O_2 indeed affects DA release rather than DA uptake (Avshalumov et al., 2003). Importantly, MCS has no effect on DA release evoked by a single stimulus pulse (Avshalumov et al., 2003), implying that even under conditions of impaired GSH peroxidase activity, basal levels of H_2O_2 in striatum are insufficient to inhibit DA release. Instead, modulatory H_2O_2 must be generated *dynamically* during the first pulse of a stimulus-train and inhibits DA released by subsequent pulses.

The central role of H_2O_2 in striatal DA release regulation by glutamate was demonstrated by the complete prevention of the effects of AMPAR blockade on evoked DA release in the presence of exogenous GSH peroxidase or the other major H_2O_2-metabolizing enzyme, catalase (Figure 17.1 (B)) (Avshalumov et al., 2003). Is this AMPAR-dependent process the primary source of modulatory H_2O_2 in the striatum? The answer appears to be yes, since the effect of GSH peroxidase inhibition by MCS is abolished when AMPARs are blocked by GYKI-52466 (Avshalumov et al., 2003).

GABA-DEPENDENT MODULATION OF DA RELEASE IN DORSAL STRIATUM ALSO REQUIRES AMPAR-DEPENDENT H_2O_2 GENERATION

Interestingly, tests of the possible involvement of conventional γ-aminobutyric acid (GABA) -ergic circuitry in the regulation of striatal DA release also gave initially surprising results: blockade of $GABA_A$ receptors ($GABA_A$Rs) by picrotoxin causes a ~50% decrease in evoked $[DA]_o$ (Figure 17.1 (C)), whereas blockade of $GABA_B$Rs by saclofen has no effect (Avshalumov et al., 2003). These data indicate that GABA, acting through $GABA_A$R, normally *enhances* DA release. The influence of GABA on DA release must be indirect, like that of glutamate, since DA axons in dorsal striatum do not express $GABA_A$Rs (Fujiyama et al., 2000). Indeed, GABAergic regulation of evoked $[DA]_o$ also requires H_2O_2 generation: the effect of picrotoxin is completely prevented by catalase (Figure 17.1 (D)). Moreover, picrotoxin has no effect on DA release when AMPARs are blocked by GYKI-52466 (Avshalumov et al., 2003). Together, these data indicate that glutamate and GABA act on the same pool of modulatory H_2O_2 that is generated downstream from AMPARs.

ACTIVATION OF K_{ATP} CHANNELS UNDERLIES H_2O_2-DEPENDENT INHIBITION OF STRIATAL DA RELEASE

The findings discussed thus far show that endogenous H_2O_2 is an inhibitory messenger that mediates the effects of glutamate and GABA on axonal DA release in dorsal striatum. How does H_2O_2 inhibit DA release? The answer is that H_2O_2 generation leads to the opening of K_{ATP} channels (Avshalumov et al., 2003; Avshalumov and Rice, 2003; Avshalumov et al., 2005). Previous physiological studies demonstrated that *exogenous* H_2O_2 can cause membrane hyperpolarization and decreased excitation by activating a K^+ conductance in many cell types, including hippocampal CA1 neurons, cardiac myocytes, and pancreatic β-cells (Seutin et al., 1995; Ichinari et al., 1996; Tokube, Kiyosue & Arita, 1998; Krippeit-Drews et al., 1999). Our studies of DA release in striatal slices provided the first evidence that endogenous H_2O_2 causes functionally relevant activation of K_{ATP} channels *in situ*. These studies showed that blockade of K_{ATP}-channels with sulfonylureas, either tolbutamide (Avshalumov et al., 2003) or glibenclamide (Figure 17.2 (A)) (Avshalumov & Rice, 2003), causes a significant increase in evoked $[DA]_o$, which indicates that DA release is normally inhibited by K_{ATP} channel activation during local pulse-train stimulation. Blockade of K_{ATP} channels also prevents the inhibitory effect of MCS on DA release and the usual effects of AMPAR blockade by GYKI-52466 and $GABA_A$R blockade by picrotoxin (Figure 17.2 (A)). These data show that K_{ATP} channels are *required* for modulation of DA release by H_2O_2, glutamate, and GABA.

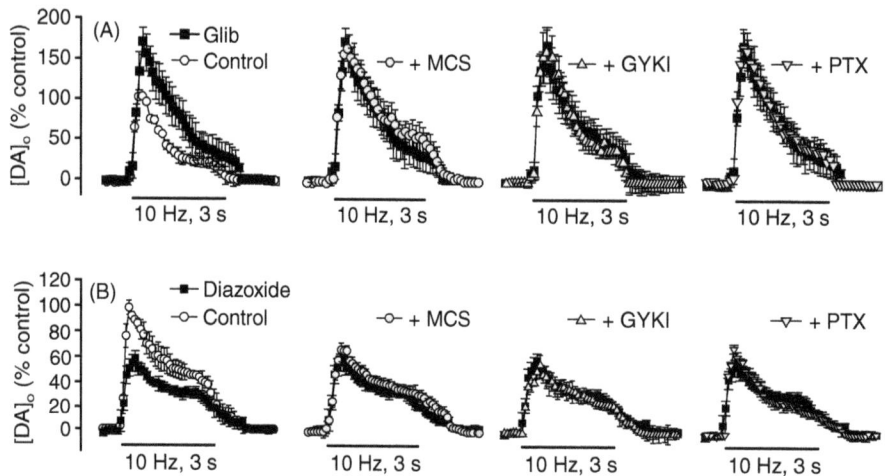

Figure 17.2. Inhibition of striatal dopamine (DA) release by endogenous H_2O_2 is mediated by K_{ATP} channels. (A) K_{ATP}-channel blockade by glibenclamide (Glib) (3 μM) increased evoked $[DA]_o$ in guinea-pig striatal slices ($p<0.01$, $n=5$) and prevented the usual modulation of DA release by MCS (1 mM), GYKI-52466 (GYKI) (50 μM), and picrotoxin (PTX) (100 μM) ($n=5$ for each). (B) Diazoxide (30 μM), a SUR1-selective K_{ATP}-channel opener, decreased evoked $[DA]_o$ ($p<0.01$, diazoxide versus control; $n=5$) and also abolished the effects of MCS, GYKI-52466, and PTX ($n=5$). Data are means (± SEM) (modified from Avshalumov & Rice, 2003).

Which K_{ATP} channels are involved? These K^+ channels are octameric proteins composed of four inwardly rectifying pore-forming subunits, typically Kir 6.2 in neurons (Karschin et al., 1997; Aschroft & Gribble, 1998) and four regulatory sulfonylurea receptor subunits (SUR1 or SUR2) (Babenko, Aguilar-Bryan & Bryan, 1998; Aguilar-Bryan et al., 1998). SUR1- and SUR2-based channels can be distinguished by their differential sensitivity to K_{ATP}-channel openers, with preferential selectivity of diazoxide for SUR1-based channels and cromakalim for SUR2-based channels (Inagaki et al., 1996; Babenko, Gonzalez & Bryan, 2000). In dorsal striatum, K_{ATP} channel opening by diazoxide causes a *suppression* of DA release and prevention of the usual pattern of H_2O_2-dependent modulation by MCS, GYKI, and picrotoxin (Figure 17.2 (B)) (Avshalumov & Rice, 2003). Although SUR2-selective cromakalim also suppresses striatal DA release, it does not alter H_2O_2-dependent DA modulation by these agents, indicating selective targeting of SUR1-based K_{ATP} channels in this process (Avshalumov & Rice, 2003). The question of whether the effect of H_2O_2 is direct or mediated by additional pathways remains open. Previous studies using inside-out membrane patches from cardiac cells have shown a direct, concentration-dependent effect of H_2O_2 on K_{ATP}-channel opening by decreasing channel sensitivity to ATP (Ichinari et al., 1996; Tokube, Kiyosue & Arita, 1998). The mechanism underlying this effect is not known; however, it cannot involve complete oxidation of redox-sensitive sites in K_{ATP}-channels, since strong oxidants cause K_{ATP}-channel closure (Coetzee, Nakamura & Faivre, 1995). On the other hand, studies in pancreatic β-cells show that H_2O_2 may also lead to K_{ATP}-channel activation by decreasing intracellular ATP (Krippeit-Drews et al., 1999).

K_{ATP} channels are expressed in dopaminergic neurons of the SNc and throughout the nigrostriatal pathway (Mourre et al., 1990; Xia & Haddad, 1991; Dunn-Meynell et al., 1997), as well as in striatal medium spiny neurons (Schwanstecher & Bassen, 1997). However, the pattern of effects of K_{ATP} channel blockers and openers on DA release in dorsal striatum (Figure 17.2) is consistent with primary localization of H_2O_2-sensitive K_{ATP} channels on dopaminergic axons rather than on spiny neurons (Avshalumov & Rice, 2003). Most dopaminergic cells in the SNc express K_{ATP} channels with either SUR1 or SUR2 subunits; intriguingly, SUR1 expression is linked to greater metabolic sensitivity of DA neurons in the SNc (Liss, Bruns & Roeper, 1999). Moreover, we have found that the SUR1 subunit also conveys sensitivity to H_2O_2 in these cells: exogenous H_2O_2 or GSH peroxidase inhibition by MCS activates glibenclamide-sensitive K_{ATP} channels in cells that hyperpolarize with SUR1-selective diazoxide, but are not affected by SUR2-selective cromakalim (Avshalumov et al., 2005).

EVIDENCE FOR MITOCHONDRIA AS THE SOURCE OF AMPAR-DEPENDENT H_2O_2 GENERATION

What cellular process generates modulatory H_2O_2? Our recent studies indicate that the primary source is mitochondrial respiration: the effect of MCS on DA release in striatum is lost when mitochondrial complex I is inhibited (Bao & Rice, 2004). This argues against significant involvement of other $\cdot O_2^-/H_2O_2$-generating pathways, including NADPH oxidase. Additional experiments confirm a lack of involvement of monoamine oxidase (MAO), which produces one molecule of H_2O_2 for each molecule of DA metabolized (Cohen, 1994): a cocktail of MAO-A and MAO-B inhibitors does not alter evoked $[DA]_o$ versus control or prevent the effect of MCS (Bao & Rice, 2004).

During local stimulation, an increase in mitochondrial activity arises naturally to meet increased energy demands following glutamate-induced depolarization and subsequent re-establishment of ion gradients. In particular, re-establishing intracellular Ca^{2+} concentration ($[Ca^{2+}]_i$) after depolarization involves energy-demanding processes and is accompanied by enhanced mitochondrial activity and O_2 consumption (Kojima et al., 1994; Bindokas et al., 1998; Jung et al., 2000; Zenisek & Matthews, 2000). Consistent with a Ca^{2+}-dependent process, exogenous H_2O_2 effectively suppresses evoked DA release in dorsal striatum when extracellular Ca^{2+} concentration ($[Ca^{2+}]_o$) is 1.5 mM, but when $[Ca^{2+}]_o$ is elevated to 2.4 mM, H_2O_2 is completely ineffective (Chen, Avshalumov & Rice, 2001). These findings suggest that the effect of exogenous H_2O_2 is occluded in higher $[Ca^{2+}]_o$ because of increased Ca^{2+} entry and enhanced Ca^{2+}-dependent H_2O_2 generation.

Importantly, rotenone, at nanomolar concentrations that are sufficient to increase H_2O_2 production in isolated mitochondria (e.g. Votyakova &

Reynolds, 2001), causes suppression of evoked DA release in striatal slices (Bao, Avshalumov & Rice, 2005). This effect is prevented by catalase or K_{ATP}-channel blockade; moreover, slice ATP levels are not altered under the conditions examined. Together, these data confirm H_2O_2-dependent K_{ATP} channel activation in the effects of rotenone on DA release. It should also be noted that rotenone-induced suppression of DA release is seen with single-pulse, as well as multiple-pulse stimulation, indicating that enhanced H_2O_2 generation during mitochondrial inhibition does not require AMPAR activation, but rather reflects a direct effect on mitochondrial respiration (Bao, Avshalumov & Rice, 2005). These findings with rotenone have potentially important implications for Parkinson's disease, since mitochondrial dysfunction appears to be a casual factor in DA neuron degeneration (Greenamyre et al., 2001; Orth & Schapira, 2002; Fiskum et al., 2003; Wood-Kaczmar, Gandhi & Wood, 2006; Fukae, Mizuno & Hattori, 2007; Dodson & Guo, 2007).

H_2O_2 GENERATED IN MEDIUM SPINY NEURONS INHIBITS DA RELEASE IN DORSAL STRIATUM

To address the cellular source of H_2O_2 requires a basic understanding of striatal circuitry and receptor localization. Although the overall circuitry of the basal ganglia is well known (Kemp & Powell, 1971; Albin, Young & Penney, 1989; Smith & Bolam, 1990), the microchemical circuitry of individual structures, including the striatum, is only beginning to be elucidated. Motor regions of the dorsal striatum receive excitatory input from cortex and thalamus and provide the major inhibitory output of the basal ganglia to subcortical structures (Albin, Young & Penney, 1989; Smith & Bolam, 1990). The principal striatal efferent cells are GABAergic medium spiny neurons (Kemp & Powell, 1971), which receive synaptic glutamatergic input to their dendrites (Smith & Bolam, 1990; Bernard & Bolam, 1998; Chen et al., 1998). These neurons also receive synaptic DA input from midbrain dopaminergic cells (Albin, Young & Penney, 1989; Smith & Bolam, 1990). The absence of ionotropic glutamate receptors (Bernard & Bolam, 1998; Chen et al., 1998) and $GABA_A$Rs (Fujiyama et al., 2000) on dopaminergic terminals suggests that glutamate-dependent H_2O_2 must be generated in non-dopaminergic cells. Prevention of the modulatory effects of endogenous glutamate and GABA by exogenously applied peroxidase enzymes (see Figure 17.1 (B, D)), which are likely to remain in the extracellular compartment, is consistent with a requirement for H_2O_2 diffusion via the extracellular space to reach pre-synaptic dopaminergic axons.

Increasing evidence implicates medium spiny neurons as the primary cellular source of modulatory H_2O_2 in dorsal striatum. Not only are these the most abundant striatal neurons (90–95%; Kemp & Powell, 1971), but the pattern of sensitivity of DA release to glutamate and GABA antagonists (Avshalumov et al., 2003) mirrors the electrophysiological responsiveness of these cells (Jiang & North, 1991; Kita, 1996). Recent studies in our laboratory have confirmed that medium spiny neurons are a key cellular source of modulatory H_2O_2 in dorsal striatum. In these studies, we used a method we developed for simultaneous whole-cell recording and fluorescence imaging of the H_2O_2-sensitive dye, 2'7'-dihydrodichlorofluorescein (H_2DCF) (Avshalumov et al., 2005; Bao, Avashalumov & Rice, 2005). When oxidized by H_2O_2 or other ROS, this dye becomes fluorescent DCF. Under control conditions, basal DCF fluorescence is seen in all striatal spiny neurons, reflecting a tonic level of intracellular H_2O_2 (Figure 17.3). Local stimulation (using the same 30 pulse, 10 Hz pulse trains that we typically use to elicit

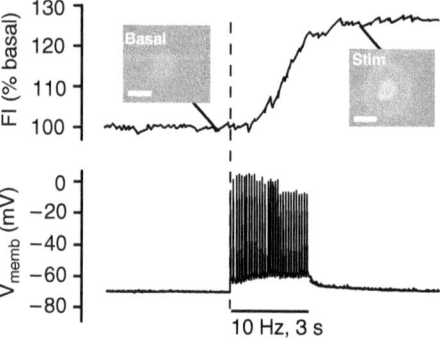

Figure 17.3. Activity-dependent H_2O_2 generation in a striatal medium spiny neuron. H_2O_2-dependent DCF fluorescence intensity (FI) under control conditions and after local stimulation (10 Hz, 30 pulses) with simultaneously recorded membrane voltage (V_{memb}); FI plateau reflects irreversible DCF activation by H_2O_2. Images are basal and stimulated (Stim) FI (scale bar is 20 μm).

concurrent release of DA and glutamate, e.g. Figure 17.1 (A), causes an increase of ~30% in DCF fluorescence in these cells (Figure 17.3), which is further enhanced when GSH peroxidase is inhibited by MCS (Patel, Avshalumov & Rice, 2004). We also found that AMPAR blockade by GYKI-52466 prevents activity-dependent H_2O_2 generation in medium spiny neurons, as well as eliminating stimulus-induced action potentials.

Importantly, glutamate synapses are closely apposed to DA synapses on medium spiny neuron dendrites (Smith & Bolam, 1990; Bernard & Bolam, 1998; Chen et al., 1998) so that they are ideally positioned to modulate DA release via post-synaptically generated H_2O_2 (Figure 17.4). Local glutamate release and AMPAR activation would be expected to increase mitochondrial respiration in spiny neuron dendrites by increasing $[Ca^{2+}]_i$ from Ca^{2+} entry via Ca^{2+}-permeable AMPARs (Carter & Sabatini, 2004), as well as via dendritic voltage-dependent Ca^{2+} channels (Stefani et al., 1998). Medium spiny neurons also express $GABA_A$Rs at dendritic sites near spines (Fujiyama et al., 2000), so that GABAergic input is well-placed to oppose AMPAR-mediated excitation and consequent H_2O_2 generation from mitochondria which are in close proximity to dendritic spines and dopaminergic synapses (Smith & Bolam, 1990; Chen et al., 1998) (Figure 17.4).

The model we propose, therefore, is that mitochondrial H_2O_2 generated in the dendrites and cell bodies of medium spiny neurons diffuses to adjacent DA axons where it inhibits DA release via opening of SUR1-containing K_{ATP}-channels. By decreasing dendritic excitability, GABA lessens H_2O_2 production (Figure 17.4, center). When $GABA_A$Rs are blocked (+PTX), however, H_2O_2 production increases, leading to greater suppression of DA release (Figure 17.4, right). When AMPARs are blocked (+GYKI), H_2O_2 generation is minimal, DA release increases, and $GABA_A$R-dependent regulation is lost (Figure 17.4, left). Like a brake when there is no motion, GABA has no direct influence on DA release, but rather counters the extent to which AMPAR activation by glutamate causes H_2O_2 generation.

This indirect, H_2O_2-dependent modulation of axonal DA release by glutamate and GABA expands possibilities for regulation of DA release by other transmitters that do not have receptors on DA axons. For example, activation of cannabinoid receptors (CB1Rs) in the striatum has been shown to inhibit the release of glutamate and GABA (Szabo et al., 1998;

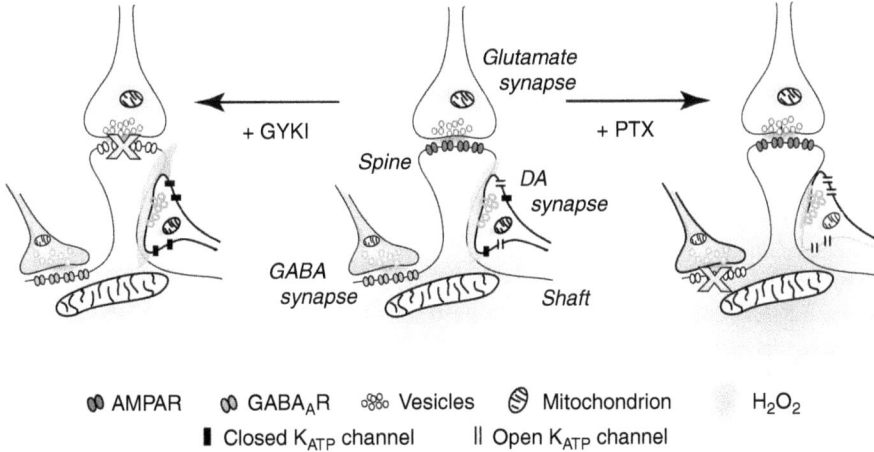

Figure 17.4. Triad of striatal dopamine (DA), glutamate and γ-aminobutyric acid (GABA) synapses on a medium spiny neuron dendrite bound together functionally by diffusible H_2O_2. Generation of modulatory H_2O_2 when GABA as well as glutamate is released (center); $GABA_A$-receptor ($GABA_A$R) blockade by picrotoxin (+PTX, right), and with α-amino-3-hydroxy-5-methyl-4-isoxazoleproprionate (AMPA) -receptor (AMPAR) blockade by GYKI-52466 (+GYKI, left) (circuitry and locations of receptors and mitochondria are from Smith & Bolam, 1990; Bernard & Bolam, 1998; Chen et al., 1998; Fujiyama et al., 2000).

Gerdeman & Lovinger, 2001; Kreitzer & Malenka, 2007). However, whether CB1R activation may also modulate striatal DA release has been controversial, given the apparent absence of CB1Rs on DA axons in dorsolateral striatum (see Sidló, Reggio & Rice, 2008 for discussion). We recently reported that pulse-train evoked DA release is indeed inhibited by CB1R activation (Sidló, Reggio & Rice, 2008). Importantly, this effect is prevented by the H_2O_2-metabolizing enzyme, catalase, by the $GABA_AR$ antagonist picrotoxin, and by the K_{ATP} channel blockers tolbutamide and glibenclamide. These findings can be understood using the model shown above in Figure 17.4. Analogous to the condition in which $GABA_ARs$ are blocked (Figure 17.4, *left*), CB1R activation inhibits GABA release, which leads to increased H_2O_2 generation and consequent K_{ATP}-channel dependent inhibition of DA release (Sidló, Reggio & Rice, 2008). Thus, CB1R-dependent regulation of DA release in dorsolateral striatum is mediated via synaptically released GABA and diffusible H_2O_2.

THE BRAIN ANTIOXIDANT NETWORK: PROTECTIVE, YET PERMISSIVE

As discussed in the Introduction, oxidative damage by endogenous ROS is prevented by the brain antioxidant network, which includes low molecular weight antioxidants, enzymes, and repair systems (Davies, 1988; Yu, 1994; Cohen, 1994; Meister, 1994; Rice, 2000; Dringen, Pawlowski & Hirrlinger, 2005; Rhee et al., 2001). Importantly, however, antioxidant regulation differs between neurons and glia, with higher levels of GSH and GSH-related enzymes in glia than in neurons (Slivka, Mytilineou & Cohen, 1987; Raps et al., 1989; Makar et al., 1994; Desagher, Glowinski & Premont, 1996; Trépanier et al., 1996; Peuchen et al., 1997; Rice & Russo-Menna, 1998; Dringen & Hirrlinger, 2003) and higher levels of the low molecular weight antioxidant ascorbate in neurons than in glia (Rice & Russo-Menna, 1998). This differentiation could reflect the need for ROS signaling in neurons, with additional protection from oxidative damage provided by surrounding glia. Consistent with this hypothesis, evidence from cultured cells suggest that glia play a critical role in protecting neurons from oxidative stress (Desahger, Glowinski & Premont, 1996; Drukarch et al., 1997; Drukarch et al., 1998; Tanaka et al., 1999; Dringen et al., 1999). Moreover, the ability of glia to protect neurons in culture is abolished by inhibition of GSH synthesis (Drukarch et al., 1997), as well as inhibition of GSH peroxidase or the major cellular peroxidase, catalase (Desagher, Glowinski & Premont, 1996; Dringen & Hamprecht, 1997).

We examined the neuroprotective role of glial antioxidants by comparing the consequences of oxidative stress caused by elevated H_2O_2 in slices of guinea-pig and rat brain, which provide a much more nearly intact neuron–glial microenvironment than is possible even using neuron–glia co-cultures (Avshalumov et al., 2004). This comparison was based on the rationale that the ratio of glia to neurons is higher in guinea-pig brain than in rat brain, because of the lower neuron density of guinea-pig brain than rat brain (Tower & Elliott, 1952) and the relative constancy of glial density across species (Friede, 1954; Bass et al., 1971; Tower & Young, 1973; Haug, 1987; Rice & Russo-Menna, 1998). We found initially that the pathophysiological consequences of H_2O_2 exposure on rat brain tissue, including epileptiform activity in hippocampal slices (Avshalumov & Rice, 2002) and edema in forebrain slices (Brahma et al., 2000), were absent in guinea-pig tissue (Avshalumov et al., 2004). We therefore examined the contributions of GSH synthesis, GSH peroxidase, and catalase to the resistance of guinea-pig brain tissue to H_2O_2 challenge and whether exogenous ascorbate could compensate for the loss of these components of the antioxidant network.

DIFFERENTIAL SENSITIVITY OF RAT AND GUINEA-PIG BRAIN SLICES TO H_2O_2 EXPOSURE

Previous studies showed that application of exogenous H_2O_2 (1.0–3.0 mM) reversibly inhibits the PS evoked by Schaffer collateral stimulation in hippocampal slices from either rat or guinea-pig (Pellmar, 1995; Avshalumov, Chen & Rice, 2000; Avshalumov & Rice, 2002). In rat hippocampal slices, recovery of the primary PS after H_2O_2 washout is accompanied by mild epileptiform activity, indicated by an additional PS after the primary spike (Avshalumov & Rice, 2002) (Figure 17.5 (A)). When guinea-pig hippocampal slices

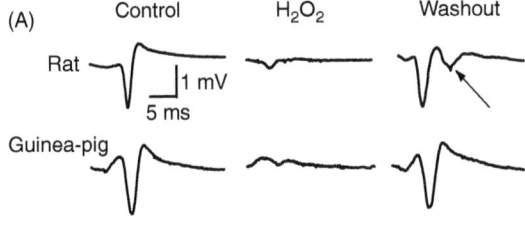
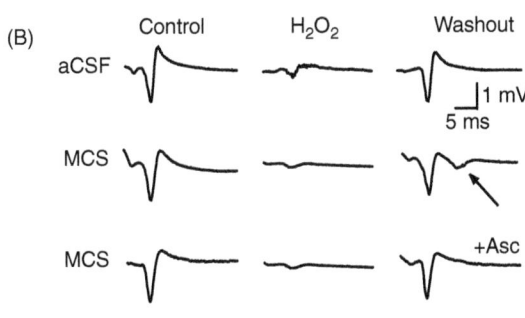
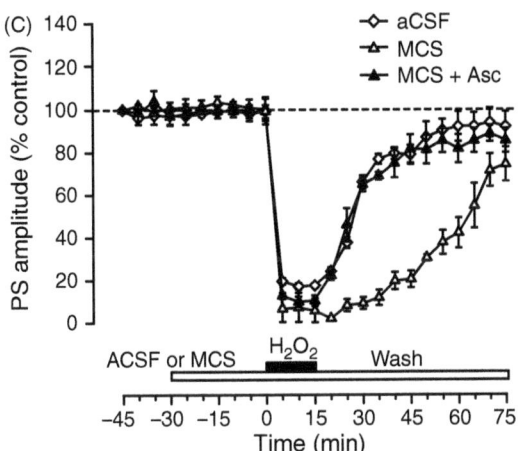

Figure 17.5. Greater tolerance of guinea-pig versus rat brain tissue to H_2O_2 elevation and role of glutathione (GSH) peroxidase. (A) Electrophysiological records showing the extracellular population spike (PS) evoked by stimulation of the Schaffer collaterals and recorded in CA1 before H_2O_2 exposure (Control), after 15 minutes of superfusion with H_2O_2 (1.5 mM), and after 30 minutes of washout of H_2O_2 (Wash) in rat and guinea-pig hippocampal slices. Recovery of the PS was accompanied by mild epileptiform activity, indicated by an additional PS after washout (arrow; $n = 9$) in rat, but not guinea-pig ($n = 8$) slices (records shown are the averaged PS responses for each species under each condition). (B) Averaged evoked PS in guinea-pig hippocampal slices under control conditions, after 15 minutes of exposure to H_2O_2 (1.5 mM) and after 30 minutes of H_2O_2 washout in artificial cerebrospinal fluid (aCSF) alone, in the presence of MCS (1 mM), and in MCS when H_2O_2 was washed out with aCSF plus ascorbate (400 µM; +Asc). Superfusion of MCS for 30 minutes had no effect on hippocampal PS amplitude; however, in the continued presence of MCS, exposure to H_2O_2 caused a larger suppression of the evoked PS than in slices with an intact antioxidant network exposed to H_2O_2 in ACSF alone ($p < 0.01$ MCS versus ACSF; $n = 7$). Recovery of the PS was accompanied by epileptiform activity (arrow, middle panel); the presence of ascorbate during H_2O_2 washout prevented this secondary pathophysiology (lower panel) ($n = 5$) (PS records are the averaged responses for each condition; $n = 5$–8). (C) Time course of PS amplitude changes during H_2O_2 exposure and washout in guinea-pig hippocampal slices with and without MCS, and with ascorbate present during H_2O_2 washout. Not only was PS suppression during H_2O_2 application more pronounced after GSH peroxidase inhibition by MCS, but recovery was also delayed compared with that seen during H_2O_2 washout in companion slices exposed to H_2O_2 in ACSF alone ($p < 0.01$ washout in MCS versus aCSF; ANOVA). When ascorbate was present during H_2O_2 washout in GSH-peroxidase inhibited slices, the time course of recovery was indistinguishable from that in aCSF alone ($n = 5$) ($p > 0.05$ washout in MCS + Asc versus aCSF; ANOVA) (modified from Avshalumov et al., 2004).

are exposed to H_2O_2 under identical conditions, the evoked PS is also reversibly depressed during H_2O_2 exposure. However, unlike rat slices, only a single PS is seen in guinea-pig hippocampal slices after H_2O_2 washout, indicating the absence of lasting pathophysiology (Figure 17.5 (A)). The higher tolerance of guinea-pig brain than rat brain tissue to H_2O_2 exposure implies that the higher glia-to-neuron ratio in this species provides additional antioxidant protection, particularly from glia.

Considerable evidence indicates that GSH synthesizing enzymes and GSH peroxidase are predominantly expressed in glia (Slivka, Mytilineou & Cohen, 1987; Raps et al., 1989; Maker et al., 1994; Desagher, Glowinski & Premont, 1996; Trépanier et al., 1996; Peuchen et al., 1997). Although GSH synthesis occurs in all cells from its substituent amino acids, glutamate, cysteine, and glycine (Meister, 1994), in the brain, synthetic enzymes for GSH are more abundant in glia than in neurons (Maker et al., 1994). Consistent with this localization, cellular GSH levels in intact brain tissue are ∼50% higher in glia (4 mM) than in neurons (2.5 mM) (Rice & Russo-Menna, 1998). Neurons also synthesize GSH (Chen & Swanson, 2003; Himi et al., 2003), however, and have additional mechanisms, including shuttling of GSH and its precursors from glia to neurons, for maintenance of neuronal GSH content (Sagara, Miura & Bannai, 1993; Wang & Cynader, 2000; Dringen & Hirrlinger, 2003).

Given the importance of GSH-based antioxidant protection in glia, we examined the effect of inhibiting GSH synthesis or GSH peroxidase on H_2O_2-induced pathophysiology in guinea-pig brain slices. Both GSH and ascorbate are readily lost from either rat or guinea-pig brain slices in vitro, with slice contents that are at least 60% lower than those in intact tissue (McIlwain, Thomas & Bell, 1956; Rice, Pérez-Pinzón & Lee, 1994; Avshalumov et al., 2004). Nonetheless, in slices of guinea-pig hippocampus, an additional small, but significant, decrease in GSH (and ascorbate) content after inhibition of GSH synthesis by buthionine sulfoximine (BSO) is accompanied by epileptiform activity in guinea-pig slices following H_2O_2 exposure (Avshalumov et al., 2004). The loss of H_2O_2 tolerance in guinea-pig hippocampal slices after GSH synthesis inhibition, despite a relatively small absolute decrease in GSH content, suggests that the protective effect of endogenous GSH is mediated through its role as a co-factor for GSH peroxidase, as well as by maintenance of ascorbate levels by recycling oxidized ascorbate (Meister, 1994; Rice, 2000; Avshalumov et al., 2004).

Consistent with this hypothesis, enhanced H_2O_2-induced pathophysiology in guinea-pig hippocampal slices also occurs when GSH-peroxidase is inhibited by MCS (Avshalumov et al., 2004) (Figure 17.5 (B)). Although superfusion of MCS for 30 minutes had no effect on hippocampal PS amplitude, exposure to H_2O_2 in the continued presence of MCS causes a larger suppression of the evoked PS than in artificial cerebrospinal fluid (aCSF) alone ($p<0.01$ versus aCSF) (Figure 17.5 (B)). The recovery of PS amplitude during H_2O_2 washout in MCS is accompanied by secondary H_2O_2-induced pathology (Figure 17.5 (B), arrow). Moreover, recovery was delayed compared with that in aCSF alone (Figure 17.5 (C)). Inclusion of ascorbate at its normal extracellular concentration (400 μM) (Rice, 2000) in the washout solution prevented these pathophysiological consequences of H_2O_2 exposure after GSH peroxidase inhibition (Figure 17.5 (B, C)).

The relative distribution of the peroxidase enzyme, catalase, between neurons and glia is less defined than for GSH and related enzymes. Although catalase is found in high levels in glial cells (Sokolova et al., 2001), the relative catalase activities in glia and neurons in culture are similar (see Dringen, Pawlowski & Hirrlinger, 2005 for review). It should be noted, however, that confinement of catalase to peroxisomes in both neurons and glia adds a diffusional component to the efficacy of this peroxidase (Dringen, Pawlowski & Hirrlinger, 2005), which might facilitate the "escape" of H_2O_2 signals from a cell of origin. In neuron–glia co-cultures, loss of the ability of glia to protect neurons from H_2O_2 toxicity after catalase inhibition (Dringen et al., 1999) reflects loss of neuronal and glial catalase (Dringen, Pawlowski & Hirrlinger, 2005). We found that guinea-pig hippocampal slices exposed to H_2O_2 after catalase inhibition showed epileptiform activity after H_2O_2 washout, indicating that catalase also provides key protection from elevated H_2O_2 in the intact neuropil of brain slices (Avshalumov et al., 2004). Washout with ascorbate was again able to prevent this secondary pathophysiology.

Prolonged H_2O_2 exposure as a model of oxidative stress may also cause other pathophysiological effects in brain tissue. One we have examined most extensively is edema formation (Brahma et al., 2000; Avshalumov et al., 2004). As noted above, significant water gain is seen in coronal slices of rat forebrain after exposure to H_2O_2 for three hours (Brahma et al., 2000); however, no increase in edema is seen in guinea-pig slices incubated under identical conditions (Avshalumov et al., 2004). Once again, guinea-pig brain tissue behaves like rat tissue when GSH synthesis, GSH peroxidase, or catalase is inhibited, with prevention of edema when ascorbate is included in the incubation medium (Avshalumov et al., 2004). Studies in neuron–glia co-cultures have shown that the glial antioxidant network may protect neurons from H_2O_2 toxicity (Desahger, Glowinski & Premont, 1996; Drukarch et al., 1997; Drukarch et al., 1998; Tanaka et al., 1999; Dringen et al., 1999). Our work indicates that this is also true in acute brain slices, in which neuronal and glial integrity is maintained (Avshalumov et al., 2004). Together, these studies indicate that glia provide protection against H_2O_2 toxicity.

UNIQUE ANTIOXIDANT PROPERTIES OF NEURONAL ASCORBATE

One other difference between rats and guinea-pigs is that rats may synthesize ascorbate, whereas guinea-pigs, like humans and non-human primates, cannot produce ascorbate endogenously and must therefore acquire it from dietary sources. However, in contrast to GSH synthesis, which occurs in all cells, ascorbate synthesis occurs only in liver (e.g. mammals) or kidney (e.g. reptiles). Regardless of its origin, therefore, ascorbate is transported throughout the body via plasma and is taken up at cell-specific levels by the sodium-dependent ascorbate (vitamin C) transporters, SVCT1 and SVCT2 (Rice, 2000); SVCT2 is the isoform expressed in brain (Tsukaguchi et al., 1999). Consequently, cellular regulation of ascorbate is species-independent, so that whether or not an animal may synthesize ascorbate should not influence other components of the antioxidant network. In the CNS, ascorbate is compartmentalized between neurons and glia, with much higher concentrations in neurons (~10 mM) than in glia (1 mM) (Rice & Russo-Menna, 1998). This distribution is consistent with expression of SVCT2 by neurons, but not glia (Tsukaguchi et al., 1999; Berger & Hediger, 2000).

Previous *in vitro* studies have shown that application of exogenous H_2O_2 has no effect on the evoked PS in hippocampal slices when ascorbate is present at its normal extracellular concentration (400 µM). Similar prevention of PS suppression is provided by deferoxamine, a metal ion chelator that prevents iron or copper-catalyzed •OH formation from H_2O_2, indicating that the ROS required for PS suppression is •OH and that protection by ascorbate under these conditions is by •OH scavenging (Avshalumov, Chen & Rice, 2000). H_2O_2-dependent PS suppression is also prevented by isoascorbate (D-ascorbate) (Avshalumov, Chen & Rice, 2000), the non-biologically active stereoisomer of L-ascorbate that is not transported by stereoselective SVCT2 (Tsukaguchi et al., 1999). This indicates an extracellular site of antioxidant action. As discussed above, ascorbate may also prevent long-lasting pathophysiological consequences of H_2O_2 exposure, indicated by the absence of epileptiform activity in rat hippocampal slices when H_2O_2 is washed out with ascorbate-containing media (Avshalumov & Rice, 2002) or in guinea-pig hippocampal slices when the antioxidant network is compromised (Avshalumov et al., 2004) (Figure 17.5 (B, C)). This protective effect of ascorbate must occur intracellularly in neurons, however, because non-transported isoascorbate is not protective under these conditions (Avshalumov & Rice, 2002).

In striking contrast to the efficacy of ascorbate in preventing pathological consequences of H_2O_2 exposure, this antioxidant and •OH scavenger has no effect on the modulation of striatal DA release by endogenously generated H_2O_2 (Avshalumov et al., 2003). This important finding indicates not only that ascorbate permits H_2O_2-dependent signaling in striatum, but also that inhibition of DA release is a direct action of H_2O_2, rather than •OH. Thus, ascorbate is ideally suited as a key neuronal antioxidant because of its ability to permit H_2O_2 signaling, yet prevent pathological consequences that could occur from unregulated H_2O_2 generation and •OH production.

SUMMARY AND CONCLUSIONS

In our first report of striatal DA release modulation by endogenous H_2O_2 (Chen, Avshalumov & Rice, 2001), we suggested that H_2O_2 might be generated presynaptically at DA synapses to serve as an autoinhibitory signal that would limit DA release after axonal activation, given the close apposition of mitochondria to presynaptic sites in DA axons (e.g. Nirenberg et al., 1997). However, our subsequent studies showed that generation of modulatory H_2O_2 requires AMPAR activation and may be "fine-tuned" by $GABA_AR$ activation. These findings argue against primary H_2O_2 generation in DA axons, because they lack AMPARs and $GABA_A$Rs (Bernard & Bolam, 1998; Chen et al., 1998; Fujiyama et al., 2000).

What we found, however, is more exciting: H_2O_2 must be generated in cells or processes other than DA axons, then diffused beyond the generating synapse to inhibit DA release. Our working hypothesis is that regulation of striatal DA release by glutamate and GABA involves a triad of DA, glutamate, and GABA synapses, separated by a few micrometers on the dendrites of medium spiny neurons (Smith & Bolam, 1990; Bernard & Bolam, 1998; Chen et al., 1998;

Fujiyama et al., 2000), and bound together *functionally* by diffusible H_2O_2 (see earlier Figure 17.4). Regardless of the source, endogenously generated H_2O_2 reverses conventional glutamatergic excitation by opening SUR1-based K_{ATP} channels to inhibit striatal DA release. This discovery represents a new and potentially important mechanism of external regulation of DA release that establishes a formerly "missing link" in the reciprocal relationship between DA and glutamate in striatum. Moreover, because DA-glutamate dysfunction has been implicated as a causal factor in Parkinson's disease (Olanow & Tatton, 1999; Chase & Oh, 2000; Greenamyre, 2001), schizophrenia (Deutsch et al., 2001; Sawa & Snyder, 2002), and addiction (Koob, 2000; Hyman & Malenka, 2001), exploration of this process may also suggest novel pathways through which dysfunction might occur.

Thus, neuromodulation by H_2O_2 may be a double-edged sword: although activity dependent H_2O_2 generation may provide important regulation of DA release, an imbalance between H_2O_2 generation and metabolism could result in oxidative stress, which has been implicated in nigrostriatal degeneration in Parkinson's disease (Cohen, 1994; Ebadi, Srinivasan & Baxi, 1996; Sonsalla et al., 1997; Olanow & Tatton, 1999; Zhang, Dawson & Dawson, 2000; Xu et al., 2002; Wood-Kaczmar, Gandhi & Wood, 2006; Fukae, Mizuno & Hattori, 2007; Dodson & Guo, 2007) and as a causal factor in schizophrenia (Do et al., 2000; Yao, Reddy & van Kammen, 2001). Loss of normal H_2O_2 regulation might therefore contribute to DA-system pathology. It is relevant to note, that inhibition of GSH peroxidase by MCS leads to DA-neuron hyperpolarization (Avshalumov et al., 2005) and suppression of somatodendritic DA release in the SNc (Chen, Avshalumov & Rice, 2002); however, MCS does not inhibit DA release in the adjacent ventral tegmental area (VTA) (Chen, Avshalumov & Rice, 2002). This difference between SNc and VTA is potentially important, because DA neurons of the SNc degenerate in Parkinson's disease, whereas those in the VTA are relatively spared (Yamada et al., 1990; Fearnley & Lees, 1991).

One final point is that regulation of DA release by endogenous H_2O_2 is similar in guinea-pig and rat striatum (Avshalumov et al., 2003). Indeed, most of our studies of glutamate-dependent H_2O_2 generation and DA release regulation have been made by use of guinea-pig brain slices, which demonstrates that H_2O_2 signaling is not impaired in the presence of the stronger antioxidant network afforded by the higher glia-to-neuron ratio of guinea-pig tissue. Thus, these findings support the hypothesis that precise antioxidant compartmentalization between neurons and glia is critical for neuronal signaling, as well as neuronal protection. Indeed, the higher glia-to-neuron ratio in guinea-pig than rat brain might better manage discrete signaling by diffusible messengers like H_2O_2, since ensheathing glia could potentially limit the effective radius of a locally broadcast ROS signal. Overall, the present findings provide encouraging news for humans, because neuron density is lower (Tower & Elliott, 1952), and thus glia-to-neuron ratio higher (Friede, 1954; Bass et al., 1971; Tower & Young 1973; Haug, 1987), than either rodent species examined previously (Avshalumov et al., 2004).

ACKNOWLEDGEMENTS

The work described here was funded by NIH/NINDS grants NS-36362 and NS-45325 and by the National Parkinson Foundation.

18 · D-serine as a putative glial neurotransmitter

Asif K. Mustafa, Paul M. Kim and Solomon H. Snyder

Stereospecificity is a crucial feature of biology, permitting the highly selective recognition of substrates by their enzymes, and of hormones and neurotransmitters by receptors. Animals typically contain D-sugars and L-amino acids. Bacteria contain D-amino acids as important constituents of peptidoglycans in their cell walls which help confer rigidity as well as resistance to antibiotics. The lack of cell walls in animals presumably obviated a need for D-amino acids.

Based on these considerations, the initial observation of Lajtha and colleagues of D-aspartate in mammalian tissues was surprising (Dunlop et al., 1986). In newborn rodents D-aspartate levels are highest in the cerebral cortex and diminish with age. Hashimoto and associates (Hashimoto et al., 1992a; Hashimoto et al., 1992b; Hashimoto & Oka, 1997) then identified a second D-amino acid, D-serine, exclusively in the brain with levels about a third of those of L-serine (Hashimoto et al., 1993a). These investigators noted variations in D-serine concentrations with highest levels in the forebrain, resembling the localization pattern of N-methyl-D-aspartate (NMDA) subtypes of glutamate receptors (Hashimoto et al., 1993a).

D-aspartate is highly localized to neuroendocrine tissues, with selective concentration in the epinephrine-containing granules of the adrenal medulla, oxytocin, or vasopressin neurons projecting from the hypothalamus to the posterior pituitary, the pineal gland, and selective neuronal populations in the brain (Schell, Cooper & Snyder, 1997a). The disposition of D-serine is even more remarkable and will be the primary focus of this chapter. D-serine is exclusively localized to protoplasmic astrocytes that ensheathe the synapse. D-serine is formed by a novel enzyme, serine racemase (SR), which occurs in the same glial cells containing D-serine (Wolosker, Blackshaw & Snyder, 1999a). D-serine potently activates the so-called "glycine" site of NMDA receptors, and D-serine localizations resemble those of NMDA receptors (NMDARs) more closely than glycine (Schell et al., 1997b). Degradation of endogenous D-serine with D-amino acid oxidase (DAO) selectively diminishes NMDA neurotransmission, establishing that D-serine is an endogenous ligand for the NMDAR (Mothet et al., 2000). D-serine is released from glia by glutamate acting at α-amino-3-hydroxyl-5-methyl-4-isoxazolepropionate (AMPA) receptors (AMPARs) via calcium-dependent exocytosis (Schell, Molliver & Snyder, 1995). Activation of AMPARs triggers the binding of glutamate receptor interacting protein (GRIP) to SR causing a major activation of SR and efflux of D-serine from glia (Kim et al., 2005). D-serine plays a major role in neuronal migration as exemplified by the progression of granule cells in the cerebellum along the processes of Bergmann glia (Kim et al., 2005). Degrading D-serine with DAO and inhibiting SR both prevent granule cell migration. Thus, D-serine appears to be a novel glial neuromodulator or neurotransmitter.

LOCALIZATIONS OF D-SERINE

To localize D-serine polyclonal antibodies to a conjugate of D-serine-glutaraldehyde and bovine serum albumin have been developed. The antibody readily detects as little as 0.01 nM of D-serine (Schell, Molliver & Snyder, 1995). It displays 100 times greater sensitivity for D- than L-serine and does not interact with a variety of other amino acids.

Immunohistochemical staining reveals high densities of D-serine in the forebrain with very low densities in the hindbrain, similar to the distribution of the NMDARs. These localizations are the inverse of DAO, first identified by Hans Krebs (1935), as an enzyme that oxidatively deaminates D- but not L-amino acids. Krebs's substrate activity was demonstrated at high pH levels, which appeared to be optimal for enzyme activity. At physiologic pH, DAO is highly selective for D-serine (Mothet et al., 2000).

The inverse localizations of D-serine and DAO, coupled with the selectivity of DAO for D-serine at physiologic pH, suggested that DAO normally degrades D-serine. This possibility fits with observations that mice lacking DAO possess elevated brain levels of D-serine (Hashimoto et al., 1993b). The greatest increases of D-serine in mice lacking DAO take place in areas which normally have negligible levels of D-serine. This suggests that very low levels of D-serine do not imply a lack of biosynthesis or physiological function but rather a rapid turnover.

Immunohistochemical studies at a cellular level reveal D-serine exclusively localized to a population of protoplasmic astrocytes that ensheathe synapses in the forebrain. DAO, like D-serine, is localized to astrocytes fitting with a role for DAO in degrading D-serine. The exact subtype of astrocytes that contain D-serine is not altogether clear. Protoplasmic astrocytes correspond to Type II astrocytes prepared in primary cultures and separable from Type I astrocytes. Type I cells are polygonal, forming confluent layers on culture dishes, whereas Type II astrocytes possess processes and overlay the Type I astrocytes (Levison & McCarthy, 1991). D-serine is selectively enriched in Type II astrocytes (Schell, Molliver & Snyder, 1995). D-serine containing cells may correspond to oligodendrocyte precursor cells (OPC) which possess many neuronal properties.

A recent study suggests that microglia activated by the amyloid-β-peptide may form D-serine (Wu et al., 2004). D-serine staining is particularly abundant in the glial processes that surround blood vessels as well as processes that pass between neuronal cell bodies that are enriched in NMDARs. At the electron microscopic level in the CA1 layer of the hippocampus, D-serine occurs in astrocytic foot processes that contact endothelial cells as well as in astrocytes which contact dendritic spines of neurons (Schell et al., 1997b).

Detailed examination of D-serine in the forebrain reveals a close relationship to NMDARs (Schell et al., 1997b). Co-staining of the forebrain for the NR2A/B types of NMDARs as well as D-serine and glycine reveals a much closer correspondence of NMDARs with D-serine than with glycine in most parts of the brain. This close relationship of D-serine and NMDARs is especially striking in deeper layers of the cerebral cortex. In the amygdala, D-serine containing astrocytes abut pyramidal cells enriched in NMDARs, whereas the very low amygdalar levels of glycine have distinctly different localizations. Highest densities of glycine in the brain occur in the hypothalamus and hindbrain areas which have negligible levels of D-serine and NMDARs. In the hippocampus, glycine staining is very low and differs markedly from the disposition of D-serine and NMDARs, which occur together in the molecular layer of the subiculum as well as pyramidal cell layers of CA1 and CA3.

The selective association of D-serine, but not glycine, with NMDARs in most parts of the brain supports the notion that in these areas D-serine is the endogenous ligand for the receptor. In some parts of the brain glycine is more closely associated with NMDARs (Schell et al., 1997b). Thus, in the adult cerebellum the very low levels of NMDARs occur selectively in the pinceau processes of basket cells which are also enriched in glycine. Granule cells of the adult cerebellum are also enriched in NMDARs and are contacted by the processes of Golgi neurons which are enriched in glycine. Thus, in the adult cerebellum glycine released from Golgi cells may regulate NMDARs of granule cells, and glycine emerging from basket cells may influence Purkinje cells.

Evaluating the function of any chemical based on its localizations may be complicated by multiple functions of the given substance. Thus all amino acids are involved in protein synthesis and intermediary metabolism. In terms of neurotransmitter roles, glycine is a major transmitter of inhibitory neurons in the spinal cord and brainstem. In the olfactory bulb, glycine is an established inhibitory neurotransmitter associated with tufted and mitral cells which possess strychnine-sensitive glycine receptors, the physiologic target for the inhibitory actions of glycine (Trombley & Shepherd, 1994). Immunohistochemical densities of glycine are concentrated around mitral cell dendrites and cell bodies.

The accessory olfactory bulb is enriched in both D-serine and glycine. D-serine levels here are the highest in the brain and occur in glia that ensheathe axons from the vomeronasal organ (VNO), an area containing NMDARs and moderate levels of glycine. The plexiform layer possesses NMDARs in mitral cell dendrites which are surrounded by astrocytes containing D-serine.

SERINE RACEMASE

There are multiple ways in which D-serine might be formed. Electron transfers that take place in transamination could lead to racemization. The glycine cleavage enzyme catalyzes the conversion of glycine to serine through keto intermediates which could also involve racemization. Dunlop and Neidle (1997) obtained evidence for the conversion of radiolabeled L- to D-serine in intact rats, but it is conceivable that this transformation might have been indirect and involved processes such as transamination.

SR activity has been detected in brain extracts and the enzyme has been purified to homogeneity about 12,000-fold (Wolosker et al., 1999b). Utilizing amino acid sequence from the purified enzyme, the full-length mouse SR has been cloned as a protein with 339 amino acids and a predicted molecular weight of 36 kDa (Wolosker et al., 1999a). SR contains a consensus sequence for pyridoxal 5'–phosphate (PLP) binding near the N-terminus where it displays homology with serine dehydratase, also a PLP-dependent enzyme. The two enzymes possess about 28% identity. SR displays 40% homology with yeast, *Caenorhabditis elegans*, and *Arabidopsis* proteins of hitherto unknown functions. Thus, SR appears to be the first member of a new family of PLP enzymes quite distinct from serine dehydratase, which resembles SR less than SR resembles its plant homolog. Cloned SR produces D-serine from L-serine abundantly and in a PLP-dependent fashion (Figure 18.1). Its ribonucleic acid (RNA) levels are highest in the liver and next-highest are in the brain, with very low levels in the kidney and other tissues. Western Blot analysis reveals a prominent single 36 kDa band in the brain with much lower levels in the liver and hardly any detectable enzyme in other tissues. With subcellular fractionation the enzyme appears to be predominantly soluble. Immunohistochemical studies reveal essentially identical localization for D-serine and SR.

More extensive studies of the enzymology of SR by Wolosker and associates (De Miranda et al., 2002) reveal novel and potentially important functions. These investigators discovered in cell extracts substances that markedly augment SR activity and which they identified as magnesium and ATP that increase SR activity 5–10-fold. The abundant SR activity demonstrable under these conditions revealed a novel enzymatic function. SR possesses robust α,β-eliminase activity converting serine to pyruvate (Figure 18.1). Because of the substantial pyruvate-forming capacity of SR and the millimolar concentrations of L-serine in the brain, SR may be a major source of pyruvate generation under some conditions. Perhaps SR switches between forming D-serine and forming pyruvate, depending on the energy status of the cell. This would fit with the importance of adenosine triphosphate (ATP) as an activator of the enzyme. In the presence of magnesium, ATP half-maximally activates SR at 10 μM – only 5% of endogenous ATP levels in most cells. Accordingly, under normal circumstances, the enzyme may be saturated with ATP. It is conceivable that in some micro-environments of the cell, ATP depletion during cell stress is sufficient to alter SR activity. In initial studies (De Miranda et al., 2002) ATP stimulated D-serine and pyruvate formation to a similar extent, but under some circumstances it might influence the two processes differentially.

Cook and associates (Cook et al., 2002) independently discovered that calcium binds to SR and activates SR, suggesting that neuronal depolarization and associated calcium entry might regulate SR activity. However, half-maximal augmentation of SR activity by calcium occurs at 26 μM, at least 100 times basal levels. Magnesium is at least 10 times more potent than calcium in activating SR, suggesting that observed calcium activation merely reflects calcium mimicking magnesium.

There may be feedback upon D-serine biosynthesis from the targets of NMDAR activation. Thus administration of the NMDA antagonist MK801 to rats leads to augmented SR transcription levels (Yoshikawa et al., 2004).

More recently Wolosker and colleagues (Foltyn et al., 2005) have shown that the α,β-elimination reaction may be exerted upon D-serine as well as L-serine (Figure 18.1). This process also influences L-threonine, whereas SR does not racemize L-threonine. A mutant of SR (Q155D), which lacks α,β-elimination but retains racemase activity, is associated with the formation of two to three times more D-serine than the wild-type enzyme both in transfected cells and glial cultures. The α,β-eliminase activity of SR exerted upon D-serine might physiologically downregulate levels of D-serine in glia and participate in synaptic inactivation of D-serine.

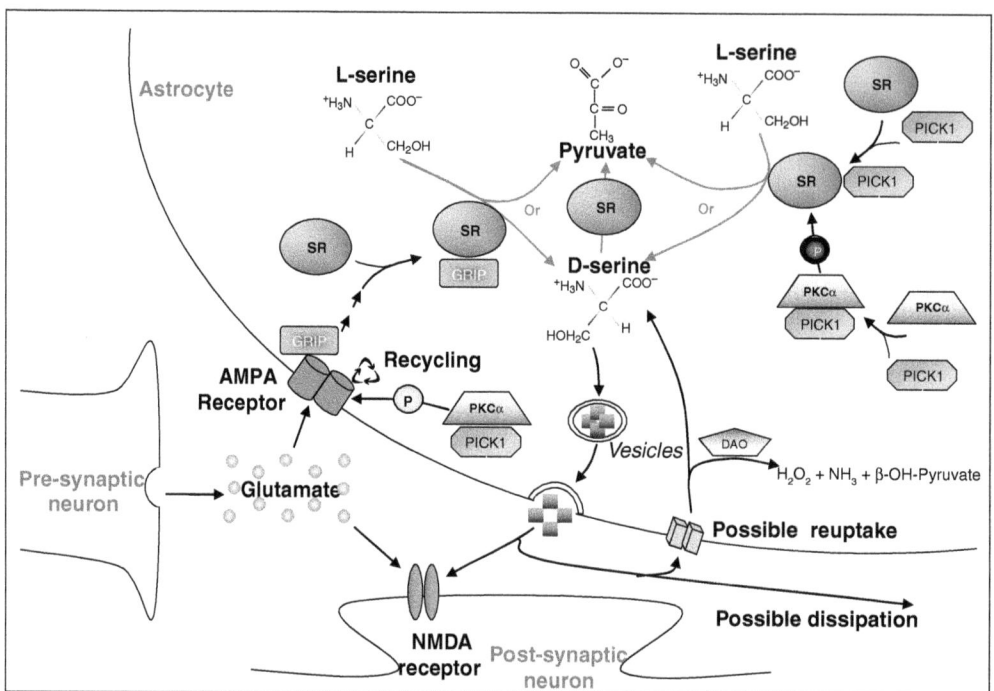

Figure 18.1. A schematic diagram of the putative D-serine and serine racemase regulatory pathway. Release of glutamate from presynaptic neurons leads to activation of glial α-amino-3-hydroxyl-5-methyl-4-isoxazolepropionate (AMPA) receptors (AMPARs) with subsequent receptor phosphorylation by protein kinase C alpha (PKCα) bound to protein interacting with c kinase (PICK1) and dissociation of receptor-bound glutamate receptor interacting protein (GRIP). The latter then binds to and activates SR promoting the formation of D-serine. Additionally, PICK1 may serve to bring PKCα into the vicinity of serine racemase (SR) thus facilitating its phosphorylation and effecting enzyme activity. The newly synthesized D-serine is stored in vesicles and released into the synapse where glutamate and D-serine co-activate N-methyl-D-aspartate receptors (NMDARs) on post-synaptic neurons. D-serine could potentially be degraded by intracellular D-amino acid oxidase (DAO) or SR after reuptake of the neurotransmitter or be dissipated by diffusion from the synaptic cleft. Reprinted from Mustafa et al. (2004), with permission.

Thus, if synaptic D-serine is inactivated by reuptake into the glial cells that make it perhaps the accumulated D-serine enters a pool that is selectively sensitive to SR's α,β-eliminase activity, which prevents the accumulation of excess levels of the amino acid.

D-SERINE REGULATION OF NMDARs

The classic studies of Ascher and colleagues (Johnson & Ascher, 1987) showed that NMDA neurotransmission in brain slices is markedly diminished when perfusion rates are increased, suggesting that perfusion washes away a critical factor. Of various amino acids examined only glycine restored neurotransmission, implying that NMDARs require the activation of two sites, one for glutamate and one for glycine. Molecular cloning of NMDARs revealed discrete "glycine" and glutamate recognition sites (Meguro et al., 1992). Why should NMDARs be the only transmitter receptors requiring two physiologic agonists? Perhaps two agonists provide a failsafe mechanism. It is well established that excess activation of NMDARs leads to neurotoxicity, including vascular stroke. Glutamate is an abundant dietary amino acid which can penetrate into the brain, suggesting a need for some system to protect NMDARs from dietary influences. However, glycine is also an abundant dietary amino acid so that it would not make much sense for nature to have selected it for such a failsafe function. By contrast, D-serine arises only from SR, which generates it selectively in the vicinity of NMDARs. The co-localization of D-serine and NMDARs supports the candidacy of D-serine as the physiologic endogenous ligand. DAO has been used to develop definitive evidence for or against D-serine as an endogenous ligand (Mothet et al., 2000). This research demonstrated that highly purified DAO is extremely selective for D-serine. At physiologic pH, the enzyme degrades only D-serine and D-alanine with levels of the latter being negligible in the brain. Under appropriate experimental conditions DAO has no effect upon glycine. In hippocampal cultures and cerebellar slices the nearly complete degradation of endogenous D-serine by DAO has been demonstrated with no influence on any other amino acid.

Utilizing the same cerebellar slices and hippocampal cultures, NMDA neurotransmission has been monitored in several ways. In cerebellar slices, the activation of neuronal nitric oxide synthase (nNOS) activity and elevation of cyclic guanosine monophosphate (GMP) levels has been examined (Bredt & Snyder, 1989). DAO treatment reduces basal nNOS activity by about 50%, suggesting that the physiologic basal levels of D-serine normally activate NMDARs. The activation of nNOS by glutamate and NMDA was also markedly reduced by DAO treatment. Additionally, DAO lowered by 70% the NMDA stimulation of cyclic GMP levels.

In hippocampal cultures, electrophysiologic techniques have been used to evaluate NMDA neurotransmission. In these experiments DAO treatment reduces transmission by about 60%, while the boiled enzyme is inactive. DAO generates hydrogen peroxide, which may be cytotoxic (Figure 18.1). If hydrogen peroxide arising from DAO activity was responsible for the loss of NMDA neurotransmission, then such effects should be irreversible. However, the reduction of nNOS activity, cyclic GMP levels, and electrophysiologically monitored NMDA neurotransmission are all reversed by addition of D-serine.

Thus, the available evidence indicates that in many, if not most, parts of the brain D-serine is the more likely endogenous ligand for the glycine site of the NMDAR than is glycine itself (Figure 18.1). Some investigators have suggested that the glycine site of NMDARs is normally saturated by endogenous glycine, whose total brain concentrations are millimolar. However, a number of neurophysiologic studies indicate that the site is not saturated *in vivo* (Wood, 1995). Moreover, total tissue concentrations are not meaningful indicators of the physiologic situation at any given synapse. Brain levels of glutamate, about 20 mM, are higher than almost any other small molecule in the body and yet it is well accepted that glutamate receptors are not saturated under basal conditions. Studies by Hashimoto and colleagues (Hashimoto, Oka & Nishikawa, 1995) indicate that extra-cellular concentrations of D-serine in the brain are about 5 μM, only about 1% of the total tissue concentrations.

D-SERINE STORAGE AND RELEASE

Classic neurotransmitters are stored in synaptic vesicles in neurons and released by exocytosis. With the advent of the atypical gaseous neurotransmitters, nitric oxide (NO) and carbon monoxide (CO), mechanisms for neurotransmitter release have been challenged. Since gases cannot be readily stored, it is assumed that release occurs immediately following synthesis by the biosynthetic enzymes that are located in close proximity to the plasma membrane. Presumably, biosynthesis of NO and CO is triggered by neuronal depolarization and associated calcium entry. This fits with observations that nNOS is activated by calcium-calmodulin (Bredt & Snyder, 1990) as is heme oxygenase-2, the biosynthetic enzyme for the neurotransmitter pools of CO (Boehning et al., 2004).

For most neurotransmitters, specific mechanisms terminate their synaptic activities, with neuronal or glial uptake being the most prevalent. In a study by Wolosker and colleagues (Ribeiro et al., 2002), [^{3}H]D-serine was accumulated into astrocytes and many other types of cells, but the processes involved appeared to be of low affinity and low specificity for D-serine. The properties of D-serine transport in astrocytes resemble the alanine–serine–cysteine transporter (ASCT) that is responsible for accumulation of small neutral amino acids. The apparently physiologic release of accumulated [^{3}H]D-serine from astrocytes by AMPAR activation suggests that accumulated D-serine labels endogenous stores.

For transmitter release, glia lack the elaborate system of vesicular machinery present in neurons, but they do possess small, clear vesicles as well as large dense-core vesicles, analogous to the same sorts of granules in neurons. Moreover, the release of ATP and glutamate from glia is thought to involve exocytosis from such granules with ATP stored in dense-core granules while glutamate occurs in small vesicles (Bezzi et al., 2004).

Our initial release studies employed glia in cultures labeled with exogenous [^{3}H]D-serine. [^{3}H]D-serine was released by substances that activate AMPA, whereas stimulation of NMDARs and potassium depolarization failed to release D-serine (Schell et al., 1995). More recently, we have demonstrated release of endogenous D-serine by AMPAR activation (Figure 18.1) (Kim et al., 2005). Thus, presumably, the [^{3}H]D-serine employed labels endogenous stores.

Recently, Mothet and colleagues (Mothet et al., 2005) developed a sensitive assay for D-serine that permits continuous monitoring of its release from cells under physiologic conditions. D-serine release was abolished in calcium-free medium and augmented by increasing extracellular levels of calcium or stimulating cells with the calcium ionophore A23187. Release appeared to arise from intracellular vesicles that utilize an electrochemical gradient of protons (Figure 18.1), because concanamycin A, an inhibitor of the vacuolar-type H^+-ATPase, reduced release of D-serine. Vesicular localization was also implied by the co-localization of a major portion of endogenous D-serine with synaptobrevin/VAMP-2, reflecting VAMP-2-bearing vesicles. Evidence for exocytotic release of D-serine was obtained utilizing tetanus toxin, which blocks exocytosis by enzymatically cleaving synaptobrevin or VAMP-2/cellubrevin/VAMP-3, which are required for neurotransmitter release. These findings suggest that D-serine is stored in small synaptic-like vesicles which are typically marked by synaptobrevin/VAMP-2. Although calcium was required for D-serine release, the release was not influenced by inhibitors of voltage-gated calcium channels, fitting with evidence that the major physiologic stimulus for D-serine release is activation of AMPARs.

Recently, novel mechanisms regulating the release of D-serine have been reported. Utilizing yeast two-hybrid techniques, these studies show that SR binds to GRIP (Figure 18.1) (Kim et al., 2005). GRIP had been previously identified by Huganir and associates (O'Brien, Lau & Huganir, 1998) as a protein that binds to AMPARs and regulates their clustering at synapses. GRIP contains seven PDZ domains, and SR binds selectively to PDZ-6. PDZ domains typically interact with the three carboxyl-terminal amino acids of partner proteins by a well-characterized consensus sequence. The C-terminal portion of SR, V-S-V, fits nicely with the PDZ domain binding consensus, and mutation of the C-terminal valine of SR abolishes interactions with GRIP. Direct protein–protein binding studies as well as co-immunoprecipitation confirm the physiologic interactions of SR and GRIP. Moreover, immunohistochemical studies show that GRIP occurs in astrocytes along with SR. The conversion of L-serine to D-serine and its release is markedly reduced in C6 glioma cells containing endogenous GRIP and transfected with SR with the C-terminal valine mutated to glycine. Thus, D-serine formation and release depends upon the binding to GRIP. GRIP also mediates influences of AMPARs on SR and D-serine release. In the absence of AMPA stimulation, D-serine levels are about 50 times higher in the medium than the cells, consistent with a basal release process. AMPA treatment triples the release of D-serine into the medium, an effect blocked by AMPA antagonists. Moreover, viral infection of primary glial cultures with GRIP augments both basal and AMPA-mediated D-serine formation and release.

Recent observations indicate binding of SR to PICK1 (protein interacting with c-kinase) (Figure 18.1). The exact role of PICK1 is unclear, but it might serve as a chaperone escorting protein kinase C alpha (PKCα) to its targets (Wang et al., 2003b). These studies report phosphorylation of SR by PKCα, which might be brought into the vicinity of SR by PICK1. In neuronal systems AMPAR activation leads to the phosphorylation of the receptor at serine-880 which causes GRIP to dissociate from the receptor and to be replaced by PICK1. Whilst GRIP causes AMPARs to cluster at the external surface of cells, PICK1 causes internalization and downregulation of AMPAR synaptic activity (Chung et al., 2003). Conceivably, GRIP and PICK1 also interact in the regulation of SR in astrocytes.

D-SERINE PHYSIOLOGICALLY REGULATES NEURONAL MIGRATION

During development, neurons in the brain migrate to their mature positions along the processes of radial glia. This process is best exemplified in the cerebellum, where granule cell neurons migrate from the external granular layer to the internal granular layer along the processes of the large Bergmann glia. Mechanisms that regulate this migration have been obscure. Rakic and Kumoro (Komuro & Rakic, 1993) showed that drugs that block the glycine site of NMDARs prevent this migratory process, a surprising finding as there are no neuronal synapses upon these granule cells during early migration. Recent experiments have investigated whether the migratory process might involve D-serine. Two approaches were adopted, selective degradation of D-serine by DAO and selective inhibition of SR (Kim et al., 2005). In mouse cerebellar slices at P8, added DAO degrades D-serine by about two-thirds with no effect on glycine and reduces granule cell migration by about 60%, an effect reversed by D-serine. The DAO inhibitor sodium benzoate reverses the effects of DAO on granule cell migration. Selective inhibitors of SR also block migration. The granule cell migration depends on augmented intracellular calcium derived from NMDAR activation, as treatment with an SR inhibitor markedly diminishes intracellular calcium.

How might D-serine influence granule cell migration? Chemokinetic influences of granule cells appear to be involved. Thus, the migration of cerebellar granule cells is augmented by overlaying them with HEK293 cells transfected with SR, but not with catalytically inactive SR or an empty vector. The physiologic influence of D-serine upon neuronal migration involves the activation of SR by GRIP. Thus, in mice infected with viruses containing the PDZ-6 domain of GRIP, movement of granule cells from the external to the internal granular layer is markedly augmented.

CLINICAL IMPLICATIONS

NMDARs have been implicated in multiple brain functions. NMDARs are critical for initiating long-term potentiation (LTP), a major model for learning and memory. D-serine appears critical for maintenance of LTP (Yang et al., 2003). Overactivation of NMDARs is associated with neurotoxicity, as is best exemplified in vascular stroke. In many animal models, NMDAR antagonists prevent stroke damage. For a variety of reasons, clinical trials of NMDAR blockers have been largely unsuccessful, usually because of toxicity associated with influences upon blood pressure or psychotomimetic actions. Drugs that block the glycine site of NMDARs also have anti-stroke influences (Kemp & McKernan, 2002). SR inhibitors might be beneficial in stroke with fewer side effects.

Decreased NMDAR activity may play a role in the pathophysiology of schizophrenia. Thus, NMDA antagonists such as phencyclidine are psychotomimetic with the effects of these drugs mimicking schizophrenic disturbances more so than actions of other psychotomimetic drugs (Javitt & Zukin, 1991). Several clinical studies have examined the influences of agents that activate the glycine site of NMDARs. Thus, administration of glycine, D-cycloserine, or D-serine itself alleviates negative as well as positive symptoms of schizophrenia (Tsai et al., 1998; Goff & Coyle, 2001).

SUMMARY

Evidence accumulating over the past decade indicates that D-serine fulfills the principal requirements of a neurotransmitter. Thus, it mimics the actions of the endogenous neurotransmitter, activating the glycine site of NMDARs with potency at least three times that of glycine. Physiologically it may be far more active than glycine because glycine is much more robustly removed from synapses by uptake systems than is D-serine. SR and D-serine are co-localized with NMDARs to a greater extent than glycine. Most important, highly selective degradation of D-serine abolishes NMDA neurotransmission in different brain regions whether monitored by neurophysiologic or neurochemical measures. D-serine is physiologically released from astrocytes by activation of AMPARs and appears to be released via exocytosis from vesicular stores, very much like classical neurotransmitters.

Assuming that D-serine is a physiologic neurotransmitter, it certainly overturns dogma. For any mammalian messenger molecule to be of the D-configuration is surprising. Equally provocative is the notion that a major neurotransmitter is made by glia, not neurons. The properties of D-serine, along with the gaseous neurotransmitters NO and CO, call for some rewriting of the "rules" of neurotransmission. NO and CO are not stored in vesicles nor are they released by exocytosis, and they do not act upon classical receptors but instead simply diffuse into adjacent cells. NO directly activates guanylyl cyclase to increase cyclic GMP levels and also nitrosylates a variety of important protein targets (Stamler, Lamas & Fang, 2001). CO also activates guanylyl cyclase.

D-aspartate, much less studied than D-serine, might also be a neurotransmitter (Schell, Cooper & Snyder, 1997a). It occurs in selected neuronal populations in the brain. Its levels are highest in early life and decline by adulthood. However, adult brain levels of D-aspartate are greater than those of acetylcholine, norepinephrine, dopamine, and serotonin (Schell, Cooper & Snyder, 1997a; Hashimoto et al., 1993c). D-aspartate is particularly concentrated in the supraoptic and paraventricular neurons of the hypothalamus and contained in the fibers from these cells which innervate the posterior pituitary (Schell, Cooper & Snyder, 1997a). Like D-serine, D-aspartate is degraded by a specific enzyme, D-aspartate oxidase, whose densities are highest in areas of the brain with lowest levels of D-aspartate. In D-aspartate oxidase knockout mice, we have found up to 100-fold elevations of D-aspartate levels. Thus, in areas with "negligible" levels of D-aspartate, the amino acid may be playing an important role with high levels of biosynthesis matched by high turnover rates.

19 · A dialogue between glia and neurons in the retina: modulation of neuronal excitability

Eric A. Newman

In recent years, glial modulation of synaptic transmission and neuronal excitability has been recognized as an important function of glial cells (Volterra, Magistretti & Haydon, 2002; Newman, 2003a). First demonstrated in astrocyte/neuron co-cultures (Parpura et al., 1994; Araque et al., 1998a; Araque et al., 1998b) and later in brain slices (Pasti et al., 1997; Kang, Goldman & Nedergaard, 1998), the release of transmitters from glial cells may have a significant influence on neuronal activity. Glial cells have also been shown to modulate neuronal function in the retina (Newman, 2004), which has proved to be a useful preparation for studying glia–neuron signaling.

Two types of macroglial cells are present in the retinas of most mammals, astrocytes and Müller cells (Figure 19.1) (Newman, 2001a). Astrocytes are confined largely to the nerve fiber layer at the inner border of the retina. They interact primarily with the axons of ganglion cells and do not contact neuronal synapses. Müller cells, in contrast, span the entire neural retina, from the inner retinal border to the photoreceptor layer. Müller cell processes ramify within the two synaptic layers of the retina, the inner and outer plexiform layers, and surround and sometimes contact neuronal synapses. Throughout most of the retina, Müller cells are the only macroglial cells present and they function as specialized astrocytes.

Glial cells may influence neuronal excitability by several mechanisms (Newman, 2005), including the uptake of glutamate and γ-aminobutyric acid (GABA) following release at the synapse, the regulation of extracellular K$^+$, and the regulation of extracellular pH. These processes are considered indirect mechanisms of neuronal modulation (Newman, 2005). Glial cells may also modulate neuronal excitability directly, by releasing transmitters or co-agonists onto neurons (Volterra, Magistretti & Haydon, 2002). Similarly, neurons can modulate glial cell function by the release of neurotransmitters (Schipke & Kettenmann, 2004). This chapter will focus on work from our laboratory which demonstrates direct communication between neurons and glial cells in the retina through the release of chemical messengers.

NEURON TO GLIA SIGNALING

Active neurons may evoke increases in Ca^{2+} within glial cells. In brain slices, electrical stimulation of neurons results in Ca^{2+} increases in neighboring astrocytes (Schipke & Kettenmann, 2004). This neuron to glia signaling is mediated by the release of transmitters from neurons. Release of glutamate (Dani, Chernjavsky & Smith, 1992; Porter & McCarthy, 1996; Pasti et al., 1997), GABA (Kang, Goldman & Nedergaard, 1998), acetylcholine (Araque et al., 2002) and adenosine triphosphate (ATP) (Bowser & Khakh, 2004) may evoke Ca^{2+} increases by activating glial metabotropic receptors. Release of nitric oxide (NO) from neurons may also produce glial Ca^{2+} increases (Matyash et al., 2001).

We have investigated whether similar neuron to glia signaling occurs in the retina (Newman, 2005b). Calcium levels in retinal Müller cells and astrocytes have been monitored with the Ca^{2+} indicator dye Fluo-4 as the retina is stimulated with light flashes. Characterizing neuron to glia signaling in the retina has the advantage over similar studies in brain slices in that a natural stimulus, light, rather than electrical stimuli, may be used to excite neurons.

Müller cell Ca^{2+} transients

Transient increases in Ca^{2+} are generated in Müller cells in retinas under constant illumination (Newman, 2005b). These transients occur at a low frequency (one to 15 transients per Müller cell per 1000 s) and do not propagate between cells. Flickering light stimulation increases the occurrence of these Ca^{2+} transients. Averaging Müller cell Ca^{2+} transients over many trials reveals that the mean Müller cell Ca^{2+} increase rises

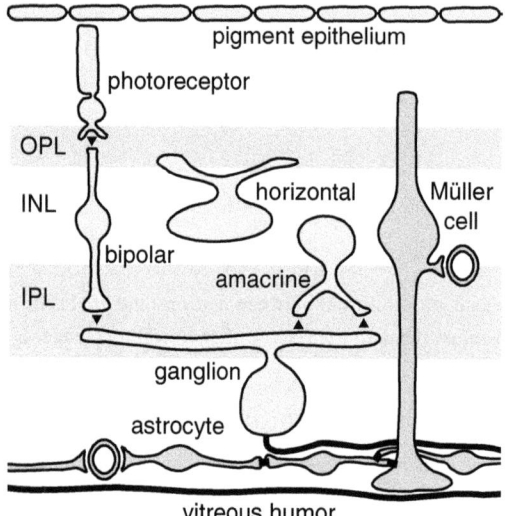

Figure 19.1. Glial cells of the mammalian retina. Astrocytes, ubiquitous central nervous system (CNS) glial cells, are confined primarily to the nerve fiber layer, adjacent to the inner (vitreal) surface of the retina. Müller cells, specialized radial glial cells found only in the retina, extend from the vitreal surface to the photoreceptor layer. Müller cells, but not astrocytes, span the two synaptic layers of the retina, the inner plexiform layer (IPL) and the outer plexiform layer (OPL). Their somata lie in the inner nuclear layer (INL). Astrocytes are coupled to each other and to Müller cells by gap junctions. The major classes of retinal neurons are also illustrated. Synapses are indicated by filled triangles. Reprinted from Newman (2004), with permission.

rapidly at the onset of a flickering stimulus and remains elevated for the duration of the stimulus (Figure 19.2). Ca^{2+} transients are not observed in astrocytes.

The light-evoked increase in Müller cell Ca^{2+} transients is largely blocked by the purinergic receptor antagonist suramin (Newman, 2005b), suggesting that neuronal release of ATP is responsible for the increase in glial Ca^{2+}. The light-evoked increase in Ca^{2+} transients is also blocked by textrodotoxin (TTX), indicating that neuron to glia signaling is mediated by neurons which generate action potentials. In the retina, only amacrine cells and ganglion cells generate action potentials, and these two types of inner retinal neurons are likely to signal glia. Cholinergic starburst amacrine cells, which are believed to co-release ATP, might mediate neuron to glia signaling.

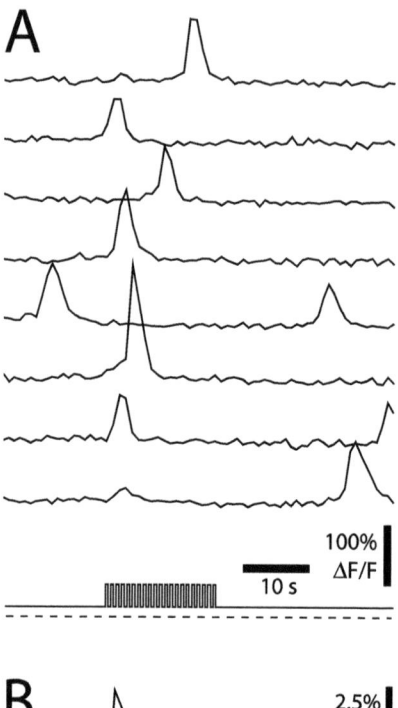

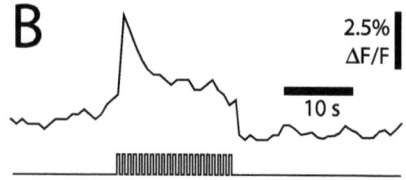

Figure 19.2. Light-evoked Ca^{2+} increases in Müller cells. (A) Calcium fluorescence measured simultaneously in eight Müller cells. Calcium transients are more likely to be generated during the flickering light stimulus. (B) Mean Ca^{2+} fluorescence increase evoked by a flickering light. The response represents transient Ca^{2+} increases averaged over 84 trials. The light stimulus is shown at the bottom in both A and B. From Newman (2005b).

Ganglion cells may also contribute to light-evoked glial Ca^{2+} increases. Antidromic stimulation of ganglion cell axons evokes a substantial increase in Müller cell Ca^{2+} transients. This increase is blocked by suramin as well, suggesting that ganglion cells may release ATP in the retina.

Adenosine potentiation of Müller cell Ca^{2+} responses

Adenosine agonists (100 μM adenosine or 2 μM NECA; 1-(6-Amino-9H-purin-9-yl)-1-deoxy-N-ethyl-b-D-ribofuranuronamide   5'-N-Ethylcarboxamidoadenosine) potentiate light-evoked Müller cell Ca^{2+} increases. In the presence of an adenosine agonist, light "on" elicits a large increase in Ca^{2+} within all Müller cells (Newman, 2005b). Adenosine-potentiated increases are approximately 100-fold larger than the increases observed in the absence of an adenosine agonist. The adenosine-potentiated, light-evoked Ca^{2+} increase begins in Müller cell processes within the inner plexiform layer and then spreads proximally into the endfeet of Müller cells at the inner surface of the retina.

In the presence of adenosine, addition of ATP evokes large increases in Müller cell Ca^{2+}. The light-evoked Ca^{2+} increase is nearly abolished by suramin and by apyrase, an ectoenzyme that hydrolyzes ATP. In contrast, the light-evoked increase is not reduced substantially by antagonists to glutamate, GABA, or acetylcholine receptors. In addition, the metabotropic glutamate agonist trans-ACPD is ineffective in evoking Müller cell Ca^{2+} increases. Together, these results indicate that neuron to glia signaling in the retina is mediated by the release of ATP from neurons.

GLIA TO NEURON SIGNALING

We have characterized glial modulation of neuronal activity in the retina in a number of studies. Initially, the effect of glial cells on light-evoked ganglion cell spike activity was characterized. Subsequent studies have been conducted to determine the mechanisms by which glial cells modulate neuronal activity.

Glial modulation of light-evoked activity in ganglion cells

Mechanical stimulation of single astrocytes evokes Ca^{2+} increases that spread from the stimulated cell to neighboring astrocytes and Müller cells to distances up to 200 μm (Newman & Zahs, 1997). The effect of these activated glial cells on neurons was assessed by recording the light-evoked spike activity of a nearby ganglion cell with an extracellular microelectrode (Newman & Zahs, 1998). We found that stimulated glial cells can either enhance or depress light-evoked neuronal activity (Figure 19.3). In some ganglion cells (9% of all cells recorded; Figure 19.3 (A)) glial cell stimulation increases cell spiking. In many more ganglion cells, however (47%, Figure 19.3 (B)), glial cell stimulation decreases cell spiking. Glial modulation of ganglion cell activity is

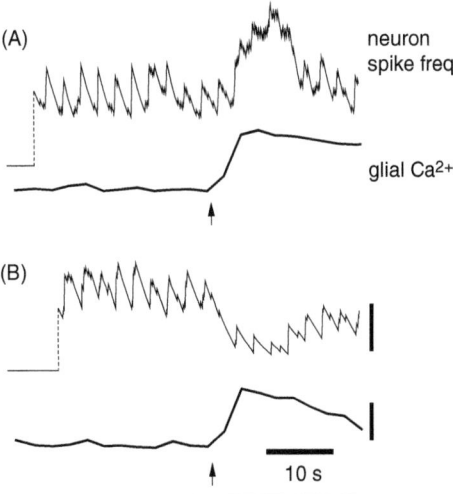

Figure 19.3 Glial modulation of light-evoked spiking in rat ganglion cells. Facilitation of neuron spiking (A) and depression of spiking (B) are illustrated in recordings from two different ganglion cells. A frequency plot of spike activity (top trace) and Ca^{2+} levels within glial cells adjacent to the neuron (bottom trace) are shown for each trial. Arrows indicate initiation of the glial Ca^{2+} wave. The bar at the bottom shows the repetitive light stimulus that evokes neuronal spiking. Calibration bars, three spikes per second; 20% ΔF/F. Modified from Newman and Zahs (1998).

blocked when glial cell Ca^{2+} increases are reduced with thapsigargin, suggesting that glia to neuron signaling is dependent on glial Ca^{2+} increases.

Mechanisms of glia to neuron signaling

The modulation of neuronal spiking by glial cells, described above, could be mediated by a number of different mechanisms. Stimulated glial cells might release a transmitter that activates receptors on pre-synaptic terminals. Glial cells have been shown, for instance, to release glutamate (Parpura et al., 1994; Jeftinija et al., 1996; Bezzi et al., 1998; Innocenti, Parpura & Haydon, 2000) and ATP (Cotrina et al., 1998; Newman, 2001b). These gliotransmitters could activate pre-synaptic glutamate, ATP or adenosine metabotropic receptors, facilitating or depressing synaptic transmission.

Gliotransmitter release might also modulate post-synaptic receptors on ganglion cells. Release of D-serine would facilitate N-methyl-D-aspartate (NMDA) synaptic transmission by binding to the NMDA co-agonist site (Wolosker, Blackshaw & Snyder, 1999; Miller, 2004). Release of glutamate, in contrast, could desensitize glutamatergic receptors, thus reducing glutamate synaptic transmission.

Gliotransmitters could directly modulate the excitability of ganglion cells as well. Release of ATP from glial cells could activate ganglion cell P2X receptors (Taschenberger, Juttner & Grantyn, 1999), depolarizing the cells. ATP, if converted to adenosine by ectoenzymes, might also activate A_1 adenosine receptors (Braas, Zarbin & Snyder, 1987), hyperpolarizing ganglion cells.

Recent studies indicate that retinal glial cells modulate ganglion cell activity by several of these mechanisms, as shown in the following experiments.

Pre-synaptic modulation of ganglion cell activity

Glial cells may modulate synaptic transmission onto ganglion cells by pre-synaptic mechanisms. When the retina is stimulated with brief light flashes repeated at five-second intervals, excitatory post-synaptic currents (EPSCs) are evoked in ganglion cells. ATPγS ejections, which elicit glial Ca^{2+} increases, result in a rise in EPSC amplitude in some ganglion cells (Figure 19.4 (A), middle). It is possible that the ATPγS ejection used to stimulate glial cells facilitates synaptic transmission, not by stimulating glial cells, but by acting directly on pre-synaptic terminals. This seems unlikely however, as activation of glial cells by mechanical stimulation produces a similar facilitation of synaptic transmission (Figure 19.4 (A), bottom).

In some ganglion cells, glial cell stimulation results in a decrease in the amplitude of light-evoked EPSCs. In the example illustrated earlier in Figure 19.4 (B), light flashes evoke a complex synaptic current composed of an initial inhibitory post-synaptic current (IPSC) followed by an EPSC. Glial stimulation reduces the EPSC but has little effect on the IPSC, suggesting that glial modulation of synaptic transmission may be selective. Additional experiments, monitoring changes in paired pulse

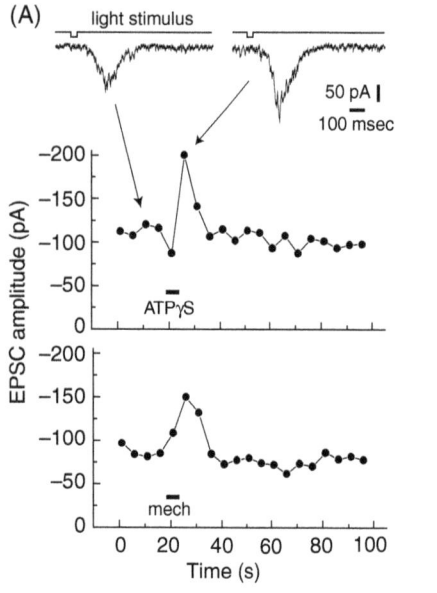
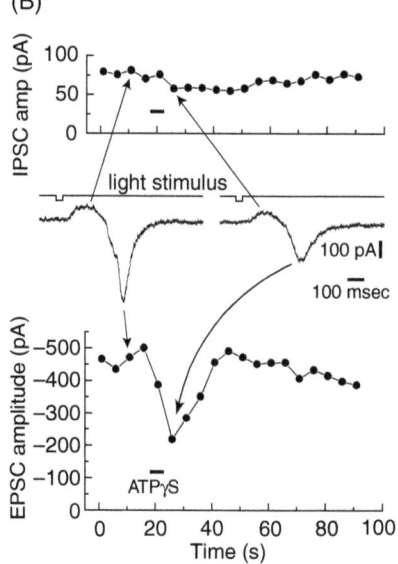

Figure 19.4 Glial facilitation and depression of synaptic transmission. Light flashes, repeated at five-second intervals, evoke excitatory post-synaptic currents (EPSCs) in ganglion cells. (A) Facilitation of synaptic transmission. Sample ganglion cell EPSCs are shown at the top. EPSC amplitudes are plotted as a function of time in the middle and bottom graphs. Stimulation of glial cells with ATPγS (middle graph) or a mechanical stimulus (bottom graph) results in a transient increase in EPSC amplitude. The two trials are from the same ganglion cell. (B) Depression of synaptic transmission. Sample ganglion cell synaptic currents (an inhibitory post-synaptic current (IPSC) followed by an EPSC) are shown in the middle. The amplitudes of IPSCs (top graph) and EPSCs (bottom graph) are shown. Stimulation of glial cells with ATPγS results in a transient decrease in EPSC amplitude but little change in IPSC amplitude. Reprinted from Newman (2004), with permission.

facilitation, indicate that glial modulation of synaptic transmission reflects a pre-synaptic mechanism.

Recordings from many ganglion cells reveal that glial modulation of synaptic transmission similar to that illustrated in Figure 19.4 occurs relatively rarely. Only a small percentage of ganglion cells show either facilitation or depression of synaptic transmission. The evidence suggests glial modulation of synaptic transmission by pre-synaptic mechanisms, although present, may play a relatively minor role in glial modulation of ganglion cell activity.

Post-synaptic modulation of ganglion cell activity

Additional experiments demonstrate that robust glial modulation of ganglion cell activity may occur by direct hyperpolarization of ganglion cells (Newman, 2003b). When glial Ca^{2+} increases are evoked by ATPγS ejection, a hyperpolarization, generated by a slow outward current, is elicited in neighboring ganglion cells (Figure 19.5 (A)). Fifty-two percent of ganglion cells monitored show moderate hyperpolarizations (< 5 mV), 12% show large hyperpolarizations (> 5 mV), and 36% show little or no response.

The hyperpolarizing response is mediated by the stimulated glial cells rather than by a direct neuronal response to the ATPγS ejection (Newman, 2003b). Diverse stimuli, including ejection of ATP, dopamine, thrombin, lysophosphatidic acid, or direct mechanical stimulation, evoke glial Ca^{2+} increases. All of these stimuli also elicit ganglion cell hyperpolarization.

Glial hyperpolarization of ganglion cells may produce substantial changes in neuronal activity. In the experiment illustrated in Figure 19.5 (B), for instance, a high rate of spontaneous spiking present in a ganglion cell is completely blocked when neighboring glial cells are stimulated with ATPγS.

Ganglion cell hyperpolarization is mediated by the release of ATP from glial cells and the activation of ganglion cell A_1 adenosine receptors, as demonstrated by several types of experiments (Newman, 2003b). Glial-mediated hyperpolarization of ganglion cells is abolished by DPCPX, an A_1 receptor antagonist (Figure 19.6 (A)). Ejection of adenosine onto ganglion cells

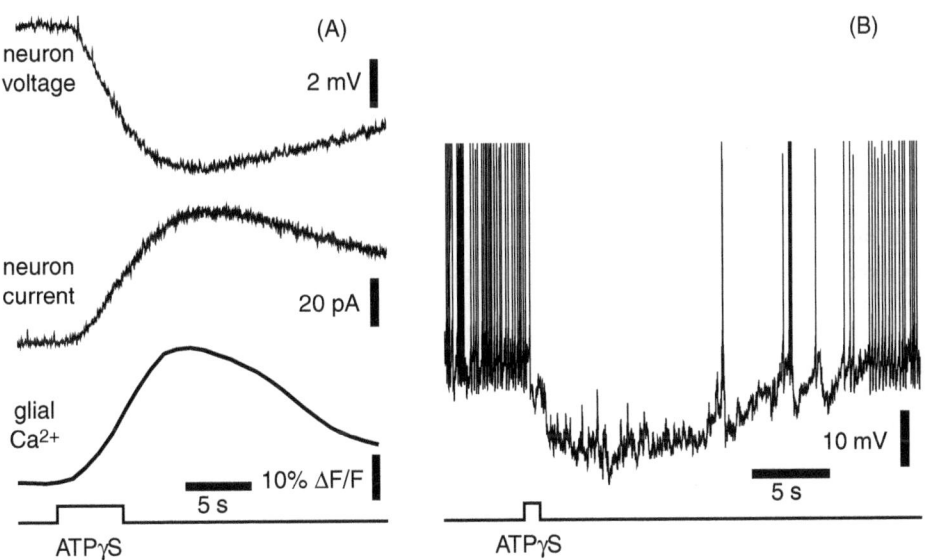

Figure 19.5 Stimulation of glial cells results in an inhibition of ganglion cells. (A) Ejection of ATPγS onto the retinal surface evokes a Ca^{2+} increase in glial cells and a hyperpolarization (current-clamp recording) and an outward current (voltage-clamp recording) in a neighboring ganglion cell. From (Newman, 2003a). (B) ATPγS stimulation of glial cells hyperpolarizes a ganglion cell and blocks all spontaneous action potentials in the cell. Reprinted from Newman (2004), with permission.

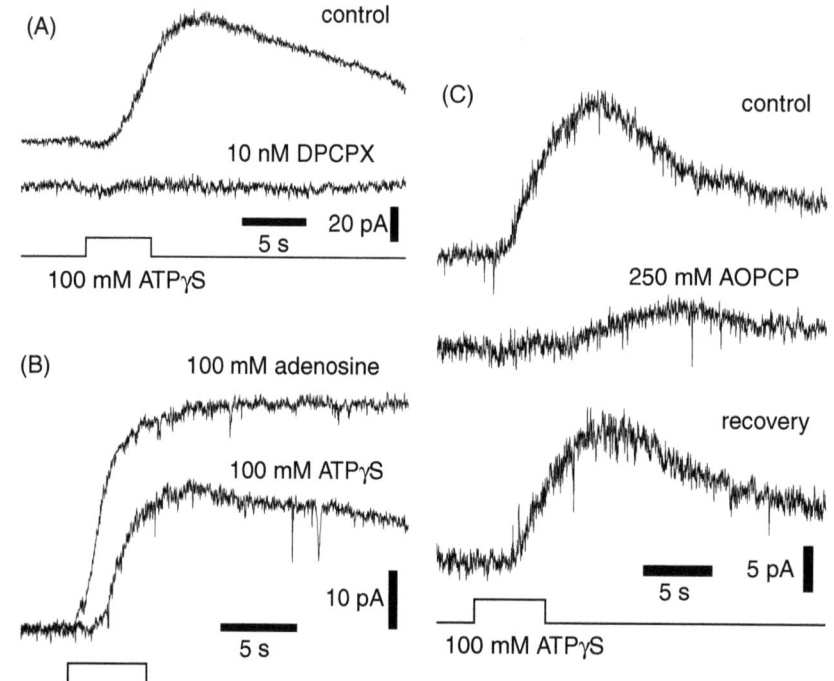

Figure 19.6 Glial inhibition of ganglion cells is mediated by glial release of adenosine triphosphate (ATP) and activation of neuronal adenosine receptors. (A) Stimulation of glial cells with ATPγS evokes an inhibitory outward current in a ganglion cell (control). Addition of DPCPX, an A_1 adenosine receptor antagonist, abolishes the outward current. (B) Adenosine ejection evokes a larger, shorter latency current in a ganglion cell than does ATPγS ejection at the same retinal location. (C) Addition of AOPCP, an ectonucleotidase inhibitor that blocks conversion of AMP to adenosine, reduces and slows the time course of outward ganglion cell current evoked by ATPγS ejection. The effect is largely reversible. From Newman, (2003b).

mimics the response produced by glial cell stimulation (Figure 19.6 (B)). Inhibition of ecto-ATPases and ectonucleotidases, both needed to convert ATP to adenosine after ATP release from glial cells, substantially reduces the glial-mediated ganglion cell response (Figure 19.6 (C)).

The luciferin–luciferase chemiluminescence assay, used to image ATP release, shows that in the presence of Cd^{2+} (which blocks transmitter release from neurons), mechanical stimulation of glial cells evokes an ATP release into the inner plexiform (synaptic) layer. This release is most likely from Müller cells. Indeed, selective stimulation of Müller cells and astrocytes demonstrates that stimulation of Müller cells, but not astrocytes, is both necessary and sufficient to elicit ganglion cell hyperpolarization (Newman, 2003b).

Activation of ganglion cell A_1 receptors evokes cell hyperpolarization by opening Ba^{2+}-sensitive K^+ channels (Newman, 2003b). The reversal potential of the slow outward current evoked by glial cell stimulation is near the K^+ equilibrium potential. The outward current is blocked by Ba^{2+}, suggesting that inwardly rectifying K^+ channels are mediating the response. Recent experiments demonstrate that the adenosine-evoked current is substantially reduced by tertiapin, which blocks G-protein coupled inwardly rectifying K^+ channels (GIRKs).

Together, these experiments demonstrate that retinal glial cells may inhibit ganglion cells by releasing ATP. The ATP is converted to adenosine by ectoenzymes, stimulating ganglion cell A_1 receptors. It is interesting to speculate that released ATP may directly excite ganglion cells as well. Ganglion cells express P2X receptors (Taschenberger, Juttner & Grantyn, 1999) and activation of these receptors would depolarize ganglion cells. Thus retinal glial cells, by releasing ATP, may be able to either excite or inhibit ganglion cells, depending on the type of receptors expressed by the neurons.

D-serine release from Müller cells

D-serine released from glial cells is believed to potentiate NMDA receptor responses in the brain (Miller, 2004). D-serine binds to the co-agonist (glycine) binding site on NMDA receptors (Wolosker, Blackshaw & Snyder, 1999) and may function as the endogenous co-agonist in the central nervous system (CNS) (Mothet et al., 2000; Schell, Molliver & Snyder, 1995).

Release of D-serine from Müller cells may potentiate NMDA receptor transmission in the retina as well (Stevens et al., 2003). Both D-serine and serine racemase, the D-serine synthesizing enzyme, are localized to Müller cells in the retina. In addition, D-serine potentiates NMDA responses in the retina. D-serine augments the NMDA component of light-evoked ganglion cell responses and potentiates ganglion cell responses to NMDA ejection (Figure 19.7 (A)). Addition of D-amino acid oxidase, which degrades D-serine, reduces both the NMDA component of light-evoked responses as well as ganglion cell responses to NMDA ejection (Figure 19.7 (B)). The results demonstrate that endogenous D-serine, which is likely released from Müller cells, modulates NMDA synaptic transmission in the retina.

GLIA TO GLIA SIGNALING

The findings outlined above demonstrate that there is bidirectional signaling between neurons and glial cells in the retina. Neuronal activity evokes Ca^{2+} increases in Müller cells. Stimulated Müller cells, in turn, release transmitters which modulate synaptic transmission and neuronal excitability. These interactions are likely to take place within a confined area, perhaps within a few microns of the synapse. Release of transmitter from the pre-synaptic terminal will evoke a Ca^{2+} increase within glial processes contacting the synapse. The stimulated processes will then release ATP which would feed back onto the pre- or post-synaptic elements of the synapse.

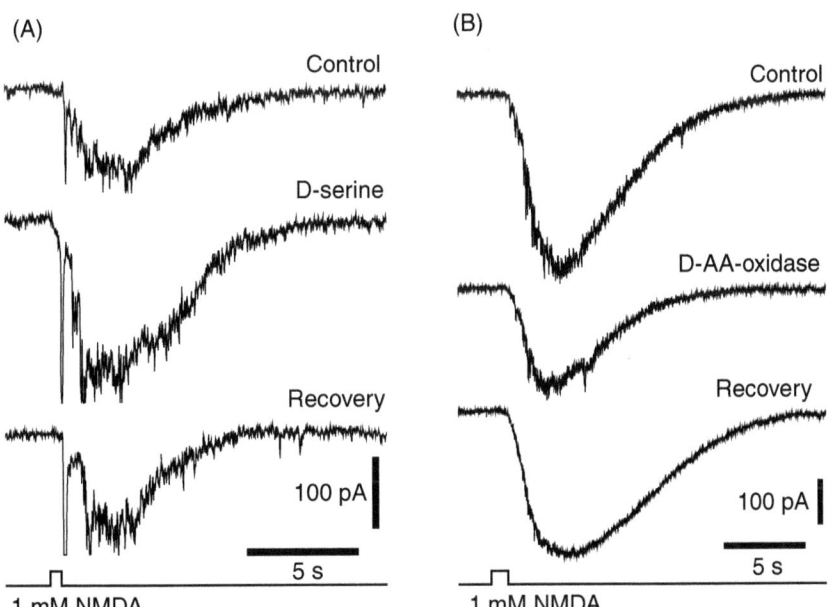

Figure 19.7 Modulation of N-methyl-D-aspartate (NMDA) receptor transmission by D-serine in the retina. (A) Addition of D-serine, an NMDA receptor co-agonist, potentiates the inward current recorded from a ganglion cell in response to NMDA ejection. (B) Addition of D-amino-acid-oxidase, an enzyme which degrades D-serine, reduces the inward current evoked by NMDA ejection. From Stevens et al. (2003).

This neuron–glia dialogue need not be restricted to a single synapse, however. Stimulation of a glial process may result in Ca^{2+} increases in other regions of the glial cell or in the initiation of a Ca^{2+} wave that propagates from the stimulated cell to other glial cells. In this manner, glial cells could influence the excitability of neurons distant from the initial site of glial cell stimulation.

Glial Ca^{2+} waves that propagate both within and between cells do indeed occur in both brain tissue (Peters et al., 2003) and the retina (Newman & Zahs, 1997). In retinal astrocytes and Müller cells, stimulation of one cell region results in a Ca^{2+} increase that propagates into other cells regions (Newman, 2001b). Focal stimulation also elicits intercellular Ca^{2+} waves (Figure 19.8) (Newman & Zahs, 1997; Newman, 2001b). These waves travel at a velocity of $\sim$23 µm/s and may propagate up to $\sim$200 µm from the stimulation site before they die out.

Ca^{2+} waves propagate between retinal glial cells by two mechanisms (Newman, 2001b). Astrocyte to astrocyte propagation is mediated principally by diffusion of an intracellular messenger between gap junction-coupled glial cells (Robinson et al., 1993; Zahs & Newman, 1997). In contrast, propagation from astrocytes to Müller cells and propagation between Müller cells is mediated by glial release of ATP, which functions as an extracellular messenger. ATP release also contributes to wave propagation between retinal astrocytes.

Glial Ca^{2+} waves, previously characterized in the retina as well as in brain slices, have been initiated by artificial stimuli such as agonist ejection, mechanical stimulation, or electrical stimulation. It has remained an open question whether neuronal activity may also initiate intercellular Ca^{2+} waves.

Recent studies provide evidence that intercellular Ca^{2+} waves may be evoked by neuronal activity in the retina (Newman, 2005b). As described above, light stimuli evoke large, transient Müller cell Ca^{2+} increases in the presence of adenosine. In some trials, the initial light-evoked Ca^{2+} increase is followed by a delayed Ca^{2+} rise in a few Müller cells. This secondary Ca^{2+} increase then propagates into neighboring Müller cells as a Ca^{2+} wave. Although these Ca^{2+} waves propagate into a limited number of glial cells and have, to date, only been observed in the presence of adenosine, their existence suggests that intercellular Ca^{2+} waves may occur in vivo.

CONCLUSION

Bidirectional communication between neurons and glial cells has been described in a number of brain slice preparations, and is believed to be responsible for glial modulation of synaptic transmission in the brain.

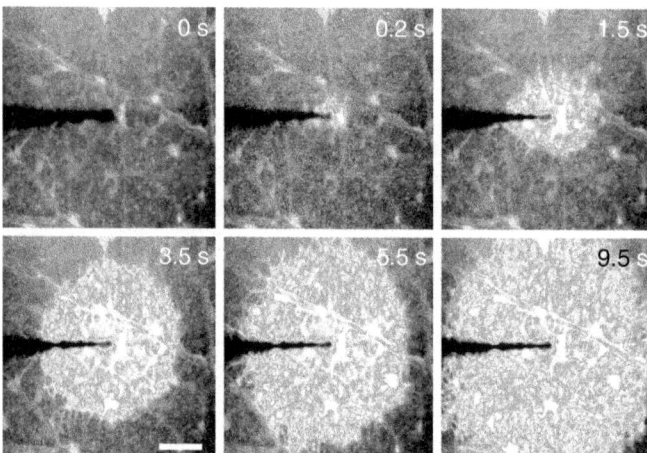

Figure 19.8 Intercellular Ca^{2+} wave in retinal glial cells. Mechanical stimulation of an astrocyte evokes an intercellular Ca^{2+} wave that propagates through both astrocytes and Müller cells. Calcium fluorescence images at the vitreal surface of the retina are shown. Numbers indicate elapsed time following stimulation. Scale bar, 50 µM. Reprinted from Newman (2004), with permission.

Similar bidirectional signaling between neurons and Müller cells is observed in the mammalian retina. Light-evoked neuronal activity leads to increases in the generation of Ca^{2+} transients in Müller cells. These light-evoked glial responses are greatly potentiated by adenosine. These experiments represent the first demonstration that a natural stimulus may evoke Ca^{2+} increases in CNS glia.

Signaling in the reverse direction, from glia to neurons, occurs by a number of mechanisms in the retina. Glial stimulation may either facilitate or depress synaptic transmission onto ganglion cells by pre-synaptic mechanisms. Glia also modulate synaptic transmission by releasing D-serine and potentiating post-synaptic NMDA receptors. Müller cells directly inhibit ganglion cells, as well, by release of ATP.

Although bidirectional neuron–glia signaling occurs in both the brain and the retina, the principal transmitters mediating the signaling appear to be different in the two CNS regions. In the brain, glutamate is the primary neurotransmitter responsible for eliciting Ca^{2+} increases in astrocytes. Similarly, glial modulation of neurons in the brain is mediated primarily by astrocytic release of glutamate. In the retina, in contrast, ATP appears to be the principal transmitter responsible for neuron to glia signaling and for glia to neuron signaling. It is unclear what accounts for this difference in signaling between the brain and the retina. One possible reason is that glutamate is tonically released from retinal neurons (Miller, 2001), and thus may not be well suited to function as the transmitter mediating neuron–glia communication in the retina.

As in the brain, the existence of signaling between neurons and glia in the retina suggests that retinal glial cells may play a role in information processing. Given the relatively slow time course of signaling that has been observed, it is likely that glial cells play a modulatory role in visual processing rather than contributing directly to the generation of rapid, light-evoked signals. However, our current understanding of neuron–glia interactions does not permit us to speculate on the precise role that retinal glial cells play in the processing of visual information.

ACKNOWLEDGEMENTS

I thank J.I. Gepner and M. Metea for helpful discussions. Dr. Newman's research is supported by NIH grant EY004077.

20 · Metabotropic glutamate receptors as a target for astrocytic control of inhibitory synaptic transmission in the hippocampus

Wing-song Liu, Qiwu Xu, Jian Kang, and Maiken Nedergaard

Astrocytes are the most abundant cell type in the central nervous system. They make extensive close contact with neurons. Thin sheets of astrocytic processes intermingle with dendrites, axons, and synapses (Ventura & Harris, 1999). This anatomic proximity forms a basis for bidirectional communication between astrocytes and neurons. Although astrocytes are unable to fire action potentials, they display a form of excitability by changing their intracellular Ca^{2+} levels ($[Ca^{2+}]_i$) (Cornell-Bell et al., 1990). Neurotransmitter release may excite astrocytes by increasing their $[Ca^{2+}]_i$ (Porter & McCarthy, 1997; Bezzi et al., 1998; Araque, Carmignoto & Haydon, 2001; Pasti et al., 2001) Ca^{2+}-dependent and -independent release of neuroactive substances (gliotransmitters) from glial cells in turn modulates excitability (Hassinger et al., 1995; Newman & Zahs, 1998; Ransom, 2000; Newman, 2003), $[Ca^{2+}]_i$ (Nedergaard, 1994; Parpura et al., 1994; Bezzi et al., 1998; Parri, Gould & Crunelli, 2001), differentiation (Blondel et al., 2000), synaptic transmission (Pfrieger & Barres, 1997; Araque et al., 1998a; Araque et al., 1998b; Kang et al., 1998; Oliet, Piet & Poulain, 2001; Beattie et al., 2002; Zhang et al., 2003; Fiacco & McCarthy, 2004), synapse number (Ullian et al., 2001), and synaptic plasticity (Kang et al., 1998; Yang et al., 2003) of neighboring neurons. Gliotransmitters may also be released at astrocyte endfeet to affect vascular tone and therefore, blood supply to neurons (Zonta et al., 2003). This feedback loop provides a powerful mechanism for coordinating neuronal activities with energy demand.

To better understand the physiological consequences of astrocyte Ca^{2+} signaling *in situ*, we have selectively stimulated astrocytes while monitoring stimulation-evoked neuronal responses in hippocampal slices. Selective intracellular Ca^{2+} elevation in astrocytes, induces by Ca^{2+} uncaging, increased the frequency of action potential-driven spontaneous inhibitory post-synaptic currents (sIPSCs) recorded from nearby interneurons. This effect may be blocked by kainate receptor antagonists. When tetrodotoxin is included to block action potentials, however, Ca^{2+} uncaging induces an unexpected decrease in the frequency of miniature IPSCs (mIPSCs), which is not affected by α-amino-3-hydroxy-5-methyl-4-propionate (AMPA)/kainate and N-methyl-D-aspartate (NMDA) receptor antagonists. The amplitude distribution of mIPSCs remained unchanged during the uncaging, consistent with a pre-synaptic inhibition of γ-aminobutyric acid (GABA) release.

It has been shown that group II/III metabotrophic glutamate receptors (mGluRs) are located at pre-synaptic terminals of many synapses (Shigemoto et al., 1997), and that agonists of group II/III mGluRs modulate synaptic transmission by a pre-synaptic mechanism (Gereau & Conn, 1995; Semyanov & Kullmann, 2000). In particular, a group III mGluR agonist induced selective inhibition of evoked IPSCs (eIPSCs) in interneurons in stratum radiatum in CA1 region of hippocampus, and this effect could be mimicked by high-frequency stimulation of neuronal pathways to trigger glutamate spillover (Semyanov & Kullmann, 2000). If glutamate released from astrocytes activates kainate receptors, it may also activate pre-synaptic mGluRs and cause pre-synaptic inhibition of transmitter release. We have tested this possibility by examining whether a potent, broad-spectrum group II/III mGluR antagonist could block the Ca^{2+} uncaging-induced pre-synaptic inhibition in interneurons.

MODULATION OF εIPSCs BY ASTROCYTIC CALCIUM ELEVATION

After a stable baseline whole-cell recording is achieved in interneurons from the stratum radiatum of CA1, an astrocyte near the interneuron process (<30 μm) may be targeted and stimulated with a train of 12 ultraviolet (UV) pulses (0.1 Hz, two minutes). The resultant fluorescence change may be monitored and eIPSCs were recorded simultaneously. Uncaging ethyleneglycol-bis-(β-aminoethylether)-N, N, N',N',-tetraacetic acid (NP-EGTA) produces a stepwise increase in $[Ca^{2+}]_i$ in the astrocyte (Figure 20.1 (A)). The astrocyte Ca^{2+} elevation is accompanied by significant reduction in the amplitude of the eIPSCs in 53% of interneurons (responder, Figure 20.1 (B, C)). During the depression the coefficient of variation (CV) (SD/mean) of the eIPSCs increases from 0.32 (±0.03) to 0.38 (±0.03), consistent with the depression of eIPSCs being mediated by a decrease in the probability of GABA release. The time-course of the decrease in the amplitude of eIPSCs in interneurons follows that of Ca^{2+} elevation in astrocytes (Figure 20.1 (B)). Ca^{2+} uncaging in astrocytes has no significant effect on the amplitude of eIPSCs in 47% of interneurons (non-responders).

Anatomical proximity of the stimulated astrocytes with the dendrites of the recorded interneurons is required for astrocyte-mediated depression of sIPSCs. In astrocytes that are 60–100 μm away from the dendrites of recorded interneurons, Ca^{2+} uncaging elevated $[Ca^{2+}]_i$ in astrocytes, but does not significantly alter the amplitude of eIPSCs.

UV flashings to astrocytes does not produce appreciable change in Ca^{2+} fluorescence when hippocampal slices from the same rat brains are loaded with fluo-4 alone ($\Delta F/F_0 = 3\%$ (±2%); $n = 9$), indicating that the astrocyte Ca^{2+} elevation during the uncaging is not caused by photo-damage or unspecific effects produced by UV laser. Under this condition, UV flashings to astrocytes do not have any significant effect on the amplitude of eIPSCs recorded from interneurons in these slices (Figure 20.1 (C)). Pre-treatment the slices with 1,2-bis (aminophenoxy) ethane-N,N', N'-tetraacetic acid (BAPTA-AM) (10 μM for 20–30 minutes) abolishes the uncaging-induced Ca^{2+} increase in astrocytes and the associated depression of eIPSCs in interneurons. These findings indicate that the depression of eIPSCs in interneurons is very likely caused by Ca^{2+} elevation in astrocytes.

In agreement with previous studies (Porter & McCarthy, 1996), we find that high frequency stimulation (50–100 Hz, one second) causes robust increases in

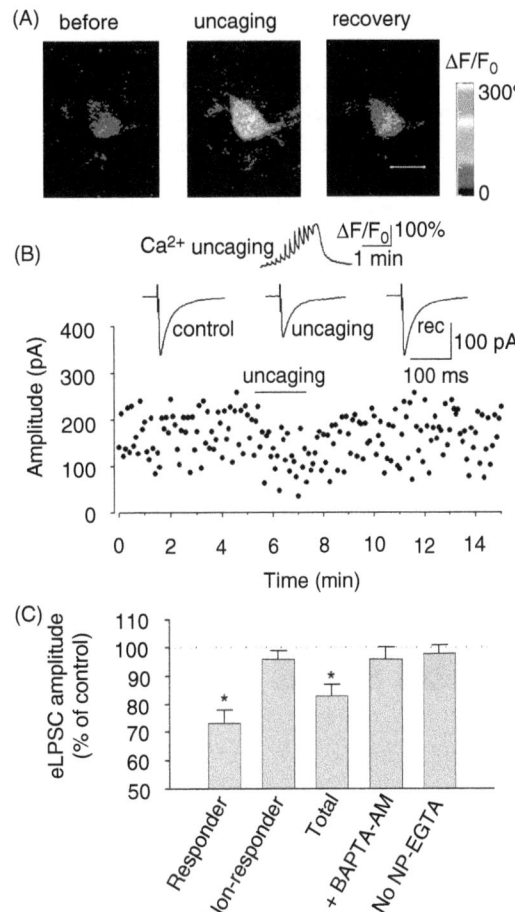

Figure 20.1 Depression of evoked inhibitory post-synaptic currents (eIPSCs) in interneurons by Ca^{2+} uncaging in astrocytes. (A) An astrocyte was stimulated with a train of 12 ultraviolet (UV) laser pulses (0.1 Hz) to uncage NP-EGTA. Images were taken before, during and after the UV pulses, as indicated. Scale bar in recovery: 10 μm. (B) Ca^{2+} uncaging produced a stepwise increase in $\Delta F/F_0$ of the astrocyte shown in A (*upper*) and a reversible decrease in the mean amplitude of eIPSCs (averaged from 24–36 traces) in an interneuron (*middle*). The amplitudes of eIPSCs over the course of the experiment (*lower*). (C) Normalized changes in the amplitude of eIPSCs by uncaging. *Responder* shows average of nine cells that responded to uncaging by decreasing the amplitude of eIPSCs. *Non-Responder* shows average of eight cells that did not respond to uncaging. *Total* shows average of all 17 cells. Preloading the slices with BAPTA-AM (10 μM), a calcium chelator, prevented the depression of eIPSCs by the uncaging ($n = 8$). *No NP-EGTA* indicates that the slices were loaded with fluo-4 alone ($n = 9$). *$p < 0.01$ compared with control, BAPTA AM, and No NP-EGTA groups by ANOVA with Dunnett's test. Reprinted from Liu et al. (2004), with permission.

$[Ca^{2+}]_i$ in a group (3–20) of astrocytes, but single or paired-pulse electrical stimulation at low frequency (0.1 Hz) does not cause appreciable change in $[Ca^{2+}]_i$. This observation indicates that $[Ca^{2+}]_i$ in astrocytes is not affected by electrical stimulation used for eliciting eIPSCs.

The fluo-4 fluorescence is present only in astrocytes, but not in neurons, from slices prepared from 11–15-day-old rats. UV flashings at pyramidal neurons fails to induce Ca^{2+}-activated K currents in NP-EGTA AM bulk-loaded slices, but uncaging NP-EGTA loaded into pyramidal neurons through conventional whole-cell recording induces such currents from pyramidal neurons in the same slices (Figure 20.2). The preferential loading of NP-EGTA in astrocytes and targeted UV stimulation ensures that only astrocytes, but not neurons, are stimulated during uncaging NP-EGTA.

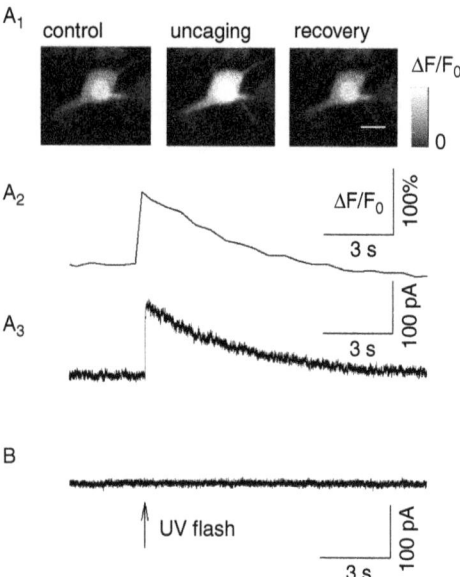

Figure 20.2 Neuronal loading of NP-EGTA in NP-EGTA-AM-loaded slices is insignificant. (A_1) A pyramidal neuron was loaded with potassium salts of NP-EGTA (2 mM, preloaded with 2 mM Ca^{2+}) and fluo-4 (0.1 mM) through conventional whole-cell recording. Uncaging NP-EGTA induced Ca^{2+} elevation as shown by change in fluo-4 fluorescence. Scale bar in "recovery": 15 μm. (A_2) Time course of $\Delta F/F_0$ of the experiment shown in A_1. (A_3) Uncaging NP-EGTA induced Ca^{2+}-activated K current in the pyramidal neuron. (B) Perforated whole-cell recording was made from a pyramidal neuron in NP-EGTA-AM-loaded slices. Ultraviolet (UV) flashing to the neuron did not activate any Ca^{2+}-activated K currents. Reprinted from Liu et al. (2004), with permission.

PHARMACOLOGICAL ACTIVATION OF GROUP II/III mGLuRs ON PRE-SYNAPTIC TERMINALS OF INTERNEURONS

The mechanism underlying the depression of eIPSCs in interneurons produced by astrocyte Ca^{2+} elevation involves group II/III mGluRs on pre-synaptic terminals of interneurons. Ca^{2+}-dependent release of glutamate has been reported to account for several neuronal responses (Araque et al., 1998a; Araque et al., 1998b; Bezzi et al., 1998; Kang et al., 1998; Pasti et al., 2001; Fiacco & McCarthy, 2004; Liu et al., 2004). It has been known that group II/III mGluRs are expressed on pre-synaptic terminals and that activation of the receptors reduces transmitter release from many excitatory and inhibitory synapses (Conn & Pin, 1997; Scanziani, Gahwiler & Charpak, 1998; Semyanov & Kullmann, 2000).

We have established that pre-synaptic group II/III mGluRs are present on pre-synaptic terminals of the interneurons by using a conventional pharmacological approach. Bath application of 0.5 μM DCG IV, a selective group II mGluR agonist (Hayashi et al., 1993), decreases the mean amplitude of the first eIPSCs to 43% (±4%) of that of baseline. Paired-pulse ratio (PPR), which was calculated as the ratio of the amplitude of the second eIPSCs over that of the first eIPSCs, is significantly increased. Co-application of 300 μM cyclopropyl-4-phosphophenylglycine (CPPG), the most potent group II/III mGluR antagonist yet described (Toms et al., 1996), with the agonist, reverses the agonist-induced inhibition. CPPG also reverses the DCG IV-induced change of PPR (Figure 20.3 (A, C)).

Similarly, application of a selective agonist of group III mGluRs, L-AP4 (50 μM) (Scanziani, Gahwiler & Charpak, 1998), reduces the mean amplitude of the first eIPSCs to 44% (±7%) of that of baseline. PPR is significantly increased by L-AP4. Co-application of 300 μM CPPG with L-AP4, reverses the agonist-induced inhibition. CPPG also reverses the L-AP4-induced change of PPR (Figure 20.3 (B, C)). Application of CPPG alone does not produce significant change in the mean amplitude of baseline eIPSCs, suggesting lack of tonic activation of group II/III mGluRs under resting condition (Figure 20.3 (C)).

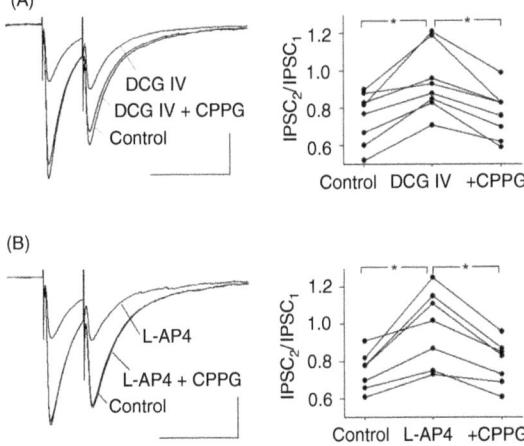

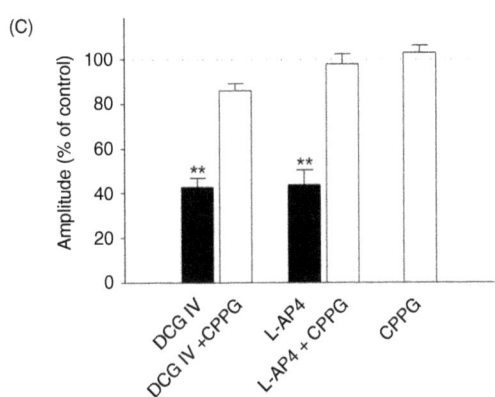

Figure 20.3 Selective group II/III metabotropic glutamate receptor (mGluR) agonists depress evoked inhibitory post-synaptic currents (eIPSCs) in interneurons. (A) Group II mGluR agonist DCG IV (0.5 μM) decreased the amplitude of eIPSCs (left), which was accompanied by an increase in the paired-pulse ratio (PPR) (right) ($n = 8$). Both effects were reversed by CPPG (300 μM), a group II/III mGluR antagonist. Scale bars: 100 pA, 100 ms. *$p < 0.05$ (paired t-test; $n = 8$). (B) Group III mGluR agonist L-AP4 (50 μM) also depressed eIPSCs. The PPR was significantly increased (*$p < 0.05$, paired t-test, $n = 7$). CPPG (300 μM) reversed the depression of eIPSCs and change of the PPR. Scale bars: 100 pA, 100 ms. Every trace in A and B is an average of 36 consecutive traces. (C) Normalized values of change of the first IPSC amplitudes by group II/III mGluR agonists and antagonist. **$p < 0.001$ compared with control, CPPG with agonists, and CPPG alone by ANOVA with Dunnett's test. Reprinted from Liu et al. (2004), with permission.

ACTIVATION OF GROUP II/III mGLuRs MEDIATES ASTROCYTE-INDUCED DEPRESSION OF eIPSCs

If astrocytes release glutamate, activate pre-synaptic group II/III mGluRs, and depress GABA release, then CPPG, a potent blocker of group II/III mGluRs, must block the astrocyte-mediated depression of eIPSCs and mIPSCs. As shown in Figure 20.4 (A), uncaging NP-EGTA in the presence of CPPG (300 μM) does not have significant effects on the amplitude of eIPSCs in any of the eight interneurons tested. CPPG does not have a significant effect on the uncaging-induced Ca^{2+} elevation in astrocytes; the peak $\Delta F/F_0$ during uncaging in the presence of CPPG is not significantly different from that in the absence of CPPG.

In order to separate eIPSCs from evoked EPSCs, AMPA/kainite and NMDA receptor antagonists CNQX (50 μM) and CPP (5 μM) are included in artificial cerebrospinal fluid (aCSF). The antagonists block the activation of all ionotropic glutamate receptors, including kainate receptors. Activation of AMPA/NMDA receptors or kainate receptors by glutamate released from astrocytes has been shown to modulate synaptic transmission (Araque et al., 1998b; Kang et al., 1998; Liu et al., 2004). We have tested the effects of Ca^{2+} uncaging in astrocytes on eIPSCs under the condition that CNQX and CPP were not added into aCSF. To separate eIPSCs from evoked EPSCs, we have clamped interneurons at the reversal potential of EPSCs, 0 mV. Uncaging NP-EGTA in astrocytes produced significant depression of amplitude of eIPSCs in five of nine interneurons tested (Figure 20.4 (B)). The mean peak $\Delta F/F_0$ during the uncaging in astrocytes is 201% ($\pm 25\%$). The depression of eIPSCs is accompanied by an increase in CV from 0.28 (± 0.02) to 0.35 (± 0.03), consistent with a pre-synaptic mechanism. In the remaining four interneurons studied, Ca^{2+} uncaging in astrocytes did not have significant effect on eIPSCs, nor did it affect the CV of eIPSCs. When experiments from the responsive and non-responsive interneurons are averaged, the depression of eIPSCs during Ca^{2+} uncaging is still significantly different from the pre-uncaging level (Figure 20.4 (B, C)). The depression of eIPSCs in the absence of CNQX and CPP is not different from that obtained in the presence of CNQX and CPP. In the absence of CNQX and CPP, the depression of eIPSCs and increases in the frequency of

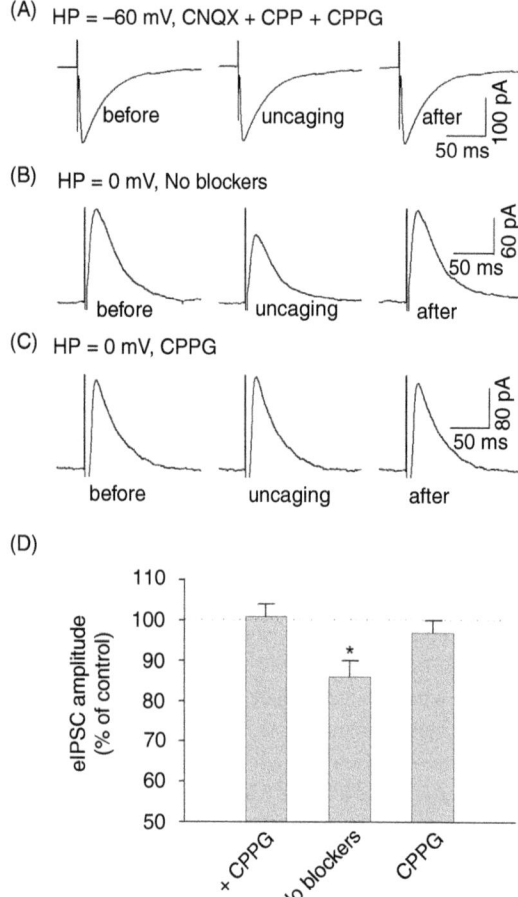

Figure 20.4 Group II/III Metabotropic glutamate receptor (mGluR) antagonist cyclopropyl-4-phosphophenylglycine (CPPG) blocked the depression of eIPSCs induced by uncaging NP-EGTA in astrocytes. (A) In the presence of AMPA/kainate and N-methyl-D-aspartate (NMDA) receptor antagonists (50 μM CNQX and 5 μM CPP) and Group II/III mGluR antagonist CPPG (300 μM), Ca^{2+} uncaging in an astrocyte had no significant effect on eIPSCs in a neighboring interneuron. The holding potential (HP) of the interneuron was -60 mV. Each sample trace of eIPSCs is an average of 24–36 traces. (B) In the absence of any glutamate receptor antagonists, Ca^{2+} uncaging in an astrocyte reduced the amplitude of eIPSCs in an interneuron. The interneuron was held at 0 mV, near the reversal potential of EPSCs. (C) In the presence of CPPG, Ca^{2+} uncaging had no effect on eIPSCs. (D) Normalized values of the amplitude of eIPSCs during uncaging NP-EGTA. +CPPG, CNQX, CPP, and CPPG were included in the bath. No blockers, the glutamate receptor antagonists were not included in the bath. CPPG, only CPPG was in the bath. * $p < 0.05$ compared with control and +CPPG group by ANOVA with Dunnett's test; $n = 8$–9 for each group. Reprinted from Liu et al. (2004), with permission.

spontaneous IPSCs may be seen from the same interneurons. The increases in the frequency of sIPSCs is attributed to activation of kainate receptors, as described in our previous study (Liu et al., 2004).

In other experiments, CPPG (300 μM) has been included in aCSF to block group II/III mGluR receptors. Uncaging NP-EGTA in astrocytes causes Ca^{2+} elevation, but does not have a significant effect on eIPSCs (Figure 20.4 (C, D)). Thus exclusion of CNQX and CPP from aCSF does not unmask any significant modulation of eIPSCs which could have been caused by astrocyte activation of AMPA/kainate and NMDA receptors. The depression of eIPSCs in interneurons by astrocytes is likely caused by Ca^{2+}-dependent release of glutamate and subsequent activation of pre-synaptic group II/III mGluRs, which is blocked by CPPG.

ACTIVATION OF mGLuRs ALSO MEDIATES THE PRE-SYNAPTIC MODULATION OF mIPSCs

We showed previously that Ca^{2+} uncaging in astrocytes caused a small but significant decrease in the frequency of mIPSCs in interneurons (Liu et al., 2004). The amplitude distribution of mIPSCs was not affected by the uncaging, suggesting that the depression of mIPSCs is of pre-synaptic origin. We demonstrated further that pharmacological blocking of AMPA/kainate and NMDA receptors by CNQX and CPP had no effect on the uncaging-induced depression of mIPSCs (Liu et al., 2004). We suspect that the depression of mIPSCs shares a similar mechanism with that of the depression of eIPSCs. We have therefore tested whether CPPG may block the effect of Ca^{2+} uncaging on mIPSCs.

We recorded mIPSCs from interneurons in the presence of tetrodotoxin (0.5 μM) to block action potentials. CNQX (50 μM) and CPP (5 μM) were also included into ACSF to block miniature EPSCs. As reported in our previous study (Liu et al., 2004), Ca^{2+} uncaging in astrocytes produces a significant depression of mIPSCs in neighboring interneurons. The frequency of mIPSCs during the uncaging is 78% (±3%) of baseline frequency (Figure 20.5 (A, C)). The amplitude distribution of mIPSCs remains unchanged (Figure 20.5 (A, D)), suggesting that a pre-synaptic mechanism was involved.

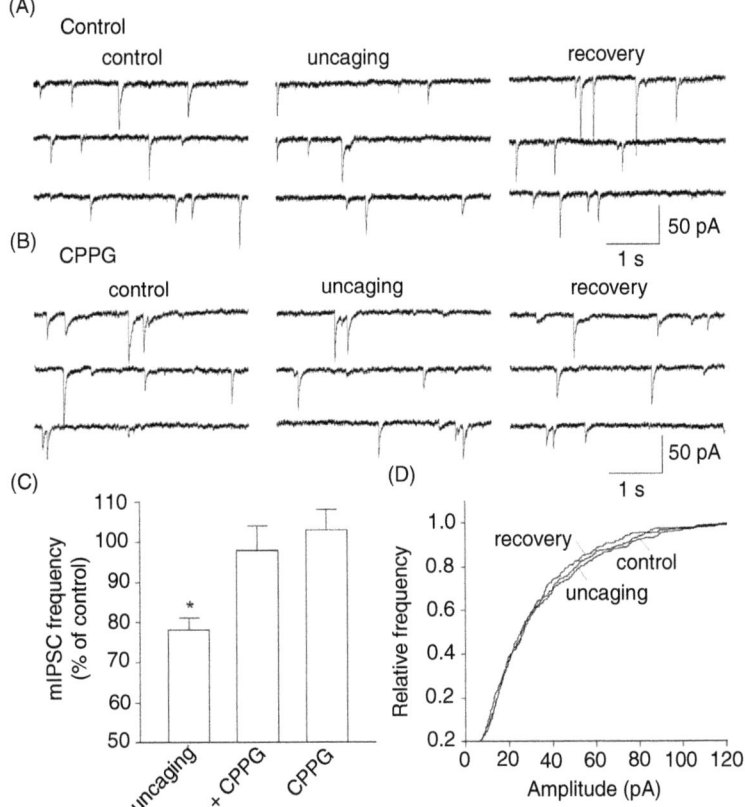

Figure 20.5 Activation of pre-synaptic group II/III metabotropic glutamate receptors (mGluRs) mediates Ca^{2+} uncaging-induced decrease in the frequency of mean inhibitory post-synaptic currents (mIPSCs). (A) Ca^{2+} uncaging in an astrocyte decreased the frequency of mIPSCs in a neighboring interneuron. (B) In the presence of cyclopropyl-4-phosphophenylglycine (CPPG) (300 μM), an antagonist of group II/III mGluRs, the uncaging had no significant effect on mIPSCs. (C) Pool data of normalized frequency of mIPSCs during Ca^{2+} uncaging. The uncaging produced a significant decrease in the frequency of mIPSCs (*uncaging*). CPPG prevented the uncaging-induced modulation of mIPSCs (*+ CPPG*). CPPG perfusion alone had no effect on baseline mIPSCs (*CPPG*). $*p < 0.01$ compared with control and $+ CPPG$ group by ANOVA with Dunnett's test, $n = 7–8$ for each group. (D) Cumulative frequency plot of amplitude distribution of the experiment shown in (A). The plot shows that uncaging had no significant effect on the amplitude distribution of mIPSCs, suggesting that a pre-synaptic mechanism was involved in the uncaging-induced depression of mIPSCs. Reprinted from Liu et al. (2004), with permission.

Bath application of CPPG has no significant effect on baseline frequency of mIPSCs (Figure 20.5 (C)), but it prevents the uncaging-induced depression of mIPSCs (Figure 20.5 (B)). The mean frequency of mIPSCs during the uncaging is 98% (±6%) of that of pre-uncaging level (Figure 20.5 (C)).

DISCUSSION

Our results show that $[Ca^{2+}]_i$ elevation in astrocytes, induced by uncaging NP-EGTA, causes a decrease in the amplitude of eIPSCs and in the frequency of mIPSCs in neighboring interneurons. The decrease in the amplitude of eIPSCs is accompanied by an increase in the coefficient of variation, and the amplitude distribution of mIPSCs remains unchanged. This pattern of changes is typical of a presynaptic effect (Bekkers & Stevens, 1990; Mitchell & Silver, 2000a). Both effects are mediated by pre-synaptic activation of group II/III mGluRs, because it may be blocked by a potent group II/III mGluR antagonist CPPG (Toms et al., 1996). Our findings are consistent with the idea that Ca^{2+}-dependent glutamate release from astrocytes (Araque et al., 1998a; Araque et al., 1998b; Bezzi et al., 1998; Kang et al., 1998; Pasti et al., 2001) activates pre-synaptic group II/III glutamate receptors and decreases GABA release from pre-synaptic terminals of interneurons.

Ca^{2+} ELEVATION IN ASTROCYTES MEDIATES THE DEPRESSION OF eIPSCs AND mIPSCs

The depression of eIPSCs and mIPSCs appears to be caused by $[Ca^{2+}]_i$ elevation in astrocytes. The duration of synaptic depression follows the time-course of $[Ca^{2+}]_i$ elevation in astrocytes, suggesting that the two events are closely linked. The depression is evident only when UV flashings were targeted on to astrocytes preloaded with NP-EGTA and fluo-4, but not with fluo-4 alone. Just like fluo-4, NP-EGTA seems to load astrocytes preferentially, leaving neurons unloaded. We find that no Ca^{2+}-activated K currents may be detected in these slices during uncaging when perforated whole-cell recordings are used to record pyramidal neurons. Such currents were readily observed when NP-EGTA is loaded directly into the pyramidal neurons through conventional whole-cell recording (Liu et al., 2004). We have also demonstrated previously that UV flashings induces Ca^{2+} elevation in astrocytes, but not in neurons in NP-EGTA AM-loaded slices, as revealed by changes of fura-2 fluorescence (Liu et al., 2004). The selective loading of NP-EGTA in astrocytes and targeted UV stimulation ensure that Ca^{2+} elevation is limited to astrocytes. Thus an increase in $[Ca^{2+}]_i$ in neurons is unlikely to contribute to the uncaging-induced depression of eIPSCs and mIPSCs.

DIFFERENTIAL NEURONAL MODULATION BY ASTROCYTES

It has been amply shown that astrocytes can release gliotransmitters to affect a diverse array of neuronal function (Nedergaard, 1994; Kang et al., 1998; Newman & Zahs, 1998; Blondel et al., 2000; Araque, Carmignoto & Haydon, 2001; Bezzi et al., 2001; Ullian et al., 2001; Yang et al., 2003; Fiacco & McCarthy, 2004). The intimate anatomical association (Ventura & Harris, 1999) and diverse functional interaction between astrocytes and neuronal synapses led to the concept "tripartite synapse," in which astrocytes play a critical role in the synaptic transmission (Araque et al., 1999). In recent years, many details of this functional interaction began to unfold in intact tissue such as brain slices. It is now generally believed that Ca^{2+}-dependent and -independent release of gliotransmitters is essential for astrocyte-mediated neuronal modulation.

Glutamate is one of the gliotransmitters that have been extensively studied (Mazzanti, Sul & Haydon, 2001). It has been well-established that there are high-affinity glutamate transporters on the astrocytes which are responsible for uptake of glutamate (Anderson & Swanson, 2000). Experimental data show that astrocytes may nevertheless release glutamate in response to intracellular Ca^{2+} elevation (Parpura et al., 1994). Vesicular exocytosis has been proposed as the mechanism for glutamate release (Calegari et al., 1999; Araque et al., 2000; Pasti et al., 2001), but this issue remains controversial (Nedergaard et al., 2002), and other release mechanisms have also been proposed (Nedergaard, Takano & Hansen, 2002; Ye et al., 2003).

Ca^{2+}-dependent glutamate released from astrocytes produces very diverse effects on different synapses and cell types (Araque et al., 1998a; Araque et al., 1998b; Bezzi et al., 1998; Kang et al., 1998; Pasti et al., 2001; Fiacco & McCarthy, 2004; Liu et al., 2004). How may these differences be explained and reconciled? We believe that the nature of astrocyte-mediated neuronal responses is determined by, first of all, specific target receptors that are activated. For example, astrocyte stimulation potentiated mIPSCs and eIPSCs in pyramidal neurons by activation of pre-synaptic AMPA/NMDA receptors (Kang et al., 1998), whereas it depressed mIPSCs and eIPSCs in interneurons by activation of pre-synaptic group II/III mGluRs. The expression of pre-synaptic mGluRs (i.e. group II/III mGluRs) is often target-specific; that is, mGluRs are present if the post-synaptic neuron is an interneuron, but not if it is a pyramidal neuron (Shigemoto et al., 1996; Scanziani, Gahwiler & Charpak, 1998). Indeed, agonists of group II and III mGluRs depress eIPSCs in interneurons, but not in pyramidal neurons (Gereau & Conn, 1995; Scanziani, Gahwiler & Charpak, 1998; Semyanov & Kullmann, 2000). This may explain why mGluR-mediated depression of eIPSCs and mIPSCs is observed in interneurons in our studies, but not in pyramidal neurons in previous study (Kang et al., 1998).

Glutamate released from astrocytes may modulate excitatory and inhibitory synapses in the same type of neurons through different mechanisms. It has been shown that astrocyte stimulation potentiates IPSCs and EPSCs in hippocampal pyramidal neurons (Kang et al., 1998; Fiacco & McCarthy, 2004). However, the potentiation of EPSCs was mediated by activation of group I mGluRs (Fiacco & McCarthy, 2004), whereas the potentiation of IPSCs was mediated by activation of AMPA/NMDA receptors (Kang et al., 1998). We have shown that astrocyte Ca^{2+} elevation increased the frequency of sIPSCs by activation of axonal or somatodendritic kainate receptors in interneurons (Liu et al., 2004), whereas it decreased the amplitude of eIPSCs and the frequency of mIPSCs by activation of pre-synaptic group II/III mGluRs. Thus astrocyte activation may also differentially modulate inhibitory synapses in the same interneurons by activation of distinct classes of glutamate receptors. Astrocytes and other glial cells may also use different gliotransmitters to affect the same types of neurons. For example, stimulation of glial cells in retinal slices either increased or decreased action potential firings in the same type of ganglion neurons (Newman & Zahs, 1998). It has been reported recently that astrocytes-mediated adenosine triphosphate (ATP) release and its conversion into adenosine accounted for the neuronal inhibition (Newman, 2003), the mechanism for glia-induced increases in action potential firings is not clear, but it is likely caused by activation of other gliotransmitters. Thus gliotransmitters released from glial cells can cause bidirectional modulation of neuronal activity and synaptic transmission, depending on the target receptors that are activated.

The spatial distribution of glutamate receptors is very critical for their activation by glutamate released from astrocytes. High-affinity glutamate transporters on astrocyte membrane limit glutamate diffusion (Anderson & Swanson, 2000). Anatomical segregation between glutamate release sites and glutamate transporters must exist so that released glutamate can reach target receptors. The anatomical proximity between astrocyte processes and neuronal glutamate receptors is required for the astrocyte-mediated neuronal modulation. It is conceivable that the astrocyte processes may contact different target glutamate receptors among different synapses. This might explain why glutamate released from astrocytes produces such diverse effects on different synapses and cell types. Electron microscopy study indicates that astrocyte processes surround the axon–spine interface, but they are not in direct contact with the post-synaptic sites (Ventura & Harris, 1999). This feature suggests that gliotransmitters released from astrocyte processes are likely to activate extra-synaptic receptors rather than post-synaptic receptors (Araque et al., 1998b). Thus astrocytes may modulate, rather than participate in, synaptic transmission.

The sensitivity of target glutamate receptors also shapes the responses to stimulating astrocytes. Different glutamate receptors have different affinity with glutamate. High-affinity glutamate receptors are more likely to be activated by glutamate. Generally, mGluRs have a high affinity for glutamate; for example, the affinity for glutamate for group II mGluRs is 4–12 μM (Pin & Duvoisin, 1995). Kainate receptors may also be activated by low micromolar concentration of glutamate (Rodriguez-Moreno, Lopez-Garcia & Lerma, 2000). The close apposition of pre-synaptic terminals with astrocytic processes (Ventura & Harris, 1999) and the high affinity for glutamate may allow glutamate to reach a sufficient concentration to activate mGluRs and kainate receptors.

FUNCTIONAL IMPLICATION

By selectively stimulating astrocytes while monitoring neuronal responses, we have found that Ca^{2+}-dependent release of glutamate from astrocytes may activate distinct classes of glutamate receptors on interneurons. Activation of kainate receptors causes an increase in the frequency of sIPSCs (Liu et al., 2004), whilst activation of group II/III mGluRs leads to a decrease in the amplitude of eIPSCs and the frequency of mIPSCs. Astrocyte-mediated modulation of synaptic transmissions through activation of mGluRs and AMPA/NMDA receptors has also been reported in hippocampal slices and hippocampal cell culture (Araque et al., 1998a; Araque et al., 1998b; Kang et al., 1998; Fiacco & McCarthy, 2004). These findings suggest that astrocyte activation could bidirectionally modulate synaptic transmission through activation of different glutamate receptors. The "net" effect of stimulating astrocytes on hippocampal circuit function is hard to predict, nevertheless, astrocytes provide a non-synaptic source of glutamate that could fine-tune the strength of excitatory and inhibitory synapses. This is particularly meaningful for inhibitory synapses. Although mGluRs and kainate receptors are expressed on pre-synaptic terminals of inhibitory synapses, how glutamate released from excitatory synapses reaches the glutamate receptors on the inhibitory terminals is unclear. Recent experimental evidence showed that under the condition of high-frequency synaptic activity, inhibition of glutamate uptake, or close apposition of excitatory and inhibitory terminals in certain brain structure, glutamate could "spillover" and activate mGluRs on inhibitory terminals to modulate inhibitory synaptic transmission (Mitchell & Silver, 2000b; Semyanov & Kullmann, 2000). The physiological significance for this heterosynaptic inhibition is not clear. Synapses are often wrapped by astrocyte processes (Ventura & Harris, 1999), Ca^{2+}-dependent release of glutamate from astrocytes could readily access glutamate receptors on inhibitory terminals. Thus, astrocyte-derived glutamate should represent a common mechanism by which inhibitory synaptic transmission may be regulated.

ACKNOWLEDGEMENTS

This work was supported by National Institutes of Health Grants NS38073 and NS41031 to M.N.

References

Abe, K. (2001) Modulation of hippocampal long-term potentiation by the amygdala: a synaptic mechanism linking emotion and memory. *Japanese Journal of Pharmacology* **86**, 18–22.

Abel, T. and Kandel, E. (1998) Positive and negative regulatory mechanisms that mediate long-term memory storage. *Brain Research Brain Research Reviews* **26**, 360–378.

Abel, T., Nguyen, P. V., Barad, M., Deuel, T. A., Kandel, E. R. and Bourtchouladze, R. (1997) Genetic demonstration of a role for PKA in the late phase of LTP and in hippocampus-based long-term memory. *Cell* **88**, 615–626.

Aberg K., Saetre, P., Lindholm, E., Ikholm, B., Pettersson, U., Adolfsson, R. and Jazin, E. (2006a) Human QKI, a new candidate gene for schizophrenia involved in myelination. *American Journal of Medical Genetics B. Neuropsychiatry Genetics* **141**, 84–90.

Aberg K., Saetre, P., Jreborg, H. and Jazin, E. (2006b) Human QKI, a potential regulator of mRNA expression of human oligodendrocyte-related genes involved in schizophrenia. *Proceedings of the National Academy of Science of the United States of America* **103**, 7482–7487.

Abraham, W. C., Mason, S. E., Demmer, J., Williams, J. M., Richardson, C. L., Tate, W. P., Lawlor, P. A. and Dragunow, M. (1993) Correlations between immediate early gene induction and the persistence of long-term potentiation. *Neuroscience* **56**, 717–727.

Adams, J. P. and Sweatt, J. D. (2002) Molecular psychology: roles for the ERK MAP kinase cascade in memory. *Annual Review of Pharmacology and Toxicology* **42**, 135–163.

Adams, M. M., Fink, S. E., Shah, R. A., Janssen, W. G. M., Hayashi, S., Milner, T. A., McEwen, B. S. and Morrison, J. H. (2002) Estrogen and aging affect the subcellular distribution of estrogen receptor-a in the hippocampus of female rats. *Journal of Neuroscience* **22**, 3608–3614.

Addis, D. R., Moscovitch, M., Crawley, A. P. and McAndrews, M. P. (2004) Recollective qualities modulate hippocampal activation during autobiographical memory retrieval. *Hippocampus* **14**, 752–762.

Aggleton, J. P. and Brown, M. W. (1999) Episodic memory, amnesia and the hippocampal–anterior thalamic axis. *Behavioral Brain Science* **22**, 425–489.

Agranoff, B. W., Davis, R. E. and Brink, J. J. (1965) Memory fixation in the goldfish. *Proceedings of the National Academy of Science of the United States of America* **54**, 788–793.

Agster, K. L., Fortin, N. J. and Eichenbaum, H. (2002) The hippocampus and disambiguation of overlapping sequences. *Journal of Neuroscience* **22**, 5760–5768.

Aguado, F., Carmona, M. A., Pozas, E., Aguilo, A., Martinez-Guijjaro, F. J., Alcantara, S., Borrell, V., Yuste, R. and Soriano, E. (2003) BDNF regulates spontaneous correlated activity at early developmental stages by increasing synaptogenesis and expression of the K^+/Cl^- co-transporter KCC2. *Development* **130**, 1267–1280.

Aguilar-Bryan, L., Clement, J. P. T., Gonzalez, G., Kunjilwar, K., Babenko, A. and Bryan, J. (1998) Toward understanding the assembly and structure of KATP channels. *Physiological Reviews* **78**, 227–245.

Ahmed, S., Bierely, R., Sheikh, J. I. and Date, E. S. (2000) Post-traumatic amnesia after closed head injury: a review of the literature and some suggestions for further research. *Brain Injury* **14**, 765–780.

Aicardi, G., Argilli, E., Cappello, S., Santi, S., Riccio, M., Thoenen, H. and Canossa, M. (2004) Induction of long-term potentiation and depression is reflected by corresponding changes in secretion of endogenous brain-derived neurotrophic factor. *Proceedings of the National Academy of Sciences of the United States of America* **101**, 15788–15792.

Akama, K. T. and McEwen, B. S. (2003) Estrogen stimulates postsynaptic density-95 rapid protein synthesis via the

Akt/protein kinase B pathway. *Journal of Neuroscience* **23**, 2333–2339.

Akaneya, Y., Tsumoto, T. and Hatanaka, H. (1996) Brain-derived neurotrophic factor blocks long-term depression in rat visual cortex. *Journal of Neurophysiology* **76**, 4198–4201.

Akirav, I. and Richter-Levin, G. (1999) Biphasic modulation of hippocampal plasticity by behavioral stress and basolateral amygdala stimulation in the rat. *Journal of Neuroscience* **19**, 10530–10535.

Akopian, G., Foy, M. R. and Thompson, R. F. (2003) 17B-estradiol enhancement of LTP involves activation of voltage-dependent calcium channels. Program No. 255.2, 2003 Abstract viewer/itinerary planner. *Society for Neuroscience Abstracts.* Society for Neuroscience, online.

Akshoomoff, N., Pierce, K. and Courchesne, E. (2002) The neurobiological basis of autism from a developmental perspective. *Development and Psychopathology* **14**, 613–634.

Alarcon, J. M., Barco, A. and Kandel, E. R. (2006) Capture of the late phase of long-term potentiation within and across the apical and basilar dendritic compartments of CA1 pyramidal neurons: synaptic tagging is compartment restricted. *Journal of Neuroscience* **26**, 256–264.

Albensi, B. C. and Mattson, M. P. (2000) Evidence for the involvement of TNF and NF-kappaB in hippocampal synaptic plasticity. *Synapse* **35**, 151–159.

Alberini, C. M. (1999) Genes to remember. *Journal of Experimental Biology* **202**, 2887–2891.

Alberini, C. M. (2005) Mechanisms of memory stabilization: are consolidation and reconsolidation similar or distinct processes? *Trends in Neurosciences* **28**, 51–56.

Alberini, C. M., Ghirardi, M., Huang, Y. Y., Nguyen, P. V. and Kandel, E. R. (1995) A molecular switch for the consolidation of long-term memory: cAMP-inducible gene expression. *Annals of the New York Academy of Sciences* **758**, 261–286.

Albin, R. L., Young, A. B. and Penney, J. B. (1989) The functional anatomy of basal ganglia disorders. *Trends in Neurosciences* **12**, 366–375.

Alderson, R. F., Curtis, R., Alterman, A. L., Lindsay, R. M. and DiStefano, P. S. (2000) Truncated TrkB mediates the endocytosis and release of BDNF and neurotrophin-4/5 by rat astrocytes and schwann cells in vitro. *Brain Research* **871**, 210–222.

Alford, S., Frenguelli, B. G., Schofield, J. G. and Collingridge, G. L. (1993) Characterization of Ca^{2+} signals induced in hippocampal CA1 neurones by the synaptic activation of NMDA receptors. *Journal of Physiology* **469**, 693–716.

Ali, S. M., Bullock, S. and Rose, S. P. R. (1988) Phosphorylation of synaptic proteins in chick forebrain: Changes with development and passive avoidance training. *Journal of Neurochemistry* **50**, 1579–1587.

Alkayed, N. J., Murphy, S. J., Traystman, R. J., Hurn, P. D. and Miller, V. M. (2000) Neuroprotective effects of female gonadal steroids in reproductively senescent female rats. *Stroke* **31**, 161–168.

Allan, S. M. (2002) Varied actions of proinflammatory cytokines on excitotoxic cell death in the rat central nervous system. *Journal of Neuroscience Research* **67**, 428–434.

Allan, S. M., and Rothwell, N. J. (2001) Cytokines and acute neurodegeneration. *Nature Reviews Neuroscience* **2**, 734–744.

Almaguer-Melian, W., Martinez-Marti, L., Frey, J. U. and Bergado, J. A. (2003) The amygdala is part of the behavioural reinforcement system modulating long-term potentiation in rat hippocampus. *Neuroscience* **119**, 319–322.

Alonso, M., Vianna, M. R., Izquierdo, I. and Medina, J. H. (2002) Signaling mechanisms mediating BDNF modulation of memory formation in vivo in the hippocampus. *Cellular and Molecular Neurobiology* **22**, 663–674.

Alsina, B. and Cohen-Cory, S. (2001) Visualizing synapse formation in arborizing optic axons in vivo: dynamics and modulation by BDNF. *Nature Neuroscience* **4**, 1093–1101.

Alsina, B., Vu, T. and Cohen-Cory, S. (2001) Visualizing synapse formation in arborizing optic axons in vivo: dynamics and modulation by BDNF. *Nature Neuroscience* **4**, 1093–1101.

Altman, J. (1962) Are new neurons formed in the brains of adult mammals? *Science* **135**, 1127–1128.

Altman, J. (1963) Autoradiographic investigation of cell proliferation in the brains of rats and cats. *Anatomical Record* **145**, 573–591.

Alvarez-Buyella, A. and Lim, D. A. (2004) For the long run: maintaining germinal niches in the adult brain. *Neuron* **41**, 683–686.

Amaral, D. G. and Witter, M. P. (1989) The three-dimensional organization of the hippocampal formation: a review of anatomical data. *Journal of Neuroscience* 31, 571–591.

Amaral, D. G., and Witter, M. P. (1995) Hippocampal formation. In Pacinos, G. (ed.) *The Rat Nervous System*. Academic Press, pp. 443–493.

Amaral, D. G., Price, J. L., Pitkänen, A. and Carmichael, S. T. (1992) Anatomical organization of the primate amygdaloid complex. In: Aggleton, J. P. (ed.) *The Amygdala: Neurobiological Aspects of Emotion, Memory, and Mental Dysfunction*. Wiley–Liss, Inc., pp 1–66.

Ambros, V. (2001) microRNAs: tiny regulators with great potential. *Cell* 107, 823–826.

Amorapanth, P., LeDoux, J. E. and Nader, K. (2000) Different lateral amygdala outputs mediate reactions and actions elicited by a fear-arousing stimulus. *Nature Neuroscience* 3, 74–79.

An, W. F., Bowlby, M. R., Betty, M., Cao, J., Ling, H. P., Mendoza, G., Hinson, J. W., Mattsson, K. I., Strassle, B. W., Trimmer, J. S. and Rhodes, K. J. (2000) Modulation of A-type potassium channels by a family of calcium sensors. *Nature* 403, 553–556.

Anderson, C. M. and Swanson, R. A. (2000) Astrocyte glutamate transport: review of properties, regulation, and physiological functions. *Glia* 32, 1–14.

Anderson, B. J., Alcantara, A. A. and Greenough, W. T. (1996) Motor-skill learning: changes in synaptic organization of the rat cerebellar cortex. *Neurobiology of Learning and Memory* 66, 221–229.

Andreasen, B. J., Li, X., Alcantara, A. A., Isaacs, K. R., Black, J. E. and Greenough, W. T. (1994) Glial hypertrophy is associated with synaptogenesis following motor-skill learning, but not with angiogenesis following exercise. *Glia* 11, 73–80.

Andrade, J. P., Madeira, M. D. & Paula-Barbosa, M. M. (2000) Sexual dimorphism in the subiculum of the rat hippocampal formation. *Brain Research* 875, 125–137.

Andreasen, N. C., Arndt, S., Swayze, V. 2nd, Cizadlo, T., Flaum, M., O'Leary, D., Ehrhardt, J. C. and Yuh, W. T. (1994) Thalamic abnormalities in schizophrenia visualized through magnetic resonance image averaging. *Science* 266, 294–298.

Andrew, R. J. (1999) The differential roles of right and left sides of the brain in memory formation *Behavioural Brain Research* 88, 289–296.

Anglada-Figueroa, D. and Quirk, G. J. (2005) Lesions of the basal amygdala block expression of conditioned fear but not extinction. *Journal of Neuroscience* 25, 9680–9685.

Anokhin, K. V., Tiunova, A. A. and Rose, S. P. R. (2002) Reminder effects – reconsolidation or retrieval deficit? Pharmacological dissection with protein synthesis inhibitors following reminder for a passive avoidance task in young chicks. *Euopean Journal of Neuroscience* 15, 1759-1765.

Anokhin, K. V., Mileusnic, R., Shamakhina, I. Y. and Rose, S. P. R. (1991) Effects of early experience on c-fos gene expression in the chick forebrain. *Brain Research* 544, 101–107.

Antonini, A. and Stryker, M. P. (1993). Rapid remodeling of axonal arbors in the visual cortex. *Science* 260, 1819–1821.

Antonny, B. and Schekman, R. (2001) ER export: public transportation by the COPII coach. *Current Opinions in Cell Biology* 13, 438–443.

Aoyagi, A., Nishikawa, K., Saito, H. and Abe, K. (1994) Characterization of basic fibroblast growth factor-mediated acceleration of axonal branching in cultured rat hippocampal neurons. *Brain Research* 661, 117–126.

Arancio, O., Zhang, H. P., Chen, X., Lin, C., Trinchese, F., Puzzo, D., Liu, S., Hegde, A., Yan, S. F., Stern, A., Luddy, J. S., Lue, L. F., Walker, D. G., Roher, A., Buttini, M., Mucke, L., Li, W., Schmidt, A. M., Kindy, M., Hyslop, P. A., Stern, D. M. and Du Yan, S. S. (2004) RAGE potentiates Abeta-induced perturbation of neuronal function in transgenic mice. *EMBO Journal* 23, 4096–4105.

Araque, A., Carmignoto, G. and Haydon, P. G. (2001) Dynamic signaling between astrocytes and neurons. *Annual Review of Physiology* 63, 795–813.

Araque, A., Parpura, V., Sanzgiri, R. P. and Haydon, P. G. (1998a) Glutamate-dependent astrocyte modulation of synaptic transmission between cultured hippocampal neurons. *European Journal of Neuroscience* 10, 2129–2142.

Araque, A., Parpura, V., Sanzgiri, R. P. and Haydon, P. G. (1999) Tripartite synapses: glia, the unacknowledged partner. *Trends in Neuroscience* 22, 208–215.

Araque, A., Sanzgiri, R. P., Parpura, V. and Haydon, P. G. (1998b) Calcium elevation in astrocytes causes an NMDA receptor-dependent increase in the frequency of miniature synaptic currents in cultured hippocampal neurons. *Journal of Neuroscience* 18, 6822–6829.

Araque, A., Li, N., Doyle, R. T. and Haydon, P. G. (2000) SNARE protein-dependent glutamate release from astrocytes. *Journal of Neuroscience* **20**, 666–673.

Araque, A., Martin, E. D., Perea, G., Arellano, J. I. and Buno, W. (2002) Synaptically released acetylcholine evokes Ca^{2+} elevations in astrocytes in hippocampal slices. *Journal of Neuroscience* **22**, 2443–2450.

Arbuthnott, G. W., Ingham, C. A. and Wickens, J. R. (2000) Dopamine and synaptic plasticity in the neostriatum. *Journal of Anatomy* **196** (Pt 4), 587–596.

Armario, A., Gil, M., Marti, J., Pol, O. and Balasch, J. (1991) Influence of various acute stressors on the activity of adult male rats in a holeboard and in the forced swim test. *Pharmacology, Biochemistry, and Behavior* **39**, 373–377.

Armony, J. L., Quirk, G. J. and LeDoux, J. E. (1998) Differential effects of amygdala lesions on early and late plastic components of auditory cortex spike trains during fear conditioning. *Journal of Neuroscience* **18**, 2592–2601.

Arnold, A. P. and Breedlove, S. M. (1985) Organizational and activational effects of sex steroids on brain and behavior: a reanalysis. *Hormones and Behavior* **19**, 469–498.

Arnold, A. P. and Gorski, R. A. (1984) Gonadal steroid induction of structural sex differences in the central nervous system. *Annual Review of Neuroscience* **7**, 413–442.

Aronica, E. M., Gorter, J. A., Paupard, M. C., Grooms, S. Y., Bennett, M. V. and Zukin, R. S. (1997) Status epilepticus-induced alterations in metabotropic glutamate receptor expression in young and adult rats. *Journal of Neuroscience* **17**, 8588–8595.

Asan, E. (1998) The catecholaminergic innervation of the rat amygdala. *Advances in Anatomy, Embryology and Cell Biology* **142**, 1–118.

Ashcroft, F. M. and Gribble, F. M. (1998) Correlating structure and function in ATP-sensitive K^+ channels. *Trends in Neurosciences* **21**, 288–294.

Asthana, S., Baker, L. D., Craft, S., Stanczyk, F. Z., Veith, R. C., Raskind, M. A. and Plymate, S. R. (2001) High-dose estradiol improves cognition for women with AD: results of a randomized study. *Neurology* **57**, 605–612.

Aston, C., Jiang, L. and Sokolov, B. P. (2005) Transcriptional profiling reveals evidence for signaling and oligodendroglial abnormalities in the temporal cortex from patients with major depressive disorder. *Molecular Psychiatry* **10**, 309–322.

Aston-Jones, G., Rajkowski, J. and Cohen, J. (1999) Role of locus coeruleus in attention and behavioral flexibility. *Biological Psychiatry* **46**, 1309–1320.

Atienza, M., Cantero, J. L. and Dominguez-Marin, E. (2002) The time course of neural changes underlying auditory perceptual learning. *Learning & Memory* **9**, 138–150.

Atienza, M., Cantero, J. L. and Stickgold, R. (2004) Posttraining sleep enhances automaticity in perceptual discrimination. *Journal of Cognitive Neuroscience* **16**, 53–64.

Atkins, C. M. and Sweatt, J. D. (1999) Reactive oxygen species mediate activity-dependent neuron-glia signaling in output fibers of the hippocampus. *Journal of Neuroscience* **19**, 7241–7248.

Atkins, C. M., Selcher, J. C., Petraitis, J. J., Trzaskos, J. M. and Sweatt, J. D. (1998) The MAPK cascade is required for mammalian associative learning. *Nature Neuroscience* **1**, 602–609.

Auerbach, J. M. and Segal, M. (1997) Peroxide modulation of slow onset potentiation in rat hippocampus. *Journal of Neuroscience* **17**, 8695–8701.

Avian Brain Nomenclature Consortium. (2005) Avian brains and a new understanding of vertebrate brain evolution. *National Review of Neuroscience* **6**, 151–158.

Avshalumov, M. V. and Rice, M. E. (2002) NMDA-receptor activation mediates hydrogen peroxide-induced pathophysiology in rat hippocampal slices. *Journal of Neurophysiology* **87**, 2896–2903.

Avshalumov, M. V. and Rice, M. E. (2003) Activation of ATP-sensitive K^+ (K_{ATP}) channels by H_2O_2 underlies glutamate-dependent inhibition of striatal dopamine release. *Proceedings of the National Academy of Sciences of the United States of America* **100**, 11,729–11,734.

Avshalumov, M. V., Chen, B. T. and Rice, M. E. (2000) Mechanisms underlying H_2O_2-mediated inhibition of synaptic transmission in rat hippocampal slices. *Brain Research* **882**, 86–94.

Avshalumov, M. V., MacGregor, D. G., Sehgal, L. M. Rice, M. E. (2004) The glial antioxidant network and neuronal ascorbate: permissive yet protective for H_2O_2 signaling? *Neuron Glia Biology*.

Avshalumov, M. V., Chen, B. T., Kóos, T., Tepper, J. M. and Rice, M. E. (2005) Endogenous hydrogen peroxide regulates the excitability of midbrain dopamine neurons via ATP-sensitive potassium channels. *Journal of Neuroscience* **25**, 4222–4231.

Avshalumov, M. V., Chen, B. T., Marshall, S. P., Peña, D. M. and Rice, M. E. (2003) Glutamate-dependent inhibition of dopamine release in striatum is mediated by a new diffusible messenger, H_2O_2. *Journal of Neuroscience* **23**, 2744–2750.

Babenko, A. P., Gonzalez, G. and Bryan, J. (2000) Pharmaco-topology of sulfonylurea receptors. Separate domains of the regulatory subunits of KATP channel isoforms are required for selective interaction with K^+ channel openers. *Journal of Biological Chemistry* **275**, 717–720.

Backstrom, T. (1976) Epileptic seizures in women related to plasma estrogen and progesterone during the menstrual cycle. *Acta Neurologica Scandinavica* **54**, 321–347.

Backstrom, P. and Hyytia, P. (2003) Attenuation of cocaine-seeking behaviour by the AMPA/kainate receptor antagonist CNQX in rats. *Psychopharmacology (Berl)* **166**, 69–76.

Bacskai, B. J., Hochner, B., Mahaut-Smith, M., Adams, S. R., Kaang, B. K., Kandel, E. R. and Tsien, R. Y. (1993) Spatially resolved dynamics of cAMP and protein kinase A subunits in Aplysia sensory neurons. *Science* **260**, 222–226.

Bading, H. (2000) Transcription-dependent neuronal plasticity the nuclear calcium hypothesis. *European Journal of Biochemistry* **267**, 5280–5283.

Bading, H., Ginty, D. D. and Greenberg, M. E. (1993) Regulation of gene expression in hippocampal neurons by distinct calcium signaling pathways. *Science* **260**, 181–186.

Bahar, A., Dorfman, N. and Dudai, Y. (2004) Amygdalar circuits required for either consolidation or extinction of taste aversion memory are not required for reconsolidation. *European Journal of Neuroscience* **19**, 1115–1118.

Bailey, C. H. and Kandel, E. R. (1993) Structural changes accompanying memory storage. *Annual Review of Physiology* **55**, 397–426.

Bailey, C. H., Guistetto, M., Huang, Y. Y., Hawkins, R. D. and Kandel, E. R. (2000) Is heterosynaptic modulation essential for stabilizing Hebbian plasticity and memory? *Nat Rev Neurosci* **1**, 11–20.

Bailey, C. H., Montarolo, P., Chen, M., Kandel, E. R. and Schacher, S. (1992) Inhibitors of protein and RNA synthesis block structural changes that accompany long-term heterosynaptic plasticity in Aplysia. *Neuron* **9**, 749–758.

Bailey, D. J., Kim, J. J., Sun, W., Thompson, R. F. Helmstetter, F. J. (1999) Acquisition of fear conditioning in rats requires the synthesis of mRNA in the amygdala. *Behavioral Neuroscience* **113**, 276–282.

Baker, G. E. and Stryker, M. P. (1990) Retinofugal fibers change conduction velocity and diameter between the optic nerve and tract in ferrets. *Nature* **344**, 342–345.

Baldelli, P., Forni, P. E. and Carbone, E. (2000) BDNF, NT-3 and NGF induce distinct new Ca^{2+} channel synthesis in developing hippocampal neurons. *European Journal of Neuroscience* **12**, 4017–4032.

Balkowiec, A. and Katz, D. M. (2002) Cellular mechanisms regulating activity-dependent release of native brain-derived neurotrophic factor from hippocampal neurons. *Journal of Neuroscience* **22**, 10,399–10,407.

Bao, L. and Rice, M. E. (2004) Mitochondrial H_2O_2 mediates glutamate-dependent inhibition of striatal dopamine release. *Abstract Viewer/Itinerary Planner.* Society for Neuroscience, 46.7.

Bao, L., Avshalumov, M. V. and Rice, M. E. (2005) Mitochondrial inhibition causes functional dopamine denervation and striatal medium spiny neurons depolarization via increased H_2O_2, not decreased ATP. *Journal of Neuroscience.*

Bao, S., Chen, L., Qiao, X. and Thompson, R. F. (1999) Transgenic brain-derived neurotrophic factor modulates a developing cerebellar inhibitory synapse. *Learning & Memory* **6**, 276–283.

Barber, T. A., Howorth, P. D., Klunk, A. M. and Cho, C. C. (1999) Lesions of the intermediate medial hyperstriatum ventrale impair sickness conditioned learning in day old chicks. *Neurobiology of Learning and Memory* **72**, 128–141.

Barbin, G., Aigrot, M. S., Charles, P., Foucher, A., Grumet, M., Schachner, M., Zalc, B. and Lubetzki, C. (2004) Axonal cell-adhesion molecule L1 in CNS myelination. *Neuron Glia Biology* **1**, 65–72.

Barco, A., Alarcon, J. M. and Kandel, E. R. (2002) Expression of constitutively active CREB protein facilitates the late phase of long-term potentiation by enhancing synaptic capture. *Cell* **108**, 689–703.

Barco, A., Patterson, S., Alarcon, J. M., Gromova, P., Mata-Roig, M., Morozov, A. and Kandel, E. R. (2005) Gene expression profiling of facilitated L-LTP in VP16-CREB mice reveals that BDNF is critical for the maintenance of LTP and its synaptic capture. *Neuron* **48**, 123–137.

Barger, S. W. and Basile, A. S. (2001) Activation of microglia by secreted amyloid precursor protein evokes release of glutamate by cystine exchange and attenuates synaptic function. *Journal of Neurochemistry* **76**, 846–854.

Barnabe-Heider, F. and Miller, F. D. (2003) Endogenously produced neurotrophins regulate survival and differentiation of cortical progenitors via distinct signaling pathways. *Journal of Neuroscience* **23**, 5149–5160.

Barnea, A. and Nottebohm, F. (1994) Seasonal recruitment of hippocampal neurons in adult free-ranging black-capped chickadees. *Proceedings of the National Academy of Sciences of the United States of America* **91**, 11,217–11,221.

Barnes, C. A. (1979) Memory deficits associated with senescence: a neurophysiological and behavioral study in the rat. *Journal of Comparative Physiological Psychology* **93**, 74–104.

Barnes, C. A. (1994) Normal aging: regionally specific changes in hippocampal synaptic transmission. *Trends in Neuroscience* **17**, 13–18.

Barnes, C. A., Rao, G., Foster, T. C. and McNaugton, B. L. (1992) Region-specific age effects on AMPA sensitivity: electrophysiological evidence for loss of synaptic contacts in hippocampal field CA1. *Hippocampus* **2**, 457–468.

Barrera, A., Imenez, L., Gonzalez, G. M., Montiel, J. and Aboitiz, F. (2001) Dendritic structure of single hippocampal neurons according to sex and hemisphere of origin in middle-aged and elderly human subjects. *Brain Research* **906**, 31–37.

Barres, B. A. and Raff, M. C. (1993) Proliferation of oligodendrocyte precursor cells depends on electrical activity in axons. *Nature* **361**, 258–260.

Barrett-Conner, E. and Kritz-Silverstein, D. (1993) Estrogen replacement therapy and cognitive function in older women. *Journal of the American Medical Association* **269**, 2637–2641.

Barria, A., Derkach, V. and Soderling, T. (1997a) Identification of the Ca^{2+}/calmodulin-dependent protein kinase II regulatory phosphorylation site in the alpha-amino-3-hydroxyl-5-methyl-4-isoxazole-propionate-type glutamate receptor. *Journal of Biological Chemistry* **272**, 32,727–32,730.

Barria, A., Muller, D., Derkach, V., Griffith, L. C. and Soderling, T. R. (1997b) Regulatory phosphorylation of AMPA-type glutamate receptors by CaM-KII during long-term potentiation. *Science* **276**, 2042–2045.

Bartel, D. P. (2004) MicroRNAs: genomics, biogenesis, mechanism, and function. *Cell* **116**, 281–297.

Bartsch, D., Ghirardi, M., Skehel, P. A., Karl, K. A., Herder, S. P., Chen, M., Bailey, C. H. and Kandel, E. R. (1995) Aplysia CREB2 represses long-term facilitation: relief of repression converts transient facilitation into long-term functional and structural change. *Cell* **83**, 979–992.

Bartzokis G., Beckson M., Lu, P. H., Nuechterlein, K. H., Edwards, N. and Mintz, J. (2001) Age-related changes in frontal and temporal lobe volumes in men: a magnetic resonance imaging study. *Archives of General Psychiatry* **58**, 461–465.

Bartzokis, G., Lu, P. H., Geschwind, D. H., Edwards, N., Mintz, J. and Cummings, J. L. (2006) Apolipoprotein E genotype and age-related myelin breakdown in healthy individuals: Implications for cognitive decline and dementia. *Archives of General Psychiatry* **63**, 63–72.

Bass, N. H., Hess, H. H., Pope, A. and Thalheimer, C. (1971) Quantitative cytoarchitechtonic distribution of neurons, glia and DNA in rat cerebral cortex. *Journal of Comparative Neurology* **143**, 481–490.

Baudry, M., Davis, J. L. and Thompson, R. F. (2000) *Advances in Synaptic Plasticity*. MIT Press.

Bauer, E. P. and LeDoux, J. E. (2004) Heterosynaptic long-term potentiation of inhibitory interneurons in the lateral amygdala. *Journal of Neuroscience* **24**, 9507–9512.

Bauer, E. P., Schafe, G. E. and LeDoux, J. E. (2002) NMDA receptors and L-type voltage-gated calcium channels contribute to long-term potentiation and different components of fear memory formation in the lateral amygdala. *Journal of Neuroscience* **22**, 5239–5249.

Bayer, S. A., Yackel, J. W. and Puri, P. S. (1982) Neurons in the rat dentate gyrus granular layer substantially increase during juvenile and adult life. *Science* **216**, 890–892.

Bayer, K. U., De Koninck, P., Leonard, A. S., Hell, J. W. and Schulman, H. (2001) Interaction with the NMDA receptor locks CaMKII in an active conformation. *Nature* **411**, 801–805.

Bear, M. F. (1996a) A synaptic basis for memory storage in the cerebral cortex. *Proceedings of the National Academy of Sciences of the United States of America* **93**, 13,453–13,459.

Bear, M. F. (1996b) Progress in understanding NMDA-receptor-dependent synaptic plasticity in the visual cortex. *Journal of Physiology (Paris)* **90**, 223–227.

Bear, M. F. and Malenka, R. C. (1994) Synaptic plasticity: LTP and LTD. *Current Opinion in Neurobiology* **4**, 389–399.

Bear, M. F., Huber, K. M. and Warren, S. T. (2004) The mGluR theory of fragile X mental retardation. *Trends in Neuroscience* **27**, 370–377.

Beattie, E. C., Carroll, R. C., Yu, X., Morishita, W., Yasuda, H., von Zastrow, M. and Malenka, R. C. (2000) Regulation of AMPA receptor endocytosis by a signaling mechanism shared with LTD. *Nature Neuroscience* **3**, 1291–1300.

Beattie, E. C., Stellwagen, D., Morishita, W., Bresnahan, J. C., Ha, B. K., Von Zastrow, M., Beattie, M. S. and Malenka, R. C. (2002a) Control of synaptic strength by glial TNFalpha. *Science* **295**, 2282–2285.

Beattie, M. S., Harrington, A. W., Lee, R., Kim, J. Y., Boyce, S. L., Longo, F. M., Bresnahan, J. C., Hempstead, B. L. and Yoon, S. O. (2002b) ProNGF induces p75-mediated death of oligodendrocytes following spinal cord injury. *Neuron* **36**, 375–386.

Beaulieu, J. M., Sotnikova, T. D., Yao, W. D., Kockeritz, L., Woodgett, J. R., Gainetdinov, R. R. and Carron, M. G. (2004) Lithium antagonizes dopamine-dependent behaviors mediated by an AKT/glycogen synthase kinase 3 signaling cascade. *Proceedings of the National Academy of Science of the United States of America* **101**, 5099–5104.

Bebchuk, J. M., Arfken, C. L., Dolan-Manji, S., Murphy, J., Hasanat, K. and Manji, H. K. (2000) A preliminary investigation of a protein kinase C inhibitor in the treatment of acute mania. *Archives of General Psychiatry* **57**, 95–97.

Bedard, K. and Krause, K. H. (2007) The NOX family of ROS-generating NADPH oxidases: physiology and pathophysiology. *Physiological Reviews* **87**, 245–313.

Beiko, J., Lander, R., Hampson, E., Boon, F. and Cain, D. P. (2004) Contribution of sex differences in the acute stress response to sex differences in water maze performance in the rat. *Behavioral Brain Research* **151**, 239–253.

Bejar, R., Yasuda, R., Krugers, H., Hood, K. and Mayford, M. (2002) Transgenic calmodulin-dependent protein kinase II activation: dose-dependent effects on synaptic plasticity, learning, and memory. *Journal of Neuroscience* **22**, 5719–5726.

Bekkers, J. M. and Stevens, C. F. (1990) Presynaptic mechanism for long-term potentiation in the hippocampus. *Nature* **346**, 724–729.

Bellgowan, P. S. F. and Helmstetter, F. J. (1996) Neural systems for the expression of hypoalgesia during nonassociative fear. *Behavioral Neuroscience* **110**, 727–736.

Belvin, M. P. and Yin, J. C. (1997) Drosophila learning and memory: recent progress and new approaches. *Bioessays* **19**, 1083–1089.

Benavides, J., Camelin, J. C., Mitrani, N., Flamand, F., Uzan, A., Legrand, J. J., Gueremy, C. and Le Fur, G. (1985) 2-Amino-6-trifluoromethoxy benzothiazole, a possible antagonist of excitatory amino acid neurotransmission – II. Biochemical properties. *Neuropharmacology* **24**, 1085–1092.

Benedito, A. B., Lehtinen, M., Massol, R., Lopes, U. G., Kirchhausen, T., Rao, A. and Bonni, A. (2004) The transcription factor NFAT3 mediates neuronal survival. *Journal of Biological Chemistry* **280**, 2818–2825.

Benes, F. M., Turtle, M., Khan, Y. and Farol, P. (1994) Myelination of a key relay zone in the hippocampal formation occurs in the human brain during childhood, adolescence, and adulthood. *Archives of General Psychiatry* **51**, 477–484.

Bengtsson, S. L., Nagy, Z., Skare, S., Forsman, L., Forssberg, H. and Ullén, F. (2005) Extensive piano practicing has regionally specific effects on white matter development. *Nature Neuroscience* **8**, 1148–1150.

Bengzon, J., Kokaia, Z., Elmer, E., Nanobashvili, A., Kokaia, M. and Lindvall, O. (1997) Apoptosis and proliferation of dentate gyrus neurons after single and intermittent limbic seizures. *Proceedings of the National Academy of Sciences of the United States of America* **94**, 10,432–10,437.

Benington, J. H. and Frank, M. G. (2003) Cellular and molecular connections between sleep and synaptic plasticity. *Progress in Neurobiology* **69**, 71–101.

Benke, T. A., Luthi, A., Isaac, J. T. and Collingridge, G. L. (1998) Modulation of AMPA receptor unitary conductance by synaptic activity. *Nature* **393**, 793–797.

Bennett, E. L., Diamond, M. C., Krech, D. and Rosenzweig, M. R. (1964) Chemical and anatomical plasticity brain. *Science* **146**, 610–619.

Bennett, J. A. and Dingledine, R. (1995) Topology profile for a glutamate receptor: three transmembrane domains and a channel-lining reentrant membrane loop. *Neuron* **14**, 373–384.

Bennett, E. L., Diamond, M. C., Krech, D. and Rosenzweig, M. R. (1964) Chemical and anatomical plasticity brain. *Science* **146**, 610–619.

Benoit, B. O., Savarese, T., Joly, M., Engstrom, C. M., Pang, L., Reilly, J., Recht, L. D., Ross, A. H. and Quesenberry, P. J. (2001) Neurotrophin channeling of neural progenitor cell differentiation. *Journal of Neurobiology* **46**, 265–280.

Benraiss, A., Chmielnicki, E., Lerner, K., Roh, D. and Goldman, S. A. (2001) Adenoviral brain-derived neurotrophic factor induces both neostriatal and olfactory neuronal recruitment from endogenous progenitor cells in the adult forebrain. *Journal of Neuroscience* **21**, 6718–6731.

Berchtold, N. C., Kesslak, J. P., Pike, C. J., Adlard, P. A. and Cotman, C. W. (2001) Estrogen and exercise interact to regulate brain-derived neurotrophic factor mRNA and protein expression in the hippocampus. *European Journal of Neuroscience* **14**, 1992–2002.

Berger, U. V. and Hediger, M. A. (2000) The vitamin C transporter SVCT2 is expressed by astrocytes in culture but not *in situ*. *Neuroreport* **11**, 1395–1399.

Berger, W., Deckert, J., Hartmann J., Krotzer, C., Kornhuber, J., Ransmayr, G., Heinsen, H., Beckmann, H. and Riederer, P. (1994) Memantine inhibits [3H]MK-801 binding to human hippocampal NMDA receptors. *Neuroreport* **5**, 1237–1240.

Bergles, D. E., Roberts, J. D., Somogyi, P. and Jahr, C. E. (2000) Glutamatergic synapses on oligodendrocyte precursor cells in the hippocampus. *Nature* **405**, 187–191.

Berman, D. E., Hazvi, S., Rosenblum, K. and Seger, R. Y. D. (1998) Specific and differential activation of mitogen-activated protein kinase cascades by unfamiliar taste in the insular cortex of the behaving rat. *Journal of Neuroscience* **18**, 10,037–10,044.

Berman, R. M., Cappiello, A., Anand, A., Oren, D. A., Heninger, G. R., Charney, D.S and Krystal, J. H. (2000) Antidepressant effects of ketamine in depressed patients. *Biological Psychiatry* **47**, 351–354.

Bernabeu, R., de Stein, M. L., Fin, C., Izquierdo, I. and Medina, J. H. (1995) Role of hippocampal NO in the acquisition and consolidation of inhibitory avoidance learning. *Neuroreport* **6**, 1498–1500.

Bernabeu, R., Schmitz, P., Faillace, M. P., Izquierdo, I. and Medina, J. H. (1996) Hippocampal cGMP and cAMP are differentially involved in memory processing of inhibitory avoidance learning. *Neuroreport* **7**, 585–588.

Bernabeu, R., Schroder, N., Quevedo, J., Cammarota, M., Izquierdo, I. and Medina, J. H. (1997a) Further evidence for the involvement of a hippocampal cGMP/cGMP-dependent protein kinase cascade in memory consolidation. *Neuroreport* **8**, 2221–2224.

Bernabeu, R., Bevilaqua, L., Ardenghi, P., Bromberg, E., Schmitz, P., Bianchin, M., Izquierdo, I. and Medina, J. H. (1997b) Involvement of hippocampal cAMP/cAMP-dependent protein kinase signaling pathways in a late memory consolidation phase of aversively motivated learning in rats. *Proceedings of the National Academy of Sciences of the United States of America* **94**, 7041–7046.

Bernard, V. and Bolam, J. P. (1998) Subcellular and subsynaptic distribution of the NR1 subunit of the NMDA receptor in the neostriatum and globus pallidus of the rat: colocalization at synapses with the GluR2/3 subunit of the AMPA receptor. *European Journal of Neuroscience* **10**, 3721–3738.

Bernard-Trifilo, J. A., Kramar, E. A., Torp, R., Lin, C. Y., Pineda, E. A. and Lynch G. (2005) Integrin signaling cascades are operational in adult hippocampal synapses and modulate NMDA receptor physiology. *Journal of Neurochemistry* **93**, 834–849.

Berninger, B., Marty, S., Zafra, F., da Penha Berzaghi, M., Thoenen, H. and Lindholm, D. (1995) GABAergic stimulation switches from enhancing to repressing BDNF expression in rat hippocampal neurons during maturation in vitro. *Development* **121**, 2327–2335.

Berridge, M. J. (1998). Neuronal calcium signaling. *Neuron* **21**, 13–26.

Berridge, M. J., Downes, C. P. and Hanley, M. R. (1982) Lithium amplifies agonist-dependent phosphatidylinositol responses in brain and salivary glands. *Biochemical Journal* **206**, 587–595.

Berry, B., McMahan, R. and Gallagher, M. (1997) Spatial learning and memory at defined points of the estrous cycle: effects on performance of a hippocampal dependent task. *Behavioral Neuroscience* **111**, 267–274.

Berthold, C.-H., Nilsson, I. and Rydmark, M. (1983) Axon diameter and myelin sheath thickness in nerve fibres of the ventral spinal root of the seventh lumbar nerve of the adult and developing cat. *Journal of Anatomy* **136**, 483–508.

Best, P. J., White, A. M. and Minai, A. (2001) Spatial processing in the brain: the activity of hippocampal place cells. *Annual Review of Neuroscience* **24**, 459–486.

Bethea, J. R., Nagashima, H., Acosta, M. C., Briceno, C., Gomez, F., Marcillo, A. E., Loor, K., Green, J. and Dietrich, W. D. (1999) Systemically administered interleukin-10 reduces tumor necrosis factor-alpha production and significantly improves functional recovery following traumatic spinal cord injury in rats. *Journal of Neurotrauma* **16**, 851–863.

Bezzi, P., Domercq, M., Vesce, S. and Volterra, A. (2001) Neuron-astrocyte cross-talk during synaptic transmission: physiological and neuropathological implications. *Progress in Brain Research* **132**, 255–265.

Bezzi, P., Vesce, S., Panzarasa, P. and Volterra, A. (1999) Astrocytes as active participants of glutamatergic function and regulators of its homeostasis. *Advances in Experimental Medicine and Biology* **468**, 69–80.

Bezzi, P., Gundersen, V., Galbete, J. L., Seifert, G., Steinhauser, C., Pilati, E. and Volterra, A. (2004) Astrocytes contain a vesicular compartment that is competent for regulated exocytosis of glutamate. *Nature Neuroscience* **7**, 613–620.

Bezzi, P., Carmignoto, G., Pasti, L., Vesce, S., Rossi, D., Lodi Rizzini, B., Pozzan, T. and Volterra, A. (1998) Prostaglandins stimulate calcium-dependent glutamate release in astrocytes. *Nature* **391**, 281–285.

Bezzi, P., Domercq, M., Brambilla, L., Galli, R., Schols, D., De Clercq, E., Vescovi, A., Bagetta, G., Kollias, G., Meldolesi, J. and Volterra, A. (2001) CXCR4-activated astrocyte glutamate release via TNFalpha: amplification by microglia triggers neurotoxicity. *Nature Neuroscience* **4**, 702–710.

Bhalla, U. S. and Iyengar, R. (1999) Emergent properties of networks of biological signaling pathways. *Science* **283**, 381–387.

Bi, G. and Poo, M. (2001) Synaptic modification by correlated activity: Hebb's postulate revisited. *Annual Review of Neuroscience* **24**, 139–166.

Bi, R., Broutman, G., Foy, M., Thompson, R. F. and Baudry, M. (2000) The tyrosine kinase and MAP kinase pathways mediate multiple effects of estrogen in hippocampus. *Proceedings of the National Academy of Sciences of the United States of America* **97**, 3602–3607.

Bi, R., Foy, M. R., Vouimba, R. M., Thompson, R. F. and Baudry, M. (2001) Cyclic changes in estrogen regulate synaptic plasticity through the MAP kinase pathway. *Proceedings of the National Academy of Sciences of the United States of America* **98**, 13,391–13,395.

Biedenkapp, J. C. and Rudy, J. W. (2004) Context memories and reactivation: constraints on the reconsolidation hypothesis. *Behavioral Neuroscience* **118**, 956–964.

Bienert G. P., Schjoerring J. K. and Jahn T. P. (2006) Membrane transport of hydrogen peroxide. *Biochimica Biophysica Acta* **1758**, 994–1003.

Bienert, G. P., Møller A. L., Kristiansen K. A., Schulz A., Møller I. M., Schjoerring J. K. and Jahn T. P. (2007) Specific aquaporins facilitate the diffusion of hydrogen peroxide across membranes. *Journal of Biological Chemistry* **282**, 1183–1192.

Bimonte, H. A. and Denenberg, V. H. (1999) Estradiol facilitates performance as working memory load increases. *Psychoneuroendocrinology* **24**, 161–173.

Bimonte-Nelson, H. A., Singleton, R. S., Nelson, M. E., Eckman, C. B., Barber, J., Scott, T. Y. and Granholm, A. C. (2003) Testosterone, but not nonaromatizable dihydrotestosterone, improves working memory and alters nerve growth factor levels in aged male rats. *Experimental Neurology* **181**, 301–312.

Bindokas, V. P., Jordan, J., Lee, C. C. and Miller, R. J. (1996) Superoxide production in rat hippocampal neurons: selective imaging with hydroethidine. *Journal of Neuroscience* **16**, 1324–1336.

Birnbaum, S. G., Yuan, P. X., Wang, M., Vijayraghaven, S., Bloom, A. K., Davis, D. J., Gobeske, K. T., Sweatt, J. D., Manji, H. K. and Arnsten, A. F. (2004) Protein kinase C overactivity impairs prefrontal cortical regulation of working memory. *Science* **306**, 882–884.

Bissiere, S., Humeau, Y. and Luthi, A. (2003) Dopamine gates LTP induction in lateral amygdala by suppressing feedforward inhibition. *Nature Neuroscience* **6**, 587–592.

Bito, H., Deisseroth, K. and Tsien, R. W. (1996) CREB phosphorylation and dephosphorylation: a Ca^{2+}- and stimulus duration-dependent switch for hippocampal gene expression. *Cell* **87**, 1203–1214.

Bittinger, M. A., McWhinnie, E., Meltzer, J., Iourgenko, V., Latario, B., Liu, X., Chen, C. H., Song, C., Garza, D. and Labow, M. (2004) Activation of cAMP response element-mediated gene expression by regulated nuclear transport of TORC proteins. *Current Biology* **14**, 2156–2161.

Bjornstrom, L. and Sjoberg, M. (2005) Mechanisms of estrogen receptor signaling: convergence of genomic and non-genomic actions on target genes. *Molecular Endocrinology* **19**, 833–842.

Black, J. E., Polinsky, M. and Greenough, W. T. (1989) Progressive failure of cerebral angiogenesis supporting neural plasticity in aging rats. *Neurobiology of Aging* **10**, 353–358.

Black, J. E., Sirevaag, A. M. and Greenough, W. T. (1987) Complex experience promotes capillary formation in young rat visual cortex. *Neuroscience Letters* **83**, 351–355.

Black, J. E., Zelazny, A. M. and Greenough, W. T. (1991) Capillary and mitochondrial support of neural plasticity in adult rat visual cortex. *Experimental Neurology* **111**, 204–209.

Black, J. E., Isaacs, K. R., Anderson, B. J., Alcantara, A. A. and Greenough, W. T. (1990) Learning causes synaptogenesis, whereas motor activity causes angiogenesis, in cerebellar cortex of adult rats. *Proceedings of the National Academy of Sciences of the United States of America* **87**, 5568–5572.

Blair, H. T., Schafe, G. E., Bauer, E. P., Rodrigues, S. M. and LeDoux, J. E. (2001) Synaptic plasticity in the lateral amygdala: a cellular hypothesis of fear conditioning. *Learning & Memory* **8**, 229–242.

Blanchard, R. J. and Blanchard, D. C. (1969) Crouching as an index of fear. *Journal of Comparative Physiological Psychology* **67**, 370–375.

Bliss, T. V. and Lomo, T. (1973) Long-lasting potentiation of synaptic transmission in the dentate area of the anaesthetized rabbit following stimulation of the perforant path. *Journal of Physiology (Lond)* **232**, 331–356.

Bliss, T. V. P. and Collingridge, G. L. (1993) A synaptic model of memory: long-term potentiation in the hippocampus. *Nature* **361**, 31–39.

Bliss, T. V. P., Collingridge, G. L. and Morris, R. G. M. (eds) (2004) *Long Term Potentiation: Enhancing Neuroscience for 30 Years*. Oxford University Press.

Blitzer, R. D., Iyengar, R. and Landau, E. M. (2005) Postsynaptic signaling networks: cellular cogwheels underlying long-term plasticity. *Biological Psychiatry* **57**, 113–119.

Blondel, O., Collin, C., McCarran, W. J., Zhu, S., Zamostiano, R., Gozes, I., Brenneman, D. E. and McKay, R. D. (2000) A glia-derived signal regulating neuronal differentiation. *Journal of Neuroscience* **20**, 8012–8020.

Blozovski, D. and Buser, P. (1988) Passive avoidance memory consolidation and reinstatement in the young rat. *Neuroscience Letters* **89**, 114–119.

Blum, S., Moore, A. N., Adams, F. and Dash, P. K. (1999) A mitogen-activated protein kinase cascade in the CA1/CA2 subfield of the dorsal hippocampus is essential for long-term spatial memory. *Journal of Neuroscience* **19**, 3535–3544.

Boccia, M. M., Acosta, G. B., Blake, M. G. and Baratti, C. M. (2004) Memory consolidation and reconsolidation of an inhibitory avoidance response in mice: effects of i.c.v. injections of hemicholinium-3. *Neuroscience* **124**, 735–741.

Boccia, M. M., Blake, M. G., Acosta, G. B. and Baratti, C. M. (2005) Memory consolidation and reconsolidation of an inhibitory avoidance task in mice: effects of a new different learning task. *Neuroscience* **135**, 19–29.

Boehning, D., Sedaghat, L., Sedlak, T. W. and Snyder, S. H. (2004) Heme oxygenase-2 is activated by calcium-calmodulin. *Journal of Biological Chemistry* **279**, 30,927–30,930.

Bohme, G. A., Bon, C., Lemaire, M., Reibaud, M., Piot, O., Stutzmann, J. M., Doble, A. and Blanchard, J. C. (1993) Altered synaptic plasticity and memory formation in nitric oxide synthase inhibitor-treated rats. *Proceedings of the National Academy of Science of the United States of America* **90**, 9191–9194.

Bolton, M. M., Lo, D. C. and Sherwood, N. T. (2000) Long-term regulation of excitatory and inhibitory synaptic transmission in hippocampal cultures by brain-derived neurotrophic factor. *Progress in Brain Research* **128**, 203–218.

Bordi, F. and LeDoux, J. (1992) Sensory tuning beyond the sensory system: an initial analysis of auditory response properties of neurons in the lateral amygdaloid nucleus and overlying areas of the striatum. *Journal of Neuroscience* **12**, 2493–2503.

Bordi, F. and LeDoux, J. E. (1994a) Response properties of single units in areas of rat auditory thalamus that project to the amygdala. I. Acoustic discharge patterns and frequency receptive fields. *Experimental Brain Research* **98**, 261–274.

Bordi, F. and LeDoux, J. E. (1994b) Response properties of single units in areas of rat auditory thalamus that project to the amygdala. II. Cells receiving convergent auditory and somatosensory inputs and cells antidromically activated by amygdala stimulation. *Experimental Brain Research* **98**, 275–286.

Bordi, F., LeDoux, J., Clugnet, M. C. and Pavlides, C. (1993) Single-unit activity in the lateral nucleus of the amygdala and overlying areas of the striatum in freely behaving rats: rates, discharge patterns, and responses to acoustic stimuli. *Behavioral Neuroscience* **107**, 757–769.

Bormann, J. (1989) Memantine is a potent blocker of N-methyl-D-aspartate (NMDA) receptor channels. *European Journal of Pharmacology* **166**, 591–592.

Boulanger, L. M. and Shatz, C. J. (2004) Immune signaling in neural development, synaptic plasticity and disease. *Nature Reviews Neuroscience* **5**, 521–531.

Boulanger, L. M., Huh, G. S. and Shatz, C. J. (2001) Neuronal plasticity and cellular immunity: shared molecular mechanisms. *Current Opinion in Neurobiology* 11, 568–578.

Bouret, S. and Sara, S. J. (2004) Reward expectation, orientation of attention and locus coeruleus–medial frontal cortex interplay during learning. *European Journal of Neuroscience* 20, 791–802.

Bourne, H. R. and Nicoll, R. (1993) Molecular machines integrate coincident synaptic signals. *Cell* 72 (Suppl.), 65–75.

Bourtchouladze, R., Abel, T., Berman, N., Gordon, R., Lapidus, K. and Kandel, E. R. (1998) Different training procedures recruit either one or two critical periods for contextual memory consolidation, each of which requires protein synthesis and PKA. *Learning & Memory* 5, 365–374.

Bourtchuladze, R., Frenguelli, B., Blendy, J., Cioffi, D., Schutz, G. and Silva, A. J. (1994) Deficient long-term memory in mice with a targeted mutation of the cAMP-responsive element-binding protein. *Cell* 79, 59–68.

Boveris, A. and Chance, B. (1973) The mitochondrial generation of hydrogen peroxide: general properties and the effect of hyperbaric oxygen. *Biochemical Journal* 134, 707–716.

Bowser, D. N. and Khakh, B. S. (2004) ATP excites interneurons and astrocytes to increase synaptic inhibition in neuronal networks. *Journal of Neuroscience* 24, 8606–8620.

Bozon, B., Davis, S. and Laroche, S. (2003) A requirement for the immediate early gene zif268 in reconsolidation of recognition memory after retrieval. *Neuron* 40, 695–701.

Braas, K. M., Zarbin, M. A. and Snyder, S. H. (1987) Endogenous adenosine and adenosine receptors localized to ganglion cells of the retina. *Proceedings of the National Academy of Sciences of the United States of America* 84, 3906–3910.

Brady, S. T., Witt, A. S., Kirkpatrick, L. L., de Waegh, S. M., Readhead, C., Tu, P. H. and Lee, V. M. (1999) Formation of compact myelin is required for maturation of the axonal cytoskeleton. *Journal of Neuroscience* 19, 7278–7288.

Braga, M. F., Aroniadou-Anderjaska, V., Manion, S. T., Hough, C. J. and Li, H. (2004) Stress impairs alpha(1A) adrenoceptor-mediated noradrenergic facilitation of GABAergic transmission in the basolateral amygdala. *Neuropsychopharmacology* 29, 45–58.

Brahma, B., Forman, R. E., Stewart, E. E., Nicholson, C. and Rice, M. E. (2000) Ascorbate inhibits brain slice edema. *Journal of Neurochemistry* 74, 1263–1270.

Braithwaite, S. P., Meyer, G. and Henley, J. M. (2000) Interactions between AMPA receptors and intracellular proteins. *Neuropharmacology* 39, 919–930.

Brake, W. G., Alves, S. E., Dunlop, J. C., Lee, S. J., Bulloch, K., Allen, P. B., Greengard, P. and McEwen, B. S. (2001) Novel target sites for estrogen action in the dorsal hippocampus: an examination of synaptic proteins. *Endocrinology* 142, 1284–1289.

Brambilla, R., Gnesutta, N., Minichiello, L., White, G., Roylance, A. J., Herron, C. E., Ramsey, M., Wolfer, D. P., Cestari, V., Rossi-Arnaud, C., Grant, S. G., Chapman, P. F., Lipp, H. P., Sturani, E. and Klein, R. (1997) A role for the Ras signaling pathway in synaptic transmission and long-term memory. *Nature* 290, 281–286.

Bramham, C. R. and Messaoudi, E. (2005) BDNF function in adult synaptic plasticity: the synaptic consolidation hypothesis. *Progress in Neurobiology* 76, 99–125.

Branchi, I., Francia, N. and Alleva, E. (2004) Epigenetic control of neurobehavioural plasticity: the role of neurotrophins. *Behavioural Pharmacology* 15, 353–362.

Brashers-Krug, T., Shadmehr, R. and Bizzi, E. (1996) Consolidation in human motor memory. *Nature* 382, 252–255.

Bredt, D. S. and Nicoll, R. A. (2003) AMPA receptor trafficking at excitatory synapses. *Neuron* 40, 361–379.

Bredt, D. S. and Snyder, S. H. (1989) Nitric oxide mediates glutamate-linked enhancement of cGMP levels in the cerebellum. *Proceeding of the National Academy of Sciences of the United States of America* 86, 9030–9033.

Bredt, D. S. and Snyder, S. H. (1990) Isolation of nitric oxide synthetase, a calmodulin-requiring enzyme. *Proceeding of the National Academy of Sciences of the United States of America* 87, 682–685.

Bredt, D. S. and Snyder, S. H. (1992) Nitric oxide, a novel neuronal messenger. *Neuron* 8, 3–11.

Bredt, D. S., Huang, P. M. and Snyder, S. H. (1990) Localization of nitric oxide synthase indicating a neural role for nitric oxide. *Nature* 347, 768–770.

Bredt, D. S., Glatt, C. E., Huang, P. M., Fotuhi, M., Dawson, T. M. and Snyder, S. H. (1991) Nitric oxide synthase protein and mRNA are discretely localized in neuronal populations of the mammalian CMS together with NADPH diaphorase. *Neuron* 7, 615–624.

Bregman, B. S. et al. (1995) Recovery from spinal cord injury mediated by antibodies to neurite growth inhibitors. *Nature* **378**, 498–501.

Brinley, F. J. (1980) Excitation and conduction in nerve fibers. In Mountcastle, V. B. (ed.) *Medical Physiology*. CV Mosby Co., pp. 46–81.

Brinley-Reed, M. and McDonald, A. J. (1999) Evidence that dopaminergic axons provide a dense innervation of specific neuronal subpopulations in the rat basolateral amygdala. *Brain Research* **850**, 127–135.

Brinton, R. D. (1994) The neurosteroid 3 alpha-hydroxy-5 alpha-pregnan-20-one induces cytoarchitectural regression in cultured fetal hippocampal neurons. *Journal of Neuroscience* **14**, 2763–2774.

Brinton, R. D., Proffitt, P., Tran, J. and Luu, R. (1997a) Equilin, a principal component of the estrogen replacement therapy, premarin, increases the growth of cortical neurons via an NMDA receptor dependent mechanism. *Experimental Neurology* **147**, 211–220.

Brinton, R. D., Tran, J., Proffitt, P. and Kihil, M. (1997b) 17B-estradiol increases the growth and survival of cultured cortical neurons. *Neurochemical Research* **22**, 1339–1351.

Briones, T. L., Klintsova, A. Y. and Greenough, W. T. (2004) Stability of synaptic plasticity in the adult rat visual cortex induced by complex environment exposure. *Brain Research* **1018**, 130–135.

Briones, T., Shah, P., Juraska, J., and Greenough, W. T. (1999) Effects of prolonged exposure to and subsequent removal from a complex environment on corpus callosum myelination in the adult rat. *Society for Neuroscience Abstracts* p. 638.

Brown, W. M. and Aiken, S. P. (1998) Felbamate: clinical and molecular aspects of a unique antiepileptic drug. *Critical Reviews in Neurobiology* **12**, 205–222.

Brown, J., Cooper-Kuhn, C. M., Kempermann, G., Van Praag, H., Winkler, J., Gage, F. H. and Kuhn, H. G. (2003) Enriched environment and physical activity stimulate hippocampal but not olfactory bulb neurogenesis. *European Journal of Neuroscience* **17**, 2042–2046.

Brown, V., Jin, P., Ceman, S., Darnell, J. C., O'Donnell, W. T., Tenenbaum, S. A., Jin, X., Feng, Y., Wilkinson, K. D., Keene, J. D., Darnell, R. B. and Warren, S. T. (2001) Microarray identification of FMRP-associated brain mRNAs and altered mRNA translational profiles in fragile X syndrome. *Cell* **107**, 477–487.

Bruce, A. J., Boling, W., Kindy, M. S., Peschon, J., Kraemer, P. J., Carpenter, M. K., Holtsberg, F. W. and Mattson, M. P. (1996) Altered neuronal and microglial responses to excitotoxic and ischemic brain injury in mice lacking TNF receptors. *Nature Medicine* **2**, 788–794.

Bruel-Jungerman, E., Laroche, S. and Rampon, C. (2005) New neurons in the dentate gyrus are involved in the expression of enhanced long-term memory following environmental enrichment. *European Journal of Neuroscience* **21**, 513–521.

Brunig, I., Penschuck, S., Berninger, B., Benson, J. and Fritschy, J. M. (2001) BDNF reduces miniature inhibitory postsynaptic currents by rapid downregulation of GABA(A) receptor surface expression. *European Journal of Neuroscience* **13**, 1320–1328.

Bryceson, Y. T., Foster, J. A., Kuppusamy, S. P., Herkenham, M. and Long, E. O. (2005) Expression of a killer cell receptor-like gene in plastic regions of the central nervous system. *Journal of Neuroimmunology* **161**, 177–182.

Bucci, D. J., Chiba, A. A. and Gallagher, M. (1995) Spatial learning in male and female Long–Evan rats. *Behavioral Neuroscience* **109**, 180–183.

Buckmaster, C. A., Eichenbaum, H., Amaral, D. G., Suzuki, W. A. and Rapp, P. (2004) Entorhinal cortex lesions disrupt the relational organization of memory in monkeys. *Journal of Neuroscience* **24**, 9811–9825.

Bull, R., Ferrera, E. and Orrego, F. (1976) Effects of anisomycin on brain protein synthesis and passive avoidance learning in newborn chicks. *Journal of Neurobiology* **7**, 37–49.

Bullock, T. H., Bennett, M. V., Johnston, D., Josephson, R., Marder, E. and Fields R. D. (2005) The neuron doctrine, redux. *Science* **310**, 791–793.

Bunsey, M. and Eichenbaum, H. (1996) Conservation of hippocampal memory function in rats and humans. *Nature* **379**, 255–257.

Buonanno, A. and Fischbach, G. D. (2001) Neuregulin and ErbB receptor signaling pathways in the nervous system. *Current Opinion in Neurobiology* **11**, 287–296.

Burchuladze, R., and Rose, S. P. R. (1992) Memory formation in day old chicks requires NMDA but not non-NMDA glutamate receptors. *European Journal of Neuroscience* **4**, 533–538.

Burchuladze, R., Potter, J. and Rose, S. P. R. (1990) Memory formation in the chick depends on membrane-bound protein kinase C. *Brain Research* **535**, 131–138.

Burne, T. H. J. and Rose, S. P. R. (1997) Effects of training procedure on memory formation using a weak passive avoidance learning paradigm. *Neurobiology of Learning and Memory* **68**, 133–139.

Burns, L. H., Everitt, B. J., Kelley, A. E. and Robbins, T. W. (1994) Glutamate–dopamine interactions in the ventral striatum: role in locomotor activity and responding with conditioned reinforcement. *Psychopharmacology (Berl)* **115**, 516–528.

Butt, A. M. and Tutton, M. (1992) Response of oligodendrocytes to glutamate and gamma-aminobutyric acid in the intact mouse optic nerve. *Neuroscience Letters* **146**, 108–110.

Buxbaum, J. D., Choi, E. K., Luo, Y., Lilliehook, C., Crowley, A. C., Merriam, D. E. and Wasco, W. (1998) Calsenilin: a calcium-binding protein that interacts with the presenilins and regulates the levels of a presenilin fragment. *Nature Medicine* **4**, 1177–1181.

Cabelli, R. J., Shelton, D. L., Segal, R. A. and Shatz, C. J. (1997) Blockade of endogenous ligands of trkB inhibits formation of ocular dominance columns. *Neuron* **19**, 63–76.

Cabelli, R. J., Allendoerfer, K. L., Radeke, M. J., Welcher, A. A., Feinstein, S. C. and Shatz, C. J. (1996) Changing patterns of expression and subcellular localization of TrkB in the developing visual system. *Journal of Neuroscience* **16**, 7965–7980.

Cahill, L., Pham, K. and Setlow, B. (2000) Impaired memory consolidation in rats produced with beta-adrenergic blockade. *Neurobiology of Learning and Memory* **74**, 259–266.

Cain, D. P. (1997) LTP, NMDA, genes and learning. *Current Opinion in Neurobiology* **7**, 235–242.

Caithness, G., Osu, R., Bays, P., Chase, H., Klassen, J., Kawato, M., Wolpert, D. M. and Flanagan, J. R. (2004) Failure to consolidate the consolidation theory of learning for sensorimotor adaptation tasks. *Journal of Neuroscience* **24**, 8662–8671.

Cajal, S.Ry. (1909) *Histologie du Systeme Nerveux de l'Homme et Des Vertebres*. Paris: A. Maloine.

Calabrese, J. R., Bowden, C. L., Sachs, G. S., Ascher, J. A., Monaghan, E. and Rudd, G. D. (1999) A double-blind placebo-controlled study of lamotrigine monotherapy in outpatients with bipolar I depression. Lamictal 602 Study Group. *Journal of Clinical Psychiatry* **60**, 79–88.

Calabresi, P., Siniscalchi, A., Pisani, A., Stefani, A., Mercuri, N. B. and Bernardi, G. (1996) A field potential analysis on the effects of lamotrigine, GP 47779, and felbamate in neocortical slices. *Neurology* **47**, 557–662.

Calegari, F., Coco, S., Taverna, E., Bassetti, M., Verderio, C., Corradi, N., Matteoli, M. and Rosa, P. (1999) A regulated secretory pathway in cultured hippocampal astrocytes. *Journal of Biological Chemistry* **274**, 22,539–22,547.

Camel, J. E., Withers, G. S. and Greenough, W. T. (1986) Persistence of visual cortex dendritic alterations induced by postweaning exposure to a "superenriched" environment in rats. *Behavioral Neuroscience* **100**, 810–813.

Cameron, H. A. and Gould, E. (1994) Adult neurogenesis is regulated by adrenal steroids in the dentate gyrus. *Neuroscience* **61**, 203–209.

Cameron, H. A. and McKay, R. D. (1998) Stem cells and neurogenesis in the adult brain. *Current Opinion in Neurobiology* **8**, 677–680.

Cameron, H. A. and McKay, R. D. (1999) Restoring production of hippocampal neurons in old age. *Nature Neuroscience* **2**, 894–897.

Cameron, H. A., Hazel, T. G. and McKay, R. D. (1998) Regulation of neurogenesis by growth factors and neurotransmitters. *Journal of Neurobiology* **36**, 287–306.

Cameron, H. A., Tanapat, P. and Gould, E. (1998) Adrenal steroids and N-methyl-D-aspartate receptor activation regulate neurogenesis in the dentate gyrus of adult rats through a common pathway. *Neuroscience* **82**, 349–354.

Cammarota, M., Bevilaqua, L. R., Medina, J. H. and Izquierdo, I. (2004) Retrieval does not induce reconsolidation of inhibitory avoidance memory. *Learning & Memory* **11**, 572–578.

Campbell, L. W., Hao, S. Y., Thibault, O., Blalock, E. M. and Landfield, P. W. (1996) Aging changes in voltage-gated calcium currents in hippocampal CA1 neurons. *Journal of Neuroscience* **16**, 6286–6295.

Campeau, S. and Davis, M. (1995) Involvement of the central nucleus and basolateral complex of the amygdala in fear conditioning measured with fear-potentiated startle in rats trained concurrently with auditory and visual conditioned stimuli. *Journal of Neuroscience* **15**, 2301–2311.

Campeau, S., Miserendino, M. J. and Davis, M. (1992) Intra-amygdala infusion of the N-methyl-D-aspartate receptor antagonist AP5 blocks acquisition but not expression of fear-potentiated startle to an auditory conditioned stimulus. *Behavioral Neuroscience* **106**, 569–574.

Cantero, J. L., Atienza, M., Stickgold, R., Kahana, M. J., Madsen, J. R. and Kocsis, B. (2003) Sleep-dependent theta oscillations in the human hippocampus and neocortex. *Journal of Neuroscience* **23**, 10,897–10,903.

Cao W., Carney, J. M., Duchon, A., Floyd, R. A. and Chevion, M. (1988) Oxygen free radical involvement in ischemia and reperfusion injury to brain. *Neuroscience Letters* **88**, 233–238.

Carew, T. J. and Kandel, E. R. (1973) Acquistion and retention of long-term habituation in aplysia: correlation of behavioral and cellular processes. *Science* **182**, 1158–1160.

Carew, T. J. and Sutton, M. A. (2001) Molecular stepping stones in memory consolidation. *Nature Neuroscience* **4**, 769–771.

Carlin, R. K. and Siekevitz, P. (1983) Plasticity in the central nervous system: do synapses divide? *Proceedings of the National Academy of Sciences of the United States of America* **80**, 3517–3521.

Carlson, N. G., Bacchi, A., Rogers, S. W. and Gahring, L. C. (1998) Nicotine blocks TNF-alpha-mediated neuroprotection to NMDA by an alpha-bungarotoxin-sensitive pathway. *Journal of Neurobiology* **35**, 29–36.

Carmignoto, G., Pizzorusso, T., Tia, S. and Vicini, S. (1997) Brain-derived neurotrophic factor and nerve growth factor potentiate excitatory synaptic transmission in the rat visual cortex. *Journal of Physiology* **498**, 153–164.

Carmona, M. A., Martinez, A., Soler, A., Blasi, J., Soriano, E. and Aguado, F. (2003) Ca^{2+}-evoked synaptic transmission and neurotransmitter receptor levels are impaired in the forebrain of trkb (-/-) mice. *Molecular and Cellular Neurosciences* **22**, 210–226.

Carr, C. E. and Konishi, M. (1990) A circuit for detection of interaural time differences in the brain stem of the barn owl. *Journal of Neuroscience* **10**, 3227–3246.

Carriedo, S. G., Sensi, S. L., Yin, H. Z. and Weiss, J. H. (2000) AMPA exposures induce mitochondrial Ca^{2+} overload and ROS generation in spinal motor neurons in vitro. *Journal of Neuroscience* **20**, 240–250.

Carrion, A. M., Link, W. A., Ledo, F., Mellstrom, B. and Naranjo, J. R. (1999) DREAM is a Ca^{2+}-regulated transcriptional repressor. *Nature* **398**, 80–84.

Carroll, R. C. and Zukin, R. S. (2002) NMDA-receptor trafficking and targeting: implications for synaptic transmission and plasticity. *Trends in Neuroscience* **25**, 571–577.

Carroll, R. C., Beattie, E. C., von Zastrow, M. and Malenka, R. C. (2001) Role of AMPA receptor endocytosis in synaptic plasticity. *Nature Reviews Neuroscience* **2**, 315–324.

Carroll, R. C., Lissin, D. V., von Zastrow, M., Nicoll, R. A. and Malenka, R. C. (1999) Rapid redistribution of glutamate receptors contributes to long-term depression in hippocampal cultures. *Nature and Neuroscience* **2**, 454–460.

Carter, A. G. and Regehr, W. G. (2002) Quantal events shape cerebellar interneuron firing. *Nature Neuroscience* **5**, 1309–1318.

Carter, A. G. and Sabatini, B. L. (2004) State-dependent calcium signaling in dendritic spines of striatal medium spiny neurons. *Neuron* **44**, 483–493.

Carter, A. R., Chen, C., Schwartz, P. M. and Segal, R. A. (2002) Brain-derived neurotrophic factor modulates cerebellar plasticity and synaptic ultrastructure. *Journal of Neuroscience* **22**, 1316–1327.

Cartford, M. C., Gould, T. and Bickford, P. C. (2004) A central role for norepinephrine in the modulation of cerebellar learning tasks. *Behav Cogn Neurosci Rev* **3**, 131–138.

Caspi, A., Roberts, B. W. and Shiner, R. L. (2005) Personality development: stability and change. *Annual Review of Psychology* **56**, 453–484.

Caspi, A., Sugden, K., Moffitt, T. E., Taylor, A., Craig, I. W., Harrington, H., McClay, J., Mill, J., Martin, J., Braithwaite, A. and Poulton, R. (2003) Influence of life stress on depression: moderation by a polymorphism in the 5-HTT gene. *Science* **301**, 386–389.

Castellucci, V., Pinsker, H., Kupfermann, I. and Kandel, E. (1970) Neuronal mechanisms of habituation and dishabituation of the gill-withdrawal reflex in Aplysia. *Science* **167**, 1745–1748.

Castren, E. (2004) Neurotrophic effects of antidepressant drugs. *Current Opinions in Pharmacology* **4**, 58–64.

Castren, E., Zafra, F., Thoenen, H. and Lindholm, D. (1992) Light regulates expression of brain-derived neurotrophic factor mRNA in rat visual cortex. *Proceedings of the National Academy of Science of the United States of America* **89**, 9444–9448.

Cato, A. C., Ponta, H. and Herrlich, P. (1992) Regulation of gene expression by steroid hormones. *Progress in Nucleic Acid Research and Molecular Biology* **43**, 1–36.

Cavus, I. and Teyler, T. J. (1996) Two forms of long-term potentiation in area CA1 activate different signal transduction cascades. *Journal of Neurophysiology* **76**, 3038–3047.

Centonze, D., Gubellini, P., Bernardi, G. and Calabresi, P. (1999) Permissive role of interneurons in corticostriatal synaptic plasticity. *Brain Research Brain Research Reviews* **31**, 1–5.

Chan, P. H. (2004) Mitochondria and neuronal death/survival signaling pathways in cerebral ischemia. *Neurochemical Research* **29**, 1943–1949.

Chan, C. S., Weeber, E. J., Kurup, S., Sweatt, J. D. and Davis, R. L. (2003) Integrin requirement for hippocampal synaptic plasticity and spatial memory. *Journal of Neuroscience* **23**, 7107–7116.

Chang, F. L. and Greenough, W. T. (1982) Lateralized effects of monocular training on dendritic branching in adult split-brain rats. *Brain Research* **232**, 283–292.

Chang, F. L. and Greenough, W. T. (1984) Transient and enduring morphological correlates of synaptic activity and efficacy change in the rat hippocampal slice. *Brain Research* **309**, 35–46.

Chang, A. Y., Huang, C. M., Chan, J. Y. and Chan, S. H. (2001) Involvement of noradrenergic innervation from locus coeruleus to hippocampal formation in negative feedback regulation of penile erection in the rat. *Hippocampus* **11**, 783–792.

Chao, M. V. (2000) Trophic factors: An evolutionary cul-de-sac or door into higher neuronal function? *Journal of Neuroscience Research* **59**, 353–355.

Chao, M. V. (2003) Neurotrophins and their receptors: a convergence point for many signalling pathways. *Nature Reviews Neuroscience* **4**, 299–309.

Chapman, P. F., Atkins, C. M., Allen, M. T., Haley, J. E. and Steinmetz, J. E. (1992) Inhibition of nitric oxide synthesis impairs two different forms of learning. *Neuroreport* **3**, 567–570.

Charifi, C., Debilly, G., Paut-Pagano, L., Cespuglio, R. and Valatx, J. L. (2001) Effect of noradrenergic denervation of medial prefrontal cortex and dentate gyrus on recovery after sleep deprivation in the rat. *Neuroscience Letters* **311**, 113–116.

Chase, T. N. and Oh, J. D. (2000) Striatal dopamine- and glutamate-mediated dysregulation in experimental Parkinsonism. *Trends in Neurosciences* **23**, S86–S91.

Chavis, P. and Westbrook, G. (2001) Integrins mediate functional pre- and postsynaptic maturation at a hippocampal synapse. *Nature* **411**, 317–321.

Chawla, S. and Bading, H. (2001) CREB/CBP and SRE-interacting transcriptional regulators are fast on-off switches: duration of calcium transients specifies the magnitude of transcriptional responses. *Journal of Neurochemistry* **79**, 849–858.

Chen, M. S., Huber, A. B., van der Haar, M. E., Frank, M., Schnell, L., Spillmann, A. A., Christ, F. and Schwab M. E. (2000) Nogo-A is a myelin-associated neurite outgrowth inhibitor and an antigen for monoclonal antibody IN-1. *Nature* **403**, 434–439.

Chen, X. and Johnston, D. (2004) Properties of single voltage-dependent K^+ channels in dendrites of CA1 pyramidal neurones of rat hippocampus. *Journal of Physiology* **559**, 187–203.

Chen, Y. and Swanson, R. A. (2003) The glutamate transporters EAAT2 and EAAT3 mediate cysteine uptake in cortical neuron cultures. *Journal of Neurochemistry* **84**, 1332–1339.

Chen, B. T., Avshalumov, M. V. and Rice, M. E. (2001) H_2O_2 is a novel, endogenous modulator of synaptic dopamine release. *Journal of Neurophysiology* **85**, 2468–2476.

Chen, B. T., Avshalumov, M. V. and Rice, M. E. (2002) Modulation of somatodendritic dopamine release by endogenous H_2O_2: susceptibility in substantia nigra but resistance in the ventral tegmental area. *Journal of Neurophysiology* **87**, 1155–1158.

Chen, J., Chopp, M. and Li, Y. (1999) Neuroprotective effects of progesterone after transient middle cerebral artery occlusion in rats. *Journal of Neurological Research* **171**, 24–30.

Chen, C., Rainnie, D. G., Greene, R. W. and Tonegawa, S. (1994) Abnormal fear response and aggressive behavior in mutant mice deficient for alpha-calcium-calmodulin kinase II. *Science* **265**, 291–294.

Chen, G., Huang, L. D., Jiang, Y. M. and Manji, H. K. (1999) The mood-stabilizing agent valproate inhibits the activity of glycogen synthase kinase-3. *Journal of Neurochemistry* **72**, 1327–1330.

Chen, H. J., Rojas-Soto, M., Oguni, A. and Kennedy, M. B. (1998a) A synaptic Ras-GTPase activating protein (p135 SynGAP) inhibited by CaM kinase II. *Neuron* **20**, 895–904.

Chen, Q., Veenman, L., Knopp, K., Yan, Z., Medina, L., Song, W. J., Surmeier, D. J. and Reiner, A. (1998b) Evidence for the preferential localization of glutamate receptor-1 subunits of AMPA receptors to the dendritic spines of medium spiny neurons in rat striatum. *Neuroscience* **83**, 749–761.

Chen, W. G., Chang, Q., Lin, Y., Meissner, A., West, A. E., Griffith, E. C., Jaenisch, R. and Greenberg, M. E. (2003) Derepression of BDNF transcription involves calcium-dependent phosphorylation of MeCP2. *Science* **31**, 885–889.

Chen, L., Chetkovich, D. M., Petralia, R. S., Sweeney, N. T., Kawasaki, Y., Wenthold, R. J., Bredt, D. S. and Nicoll, R. A. (2000) Stargazin regulates synaptic targeting of AMPA receptors by two distinct mechanisms. *Nature* **408**, 936–943.

Chen, X., Yuan, L. L., Zhao, C., Birnbaum, S. G., Frick, A., Jung, W. E., Schwarz, T. L., Sweatt, J. D. and Johnston, D. (2006) Deletion of Kv4.2 gene eliminates dendritic A-type K^+ current and enhances induction of long-term potentiation in hippocampal CA1 pyramidal neurons. *Journal of Neuroscience* **26**, 12143–12151.

Chenard, B. L., Menniti, F. S. (1999) Antagonists selective for NMDA receptors containing the NR2B subunit. *Curr Pharm Des* **5**, 381–404.

Cheng, B., Christakos, S. and Mattson, M. P. (1994) Tumor necrosis factors protect neurons against metabolic-excitotoxic insults and promote maintenance of calcium homeostasis. *Neuron* **12**, 139–153.

Cheng, H. Y., Pitcher, G. M., Laviolette, S. R., Whishaw, I. Q., Tong, K. I., Kockeritz, L. K., Wada, T., Joza, N. A., Crackower, M., Goncalves, J., Sarosi, I., Woodgett, J. R., Oliveira-dos-Santos, A. J., Ikura, M., van der Kooy, D., Salter, M. W. and Penninger, J. M. (2002) DREAM is a critical transcriptional repressor for pain modulation. *Cell* **108**, 31–43.

Cheramy, A., Romo, R., Godeheu, G., Baruch, P. and Glowinski, J. (1986) *In vivo* presynaptic control of dopamine release in the cat caudate nucleus. II. Facilitatory or inhibitory influence of L-glutamate. *Neuroscience* **19**, 1081–1090.

Cherkin, A. (1969) Kinetics of memory consolidation; role of amnestic treatment parameters. *Proceedings of the National Academy of Sciences of the United States of America* **63**, 1094–1101.

Cherrier, M. M., Craft, S. and Matsumoto, A. H. (2003) Cognitive changes associated with supplementation of testosterone or dihydrotestosterone in mildly hypogonadal men: a preliminary report. *Journal of Andrology* **24**, 568–576.

Child, F. M., Epstein, H. T., Kuzirian, A. M. and Alkon, D. L. (2003) Memory reconsolidation in Hermissenda. *Biological Bulletin* **205**, 218–219.

Chittajallu, R., Braithwaite, S. P., Clarke, V. R. and Henley, J. M. (1999) Kainate receptors: subunits, synaptic localization and function. *Trends in Pharmacological Science* **20**, 26–35.

Choi, D. W. (1992) Excitotoxic cell death. *Journal of Neurobiology* **23**, 1261–1276.

Choi, D. W. (1994) Glutamate receptors and the induction of excitotoxic neuronal death. *Progress in Brain Research* **100**, 47–51.

Choi, J. M., Romeo, R. D., Brake, W. G., Bethea, C. L., Rosenwaks, Z. and McEwen, B. S. (2003) Estradiol increases pre- and post-synaptic proteins in the CA1 region of the hippocampus in female rhesus macaques (*Macaca mulatta*). *Endocrinology* **144**, 4734–4738.

Christensen, L. W., Nance, D. M. and Gorski, R. A. (1977) Effects of hypothalamic and preoptic lesions on reproductive behavior in male rats. *Brain Research Bulletin* **2**, 137–141.

Christianson, S.-A. (1992) Preface. In Christianson, S.-A. (ed.) *Handbook of Emotion and Memory: Research and Theory*. Erlbaum Associates Inc.

Christopherson, K. S., Ullian, E. M., Stokes, C. C. A., Mullowney, C. E., Hell, J. W., Agah, A., Lawler, J., Mosher, D. F., Bornstein, P. and Barres, B. A. (2005) Thrombospondins are astrocyte-secreted proteins that promote CNS synaptogenesis. *Cell* **120**, 421–433.

Chrivia, J. C., Kwok, R. P., Lamb, N., Hagiwara, M., Montminy, M. R. and Goodman, R. H. (1993) Phosphorylated CREB binds specifically to the nuclear protein CBP. *Nature* **365**, 855–859.

Chun, M. M. (2005) Drug-induced amnesia impairs implicit relational memory. *Trends in Cognitive Sciences* **9**, 355–357.

Chun, D., Gall, C. M., Bi, X. and Lynch, G. (2001) Evidence that integrins contribute to multiple stages in the consolidation of long term potentiation in rat hippocampus. *Neuroscience* **105**, 815–829.

Chung, H. J., Steinberg, J. P., Huganir, R. L. and Linden, D. J. (2003) Requirement of AMPA receptor GluR2 phosphorylation for cerebellar long-term depression. *Science* **300**, 1751–1755.

Cirelli, C. and Tononi, G. (1998) Differences in gene expression between sleep and waking as revealed by mRNA differential display. *Brain Research Molecular Brain Research* **56**, 293–305.

Cirelli, C. and Tononi, G. (2000) Differential expression of plasticity-related genes in waking and sleep and their regulation by the noradrenergic system. *Journal of Neuroscience* **20**, 9187–9194.

Cirelli, C. and Tononi, G. (2000a). Gene expression in the brain across the sleep–waking cycle. *Brain Research* **885**, 303–321.

Cirelli, C. and Tononi, G. (2000b) On the functional significance of c-fos induction during the sleep–waking cycle. *Sleep* **23**, 453–469.

Cirelli, C., Gutierrez, C. M. and Tononi, G. (2004) Extensive and divergent effects of sleep and wakefulness on brain gene expression. *Neuron* **41**, 35–43.

Clayton, D. F. (2000) The genomic action potential. *Neurobiology of Learning and Memory* **74**, 185–216.

Clayton, N. S. (2004) Do animals think? *Science* **305**, 344.

Clements, M. P. and Rose, S. P. R. (1996) Time dependent increase in release of arachidonic acid following passive avoidance training in the day-old chick. *Journal of Neurochemistry* **67**, 1317–1323.

Clements, M. P., Rose, S. P. R. and Tiunova, A. (1995). ω-Ω Conotoxin GVIA disrupts memory formation in the day old chick *Neurobiology of Learning and Memory* **64**, 276–284.

Clugnet, M. C., LeDoux, J. E. and Morrison, S. F. (1990) Unit responses evoked in the amygdala and striatum by electrical stimulation of the medial geniculate body. *Journal of Neuroscience* **10**, 1055–1061.

Coetzee, W. A., Nakamura, T. Y. and Faivre, J. F. (1995) Effects of thiol-modifying agents on KATP channels in guinea pig ventricular cells. *American Journal of Physiology* **269**, H1625–H1633.

Cohen, G. (1994) Enzymatic/nonenzymatic sources of oxyradicals and regulation of antioxidant defenses. *Annals of the New York Academy of Sciences* **738**, 8–14.

Cohen, N. J. (1984) Preserved learning capacity in amnesia: evidence for multiple memory systems. In Butters, N. and Squire, L. R. (eds) *The Neuropsychology of Memory*. Guilford Press, pp. 83–103.

Cohen, N. J., Ryan, J., Hunt, C., Romine, L., Wszalek, T. and Nash, C. (1999) Hippocampal system and declarative (relational) memory: summarizing the data from functional neuroimaging studies. *Hippocampus* **9**, 83–98.

Cohen-Cory, S. (2002) The developing synapse: construction and modulation of synaptic structures and circuits. *Science* **298**, 770–776.

Cohen-Cory, S. and Fraser, S. E. (1995) Effects of brain-derived neurotrophic factor on optic axon branching and remodelling in vivo. *Nature* **378**, 192–196.

Cole, J. S., Messing, A., Trojanowski, J. Q. and Lee, V. M.-Y. (1994) Modulation of axon diameter and neurofilaments by hypomyelinating Schwann cells in transgenic mice. *Journal of Neuroscience* **14**, 6956–6966.

Colledge, M., Dean, R. A., Scott, G. K., Langeberg, L. K., Huganir, R. L. and Scott, J. D. (2000) Targeting of PKA to glutamate receptors through a MAGUK–AKAP complex. *Neuron* **27**, 107–119.

Collin, C., Vicario-Abejon, C., Rubio, M. E., Wenthold, R. J., McKay, R. D. and Segal, M. (2001) Neurotrophins act at presynaptic terminals to activate synapses among cultured hippocampal neurons. *European Journal of Neuroscience* **13**, 1273–1282.

Collingridge, G. L., Isaac, J. T. and Wang, Y. T. (2004) Receptor trafficking and synaptic plasticity. *Nature Reviews Neuroscience* **5**, 952–962.

Collins, D. R. and Pare, D. (2000) Differential fear conditioning induces reciprocal changes in the sensory responses of lateral amygdala neurons to the CS(+) and CS(–). *Learning & Memory* **7**, 97–103.

Comery, T. A., Shah, R. and Greenough, W. T. (1995) Differential rearing alters spine density on medium-sized spiny neurons in the rat corpus striatum: evidence for association of morphological plasticity with early response gene expression. *Neurobiology of Learning and Memory* **63**, 217–219.

Comery, T. A., Stamoudis, C. X., Irwin, S. A. and Greenough, W. T. (1996) Increased density of multiple-head dendritic spines on medium-sized spiny neurons of the striatum in rats reared in a complex environment. *Neurobiology of Learning and Memory* **66**, 93–96.

Condes-Lara, M. (1998) Different direct pathways of locus coeruleus to medial prefrontal cortex and centrolateral thalamic nucleus: electrical stimulation effects on the evoked responses to nociceptive peripheral stimulation. *European Journal of Pain* **2**, 15–23.

Condorelli, D. F., Dell'Albani, P., Mudo, G., Timmusk, T. and Belluardo, N. (1994) Expression of neurotrophins and their receptors in primary astroglial cultures: induction by cyclic AMP-elevating agents. *Journal of Neurochemistry* **63**, 509–516.

Conejo, N. M., Gonzalez-Pardo, H., Pedraza, C., Navarro, F. F., Vallego, G. and Arias, J. L. (2003) Evidence for sexual difference in astrocytes of adult hippocampus. *Neuroscience Letters* **339**, 119–122.

Conkright, M. D., Canettieri, G., Screaton, R., Guzman, E., Miraglia, L., Hogenesch, J. B. and Montminy, M. (2003) TORCs: transducers of regulated CREB activity. *Molecular Cell* **12**, 413–423.

Conn, P. and Sweatt, J. (1994) Protein kinase C in the nervous system. In Kuo, J. (ed.) *Protein Kinase C*. Oxford University Press, pp. 199–235.

Conn, P. J. and Pin, J. P. (1997) Pharmacology and functions of metabotropic glutamate receptors. *Annual Review of Pharmacology and Toxicology* **37**, 205–237.

Conner, R. L., Vernikos-Danellis, J. and Levine, S. (1971) Stress, fighting and neuroendocrine function. *Nature* **234**, 564–566.

Conrad, C. D., Lupien, S. J. and McEwen, B. S. (1999) Support for a bimodal role for type II adrenal steroid receptors in spatial memory. *Neurobiology of Learning and Memory* **72**, 39–46.

Conrad, C. D., Grote, K. A., Hobbs, R. J. and Ferayorni, A. (2003) Sex differences in spatial and non-spatial Y-maze performance after chronic stress. *Neurobiology of Learning and Memory* **79**, 32–40.

Conrad, C. D., Jackson, J. L., Wieczorek, L., Baran, S. E., Harman, J. S., Wright, R. L. and Korol, D. L. (2004) Acute stress impairs spatial memory in male but not female rats: influence of estrous cycle. *Pharmacology, Biochemistry, and Behavior* **78**, 569–579.

Contreras, D., Destexhe, A. and Steriade, M. (1997) Intracellular and computational characterization of the intracortical inhibitory control of synchronized thalamic inputs in vivo. *Journal of Neurophysiology* **78**, 335–350.

Cook, S. P., Galve-Roperh, I., Martinez del Pozo, A. and Rodriguez-Crespo, I. (2002) Direct calcium binding results in activation of brain serine racemase. *Journal of Biological Chemistry* **277**, 27,782–27,792.

Coq, J. O. and Xerri, C. (1998) Environmental enrichment alters organizational features of the forepaw representation in the primary somatosensory cortex of adult rats. *Experimental Brain Research* **121**, 191–204.

Cordoba Montoya, D. A. and Carrer, H. F. (1997) Estrogen facilitates induction of long term potentiation in the hippocampus of awake rats. *Brain Research* **778**, 430–438.

Cornell-Bell, A. H., Finkbeiner, S. M., Cooper, M. S. and Smith, S. J. (1990) Glutamate induces calcium waves in cultured astrocytes: long-range glial signaling. *Science* **247**, 470–473.

Corriveau, R. A., Huh, G. S. and Shatz, C. J. (1998) Regulation of class I MHC gene expression in the developing and mature CNS by neural activity. *Neuron* **21**, 505–520.

Cosgaya, J. M., Chan, J. R. and Shooter, E. M. (2002) The neurotrophin receptor p75NTR as a positive modulator of myelination. *Science* **298**, 1245–1248.

Cotman, C. W. and Berchtold, N. C. (2002) Exercise: a behavioral intervention to enhance brain health and plasticity. *Trends in Neurosciences* **25**, 295–301.

Cotrina, M. L., Lin, J. H. C., Alves-Rodriques, A., Liu, S., Li, J., Azmi-Ghadimi, H., Kang, J., Naus, C. C. G. and Nedergaard, M. (1998) Connexins regulate calcium signaling by controlling ATP release. *Proceedings of the National Academy of Science of the United States of America* **95**, 15,735–15,740.

Couldwell, W. T., Weiss, M. H., DeGiorgio, C. M., Weiner, L. P., Hinton, D. R., Ehresmann, G. R., Conti, P. S. and Apuzzo, M. L. (1993) Clinical and radiographic response in a minority of patients with recurrent malignant gliomas treated with high-dose tamoxifen. *Neurosurgery* **32**, 485–489; discussion 489–490.

Crabtree, G. R. and Olson, E. N. (2002) NFAT signaling: choreographing the social lives of cells. *Cell* **109**, S67–79.

Cracco, J. B., Serrano, P., Moskowitz, S. I., Bergold, P. J. and Sacktor, T. C. (2005) Protein synthesis-dependent LTP in isolated dendrites of CA1 pyramidal cells. *Hippocampus* **15**, 551–556.

Crane, G. (1959) Cycloserine as an antidepressant agent. *American Journal of Psychiatry* **115**, 1025–1026.

Crane, G. (1961) The psychotropic effect of cycloserine: a new use of an antibiotic. *Comprehensive Psychiatry* **2**, 51–59.

Crick, F. and Mitchison, G. (1983) The function of dream sleep. *Nature* **304**, 111–114.

Crow, T., Xue-Bian, J. J. and Siddiqi, V. (1999) Protein synthesis-dependent and mRNA synthesis-independent intermediate phase of memory in Hermissenda. *Journal of Neurophysiology* **82**, 495–500.

Crowe, M. J., Bresnahan, J. C., Shuman, S. L., Masters, J. N. and Beattie, M. S. (1997) Apoptosis and delayed degeneration after spinal cord injury in rats and monkeys. *Nature Medicine* **3**, 73–76.

Cruzalegui, F. H. and Bading, H. (2000) Calcium-regulated protein kinase cascades and their transcription factor targets. *Cellular and Molecular Life Sciences* **57**, 402–410.

Cummins, R. A., Walsh, R. N., Budtz-Olsen, O. E., Konstantinos, T. and Horsfall, C. R. (1973) Environmentally-induced changes in the brains of elderly rats. *Nature* **243**, 516–518.

D'Sa, C. and Duman, R. S. (2002) Antidepressants and neuroplasticity. *Bipolar Disord* **4**, 183–194.

Daisley, J. N. and Rose, S. P. R. (2001) Amino acid release from the IMHV of day-old chicks following a one-trial passive avoidance task *Neurobiology of Learning and Memory* **77**, 185–201.

Dalrymple, A. M. (1995) Treatment of trauma-induced amnesia. *Journal of Neurosurgery* **82**, 518.

Dalrymple-Alford, J. C. and Benton, D. (1984) Preoperative differential housing and dorsal hippocampal lesions in rats. *Behavioral Neuroscience* **98**, 23–34.

Dalva, M. B., Takasu, M. A., Lin, M. Z., Shamah, S. M., Hu, L., Gale, N. W. and Greenberg, M. E. (2000) EphB receptors interact with NMDA receptors and regulate excitatory synapse formation. *Cell* **103**, 945–956.

Damasio, A. R. (1994) *Descartes' Error: Emotion, Reason and the Human Brain*. Macmillan.

Dan, Y. and Poo, M. M. (2004) Spike timing-dependent plasticity of neural circuits. *Neuron* **44**, 23–30.

Dani, J. W., Chernjavsky, A. and Smith, S. J. (1992) Neuronal activity triggers calcium waves in hippocampal astrocyte networks. *Neuron* **8**, 429–440.

Daniel, J. M. and Dohanich, G. P. (2001) Acetylcholine mediates the estrogen-induced increase in NMDA receptor binding in CA1 of the hippocampus and the associated improvement in working memory. *Journal of Neuroscience* **21**, 6949–6956.

Daniel, J. M., Fader, A. J., Spencer, A. L. and Dohanich, G. P. (1998) Estrogen enhances performance of female rats during acquisition of a radial arm maze. *Hormones and Behavior* **32**, 217–225.

Datta, S. (2000) Avoidance task training potentiates phasic pontine-wave density in the rat: a mechanism for sleep-dependent plasticity. *Journal of Neuroscience* **20**, 8607–8613.

Datta, S. R., Brunet, A. and Greenberg, M. E. (1999) Cellular survival: a play in three Akts. *Genes Development* **13**, 2905–2927.

Davachi, L. and Wagner, A. G. (2002) Hippocampal contributions to episodic encoding, insights from relational and item-based learning. *Journal of Neurophysiology* **88**, 982–990.

Davachi, L., Mitchell, J. P. and Wagner, A. D. (2003) Multiple routes to memory: distinct medial temporal lobe processes build item and source memories. *Proceedings of the National Academy of Sciences of the United States of America* **100**, 2157–2162.

Dave, A. S. and Margoliash, D. (2000) Song replay during sleep and computational rules for sensorimotor vocal learning. *Science* **290**, 812–816.

Dave, A. S., Yu, A. C. and Margoliash, D. (1998) Behavioral state modulation of auditory activity in a vocal motor system. *Science* **282**, 2250–2254.

Davies, K. J. A. (1988) Proteolytic systems as secondary antioxidant defenses. In Chow, C. K. (ed.), *Cellular Antioxidant Defense Mechanisms*. CRC Press, pp 25–67.

Davies, H. P. and Squire, L. R. (1984) Protein synthesis and memory: a review *Psychological Bulletin* **96**, 518–559.

Davis, M. (1986) Pharmacological and anatomical analysis of fear conditioning using the fear-potentiated startle paradigm. *Behavioral Neuroscience* **100**, 814–824.

Davis, M. (1998) Anatomic and physiologic substrates of emotion in an animal model. *Journal of Clinical Neurophysiology* **15**, 378–387.

Davis, H. P. and Squire, L. R. (1984) Protein synthesis and memory: a review. *Psychological Bulletin* **96**, 518–559

Davis, M. and Shi, C. (1999) The extended amygdala: are the central nucleus of the amygdala and the bed nucleus of the stria terminalis differentially involved in fear versus anxiety? *Annals of the New York Academy of Sciences* **877**, 281–291.

Davis, R. E. and Klinger, P. D. (1994) NMDA receptor antagonist MK-801 blocks learning of conditioned stimulus-unconditioned sitmulus contiguity but not fear of conditioned stimulus in goldfish (*Carassius auratus L.*). *Behavioral Neuroscience* **108**, 935–940.

Day, M., Langston, R. and Morris, R. G. M. (2003) Glutamate receptor mediated encoding and retrieval of paired associate learning. *Nature* **424**, 205–209.

De, A., Krueger, J. M. and Simasko, S. M. (2003) Tumor necrosis factor alpha increases cytosolic calcium responses to AMPA and KCl in primary cultures of rat hippocampal neurons. *Brain Research* **981**, 133–142.

De Kloet, E. R., Oitzl, M. S. and Joels, M. (1993) Functional implications of brain corticosteroid receptor diversity. *Cellular and Molecular Neurobiology* **13**, 433–455.

De Miranda, J., Panizzutti, R., Foltyn, V. N. and Wolosker, H. (2002) Cofactors of serine racemase that physiologically stimulate the synthesis of the N-methyl-D-aspartate (NMDA) receptor coagonist D-serine.

Proceeding of the National Academy of Sciences of the United States of America **99**, 14,542–14,547.

De Oca, B. M., DeCola, J. P., Maren, S. and Fanselow, M. S. (1998) Distinct regions of the periaqueductal gray are involved in the acquisition and expression of defensive responses. *Journal of Neuroscience* **18**, 3426–3432.

de Waegh, S. M., Lee, V. M. and Brady, S. T. (1992) Local modulation of neurofilament phosphorylation, axonal caliber, and slow axonal transport by myelinating Schwann cells. *Cell* **68**, 451–463.

Debiec, J. and LeDoux, J. E. (2004) Disruption of reconsolidation but not consolidation of auditory fear conditioning by noradrenergic blockade in the amygdala. *Neuroscience* **129**, 267–272.

Debiec, J., LeDoux, J. E. and Nader, K. (2002) Cellular and systems reconsolidation in the hippocampus. *Neuron* **36**, 527–538.

Deisseroth, K. and Tsien, R. W. (2002) Dynamic multiphosphorylation passwords for activity-dependent gene expression. *Neuron* **34**, 179–182.

Deisseroth, K., Bito, H. and Tsien, R. W. (1996) Signaling from synapse to nucleus: postsynaptic CREB phosphorylation during multiple forms of hippocampal synaptic plasticity. *Neuron* **16**, 89–101.

Deisseroth, K., Heist, E. K. and Tsien, R. W. (1998) Translocation of calmodulin to the nucleus supports CREB phosphorylation in hippocampal neurons. *Nature* **392**, 198–202.

Deisseroth, K., Mermelstein, P. G., Xia, H. and Tsien, R. W. (2003) Signaling from synapse to nucleus: the logic behind the mechanisms. *Current Opinion in Neurobiology* **13**, 354–365.

Deisseroth, K., Singla, S., Toda, H., Monje, M., Palmer, T. D. and Malenka, R. C. (2004) Excitation-neurogenesis coupling in adult neural stem/progenitor cells. *Neuron* **42**, 535–552.

Demerens, C., Stankoff, B., Logak, M., Anglade, P., Allinquant, B., Couraud, F., Zalc, B. and Lubetzki, C. (1996) Induction of myelination in teh central nervous system by electrical activity. *Proceedings of the National Academy of Science of the United States of America* **93**, 9887–9892.

Derkach, V., Barria, A. and Soderling, T. R. (1999) $Ca^{2+}/$ calmodulin-kinase II enhances channel conductance of alpha-amino-3-hydroxy-5-methyl-4-isoxazolepropionate type glutamate receptors. *Proceedings of the National Academy of Science of the United States of America* **96**, 3269–3274.

Desagher, S., Glowinski, J. and Premont, J. (1996) Astrocytes protect neurons from hydrogen peroxide toxicity. *Journal of Neuroscience* **16**, 2553–2562.

Deutsch, S. I., Rosse, R. B., Schwartz, B. L. and Mastropaolo, J. (2001) A revised excitotoxic hypothesis of schizophrenia: therapeutic implications, *Clinical Neuropharmacology* **24**, 43–49.

Devauges, V. and Sara, S. J. (1990) Activation of the noradrenergic system fascilitates an attentional shift in the rat. *Behavioral Brain Research* **39**, 19–28.

DeZazzo, T. and Tully, T. (1995) Dissection of memory formation: from behavioral pharmacology to molecular genetics. *Trends in Neurosciences* **18**, 212–218.

Diamond, M. C., Krech, D. and Rosenzweig, M. R. (1964) The effects of an enriched environment on the histology of the rat cerebral cortex. *Journal of Comparative Neurology* **123**, 111–120.

Diamond, M. I., Miner, J. N., Yoshinaga, S. K. and Yamamoto, K. R. (1990) Transcription factor interactions: selectors of positive or negative regulation from a single DNA element. *Science* **249**, 1266–1272.

Diamond, M. C., Lindner, B., Johnson, R., Bennett, E. L. and Rosenzweig, M. R. (1975) Differences in occipital cortical synapses from environmentally enriched, impoverished, and standard colony rats. *Journal of Neuroscience Research* **1**, 109–119.

Diamond, M. C., Law, F., Rhodes, H., Lindner, B., Rosenzweig, M. R., Krech, D. and Bennett, E. L. (1966) Increases in cortical depth and glia numbers in rats subjected to enriched environment. *Journal of Comparative Neurology* **128**, 117–126.

Disterhoft, J. F., Wu, W. W. and Ohno, M. (2004) Biophysical alterations of hippocampal pyramidal neurons in learning, ageing and Alzheimer's disease. *Ageing Research Reviews* **3**, 383–406.

Dityatev, A. and Schachner, M. (2003) Extracellular matrix molecules and synaptic plasticity. *Nature Reviews Neuroscience* **4**, 456–468.

Dixon, J. F. and Hokin, L. E. (1998) Lithium acutely inhibits and chronically up-regulates and stabilizes glutamate uptake by presynaptic nerve endings in mouse cerebral cortex. *Proceedings of the National Academy of Science of the United States of America* **95**, 8363–8368.

Do, K. Q., Trabesinger, A. H., Kirsten-Kruger, M., Lauer, C. J., Dydak, U., Hell, D., Holsboer, F., Boesiger, P. and Cuenod, M. (2000) Schizophrenia: glutathione deficit in cerebrospinal fluid and prefrontal cortex in vivo. *European Journal of Neuroscience* **12**, 3721–3728.

Dobrunz, L. E. and Stevens, C. F. (1999) Response of hippocampal synapses to natural stimulation patterns. *Neuron* **22**, 157–166.

Dodson, M. W. and Guo, M. (2007) Pink1, Parkin, DJ-1 and mitochondrial dysfunction in Parkinson's disease. *Current Opinion in Neurobiology* **17**, 331–337.

Dolan, R. (2002) Emotion, cognition and behavior. *Science* **298**, 1191–1197.

Dolmetsch, R. E., Xu, K. and Lewis, R. S. (1998) Calcium oscillations increase the efficiency and specificity of gene expression. *Nature* **392**, 933–936.

Dolmetsch, R. E., Pajvani, U., Fife, K., Spotts, J. M. and Greenberg, M. E. (2001) Signaling to the nucleus by an L-type calcium channel-calmodulin complex through the MAP kinase pathway. *Science* **294**, 333–339.

Donaldson, H. H. and Hoke, G. W. (1905) On the areas of the axis cylinder and medullary sheath as seen in cross sections of the spinal nerves of vertebrates. *Journal of Comparative Neurology* **15**, 1–16.

Dorfman, J. R., Zerrahn, J., Coles, M. C. and Raulet, D. H. (1997) The basis for self-tolerance of natural killer cells in beta2-microglobulin- and TAP-1- mice. *Journal of Immunology* **159**, 5219–5225.

Drevets, W. C. (2001) Neuroimaging and neuropathological studies of depression: implications for the cognitive-emotional features of mood disorders. *Current Opinions in Neurobiology* **11**, 240–249.

Drew, P. D., Lonergan, M., Goldstein, M. E., Lampson, L. A., Ozato, K. and McFarlin, D. E. (1993) Regulation of MHC class I and beta 2-microglobulin gene expression in human neuronal cells. Factor binding to conserved cis-acting regulatory sequences correlates with expression of the genes. *Journal of Immunology* **150**, 3300–3310.

Dringen, R. and Hamprecht, B. (1997) Involvement of glutathione peroxidase and catalase in the disposal of exogenous hydrogen peroxide by cultured astroglial cells. *Brain Research* **759**, 67–75.

Dringen, R. and Hirrlinger, J. (2003) Glutathione pathways in the brain. *Biological Chemistry* **384**, 505–516.

Dringen R., Pawlowski, P. G. and Hirrlinger, J. (2005) Peroxide detoxification by brain cells. *Journal of Neuroscience Research* **79**, 157–165.

Dringen, R., Kussmaul, L., Gutterer, J. M., Hirrlinger, J. and Hamprecht, B. (1999) The glutathione system of peroxide detoxification is less efficient in neurons than in astroglial cells. *Journal of Neurochemistry* **72**, 2523–2530.

Drukarch, B., Schepens, E., Jongenelen, C. A., Stoof, J. C., Langeveld, C. H. (1997) Astrocyte-mediated enhancement of neuronal survival is abolished by glutathione deficiency. *Brain Research* **770**, 123–130.

Drukarch, B., Schepens, E., Stoof, J. C., Langeveld, C. H. and Van Muiswinkel, F. L. (1998) Astrocyte-enhanced neuronal survival is mediated by scavenging of extracellular reactive oxygen species. *Free Radical Biology and Medicine* **25**, 217–220.

Dryer, C. A. and Benjamins, J. A. (1989) Organization of oligodendroglial membrane sheets: II. Galactocerebroside: Antibody interactions signal changes in cytoskeleton and myelin basic protein. *Journal of Neuroscience Research* **24**, 212–221.

Du, J., Feng, L., Yang, F. and Lu, B. (2000) Activity- and Ca^{2+}-dependent modulation of surface expression of brain-derived neurotrophic factor receptors in hippocampal neurons. *Journal of Cell Biology* **150**, 1423–1434.

Du, J., Szabo, S. T., Gray, N. A. and Manji, H. K. (2004) Focus on CaMKII: a molecular switch in the pathophysiology and treatment of mood and anxiety disorders. *International Journal of Neuropsychopharmacology* **7**, 243–248.

Du, J., Creson, T. K., Wu, L. J., Ren, M., Gray, N. A., Falke, C. (2008) The role of hippocampal GluR1 and GluR2 receptors in manic-like behavior. *Journal of Neuroscience* **28**, 68–79.

Du, J., Feng, L., Zaitsev, E., Je, H., Liu, X. and Lu, B. (2003a) Regulation of TrkB receptor tyrosine kinase and its internalization by neuronal activity and calcium influx. *Journal of Cell Biology* **163**, 385–395.

Du, J., Gray, N., Falke, C., Yuan, P., Szabo, S. and Manji, H. (2003b) Structurally dissimilar antimanic agents modulate synaptic plasticity by regulating AMPA glutamate receptor subunit GluR1 synaptic expression. *Annals of the New York Academy of Sciences* **1003**, 378–380.

Du, J., Gray, N. A., Falke, C. A., Chen, W., Yuan, P., Szabo, S. T., Einat, H. and Manji, H. K. (2004) Modulation of synaptic plasticity by antimanic agents: the role of AMPA glutamate receptor subunit 1 synaptic expression. *Journal of Neuroscience* **24**, 6578–6589.

Dubal, D. B., Zhu, H., Yu, J., Rau, S. W., Shughrue, P. J., Merchenthaler, I., Kindy, M. S. and Wise, P. M. (2001) Estrogen receptor alpha, not beta, is a critical link in estradiol-mediated protection against brain injury. *Proceedings of the National Academy of Sciences of the United States of America* **98**, 1952–1957.

Dubnau, J., Chiang, A. S. and Tully, T. (2003) Neural substrates of memory: from synapse to system. *Journal of Neurobiology* **54**, 238–253.

Duda, R. O. and Hart, P. E. (1973) *Pattern Classification and Scene Analysis*. Wiley.

Dudai, Y. (2002) Molecular bases of long-term memories: a question of persistence. *Current Opinion in Neurobiology* **12**, 211–216.

Dudai, Y. (2004) The neurobiology of consolidations, or, how stable is the engram? *Annual Reviews of Psychology* **55**, 51–86.

Dudek, S. M. and Fields, R. D. (2001) Mitogen-activated protein kinase/extracellular signal-regulated kinase activation in somatodendritic compartments: roles of action potentials, frequency, and mode of calcium entry. *Journal of Neuroscience* **21**, RC122.

Dudek, S.M and Fields, R. D. (2002) Somatic action potentials are sufficient for late-phase LTP-related cell signaling. *Proceedings of the National Academy of Science of the United States of America* **99**, 3962–3967.

Duffy, S. N., Craddock, K. J., Abel, T. and Nguyen, P. V. (2001) Environmental enrichment modifies the PKA-dependence if hippocampal LTP and improves hippocampus-dependent memory. *Learning & Memory* **8**, 26–34.

Dugan, L. L., Sensi, S. L., Canzoniero, L. M., Handran, S. D., Rothman, S. M., Lin, T. S., Goldberg, M. P. and Choi, D. W. (1995) Mitochondrial production of reactive oxygen species in cortical neurons following exposure to N-methyl-D-aspartate. *Journal of Neuroscience* **15**, 6377–6388.

Dumay, N. and Gaskell, M. G. (2007) Sleep-associated changes in the mental representation of spoken words. *Psychological Sciences* **18**(1), 35–9.

Duncan, C. P. (1949) The retroactive effect of electroshock on learning. *Journal of Comparative Physiology and Psychology* **42**, 32–44.

Dunlop, D. S. and Neidle, A. (1997) The origin and turnover of D-serine in brain. *Biochemical and Biophysical Research Communications* **235**, 26–30.

Dunlop, D. S., Neidle, A., McHale, D., Dunlop, D. M. and Lajtha, A. (1986) The presence of free D-aspartic acid in rodents and man. *Biochemical and Biophysical Research Communications* **141**, 27–32.

Dunn, A. J., Gray, H. E. and Iuvone, P. M. (1977) Protein synthesis and amnesia: studies with emetine and pactamycin. *Pharmacology, Biochemistry and Behavior* **6**, 1–4.

Dunn-Meynell, A. A., Routh, V. H., McArdle, J. J. and Levin, B. E. (1997) Low-affinity sulfonylurea binding sites reside on neuronal cell bodies in the brain. *Brain Research* **745**, 1–9.

Dupree, J. L., Mason, J. L., Marcus, J. R., Stull, M., Levinson, R., Matsushima, G. K. and Popko, B. (2004) Oligodendrocytes assist in the maintenance of sodium channel clusters independent of the myelin sheath. *Neuron Glia Biology* **1**, 179–192.

Dusek, J. A. and Eichenbaum, H. (1997) The hippocampus and memory for orderly stimulus relations. *Proceedings of the National Academy of Science of the United States of America* **94**, 7109–7114.

Duvarci, S. and Nader, K. (2004) Characterization of fear memory reconsolidation. *Journal of Neuroscience* **24**, 9269–9275.

Duvarci, S., Nader, K. and LeDoux, J. E. (2005) Activation of extracellular signal-regulated kinase-mitogen-activated protein kinase cascade in the amygdala is required for memory reconsolidation of auditory fear conditioning. *European Journal of Neuroscience* **21**, 283–289.

Dyer, C. A., Philibotte, T. M., Wolf, M. K. and Billings-Gagliardi, S. (1994) Myelin basic protein mediates extreculualr signals that regulate microtubule stability in oligodendrocyte membrane sheets. *Journal of Neuroscience Research* **39**, 97–107.

Dyson, S. E. and Jones, D. G. (1980) Quantitation of terminal parameters and their inter-relationships in maturing central synapses: a perspective for experimental studies. *Brain Research* **183**, 43–59.

Eacott, M. J. and Norman, G. (2004) Integrated memory for object, place, and context in rats: a possible model of episodic memory? *Journal of Neuroscience* **24**, 1948–1953.

Ebadi, M., Srinivasan, S. K. and Baxi, M. D. (1996) Oxidative stress and antioxidant therapy in Parkinson's disease. *Progress in Neurobiology* **48**, 1–19.

Ebbinghaus, H. (1885) *Ueber das Gedaechtnis*. Leipzig: Duncker & Humbolt.

Ebinu, J.O and Yankner, B. A. (2002) A RIP tide in neuronal signal transduction. *Neuron* **34**, 499–502.

Edelman, G. M. (1985) Cell adhesion and the molecular process of morphogenesis *Annual Review of Biochemistry* **54**, 135–169.

Edstrom, E., Kullberg, S., Ming, Y., Zheng, H. and Ulfhake, B. (2004) MHC class I, beta2 microglobulin, and the INF-gamma receptor are upregulated in aged motoneurons. *Journal of Neuroscience Research* **78**, 892–900.

Ehlers, M. D. (2000) Reinsertion or degradation of AMPA receptors determined by activity-dependent endocytic sorting. *Neuron* **28**, 511–525.

Ehrlich, I. and Malinow, R. (2004) Postsynaptic density 95 controls AMPA receptor incorporation during long-term potentiation and experience-driven synaptic plasticity. *Journal of Neuroscience* **24**, 916–927.

Eichenbaum, H. (1997) Declarative memory: insights from cognitive neurobiology. *Annual Review of Neuroscience* **48**, 547–572.

Eichenbaum, H. and Cohen, N. J. (2001) *From Conditioning to Conscious Recollection: Memory Systems of the Brain*. Oxford University Press.

Eichenbaum, H., Dudchencko, P., Wood E., Shapiro, M., and Tanila, H. (1999) The hippocampus, memory, and place cells: is it spatial memory or a memory space? *Neuron* **23**, 209–226.

Einat, H., Yuan, P., Gould, T. D., Li, J., Du, J., Zhang, L., Manji, H. K. and Chen, G. (2003) The role of the extracellular signal-regulated kinase signaling pathway in mood modulation. *Journal of Neuroscience* **23**, 7311–7316.

Eisenberg, M., Kobilo, T., Berman, D. E. and Dudai, Y. (2003) Stability of retrieved memory: inverse correlation with trace dominance. *Science* **301**, 1102–1104.

Ekstrom, A. D., Kahana, M. J., Caplan, J. B., Fields, T. A., Isham, E. A., Newman, E. L. and Fried, I. (2003) Cellular networks underlying human spatial navigation. *Nature* **425**, 184–187.

Eldridge, L. L., Knowlton, B. J., Furmanski, C. S., Brookheimer, S. Y. and Engel, S. A. (2000) Remembering episodes: a selective role for the hippocampus during retrieval. *Nature Neuroscience* **3**, 1149–1152.

Elgersma, Y., Fedorov, N. B., Ikonen, S., Choi, E. S., Elgersma, M., Carvalho, O. M., Giese, K. P. and Silva, A. J. (2002) Inhibitory autophosphorylation of CaMKII controls PSD association, plasticity, and learning. *Neuron* **36**, 493–505.

Ellis-Davies, G. C. and Kaplan, J. H. (1994) Nitrophenyl-EGTA, a photolabile chelator that selectively binds Ca^{2+} with high affinity and releases it rapidly upon photolysis. *Proceedings of the National Academy of Science of the United States of America* **91**, 187–191.

Elmariah, S. B., Crumling, M. A., Parsons, T. D. and Balice-Gordon, R. J. (2004) Postsynaptic TrkB-mediated signaling modulates excitatory and inhibitory neurotransmitter receptor clustering at hippocampal synapses. *Journal of Neuroscience* **24**, 2380–2393.

Elmariah, S. B., Oh, E. J., Hughes, E. M. and Balice-Gordon, R. J. (2005) Astrocytes regulate inhibitory synapse formation via Trk-mediated modulation of postsynaptic GABAA receptors. *Journal of Neuroscience* **25**, 3638–3650.

Emamghoreishi, M., Schlichter, L., Li, P. P., Parikh, S., Sen, J., Kamble, A. and Warsh, J. J. (1997) High intracellular calcium concentrations in transformed lymphoblasts from subjects with bipolar I disorder. *American Journal of Psychiatry* **154**, 976–982.

English, J. D. and Sweatt, J. D. (1997) A requirement for the mitogen-activated protein kinase cascade in hippocampal long term potentiation. *Journal of Biological Chemistry* **272**, 19,103–19,106.

Eriksson, P. S., Perfilieva, E., Bjork-Eriksson, T., Alborn, A. M., Nordborg, C., Peterson, D. A. and Gage, F. H. (1998) Neurogenesis in the adult human hippocampus. *Nature Medicine* **4**, 1313–1317.

Ernfors, P., Bengzon, J., Kokaia, Z., Persson, H. and Lindvall, O. (1991) Increased levels of messenger RNAs for neurotrophic factors in the brain during kindling epileptogenesis. *Neuron* **7**, 165–176.

Eshete, F. and Fields, R. D. (2001) Spike frequency decoding and autonomous activation of Ca^{2+}-calmodulin-dependent protein kinase II in dorsal root ganglion neurons. *Journal of Neuroscience* **21**, 6694–6705.

Eshhar, N., Petralia, R. S., Winters, C. A., Niedzielski, A. S. and Wenthold, R. J. (1993) The segregation and expression of glutamate receptor subunits in cultured hippocampal neurons. *Neuroscience* **57**, 943–964.

Esposito, F., Ammendola, R., Faraonio, R., Russo, T. and Cimino, F. (2004) Redox control of signal transduction, gene expression and cellular senescence. *Neurochemical Research* **29**, 617–628.

Esteban, J. A., Shi, S. H., Wilson, C., Nuriya, M., Huganir, R. L. and Malinow, R. (2003) PKA phosphorylation of AMPA receptor subunits controls synaptic trafficking underlying plasticity. *Nature Neuroscience* **6**, 136–143.

Eyre, J. A., Miller, S. and Ramesh, V. (1991) Constancy of central conduction delays during development in man: investigation of motor and somatosensory pathways. *Journal of Physiology* **343**, 441–452.

Falkenberg, T., Mohammed, A. K., Henriksson, B., Persson, H., Winblad, B. and Lindefors, N. (1992) Increased expression of brain-derived neurotrophic factor mRNA in rat hippocampus is associated with improved spatial memory and enriched environment. *Neuroscience Letters* **138**, 153–156.

Fallon, J. H. and Ciofi, P. (1992) Distribution of monoamines within the amygdala. In: Aggleton, J. P. (ed.) *The Amygdala: Neurobiological Aspects of Emotion, Memory, and Mental Dysfunction.* Wiley-Liss, pp 97–114.

Falls, W. A., Miserendino, M. J. D. and Davis, M. (1992) Extinction of fear-potentiated startle: blockade by infusion of an NMDA antagonist into the amygdala. *Journal of Neuroscience* **12**, 854–863.

Fanselow, M. S. (2000) Contextual fear, gestalt memories, and the hippocampus. *Behavioral Brain Research* **110**, 73–81.

Fanselow, M. S. and Bolles, R. C. (1979) Naloxone and shock-elicited freezing in the rat. *Journal of Comparative Physiological Psychology* **93**, 736–744.

Fanselow, M. S. and Kim, J. J. (1994) Acquisition of contextual Pavlovian fear conditioning is blocked by application of an NMDA receptor antagonist D,L-2-amino-5-phosphonovaleric acid to the basolateral amygdala. *Behavioral Neuroscience* **108**, 210–212.

Fanselow, M. S. and LeDoux, J. E. (1999) Why we think plasticity underlying Pavlovian fear conditioning occurs in the basolateral amygdala. *Neuron* **23**, 229–232.

Fanselow, M. S. and Poulos, A. M. (2005) The neuroscience of mammalian associative aearning. *Annual Review of Psychology* **56**, 207–234.

Farb, C. R. and LeDoux, J. E. (1997) NMDA and AMPA receptors in the lateral nucleus of the amygdala are postsynaptic to auditory thalamic afferents. *Synapse* **27**, 106–121.

Farb, C. R., Aoki, C. and LeDoux, J. E. (1995) Differential localization of NMDA and AMPA receptor subunits in the lateral and basal nuclei of the amygdala: a light and electron microscopic study. *Journal of Comparative Neurology* **362**, 86–108.

Fearnley, J. and Lees, A. J. (1991) Aging and Parkinson's disease: substantia nigra regional selectivity. *Brain* **114**, 2283–2301.

Federmeier, K. D., Kleim, J. A. and Greenough, W. T. (2002) Learning-induced multiple synapse formation in rat cerebellar cortex. *Neuroscience Letters* **332**, 180–184.

Feldman, D. E., Nicoll, R. A. and Malenka, R. C. (1999) Synaptic plasticity at thalamocortical synapses in developing rat somatosensory cortex: LTP, LTD, and silent synapses. *Journal of Neurobiology* **41**, 92–101.

Fendt, M. and Schmid, S. (2002) Metabotropic glutamate receptors are involved in amygdaloid plasticity. *European Journal of Neuroscience* **15**, 1535–1541.

Feng, R., Rampon, C., Tang, Y. P., Shrom, D., Jin, J., Kyin, M., Sopher, B., Miller, M. W., Ware, C. B., Martin, G. M., Kim, S. H., Langdon, R. B., Sisodia, S. S. and Tsien, J. Z. (2001) Deficient neurogenesis in forebrain-specific presenilin-1 knockout mice is associated with reduced clearance of hippocampal memory traces. *Neuron* **32**, 911–926.

Fenn, K. M., Nusbaum, H. C. and Margoliash, D. (2003) Consolidation during sleep of perceptual learning of spoken language. *Nature* **425**, 614–616.

Ferbinteanu J. and Shapiro M. L. (2003) Prospective and retrospective memory coding in the hippocampus. *Neuron* **40**, 1227–1239.

Ferchmin, P. A. and Bennett, E. L. (1975) Direct contact with enriched environment is required to alter cerebral weights in rats. *Journal of Comparative and Physiological Psychology* **88**, 360–367.

Ferry, B. and McGaugh, J. L. (2000) Role of amygdala norepinephrine in mediating stress hormone regulation of memory storage. *Acta Pharmacologica Sinica* **21**, 481–493.

Ferry, B., Roozendaal, B. and McGaugh, J. L. (1999) Role of norepinephrine in mediating stress hormone regulation of long-term memory storage: a critical involvement of the amygdala. *Biological Psychiatry* **46**, 1140–1152.

Fiacco, T. A. and McCarthy, K. D. (2004) Intracellular astrocyte calcium waves in situ increase the frequency of spontaneous AMPA receptor currents in CA1 pyramidal neurons. *Journal of Neuroscience* **24**, 722–732.

Fiala, B. A., Joyce, J. N. and Greenough, W. T. (1978) Environmental complexity modulates growth of granule cell dendrites in developing but not adult hippocampus of rats. *Experimental Neurology* **59**, 372–383.

Fiala, J. C., Allwardt, B. and Harris, K. M. (2002) Dendritic spines do not split during hippocampal LTP or maturation. *Nature Neuroscience* 5, 297–298.

Fields, R. D. (2005) Myelination: an overlooked mechanism of synaptic plasticity? *The Neuroscientist* 11, 528–531.

Fields, R. D. (2008a) White Matter Matters. *Scientific American* 298, 51–61.

Fields, R. D. (2008b) White matter in learning, cognition, and psychiatric disorders. *Trends in Neuroscience* (in press).

Fields, R. D. and Burnstock, G. (2006) Purinergic signaling in neuron–glia interactions. *Nature Review Neuroscience* 7, 423–436.

Fields, R. D. and Itoh, K. (1996) Neural cell adhesion molecules in activity-dependent development and synaptic plasticity. *Trends in Neuroscience* 19, 473–480.

Fields, R. D. and Stevens-Graham, B. (2002) New views of neuron–glia communication. *Science* 298, 556–562.

Fields, R. D., Lee, P. R. and Cohen, J. E. (2005) Temporal integration of intracellular Ca^{2+} signaling networks in regulating gene expression by action potentials. *Cell Calcium* 37, 433–442.

Fink, C. C. and Meyer, T. (2002) Molecular mechanisms of CaMKII activation in neuronal plasticity. *Current Opinions in Neurobiology* 12, 293–299.

Fischer, M., Kaech, S., Knutti, D. and Matus, A. (1998) Rapid actin-based plasticity in dendritic spines. *Neuron* 20, 847–854.

Fischer, S., Hallschmid, M., Elsner, A. L. and Born, J. (2002) Sleep forms memory for finger skills. *Proceedings of the National Academy of Science* 99, 11,987–11,991.

Fiskum, G., Starkov, A., Polster, B. M. and Chinopoulos, C. (2003) Mitochondrial mechanisms of neural cell death and neuroprotective interventions in Parkinson's disease. *Annals of the New York Academy of Sciences* 991, 111–119.

Fitzgerald, H. E. and Brackbill, Y. (1976) Classical conditioning in infancy: development and constraints. *Psychological Bulletin* 83, 353–376.

Flexner, L. B., Flexner, J. B., De La Haba, G. and Roberts, R. B. (1965) Loss of memory as related to inhibition of cerebral protein synthesis. *Journal of Neurochemistry* 12, 535–541.

Floeter, M. K. and Greenough, W. T. (1979) Cerebellar plasticity: modification of Purkinje cell structure by differential rearing in monkeys. *Science* 206, 227–229.

Flynn, S. W., Lang, D. J., Mackay, A. L., Goghari, V., Vavasour, I. M., Whittall, K. P., Smith, G. N., Arango, V., Mann, J. J., Dwork, A. J., Falkai, P. and Honer, W. G. (2003) Abnormalities of myelination in schizophrenia detected in vivo with MRI, and post-mortem with analysis of oligodendorcyte proteins. *Molecular Psychiatry* 8, 811–820.

Fogel, S., Jacob, J. and Smith, C. (2001) Increased sleep spindle activity following simple motor procedural learning in humans. *Congress Physiological Basis for Sleep Medicine*, Uruguay, Actas de Fisiología.

Foltyn, V. N., Bendikov, I., De Miranda, J., Panizzutti, R., Dumin, E., Shleper, M., Li, P., Toney, M. D., Kartvelishvily, E. and Wolosker, H. (2005) Serine racemase modulates intracellular D-serine levels through an alpha,beta-elimination bctivity. *Journal of Biological Chemistry* 280, 1754–1763.

Fontaine, V., Mohand-Said, S., Hanoteau, N., Fuchs, C., Pfizenmaier, K. and Eisel, U. (2002) Neurodegenerative and neuroprotective effects of tumor necrosis factor (TNF) in retinal ischemia: opposite roles of TNF receptor 1 and TNF receptor 2. *Journal of Neuroscience* 22, RC216.

Forgays, D. G. and Forgays, J. W. (1952) The nature of the effect of free-environmental experience in the rat. *Journal of Comparative and Physiological Psychology* 45, 322–328.

Fortin, N. J., Agster, K. L. and Eichenbaum, H. (2002) Critical role of the hippocampus in memory for sequences of events. *Nature Neuroscience* 5, 458–462.

Foster, T. C. (1999) Involvement of hippocampal synaptic plasticity in age-related memory decline. *Brain Research Reviews* 30, 236–249.

Foster, T. C. and Norris, C. M. (1997) Age-associated -changes in Ca^{2+}-dependent processes: relation to hippocampal synaptic plasticity. *Hippocampus* 7, 602–612.

Foster, J. A., Quan, N., Stern, E. L., Kristensson, K. and Herkenham, M. (2002) Induced neuronal expression of class I major histocompatibility complex mRNA in acute and chronic inflammation models. *Journal of Neuroimmunology* 131, 83–91.

Foy, M. R. (1983) Neuromodulation: effects of estradiol and THC on brain excitability. In: Neurobiology Program. Kent: Kent State University.

Foy, M. R. and Teyler, T. J. (1983) 17-alpha-estradiol and 17-beta-estradiol in hippocampus. *Brain Research Bulletin* 10, 735–739.

Foy, M. R., Baudry, M. and Thompson, R. (2004) Estrogen and hippocampal synaptic plasticity. *Neuron Glia Biology* 1, 327–338.

Foy, M. R., Xu, J., Xie, X., Brinton, R. D., Thompson, R. F. and Berger, T. W. (1999) 17b-estradiol enhances NMDA receptor-mediated EPSPs and long-term potentiation. *Journal of Neurophysiology* **81**, 925–929.

Frank, L. M., Brown, E. N. and Wilson, M. (2000) Trajectory encoding in the hippocampus and entorhinal cortex. *Neuron* **27**, 169–178.

Frank, L. M., Stanley, G. B. and Brown, E. N. (2004) Hippocampal plasticity across multiple days of exposure to novel environments. *Journal of Neuroscience* **24**, 7681–7689.

Frank, M. G., Issa, N. P. and Stryker, M. P. (2001) Sleep enhances plasticity in the developing visual cortex. *Neuron* **30**, 275–287.

Frankfurt, M., Gould, E., Woolley, C. S. and McEwen, B. S. (1990) Gonadal steroids modify dendritic spine density in ventromedial hypothalamic neurons: a Golgi study in the adult rat. *Neuroendocrinology* **51**, 530–535.

Frankland, P. W. and Bontempi, B. (2005) The organization of recent and remote memories. *Nature Reviews Neuroscience* **6**, 119–130.

Frankland, P. W., Bontempi, B., Talton, L. E., Kaczmarek, L. and Silva A. J. (2004) The involvement of the anterior cingulate cortex in remote contextual fear memory. *Science* **304**, 881–883.

Frankland, P. W., O'Brien, C., Ohno, M., Kirkwood, A. and Silva, A. J. (2001) Alpha-CaMKII-dependent plasticity in the cortex is required for permanent memory. *Nature* **411**, 309–313.

Freeman, B. (1978) Myelin sheath thickness and conduction latency groups in the cat optic nerve. *Journal of Comparative Neurology* **181**, 183–196.

Freeman, F. M. and Rose, S. P. R. (1995) Two time windows of anisomycin-induced amnesia for passive avoidance training in the day old chick *Neurobiology of Learning and Memory* **63**, 291–295.

Frenkel, L., Maldonado, H. and Delorenzi, A. (2005) Memory strengthening by a real-life episode during reconsolidation: an outcome of water deprivation via brain angiotensin II. *European Journal of Neuroscience* **22**, 1757–1766.

Freudenthal, R., Romano, A. and Routtenberg, A. (2004) Transcription factor NF-kappaB activation after in vivo perforant path LTP in mouse hippocampus. *Hippocampus* **14**, 677–683.

Frey, S., Bergado-Rosado, J., Seidenbecher, T., Pape, H. C. and Frey, J. U. (2001) Reinforcement of early long-term potentiation (early-LTP) in dentate gyrus by stimulation of the basolateral amygdala: heterosynaptic induction mechanisms of late-LTP. *Journal of Neuroscience* **21**, 3697–3703.

Frey, U. and Morris, R. G. (1997) Synaptic tagging and long-term potentiation. *Nature* **385**, 533–536.

Frey, U. and Morris, R. G. (1998) Synaptic tagging: implications for late maintenance of hippocampal long-term potentiation. *Trends in Neuroscience* **21**, 181–188.

Frey, U. and Morris, R. G. (1998b) Weak before strong: dissociating synaptic tagging and plasticity-factor accounts of late LTP. *Neuropharmacology* **37**, 545–552.

Frey, U., Hartmann, S. and Matthies, H. (1989) Domperidone, an inhibitor of the D2-receptor, blocks a late phase of an electrically induced long-term potentiation in the CA1-region in rats. *Biomedica Biochimica Acta* **48**, 473–476.

Frey, U., Huang, Y.-Y. and Kandel, E. R. (1993) Effects of cAMP simulate a late stage of LTP in hippocampal CA1 neurons. *Science* **260**, 1661–1664.

Frey, S., Bergado, J. A. and Frey, J. U. (2003) Modulation of late phases of long-term potentiation in rat dentate gyrus by stimulation of the medial septum. *Neuroscience* **118**, 1055–1062.

Frey, U., Schröder, H. and Matthies, H. (1990) Dopaminergic antagonists prevent long-term maintenance of posttetanic LTP in the CA1 region of rat hippocampal slices. *Brain Research* **522**, 69–75.

Frey, U., Krug, M., Reymann, K. G. and Matthies, H. (1988) Anisomycin, an inhibitor of protein synthesis, blocks late phases of LTP phenomena in the hippocampal CA1 region in vitro. *Brain Research* **452**, 57–65.

Frey, U., Krug, M., Brodemann, R., Reymann, K. and Matthies, H. (1989) Long-term potentiation induced in dendrites separated from rat's CA1 pyramidal somata does not establish a late phase. *Neuroscience Letters* **97**, 135–139.

Friede, R. (1954) Der quantitative Anteil der Glia an der Cortexentwicklung. *Acta Anatomica* **20**, 290–296.

Friede, R. L. (1972) Control of myelin formation by axon caliber (with a model of the control mechanism). *Journal of Comparative Neurology* **144**, 233–252.

Friede, R. L., Brzoska, J. and Hartmann, U. (1985) Changes in myelin sheath thickness and internode geometry in rabbit phrenic nerve during growth. *Journal of Anatomy* **143**, 103–113.

Friedman, W. J., Ernfors, P. and Persson, H. (1991) Transient and persistent expression of NT-3/HDNF mRNA in the rat brain during postnatal development. *Journal of Neuroscience* 11, 1577–1584.

Friedman, E., Hoau Yan, W., Levinson, D., Connell, T. A. and Singh, H. (1993) Altered platelet protein kinase C activity in bipolar affective disorder, manic episode. *Biological Psychiatry* 33, 520–525.

Frye, C. A. and Lacey, E. H. (2001) Posttraining androgens' enhancement of cognitive performance is temporally distinct from androgens' increases in affective behavior. *Cognitive Affective and Behavioral Neuroscience* 1, 172–182.

Frye, C. A., Edinger, K. L., Seliga, A. M. and Wawrzycki, J. M. (2004) 5a-reduced androgens may have actions in the hippocampus to enhance cognitive performance of male rats. *Psychoneuroendocrinology* 29, 1019–1027.

Fuchs, E. and Flugge, G. (1998) Stress, glucocorticoids and structural plasticity of the hippocampus. *Neuroscience and Biobehavioral Reviews* 23, 295–300.

Fujimaki, H., Hikawa, N., Nagoya, M., Nagata, T. and Minami, M. (1996) IFN-gamma induces expression of MHC class I molecules in adult mouse dorsal root ganglion neurones. *Neuroreport* 7, 2951–2955.

Fujiyama, F., Fritschy, J. M., Stephenson, F. A. and Bolam, J. P. (2000) Synaptic localization of GABAA receptor subunits in the striatum of the rat. *Journal of Comparative Neurology* 416, 158–172.

Fukae, J., Mizuno, Y. and Hattori, N. (2007) Mitochondrial dysfunction in Parkinson's disease. *Mitochondrion* 7, 58–62.

Fukui, Y., Hayasaka, S., Bedi, K. S., Ozaki, H. S. and Takeuchi, Y. (1991) Quantitative study of the development of the optic nerve in rats reared in the dark during early postnatal life. *Journal of Anatomy* 174, 37–47.

Furukawa, K. and Mattson, M. P. (1998) The transcription factor NF-kappaB mediates increases in calcium currents and decreases in NMDA- and AMPA/kainate-induced currents induced by tumor necrosis factor-alpha in hippocampal neurons. *Journal of Neurochemistry* 70, 1876–1886.

Furukawa, S., Furukawa, Y., Satoyoshi, E. and Hayashi, K. (1987) Synthesis/secretion of nerve growth factor is associated with cell growth in cultured mouse astroglial cells. *Biochemical Biophysical Research Communications* 142, 395–402.

Gaab, N., Paetzold, M., Becker, M., Walker, M. P. and Schlaug, G. (2004) The influence of sleep on auditory learning – a behavioral study. *Neuroreport* 15, 731–734.

Gabor, R., Nagle, R., Johnston, D. A. and Gibbs, R. B. (2003) Estrogen enhances potassium-stimulated acetylcholine release in the rat hippocampus. *Brain Research* 962, 244–247.

Gage, F. H. (2000) Mammalian neural stem cells. *Science* 287, 1433–1438.

Gage, F. H., Coates, P. W., Palmer, T. D., Kuhn, H. G., Fisher, L. J., Suhonen, J. O., Peterson, D. A., Suhr, S. T. and Ray, J. (1995) Survival and differentiation of adult neuronal progenitor cells transplanted to the adult brain. *Proceedings of the National Academy of Sciences of the United States of America* 92, 11,879–11,883.

Gais, S. and Born, J. (2004) Low acetylcholine during slow-wave sleep is critical for declarative memory consolidation. *Proceedings of the National Academy of Science* 101, 2140–2144.

Gais, S., Molle, M., Helms, K. and Born, J. (2002) Learning-dependent increases in sleep spindle density. *Journal of Neuroscience* 22, 6830–6834.

Gais, S., Plihal, W., Wagner, U. and Born, J. (2000) Early sleep triggers memory for early visual discrimination skills. *Nature Neuroscience* 3, 1335–1339.

Galani, R., Coutureau, E. and Kelche, C. (1998) Effects of enriched postoperative housing conditions on spatial memory deficits in rats with selective lesions of either the hippocampus, subiculum or entorhinal cortex. *Restorative Neurology and Neuroscience* 13, 173–184.

Galani, R., Jarrard, L. E., Will, B. E. and Kelche, C. (1997) Effects of postoperative housing conditions on functional recovery in rats with lesions of the hippocampus, subiculum, or entorhinal cortex. *Neurobiology of Learning and Memory* 67, 43–56.

Gale, G. D., Anagnostaras, S. G., Godsil, B. P., Mitchell, S., Nozawa, T., Sage, J. R., Wiltgen, B. and Fanselow, M. S. (2004) Role of the basolateral amygdala in the storage of fear memories across the adult lifetime of rats. *Journal of Neuroscience* 24, 3810–3815.

Galea, L. A. M., Kavakiers, M., Ossenkopp, K. P. and Hampson, E. (1995) Gonadal hormone levels and spatial learning performance in the Morris water maze in male and female meadow voles, Microtus pennsylvanicus. *Hormones and Behavior* 29, 106–125.

Gall, C. M., Pinkstaff, J. K., Lauterborn, J. C., Xie, Y. and Lynch, G. (2003) Integrins regulate neuronal neurotrophin gene expression through effects on voltage-sensitive calcium channels. *Neuroscience* **118**, 925–940.

Galofre, E. and Ferrer, I. (1987) Development of dendritic spines in the Vth's layer pyramidal neurons of the rat's somatosensory cortex. A qualitative and quantitative study with the Golgi method. *Journal fur Hirnforschung* **28**, 653–659.

Ganeshina, O., Berry, R. W., Petralia, R. S., Nicholson, D. A. and Geinisman, Y. (2004) Differences in the expression of AMPA and NMDA receptors between axospinous perforated and nonperforated synapses are related to the configuration and size of postsynaptic densities. *Journal of Comparative Neurology* **468**, 86–95.

Garber, K., Smith, K. T., Reines, D. and Warren, S. T. (2006) Transcription, translation and fragile X syndrome. *Current Opinion in Genetics and Development* **16**, 270–275.

Garner, C. C., Zhai, R. G., Gundelfinger, E. D. and Ziv, N. E. (2002) Molecular mechanisms of CNS synaptogenesis. *Trends in Neuroscience* **25**, 243–251.

Gartner, A. and Staiger, V. (2002) Neurotrophin secretion from hippocampal neurons evoked by long-term-potentiation-inducing electrical stimulation patterns. *Proceedings of the National Academy of Sciences of the United States of America* **99**, 6386–6391.

Gazzaley, A. H., Weiland, N. G., McEwen, B. S. and Morrison, J. H. (1996) Differential regulation of NMDAR1 mRNA and protein by estradiol in the rat hippocampus. *Journal of Neuroscience* **16**, 6830–6838.

Geinisman, Y. (2000) Structural synaptic modifications associated with hippocampal LTP and behavioral learning. *Cerebral Cortex* **10**, 952–962.

Geinisman, Y., Detoledo-Morell, F. and Heller, R. E. (1995) Hippocampal markers of age-related memory dysfunction: behavioral, electrophysiological and morphological perspectives. *Progress in Neurobiology* **45**, 223–252.

Geinisman, Y., deToledo-Morrell, L. and Morrell, F. (1991) Induction of long-term potentiation is associated with an increase in the number of axospinous synapses with segmented postsynaptic densities. *Brain Research* **566**, 77–88.

Geinisman, Y., Morrell, F. and de Toledo-Morrell, L. (1990) Increase in the relative proportion of perforated axospinous synapses following hippocampal kindling is specific for the synaptic field of stimulated axons. *Brain Research* **507**, 325–331.

Genoux, D., Haditsch, U., Knobloch, M., Michalon, A., Storm, D. and Mansuy, I. M. (2002) Protein phosphatase 1 is a molecular constraint on learning and memory. *Nature* **418**, 970–975.

Georgieva, L., Moskvina, V., Peirce, T., Norton, N., Bray, N. J., Jones, L., Holmans, P., Macgregor, S., Zammit, S., Wilkinson, J., Williams, H., Nikolov, I., Williams, N., Ivanov, D., Davis, K. L., Haroutunian, V., Buxbaum, J. D., Craddock, N., Kirov, G., Owen, M. J. and O'Donovan, M. C. (2006) Convergent evidence that oligodendrocyte lineage transcription factor 2 (OLIG2) and interacting genes influence susceptibility to schizophrenia. *Proceedings of the National Academy of Sciences of the United States of America* **103**, 12,469–12,474.

Gerard, R. W. (1949) Physiology and psychiatry. *American Journal of Psychiatry* **106**, 161–173.

Gerdeman G. L. and Lovinger D. M. (2001) CB1 cannabinoid receptor inhibits synaptic release of glutamate in rat dorsolateral striatum. *Journal of Neurophysiology* **85**, 468–471.

Gereau, R.Wt. and Conn, P. J. (1995) Multiple presynaptic metabotropic glutamate receptors modulate excitatory and inhibitory synaptic transmission in hippocampal area CA1. *Journal of Neuroscience* **15**, 6879–6889.

Gerlai, R. and McNamara, A. (2000) Anesthesia induced retrograde amnesia is ameliorated by ephrinA5-IgG in mice: EphA receptor tyrosine kinases are involved in mammalian memory. *Behavioural Brain Research* **108**, 133–143.

Gesing, A., Bilang-Bleuel, A., Droste, S. K., Linthorst, A. C., Holsboer, F. and Reul, J. M. (2001) Psychological stress increases hippocampal mineralocorticoid receptor levels: involvement of corticotropin-releasing hormone. *Journal of Neuroscience* **21**, 4822–4829.

Ghirardi, M., Montarolo, P. G. and Kandel, E. R. (1995) A novel intermediate stage in the transition between short- and long-term facilitation in the sensory to motor neuron synapse of aplysia. *Neuron* **14**, 413–420.

Ghooray, G. T. and Martin, G. F. (1993) The development of myelin in the spinal cord of the North American opossum and its possible role in loss of rubrospinal plasticity. *Brain Research, Developmental Brain Research* **72**, 67–74.

Ghosh, A. and Greenberg, M. E. (1995) Distinct roles for bFGF and NT-3 in the regulation of cortical neurogenesis. *Neuron* **15**, 89–103.

Ghosh, S. and Karin, M. (2002) Missing pieces in the NF-kappaB puzzle. *Cell* **109**, S81–96.

Giancotti, F. G. and Ruoslahti, E. (1999) Integrin signaling. *Science* **285**, 1028–1032.

Gibbs, M. E. and Ng, K. T. (1977) Psychobiology of memory: Towards a model of memory formation. *Behav, Rev.* **1**, 113–136.

Gibbs, M. E. and Summers, R. J. (2002) Role of adrenoceptor subtypes in memory consolidation. *Progress in Neurobiology* **67**, 345–391.

Gibbs, M. E., Richdale, A. L. and Ng, K. T. (1979) Biochemical aspects of protein synthesis inhibition by cycloheximide in one or both hemispheres of the chick brain. *Pharmacology, Biochemistry and Behavior* **10**, 929–931.

Gibbs, R. B. (1996) Fluctuations in relative levels of choline acetyltransferase mRNA in different regions of the rat basal forebrain across the estrous cycle: effects of estrogen and progesterone. *Journal of Neuroscience* **16**, 1049–1055.

Gibbs, R. B. (1997) Effects of estrogen on basal forebrain cholinergic neurons vary as a function of dose and duration of treatment. *Brain Research* **757**, 10–16.

Gibbs, R. B. (1999) Estrogen replacement enhances acquisition of a spatial memory task and reduces deficits associated with hippocampal muscarinic receptor inhibition. *Hormones and Behavior* **36**, 222–233.

Gibbs, R. B. (2000) Effects of gonadal hormone replacement on measures of basal forebrain cholinergic function. *Neuroscience* **101**, 931–938.

Gibbs, R. B., Gabor, R., Cox, T. and Johnson, D. A. (2004) Effects of raloxifene and estradiol on hippocampal acetylcholine release and spatial learning in the rat. *Psychoneuroendocrinology* **29**, 741–748.

Gibbs, R. B., Wu, D., Hersh, L. B. and Pfaff, D. W. (1994) Effects of estrogen replacement on the relative levels of choline acetyltransferase, trkA, and nerve growth factor messenger RNAs in the basal forebrain and hippocampal formation of adult rats. *Experimental Neurology* **129**, 70–80.

Gibson, C. L. and Murphy, S. P. (2004) Progesterone enhances functional recovery after middle cerebral artery occlusion in male mice. *Journal of Cerebral Blood Flow and Metabolism* **24**, 805–813.

Giedd, J. N. (2004) Structural magnetic resonance imaging of the adolescent brain. *Annals of the New York Academy of Science* **1021**, 105–109.

Gilbert, P. E. and Kesner, R. P. (2003) Recognition memory for complex visual discriminations is influenced by stimulus interference in rodents with perirhinal cortex damage. *Learning & Memory* **10**, 525–530.

Giovanello, K. S., Schnyer, D. M. and Verfaellie, M. (2003) A critical role for the anterior hippocampus in relational memory: evidence from an fMRI study comparing associative and item recognition. *Hippocampus* **14**, 5–8.

Glanzer, J. G. and Eberwine, J. H. (2003) Mechanisms of translational control in dendrites. *Neurobiology and Aging* **24**, 1105–1111.

Globus, A., Rosenzweig, M. R., Bennett, E. L. and Diamond, M. C. (1973) Effects of differential experience on dendritic spine counts in rat cerebral cortex. *Journal of Comparative and Physiological Psychology* **82**, 175–181.

Goddard, C. A., Butts, D. A. and Shatz, C. J. (2007) Regulation of CNS synapses by neuronal Class I. *Proceedings of the National Academy of Science of the United States of America* **104**, 6828–6833.

Goelet, P., Castellucci, V. F., Schacher, S. and Kandel, E. R. (1986) The long and the short of long-term memory – a molecular framework. *Nature* **322**, 419–422.

Goff, D. C. and Coyle, J. T. (2001) The emerging role of glutamate in the pathophysiology and treatment of schizophrenia. *American Journal of Psychiatry* **158**, 1367–1377.

Gold, P. E. and McGaugh, J. L. (1975) A single-trace, two process view of memory storage processes. In Deutsch, D. And Deutsch, J. A. (eds) *Short-term Memory*. Academic Press, pp 355–378.

Goldberg, J. F. and Harrow, M. (2004) Consistency of remission and outcome in bipolar and unipolar mood disorders: a 10-year prospective follow-up. *Journal of Affective Disorders* **81**, 123–131.

Goldman, J. E. (1995) Lineage, migration, and fate determination of postnatal subventricular zone cells in the mammalian CNS. *Journal of Neurooncology* **24**, 61–64.

Goldman, S. A., Kirschenbaum, B., Harrison-Restelli, C. and Thaler, H. T. (1997) Neuronal precursors of the adult rat subependymal zone persist into senescence, with no decline in spatial extent or response to BDNF. *Journal of Neurobiology* **32**, 554–566.

Gomez-Ospina, N., Tsuruta, F., Barreto-Chang, O., Hu, L. and Dolmetsch, R. (2006) The C terminus of the L-type voltage-gated calcium channel Ca(V)1.2 encodes a transcription factor. *Cell* **127**, 591–606.

Gonzalez, M., Ruggiero, F. P., Chang, Q., Shi, Y. J., Rich, M. M., Kraner, S. and Balice-Gordon, R. J. (1999) Disruption of Trkb-mediated signaling induces disassembly of postsynaptic receptor clusters at neuromuscular junctions. *Neuron* 24, 567–583.

Good, M., Day, M. and Muir, J. L. (1999) Cyclical changes in endogenous levels of oestrogen modulate the induction of LTD and LTP in the hippocampal CA1 region. *European Journal of Neuroscience* 11, 4476–4480.

Goodwin, F. K. and Jamison, K. R. (1990) *Manic-depressive Illness*. Oxford University Press.

Goosens, K. A. and Maren, S. (2001) Contextual and auditory fear conditioning are mediated by the lateral, basal, and central amygdaloid nuclei in rats. *Learning & Memory* 8, 148–155.

Goosens, K. A. and Maren, S. (2004) NMDA receptors are essential for the acquisition, but not expression, of conditional fear and associative spike firing in the lateral amygdala. *European Journal of Neuroscience* 20, 537–548.

Gorski, R. A., Gordon, J. H., Shryne, J. E. and Southam, A. M. (1978) Evidence for a morphological sex difference within the medial preoptic area of the rat brain. *Brain Research* 148, 333–346.

Gothard, K. M., Skaggs, W. E., Moore, K. M., and McNaughton, B. L. (1996) Binding of hippocampal CA1 neural activity to multiple reference frames in a landmark-based navigation task. *Journal of Neuroscience* 16, 823–835.

Gould, E. and Gross, C. G. (2002) Neurogenesis in adult mammals: some progress and problems. *Journal of Neuroscience* 22, 619–623.

Gould, E. and Tanapat, P. (1999) Stress and hippocampal neurogenesis. *Biological Psychiatry* 46, 1472–1479.

Gould, T. D. and Manji, H. K. (2002) Signaling networks in the pathophysiology and treatment of mood disorders. *Journal of Psychosometric Research* 53, 687–697.

Gould, E., Woolley, C. S. and McEwen, B. S. (1991) The hippocampal formation: morphological changes induced by thyroid, gonadal, and adrenal hormones. *Psychoneuroendocrinology* 16, 67–84.

Gould, T. D., Chen, G. and Manji, H. K. (2004) In vivo evidence in the brain for lithium inhibition of glycogen synthase kinase-3. *Neuropsychopharmacology* 29, 32–38.

Gould, E., Tanapat, P., Hastings, N. B. and Shors, T. J. (1999a) Neurogenesis in adulthood: a possible role in learning. *Trends in Cognitive Sciences* 3, 186–192.

Gould, E., Vail, N., Wagers, M. and Gross, C. G. (2001) Adult-generated hippocampal and neocortical neurons in macaques have a transient existence. *Proceedings of the National Academy of Sciences of the United States of America* 98, 10,910–10,917.

Gould, E., Westlind-Danielsson, A., Frankfurt, M. and McEwen, B. S. (1990a) Sex differences and thyroid hormone sensitivity of hippocampal pyramidal cells. *Journal of Neuroscience* 10, 996–1003.

Gould, E., Woolley, C. S., Frankfurt, M. and McEwen, B. S. (1990b) Gonadal steroids regulate dendritic spine density in hippocampal pyramidal cells in adulthood. *Journal of Neuroscience* 10, 1286–1291.

Gould, E., Reeves, A. J., Graziano, M. S. and Gross, C. G. (1999b) Neurogenesis in the neocortex of adult primates. *Science* 286, 548–552.

Gould, E., Beylin, A., Tanapat, P., Reeves, A. and Shors, T. J. (1999a) Learning enhances adult neurogenesis in the hippocampal formation. *Nature Neuroscience* 2, 260–265.

Gould, E., McEwen, B. S., Tanapat, P., Galea, L. A. and Fuchs, E. (1997) Neurogenesis in the dentate gyrus of the adult tree shrew is regulated by psychosocial stress and NMDA receptor activation. *Journal of Neuroscience* 17, 2492–2498.

Gould, E., Tanapat, P., McEwen, B. S., Flugge, G. and Fuchs, E. (1998) Proliferation of granule cell precursors in the dentate gyrus of adult monkeys is diminished by stress. *Proceedings of the National Academy of Sciences of the United States of America* 95, 3168–3171.

Gould, T. D., Quiroz, J. A., Singh, J., Zarate, C. A. and Manji, H. K. (2004) Emerging experimental therapeutics for bipolar disorder: insights from the molecular and cellular actions of current mood stabilizers. *Molecular Psychiatry* 9, 734–755.

Graef, I. A., Wang, F., Charron, F., Chen, L., Neilson, J., Tessier-Lavigne, M. and Crabtree, G. R. (2003) Neurotrophins and netrins require calcineurin/NFAT signaling to stimulate outgrowth of embryonic axons. *Cell* 113, 657–670.

Graef, I. A., Mermelstein, P. G., Stankunas, K., Neilson, J. R., Deisseroth, K., Tsien, R. W. and Crabtree, G. R. (1999) L-type calcium channels and GSK-3 regulate the activity of NF-ATc4 in hippocampal neurons. *Nature* 401, 703–708.

GrandPré, T., Nakamura, F., Vartanian, T., and Strittmatter S. M. (2000) Identification of the Nogo inhibitor of axon regeneration as a Reticulon protein. *Nature* 403, 439–444.

Gray, N., Du, J., Falke, C., Yuan, P. and Manji, H. (2003) Lithium regulates total and synaptic expression of the AMPA glutamate receptor GluR2 in vitro and in vivo. *Annals of the New York Academy of Sciences* **1003**, 402–404.

Greba, Q. and Kokkonidis, L. (2000) Peripheral and intraamygdalar administration of the dopamine D1 receptor antagonist SCH 23390 blocks fear-potentiated startle but not shock reactivity or the shock sensitization of acoustic startle. *Behavioral Neuroscience* **114**, 262–272.

Greba, Q., Grifkins, A. and Kokkonidis, L. (2001) Inhibition of amygdaloid dopamine D2 receptors impairs emotional learning measured with fear-potentiated startle. *Brain Research* **899**, 218–226.

Grecksch, G. and Matthies, H. (1980) Two sensitive periods for the amnesic effect of anisomycin. *Pharmacology, Biochemistry, and Behavior* **12**, 663–665.

Green, E. J., Greenough, W. T. and Schlumpf, B. E. (1983) Effects of complex or isolated environments on cortical dendrites of middle-aged rats. *Brain Research* **264**, 233–240.

Greenamyre, J. T. (2001) Glutamatergic influences on the basal ganglia, *Clinical Neuropharmacology* **24**, 65–70.

Greenamyre, J. T., Sherer, T. B., Betarbet, R. and Panov, A. V. (2001) Complex I and Parkinson's disease. *IUBMB Life* **52**, 135–141.

Greenough, W. T. (1975) Experiential modification of the developing brain. *American Scientist* **63**, 37–46.

Greenough, W. T. and Chang, F. L. (1988) Plasticity of synapse structure and pattern in the cerebral cortex. In Peters, A. and Jones, E. G. (eds) *Cerebral Cortex: Development and Maturation of Cerebral Cortex*. Plenum Publishing Corporation, pp. 391–440.

Greenough, W. T., Larson, J. R. and Withers, G. S. (1985) Effects of unilateral and bilateral training in a reaching task on dendritic branching of neurons in the rat motor-sensory forelimb cortex. *Behavioral and Neural Biology* **44**, 301–314.

Greenough, W. T., Volkmar, F. R. and Juraska, J. M. (1973) Effects of rearing complexity on dendritic branching in frontolateral and temporal cortex of the rat. *Experimental Neurology* **41**, 371–378.

Greenough, W. T., West, R. W. and DeVoogd, T. J. (1978) Subsynaptic plate perforations: changes with age and experience in the rat. *Science* **202**, 1096–1098.

Greenough, W. T., McDonald, J. W., Parnisari, R. M. and Camel, J. E. (1986) Environmental conditions modulate degeneration and new dendrite growth in cerebellum of senescent rats. *Brain Research* **380**, 136–143.

Grillon, C., Cordova, J., Morgan, C. A. IIIrd, Charney, D. S. and Davis, M. (2004) Effects of the beta-blocker propranolol on cued and contextual fear conditioning in humans. *Psychopharmacology (Berl)* **175**, 342–352.

Grollman, A. P. (1967) Inhibitors of protein biosynthesis. II. Mode of action of anisomycin. *Journal of Biological Chemistry* **242**, 3226–3233.

Grosche, J., Kettenmann, H. and Reichenbach, A. (2002) Bergmann glial cells form distinct morphological structures to interact with cerebellar neurons. *Journal of Neuroscience Research* **68**, 138–149.

Grosche, J., Matyash, V., Moller, T., Verkhratsky, A., Reichenbach, A. and Kettenmann, H. (1999) Microdomains for neuron-glia interaction: parallel fiber signaling to Bergmann glial cells. *Nature Neuroscience* **2**, 139–143.

Grossman, S. D., Wolfe, B. B., Yasuda, R. P. and Wrathall, J. R. (1999) Alterations in AMPA receptor subunit expression after experimental spinal cord contusion injury. *Journal of Neuroscience* **19**, 5711–5720.

Grossman, A. W., Churchill, J. D., Bates, K. E., Kleim, J. A. and Greenough, W. T. (2002) A brain adaptation view of plasticity: is synaptic plasticity an overly limited concept? *Progress in Brain Research* **138**, 91–108.

Grossman, A. W., Churchill, J. D., McKinney, B. C., Kodish, I. M., Otte, S. L. and Greenough, W. T. (2003) Experience effects on brain development: possible contributions to psychopathology. *Journal of Child Psychology and Psychiatry and Allied Disciplines* **44**, 33–63.

Groth, R. D. and Mermelstein, P. G. (2003) Brain-derived neurotrophic factor activation of NFAT (nuclear factor of activated T-cells)-dependent transcription: a role for the transcription factor NFATc4 in neurotrophin-mediated gene expression. *Journal of Neuroscience* **23**, 8125–8134.

Grover, L. M. and Teyler, T. J. (1990) Two components of long-term potentiation induced by different patterns of afferent activation. *Nature* **347**, 477–479.

Gruest, N., Richer, P. and Hars, B. (2004) Memory consolidation and reconsolidation in the rat pup require protein synthesis. *Journal of Neuroscience* **24**, 10,488–10,492.

Gruest, N., Richer, P. and Hars, H. (2004) Memory consolidation and reconsolidation in the rat pup require protein synthesis. *Journal of Neuroscience* **24**, 10,488–10,492.

Gu, Q. and Moss, R. L. (1996) 17B-estradiol potentiates kainite-induced currents via activation of the camp cascade. *Journal of Neuroscience* **16**, 3620–3629.

Guarraci, F. A., Frohardt, R. J. and Kapp, B. S. (1999) Amygdaloid D1 dopamine receptor involvement in Pavlovian fear conditioning. *Brain Research* **827**, 28–40.

Guarraci, F. A., Frohardt, R. J., Falls, W. A. and Kapp, B. S. (2000) The effects of intra-amygdaloid infusions of a D2 dopamine receptor antagonist on Pavlovian fear conditioning. *Behavioral Neuroscience* **114**, 647–651.

Gulinello, M., Gong, Q. H. and Smith, S. S. (2002) Progesterone withdrawal increases the alpha4 subunit of the GABA(A) receptor in male rats in association with anxiety and altered pharmacology-a comparison with female rats. *Neuropharmacology* **43**, 701–714.

Gubellini, P., Pisani, A., Centonze, D., Bernardi, G. and Calabresi, P. (2004) Metabotropic glutamate receptors and striatal synaptic plasticity: implications for neurological diseases. *Progress in Neurobiology* **74**, 271–300.

Guthrie, K. M. and Gall, C. M. (1991) Differential expression of mRNAs for the NGF family of neurotrophic factors in the adult rat central olfactory system. *Journal of Comparative Neurology* **313**, 95–102.

Gutlerner, J. L., Penick, E. C., Snyder, E. M. and Kauer, J. A. (2002) Novel protein kinase A-dependent long-term depression of excitatory synapses. *Neuron* **36**, 921–931.

Guzowski, J. F. (2002) Insights into immediate-early gene function in hippocampal memory consolidation using antisense oligonucleotide and fluorescent imaging approaches. *Hippocampus* **12**, 86–104.

Guzowski, J. F. and McGaugh, J. L. (1997) Antisense oligodeoxynucleotide-mediated disruption of hippocampal cAMP response element binding protein levels impairs consolidation of memory for water maze training. *Proceedings of the National Academy of Sciences of the United States of America* **94**, 2693–2698.

Gyllensten, L. and Malmfors, T. (1963) Myelinization of the optic nerve and its dependence on visual function – a quantitative investigation in mice. *Journal of Embryology and Experimental Morphology* **11**, 255–266.

Ha, B. K., Vicini, S., Rogers, R. C., Bresnahan, J. C., Burry, R. W. and Beattie, M. S. (2002) Kainate-induced excitotoxicity is dependent upon extracellular potassium concentrations that regulate the activity of AMPA/KA type glutamate receptors. *Journal of Neurochemistry* **83**, 934–945.

Haddad, J. J. (2002) Antioxidant and prooxidant mechanisms in the regulation of redox(y)-sensitive transcription factors. *Cellular Signalling* **14**, 879–897.

Hahn, C. G. and Friedman, E. (1999) Abnormalities in protein kinase C signaling and the pathophysiology of bipolar disorder. *Bipolar Disord* **1**, 81–86.

Hakak, Y., Walker, J. R., Li, C., Wong, W. H., Davis, K. L., Buxbaum, J. D., Haroutunian, V. and Fienberg, A. A. (2001) Genome-wide expression analysis reveals dysregulation of myeolination-related genes in chronic schizophrenia. *Proceedings of the National Academy of Sciences of the United States of America* **98**, 4747–4751.

Hallcher, L. M. and Sherman, W. R. (1980) The effects of lithium ion and other agents on the activity of myo-inositol-1-phosphatase from bovine brain. *Journal of Biological Chemistry* **255**, 10,896–10,901.

Hama, H., Hara, C., Yamaguchi, K. and Miyawaki, A. (2004) PKC signaling mediates global enhancement of excitatory synaptogenesis in neurons triggered by local contact with astrocytes. *Neuron* **41**, 405–415.

Hamm, R. J. (1981) Hypothermia-induced retrograde amnesia in mature and aged rats. *Developmental Psychobiology* **14**, 357–364.

Hampson, R. E., Heyser, C. J. and Deadwyler, S. A. (1993) Hippocampal cell firing correlates of delayed-match-to-sample performance in the rat. *Behavioral Neuroscience* **107**, 715–739.

Hampson, R. E., Pons, T. P., Stanford, T. R., Deadwyler, S. A. (2004) Categorization in the monkey hippocampus: a possible mechanism for encoding information into memory. *Proceedings of the National Academy of Sciences of the United States of America* **101**, 3184–3189.

Handa, R. J., Kerr, J. E., DonCarlos, L. L., McGivern, R. F. and Hejna, G. (1996) Hormonal regulation of androgen receptor messenger RNA in the medial preoptic area of the male rat. *Molecular Brain Research* **39**, 57–67.

Hansel, C., Linden, D. J. and D'Angelo, E. (2001) Beyond parallel fiber LTD: the diversity of synaptic and non-synaptic plasticity in the cerebellum. *Nature Neuroscience* **4**, 467–475.

Hara, Y., Shiga, T., Abe, I., Tsujino, A., Ichimura, H., Okado, N. and Ochiai, N. (2003) P0 mRNA expression increases during gradual nerve elongation in adult rats. *Experimental Neurology* **103**, 428–435.

Hardingham, G. E., Fukunaga, Y. and Bading, H. (2002). Extrasynaptic NMDARs oppose synaptic NMDARs by triggering CREB shut-off and cell death pathways. *Nature Neuroscience* 5, 405–414.

Hardingham, G. I., Arnold, F. J. and Bading, H. (2001) Nuclear calcium signaling controls CREB-mediated gene expression triggered by synaptic activity. *Nature Neuroscience* 4, 261–267.

Hardingham, G. E., Chawla, S., Johnson, C. M. and Bading, H. (1997) Distinct functions of nuclear and cytoplasmic calcium in the control of gene expression. *Nature* 385, 260–265.

Harik, S. I., Hritz, M. A. and LaManna, J. C. (1995) Hypoxia-induced brain angiogenesis in the adult rat. *Journal of Physiology* 485 (Pt 2), 525–530.

Harley, C. (1991) Noradrenergic and locus coeruleus modulation of the perforant path-evoked potential in rat dentate gyrus supports a role for the locus coeruleus in attentional and memorial processes. *Progress in Brain Research* 88, 307–321.

Harley, C. W., Malsbury, C. W., Squires, A. and Brown, R. A. M. (2000) Testosterone decreases CA1 plasticity in vivo in gonadectomized male rats. *Hippocampus* 10, 693–697.

Haroutunian, V., Katsel, P., Dracheva, S. and Davis, K. L. (2006) The human homolog of the QKI gene affected in teh severe dysmyelinatin "quaking" mouse phenotype: downregulated in multiple brain regions in schizophrenia. *American Journal of Psychiatry* 163, 1834–1837.

Harrison, N. L. and Gibbons, S. J. (1994) Zn^{2+}: an endogenous modulator of ligand- and voltage-gated ion channels. *Neuropharmacology* 33, 935–952.

Hartell, N. A. (2002) Parallel fiber plasticity. *Cerebellum* 1, 3–18.

Hartmann, M., Heumann, R. and Lessmann, V. (2001) Synaptic secretion of BDNF after high-frequency stimulation of glutamatergic synapses. *EMBO Journal* 20, 5887–5897.

Hashimoto, A. and Oka, T. (1997) Free D-aspartate and D-serine in the mammalian brain and periphery. *Progress in Neurobiology* 52, 325–353.

Hashimoto, A., Oka, T. and Nishikawa, T. (1995) Extracellular concentration of endogenous free D-serine in the rat brain as revealed by in vivo microdialysis. *Neuroscience* 66, 635–643.

Hashimoto, A., Nishikawa, T., Oka, T. and Takahashi, K. (1993a) Endogenous D-serine in rat brain: N-methyl-D-aspartate receptor-related distribution and aging. *Journal of Neurochemistry* 60, 783–786.

Hashimoto, A., Nishikawa, T., Oka, T., Takahashi, K. and Hayashi, T. (1992a) Determination of free amino acid enantiomers in rat brain and serum by high-performance liquid chromatography after derivatization with N-tert-butyloxycarbonyl-L-cysteine and o-phthaldialdehyde. *Journal of Chromatography* 582, 41–48.

Hashimoto, A., Nishikawa, T., Hayashi, T., Fujii, N., Harada, K., Oka, T. and Takahashi, K. (1992b) The presence of free D-serine in rat brain. *FEBS Letters* 296, 33–36.

Hashimoto, A., Nishikawa, T., Konno, R., Niwa, A., Yasumura, Y., Oka, T. and Takahashi, K. (1993b) Free D-serine, D-aspartate and D-alanine in central nervous system and serum in mutant mice lacking D-amino acid oxidase. *Neuroscience Letters* 152, 33–36.

Hashimoto, A., Kumashiro, S., Nizhikawa, T., Oka, T., Takahashi, K., Mito, T., Takashima, S., Doi, N., Mizutani, Y. and Yamazaki, T. (1993c) Embryonic development and postnatal changes in free D-aspartate and D-serine in the human prefrontal cortex. *Journal of Neurochemistry* 61, 348–351.

Hasselmo, M. E. (1999) Neuromodulation: acetylcholine and memory consolidation. *Trends in Cognitive Science* 3, 351–359.

Hasselmo, M. E. and Eichenbaum H. B. (2005) Hippocampal mechanisms for the context-dependent retrieval of episodes. *Neural Networks* 18, 1172–1190. doi:10.1016/j.neunet.2005.08.007.

Hassinger, T. D., Atkinson, P. B., Strecker, G. J., Whalen, L. R., Dudek, F. E., Kossel, A. H. and Kater, S. B. (1995) Evidence for glutamate-mediated activation of hippocampal neurons by glial calcium waves. *Journal of Neurobiology* 28, 159–170.

Haug, H. (1987) Brain sizes, surfaces, and neuronal sizes of the cortex cerebri: a stereological investigation of man and his variability and a comparison with some mammals (primates, whales, marsupials, insectivores, and one elephant). *American Journal of Anatomy (now Developmental Dynamics)* 180, 126–142.

Hawkins, R. D. (1996) NO honey, I don't remember. *Neuron* 16, 465–467.

Hawkins, R. D., Kandel, E. R. and Siegelbaum, S. A. (1993) Learning to modulate transmitter release: themes and variations in synaptic plasticity. *Annual Review of Neuroscience* 16, 635–665.

Hawkins, R. D., Son, H. and Arancio, O. (1998) Nitric oxide as a retrograde messenger during long-term potentiation in hippocampus. *Progress in Brain Research* **118**, 155–172.

Hayashi, Y., Shi, S. H., Esteban, J. A., Piccini, A., Poncer, J. C. and Malinow, R. (2000) Driving AMPA receptors into synapses by LTP and CaMKII: requirement for GluR1 and PDZ domain interaction. *Science* **287**, 2262–2267.

Hayashi, Y., Momiyama, A., Takahashi, T., Ohishi, H., Ogawa-Meguro, R., Shigemoto, R., Mizuno, N. and Nakanishi, S. (1993) Role of a metabotropic glutamate receptor in synaptic modulation in the accessory olfactory bulb. *Nature* **366**, 687–690.

Haydon, P. G. (2001) Glia: listening and talking to the synapse. *Nature Review Neuroscience* **2**, 185–193.

Head, J. R. and Griffin, W. S. (1985) Functional capacity of solid tissue transplants in the brain: evidence for immunological privilege. *Proceedings of the Royal Society of London: B Biological Sciences* **224**, 375–387.

Healy, S. D., Braham, S. R. and Braithwaite, V. A. (1999) Spatial working memory in rats: no differences between the sexes. *Proceedings of the Royal Society of London B: Biological Sciences* **266**, 2303–2308.

Hebb, D. O. (1949) *The Organization of Behavior*. John Wiley & Sons.

Heckers, S., Zalezak, M., Weiss, A. P., Ditman, T. and Titone, D. (2004) Hippocampal activation during transitive inference in humans. *Hippocampus* **14**, 153–162.

Heikkinen, T., Puolivali, J., Liu, L., Rissanen, A. and Heikki, T. (2002) Effects of ovariectomy and estrogen treatment on learning and hippocampal neurotransmitters in mice. *Hormones and Behavior* **41**, 22–32.

Heim, C., Pardowitz, I., Sieklucka, M., Kolasiewicz, W., Sontag, T. and Sontag, K. H. (2002) The analysis system COGITAT for the study of cognitive deficiencies in rodents. *Behavior Research Methods, Instruments, and Computers: a Journal of the Psychonomic Society, Inc.* **32**, 140–156.

Hempstead, B. L. and Salzer, J. L. (2002) Neurobiology. A glial spin on neurotrophins. *Science* **298**, 1184–1186.

Henderson, V. W. (1997) Estrogen replacement therapy for the prevention and treatment of Alzheimer's disease. *CNS Drugs* **8**, 343–351.

Henderson, V. W. (2000) *Hormone Therapy and the Brain: A Clinical Perspective on the Role of Estrogen*. Parthenon Publishing.

Henderson, V. W., Watt, L. and Buckwalter, J. G. (1996) Cognitive skills associated with estrogen replacement in women with Alzheimer's disease. *Psychoneuroendocrinology* **21**, 421–430.

Henderson, V. W., Benke, K. S., Green, R. C., Cupples, L. A., Farrer, L. A. and Group, M. S. (2005) Postmenopausal hormone therapy and Alzheimer's disease risk: interaction with age. *Journal of Neurology, Neurosurgery and Neuropsychiatry* **76**, 103–105.

Henke, K., Buck, A., Weber, B. and Wieser, H. G. (1997) Human hippocampus establishes associations in memory. *Hippocampus* **7**, 249–256.

Heresco-Levy, U. and Javitt, D. C. (1998) The role of N-methyl-D-aspartate (NMDA) receptor-mediated neurotransmission in the pathophysiology and therapeutics of psychiatric syndromes. *European Neuropsychopharmacology* **8**, 141–152.

Hering, H. and Sheng, M. (2001) Dendritic spines: structure, dynamics and regulation. *Nature Reviews Neuroscience* **2**, 880–888.

Hermann, G. E., Rogers, R. C., Bresnahan, J. C. and Beattie, M. S. (2001) Tumor necrosis factor-alpha induces cFOS and strongly potentiates glutamate-mediated cell death in the rat spinal cord. *Neurobiological Diseases* **8**, 590–599.

Hermans, E. and Challiss, R. A. (2001) Structural, signalling and regulatory properties of the group I metabotropic glutamate receptors: prototypic family C G-protein-coupled receptors. *Biochemical Journal* **359**, 465–484.

Hernandez, P. J. and Kelley, A. E. (2004) Long-term memory for instrumental responses does not undergo protein synthesis-dependent reconsolidation upon retrieval. *Learning & Memory* **11**, 748–754.

Hernandez, R. V., Garza, J. M., Graves, M. E., Martinez, J. L. Jr. and LeBaron, R. G. (2001) The process of reducing CA1 long-term potentiation by the integrin peptide, GRGDSP, occurs within the first few minutes following theta-burst stimulation. *Biological Bulletin* **201**, 236–237.

Hernandez, R. V., Navarro, M. M., Rodriguez, W. A., Martinez, J. L. Jr and LeBaron, R. G. (2005) Differences in the magnitude of long-term potentiation produced by theta burst and high frequency stimulation protocols matched in stimulus number. *Brain Research Protocols* **15**, 6–13.

Herron, C. E., Lester, R. A., Coan, E. J. and Collingridge, G. L. (1986) Frequency-dependent involvement of NMDA receptors in the hippocampus: a novel synaptic mechanism. *Nature* **322**, 265–268.

Hidalgo C., Donoso P. and Carrasco M. A. (2005) The ryanodine receptors Ca^{2+} release channels: cellular redox sensors? *IUBMB Life* **57**, 315–322.

Himi, T., Ikeda, M., Yasuhara, T., Nishida, M. and Morita, I. (2003) Role of neuronal glutamate transporter in the cysteine uptake and intracellular glutathione levels in cultured cortical neurons. *Journal of Neural Transmission* **110**, 1337–1348.

Hirbec, H., Francis, J. C., Lauri, S. E., Braithwaite, S. P., Coussen, F., Mulle, C., Dev, K. K., Coutinho, V., Meyer, G., Isaac, J. T., Collingridge, G. L., Henley, J. M. and Couthino, V. (2003) Rapid and differential regulation of AMPA and kainate receptors at hippocampal mossy fibre synapses by PICK1 and GRIP. *Neuron* **37**, 625–638.

Hitchcock, J. and Davis, M. (1986) Lesions of the amygdala but not of the cerebellum or red nucleus block conditioned fear as measured with the potentiated startle paradigm. *Behavioral Neuroscience* **100**, 11–22.

Hof, P. R., Haroutunian, V., Friedrich, V. L. Jr., Byne, W., Buitron, C., Perl, D. P. and Davis K. L. (2003) Loss and altered spatial distribution of oligodendrocytes in the superior frontal gyrus in schizophrenia. *Biological Psychiatry* **53**, 1075–1085.

Hoffman, D. A. and Johnston, D. (1998) Downregulation of transient K^+ channels in dendrites of hippocampal CA1 pyramidal neurons by activation of PKA and PKC. *Journal of Neuroscience* **18**, 3521–3528.

Hofmann, B., Hecht, H. J. and Flohe, L. (2002) Peroxiredoxins. *Biological Chemistry* **383**, 347–364.

Hogan, P. G., Chen, L., Nardone, J. and Rao, A. (2003) Transcriptional regulation by calcium, calcineurin, and NFAT. *Genes and Development* **17**, 2205–2232.

Hogervorst, E., Williams, J., Budge, M., Barnetson, L., Combrinck, M. and Smith, A. D. (2001) Serum total testosterone is lower in men with Alzheimer's disease. *Neuroendocrinology Letters* **22**, 163–168.

Hollmann, M. and Heinemann, S. (1994) Cloned glutamate receptors. *Annuals Reviews in Neuroscience* **17**, 31–108.

Hollup, S. A., Molden, S., Donnett, J. G., Moser, M. B. and Moser, E. I. (2001) Accumulation of hippocampal place fields at the goal location in an annular watermaze task. *Journal of Neuroscience* **21**, 1635–1644.

Holscher, C. and Rose, S. P. R. (1992) An inhibitor of nitric oxide synthase prevents memory formation in the chick. *Neuroscience Letters* **145**, 165–167.

Holscher, C. and Rose, S. P. R. (1993) Inhibiting synthesis of the putative retrograde messenger nitric oxide results in amnesia in a passive avoidance task in the chick. *Brain Research* **619**, 189–194.

Holscher, C. and Rose, S. P. R. (1994) Inhibitors of phospholipase A2 produce amnesia for a passive avoidance task in the chick. *Behavioral & Neuralogical Biology* **61**, 225–232.

Holscher, C., Anwyl, R. and Rowan, M. J. (1997) Stimulation on the positive phase of hippocampal theta rhythm induces long-term potentiation that can be depotentiated by stimulation on the negative phase in area CA1 in vivo. *Journal of Neuroscience* **17**, 6470–6477.

Honey, R. C., Eatt, A. and Good, M. (1998) Hippocampal lesions disrupt an associative mismatch process. *Journal of Neuroscience* **18**, 2226–2230.

Hotsenpiller, G., Giorgetti, M. and Wolf, M. E. (2001) Alterations in behaviour and glutamate transmission following presentation of stimuli previously associated with cocaine exposure. *European Journal of Neuroscience* **14**, 1843–1855.

Houlgatte, R., Mallat, M., Brachet, P. and Prochiantz, A. (1989) Secretion of nerve growth factor in cultures of glial cells and neurons derived from different regions of the mouse brain. *Journal of Neuroscience Research* **24**, 143–152.

Hsieh, S.-T., Kidd, G. J., Crawford, T. O., Xu, Z., Lin, W.-M., Trapp, B. D., Cleveland, D. W. and Griffin, J. W. (1994) Regional modulation of neurofilament organization by myelination in normal axons. *Journal of Neuroscience* **14**, 6392–6401.

Hu, P., Stylos-Allen, M. and Walker, M. P. (2006) Sleep facilitates consolidation of emotionally arousing declarative memory. *Psychological Science* **17**, 891–898.

Huang, E. J. and Reichardt, L. F. (2001) Neurotrophins: roles in neuronal development and function. *Annual Reviews of Neuroscience* **24**, 677–736.

Huang, Y. Y. and Kandel, E. R. (1998) Postsynaptic induction and PKA-dependent expression of LTP in the lateral amygdala. *Neuron* **21**, 169–178.

Huang, Y. Y., Martin, K. C. and Kandel, E. R. (2000) Both protein kinase A and mitogen-activated protein kinase are required in the amygdala for the macromolecular synthesis-dependent late phase of long-term potentiation. *Journal of Neuroscience* **20**, 6317–6325.

Huang, Z. J., Kirkwood, A., Pizzorusso, T., Porciatti, V., Morales, B., Bear, M. F., Maffei, L. and Tonegawa, S. (1999) BDNF regulates the maturation of inhibition and the critical period of plasticity in mouse visual cortex. *Cell* **98**, 739–755.

Hubel, D. H. and Wiesel, T. N. (1962) Receptive fields, binocular interaction and functional architecture in the cat's visual system. *Journal of Physiology* **160**, 106–154.

Huber, K. M., Roder, J. C. and Bear, M. F. (2001) Chemical induction of mGluR5- and protein synthesis-dependent long-term depression in hippocampal area CA1. *Journal of Neurophysiology* **86**, 321–325.

Huber, K. M., Gallagher, S. M., Warren, S. T. and Bear, M. F. (2002) Altered synaptic plasticity in a mouse model of fragile X mental retardation. *Proceedings of the National Academy of Sciences of the United States of America* **99**, 7746–7750.

Huber, R., Ghilardi, M. F., Massimini, M. and Tononi, G. (2004) Local sleep and learning. *Nature* **430**, 78–81.

Huerta, P. T. and Lisman, J. E. (1995) Bidirectional synaptic plasticity induced by a single burst during cholinergic theta oscillation in CA1 in vitro. *Neuron* **15**, 1053–1063.

Hugin-Flores, M. E., Steimer, T., Aubert, M. L. and Schulz, P. (2004) Mineralo- and glucocorticoid receptor mrnas are differently regulated by corticosterone in the rat hippocampus and anterior pituitary. *Neuroendocrinology* **79**, 174–184.

Huh, G. S., Boulanger, L. M., Du, H., Riquelme, P. A., Brotz, T. M. and Shatz, C. J. (2000) Functional requirement for class I MHC in CNS development and plasticity. *Science* **290**, 2155–2159.

Hyde, T. M., Ziegler, J. C. and Weinberger, D. R. (1992) Psychiatric disturbances in metachromatic leukodystrophy. *Insights into the neurobiology of psychosis. Archives in Neurology* **489**, 401–406.

Hyman, S. E. and Malenka, R. C. (2001) Addiction and the brain: the neurobiology of compulsion and its persistence *Nature Reviews Neuroscience* **2**, 695–703.

Hymovitch, B. (1952) The effects of experimental variations on problem solving in the rat. *Journal of Comparative and Physiological Psychology* **45**, 313–321.

Hyslop, P. A., Zhang, Z., Pearson, D. V., Phebus, L. A. (1995) Measurement of striatal H_2O_2 by microdialysis following global forebrain ischemia and reperfusion in the rat: correlation with the cytotoxic potential of H_2O_2 in vitro. *Brain Research* **671**, 181–186.

Ibanez, C., Guennoun, R., Liere, P., Eychenne, B., Pianos, A., El-Etr, M., Baulieu, E. E. and Schumacher, M. (2003) Developmental expression of genes involved in neurosteroidogenesis: 3beta-hydroxysteroid dehydrogenase/delta5-delta4 isomerase in the rat brain. *Endocrinology* **144**, 2902–2911.

Ichinari, K., Kakei, M., Matsuoka, T., Nakashima, H. and Tanaka, H. (1996) Direct activation of the ATP-sensitive potassium channel by oxygen free radicals in guinea-pig ventricular cells: its potentiation by MgADP. *Journal of Molecular and Cellular Cardiology* **28**, 1867–1877.

Ickes, B. R., Pham, T. M., Sanders, L. A., Albeck, D. S., Mohammed, A. H. and Granholm, A. C. (2000) Long-term environmental enrichment leads to regional increases in neurotrophin levels in rat brain. *Experimental Neurology* **164**, 45–52.

Inagaki, N., Gonoi, T., Clement, J. P., Wang, C. Z., Aguilar-Bryan, L., Bryan, J. and Seino, S. (1996) A family of sulfonylurea receptors determines the pharmacological properties of ATP-sensitive K^+ channels. *Neuron* **16**, 1011–1017.

Inda, M. C., Delgado-Garcia, J. M. and Carrion, A. M. (2005) Acquisition, consolidation, reconsolidation, and extinction of eyelid conditioning responses require de novo protein synthesis. *Journal of Neuroscience* **25**, 2070–2080.

Infanger, D. W., Sharma, R. V. and Davisson, R. L. (2006) NADPH oxidases of the brain: distribution, regulation, and function. *Antioxidants and Redox Signaling* **8**, 1583–1596.

Inglis, F. M. and Moghaddam, B. (1999) Dopaminergic innervation of the amygdala is highly responsive to stress. *Journal of Neurochemistry* **72**, 1088–1094.

Innocenti, B., Parpura, V. and Haydon, P. G. (2000) Imaging extracellular waves of glutamate during calcium signaling in cultured astrocytes. *Journal of Neuroscience* **20**, 1800–1808.

Iourgenko, V., Zhang, W., Mickanin, C., Daly, I., Jiang, C., Hexham, J. M., Orth, A. P., Miraglia, L., Meltzer, J., Garza, D., Chirn, G. W., McWhinnie, E., Cohen, D., Skelton, J., Terry, R., Yu, Y., Bodian, D., Buxton, F. P., Zhu, J., Song, C. and Labow, M. A. (2003) Identification of a family of cAMP response element-binding protein coactivators by genome-scale functional analysis in mammalian cells. *Proceedings of the National Academy of Sciences of the United States of America* **100**, 12,147–12,152.

Isaac, J. T., Nicoll, R. A. and Malenka, R. C. (1995) Evidence for silent synapses: implications for the expression of LTP. *Neuron* 15, 427–434.

Isaacs, K. R., Anderson, B. J., Alcantara, A. A., Black, J. E. and Greenough, W. T. (1992) Exercise and the brain: angiogenesis in the adult rat cerebellum after vigorous physical activity and motor skill learning. *Journal of Cerebral Blood Flow and Metabolism* 12, 110–119.

Isgor, C. and Sengelaub, D. R. (1998) Prenatal gonadal steroids affect adult spatial behavior, CA1 and CA3 pyramidal cell morphology in rats. *Hormones and Behavior* 34, 183–198.

Isgor, C. and Sengelaub, D. R. (2003) Effects of neonatal gonadal steroids on adult CA3 pyramidal neuron dendritic morphology and spatial memory in rats. *Journal of Neurobiology* 55, 179–190.

Ishibashi, T., Dakin, K. A., Stevens, B., Lee, P. R., Kozlov, S. V., Stewart, C. L. and Fields, R. D. (2006) Astrocytes promote myelination in response to electrical impulses. *Neuron* 49, 823–832.

Ishii, T. and Mombaerts, P. (2008) Expression of non-classical class I major histocompatibility genes defines a tripartite organization of the mouse vomeronasal system. *Journal of Neuroscience* 28, 2332–2341.

Ishii, T., Hirota, J. and Mombaerts, P. (2003) Combinatorial coexpression of neural and immune multigene families in mouse vomeronasal sensory neurons. *Current Biology* 13, 394–400.

Itoh, K., Ozaki, M., Stevens, B. and Fields, R. D. (1995) Regulated expression of the neural cell adhesion molecule L1 by specific patterns of neural impulses. *Science* 270, 1369–1372.

Itoh, K., Stevens, B., Schachner, M. Fields, R. D. (1997) Activity-dependent regulation of N-cadherin in DRG neurons: differential regulation of N-cadherin, NCAM, and L1 by distinct patterns of action potentials. *Journal of Neurobiology* 33, 735–748.

Izquierdo, I. and Medina, J. H. (1997) Memory formation: the sequence of biochemical events in the hippocampus and its connection to activity in other brain structures. *Neurobiology of Learning and Memory* 68, 285–316.

Izquierdo, L. A., Barros, D. M., Vianna, M. R., Coitinho, A., deDavid, E., Silva, T., Choi, H., Moletta, B., Medina, J. H. and Izquierdo, I. (2002) Molecular pharmacological dissection of short- and long-term memory. *Cellular and Molecular Neurobiology* 22, 269–287.

Jackson, M. R. and Peterson, P. A. (1993) Assembly and intracellular transport of MHC class I molecules. *Annual Review of Cellular Biology* 9, 207–235.

Jacobs, J. M. and Meyer, T. (1997) Control of action potential-induced Ca^{2+} signaling in the soma of hippocampal neurons by Ca^{2+} release from intracellular stores. *Journal of Neuroscience* 17, 4129–4135.

Janeway, C. A., Travers, P., Walport, M., Shlomchik, M. (eds) (2001) *Immunobiology*. Garland Publishing.

Javitt, D. C. and Zukin, S. R. (1991) Recent advances in the phencyclidine model of schizophrenia. *American Journal of Psychiatry* 148, 1301–1308.

Jeftinija, S. D., Jeftinija, K. V., Stefanovic, G. and Liu, F. (1996) Neuroligand-evoked calcium-dependent release of excitatory amino acids from cultured astrocytes. *Journal of Neurochemistry* 66, 676–684.

Jiang, Z. G. and North, R. A. (1991) Membrane properties and synaptic responses of rat striatal neurones *in vitro*. *Journal of Physiology (London)* 443, 533–553.

Jin, P., Alisch, R. S., Warren, S. T. (2004) RNA and microRNAs in fragile X mental retardation. *Nature Cell Biology* 6, 1048–1053.

Jodo, E. and Aston-Jones, G. (1997) Activation of locus coeruleus by prefrontal cortex is mediated by excitatory amino acid inputs. *Brain Research* 768, 327–332.

Johenning, F. W. and Holthoff, K. (2007) Nuclear calcium signals during L-LTP induction do not predict the degree of synaptic potentiation. *Cell Calcium* 41, 271–283.

Johnson, J. W. and Ascher, P. (1987) Glycine potentiates the NMDA response in cultured mouse brain neurons. *Nature* 325, 529–531.

Johnson, R. S., Spiegelman, B. M. and Papaioannou, V. (1992) Pleiotropic effects of a null mutation in the c-fos proto-oncogene. *Cell* 71, 577–586.

Johnson, L. R., Farb, C., Morrison, J. H., McEwen, B. S. and LeDoux, J. E. (2005) Localization of glucocorticoid receptors at postsynaptic membranes in the lateral amygdala. *Neuroscience* 136, 289–299.

Johnston, A. N. B. and Rose, S. P. R. (2001) Memory consolidation in day-old chicks requires BDNF but not NGF or NT-3; an antisense study. *Molecular Brain Research* 88, 26–36.

Johnston, D., Hoffman, D. A., Colbert, C. M. and Magee, J. C. (1999) Regulation of back-propagating action potentials in hippocampal neurons. *Current Opinion in Neurobiology* 9, 288–292.

Joly, E. and Oldstone, M. B. (1992) Neuronal cells are deficient in loading peptides onto MHC class I molecules. *Neuron* **8**, 1185–1190.

Joly, E., Mucke, L. and Oldstone, M. B. (1991) Viral persistence in neurons explained by lack of major histocompatibility class I expression. *Science* **253**, 1283–1285.

Jones, R. S. (1993) Entorhinal-hippocampal connections: a speculative view of their function. *Trends in Neuroscience* **16**, 58–64.

Jones, T. A. (1999) Multiple synapse formation in the motor cortex opposite unilateral sensorimotor cortex lesions in adult rats. *Journal of Comparative Neurology* **414**, 57–66.

Jones, D. G. and Calverley, R. K. (1991) Frequency of occurrence of perforated synapses in developing rat neocortex. *Neuroscience Letters* **129**, 189–192.

Jones, T. A. and Greenough, W. T., (2002) Behavioural experience-dependent plasticity of glial-neuronal interactions. In Volterra, A., Magistretti, P. and Hayden, P. G. (eds) *The Tripartite Synapse: Glia in Synaptic Transmission*. Oxford University Press, pp. 248–265.

Jones, T. A. and Greenough, W. T. (1996) Ultrastructural evidence for increased contact between astrocytes and synapses in rats reared in a complex environment. *Neurobiology of Learning and Memory* **65**, 48–56.

Jones, T. A., Hawrylak, N. and Greenough, W. T. (1996) Rapid laminar-dependent changes in GFAP immunoreactive astrocytes in the visual cortex of rats reared in a complex environment. *Psychoneuroendocrinology* **21**, 189–201.

Jones, T. A., Klintsova, A. Y., Kilman, V. L., Sirevaag, A. M. and Greenough, W. T. (1997) Induction of multiple synapses by experience in the visual cortex of adult rats. *Neurobiology of Learning and Memory* **68**, 13–20.

Jones, M. W., Errington, M. L., French, P. J., Fine, A., Bliss, T. V., Garel, S., Charnay, P., Bozon, B., Laroche, S. and Davis, S. (2001) A requirement for the immediate early gene Zif268 in the expression of late LTP and long-term memories. *Nature Neuroscience* **4**, 289–296.

Jope, R. S., Song, L., Li, P. P., Young, L. T., Kish, S. J., Pacheco, M. A. and Warsh, J. J. (1996) The phosphoinositide signal transduction system is impaired in bipolar affective disorder brain. *Journal of Neurochemistry* **66**, 2402–2409.

Josselyn, S. A., Shi, C., Carlezon, W. A., Jr., Neve, R. L., Nestler, E. J. and Davis, M. (2001) Long-term memory is facilitated by cAMP response element-binding protein overexpression in the amygdala. *Journal of Neuroscience* **21**, 2404–2412.

Jovanovic, J. A., Thomas, P., Kittler, J. T., Smart, T. G. and Moss, S. J. (2004) Brain-derived neurotrophic factor modulates fast synaptic inhibition by regulating GABA (A) receptor phosphorylation, activity, and cell-surface stability. *Journal of Neuroscience* **24**, 522–530.

Ju, W., Morishita, W., Tsui, J., Gaietta, G., Deerinck, T. J., Adams, S. R., Garner, C. C., Tsien, R. Y., Ellisman, M. H. and Malenka, R. C. (2004) Activity-dependent regulation of dendritic synthesis and trafficking of AMPA receptors. *Nature Neuroscience* **7**, 244–253.

Judge, M. E. and Quartermain, D. (1982) Characteristics of retrograde amnesia following reactivation of memory in mice. *Physiology and Behavior* **28**, 585–590.

Jung, S. K., Kauri, L. M., Qian, W. J. and Kennedy, R. T. (2000) Correlated oscillations in glucose consumption, oxygen consumption, and intracellular free Ca^{2+} in single islets of Langerhans. *Journal of Biological Chemistry* **275**, 6642–6650.

Juraska, J. M. and Kopcik, J. R. (1988) Sex and environmental influences on the size and ultrastructure of the rat corpus callosum. *Brain Research* **450**, 1–8.

Juraska, J. M., Fitch, J. M. and Washburne, D. L. (1989) The dendritic morphology of pyramidal neurons in the rat hippocampal CA3 area. II. Effects of gender and the environment. *Brain Research* **479**, 115–119.

Juraska, J. M., Fitch, J. M., Henderson, C. and Rivers, N. (1985) Sex differences in the dendritic branching of dentate granule cells following differential experience. *Brain Research* **333**, 73–80.

Kalra, P. S. and Kalra, S. P. (1977) Circadian properties of serum androgens, progesterone, gonadotropins and luteinizing hormone-releasing hormone in male rats: the effects of hypothalamic deafferentation, castration, and adrenalectomy. *Endocrinology* **101**, 1821–1827.

Kamenetz, F., Tomita, T., Hsieh, H., Seabrook, G., Borchelt, D., Iwatsubo, T., Sisodia, S. and Malinow, R. (2003) APP processing and synaptic function. *Neuron* **37**, 925–937.

Kampen, D. L. and Sherwin, B. B. (1994) Estrogen use and verbal memory in healthy postmenopausal women. *Obstetrics and Gynecology* **83**, 979–983.

Kamsler, A. and Segal, M. (2003) Hydrogen peroxide modulation of synaptic plasticity. *Journal of Neuroscience* **23**, 269–276.

Kamsler, A. and Segal, M. (2004) Hydrogen peroxide as a diffusible signal molecule in synaptic plasticity. *Molecular Neurobiology* **29**, 167–178.

Kandel, E. R. (1997) Genes, synapses, and long-term memory. *Journal of Cellular Physiology* **173**, 124–125.

Kandel, E. R. (2001) The molecular biology of memory storage: a dialogue between genes and synapses. *Science* **294**, 1030–1038.

Kandel, E. R., Klein, M., Castellucci, V. F., Schacher, S. and Goelet, P. (1986) Some principles emerging from the study of short- and long-term memory. *Neuroscience Research* **3**, 498–520.

Kang, H. and Schuman, E. M. (1995) Long-lasting neurotrophin-induced enhancement of synaptic transmission in the adult hippocampus. *Science* **267**, 1658–1662.

Kang, H. and Schuman, E. M. (1996) A requirement for local protein synthesis in neurotrophin-induced hippocampal synaptic plasticity. *Science* **273**, 1402–1406.

Kang, J., Goldman, S. A. and Nedergaard, M. (1998) Astrocyte-mediated potentiation of inhibitory synaptic transmission. *Nature Neuroscience* **1**, 683–692.

Kang, J., Jiang, L., Goldman, S. A. and Nedergaard, M. (1998) Astrocyte-mediated potentiation of inhibitory synaptic transmission. *Nature Neuroscience* **1**, 683–692.

Kanit, L., Taskirna, D., Yilmaz, O. A., Balkan, B., Demirgoren, S., Furedy, J. J. and Pogun, S. (2000) Sexually dimorphic cognitive style in rats emerges after puberty. *Brain Research Bulletin* **52**, 243–248.

Kaplan, M. S. (1981) Neurogenesis in the 3-month-old rat visual cortex. *Journal of Comparative Neurology* **195**, 323–338.

Kaplan, M. S. and Bell, D. H. (1984) Mitotic neuroblasts in the 9-day-old and 11-month-old rodent hippocampus. *Journal of Neuroscience* **4**, 1429–1441.

Kapp, B. S., Whalen, P. J., Supple, W. F. and Pascoe, J. P. (1992) Amygdaloid contributions to conditioned arousal and sensory information processing. In Aggleton, J. P. (ed.) *The Amygdala: Neurobiological Aspects of Emotion, Memory, and Mental Dysfunction*. Wiley–Liss, pp 229–254.

Karandrea, D., Kittas, C. and Kitraki, E. (2002) Forced swimming differentially affects male and female brain corticosteroid receptors. *Neuroendocrinology* **75**, 217–226.

Karashima, A., Nakamura, K., Horiuchi, M., Nakao, M., Katayama, N. and Yamamoto, M. (2002) Elicited ponto-geniculo-occipital waves by auditory stimuli are synchronized with hippocampal theta-waves. *Psychiatry and Clinical Neurosciences* **56**, 343–344.

Karashima, A., Nakamura, K., Sato, N., Nakao, M., Katayama, N., Yamamoto, M. and Horiuchi, M. (2002) Phase-locking of spontaneous and elicited ponto-geniculo-occipital waves is associated with acceleration of hippocampal theta waves during rapid eye movement sleep in cats. *Brain Research* **958**, 347–358.

Karni, A., Tanne, D., Rubenstein, B. S., Askenasy, J. J. and Sagi, D. (1994) Dependence on REM sleep of overnight improvement of a perceptual skill. *Science* **265**, 679–682.

Karschin, C., Ecke, C., Ashcroft, F. M. and Karschin, A. (1997) Overlapping distribution of KATP channel-forming Kir6.2 subunit and the sulfonylurea receptor SUR1 in rodent brain, *FEBS Letters* **401**, 59–64.

Kasa, P., Rakonczay, Z. and Gulya, K. (1997) The cholinergic system in Alzheimer's disease. *Progress in Neurobiology* **52**, 511–535.

Kass, D. (1994) Differential response to anticonvulsants among bipolar patients. *Curr Affect Illness* **13**.

Katona, I., Rancz, E. A., Acsady, L., Ledent, C., Mackie, K., Hajos, N. and Freund, T. F. (2001) Distribution of CB1 cannabinoid receptors in the amygdala and their role in the control of GABAergic transmission. *Journal of Neuroscience* **21**, 9506–9518.

Kawas, C., Resnick, S., Morrison, A., Brookmeyer, R., Corrada, M., Zonderman, A., Bacal, C., Lingle, D. D. and Metter, E. (1997) A prospective study of estrogen replacement therapy and the risk of developing Alzheimer's disease: the Baltimore Longitudinal Study of Aging. *Neurology* **48**, 1517–1521.

Keefe, K. A., Zigmond, M. J. and Abercrombie, E. D. (1993) *In vivo* regulation of extracellular dopamine in the neostriatum: influence of impulse activity and local excitatory amino acids. *Journal of Neural Transmission [General Section]* **91**, 223–240.

Kelleher, R. J. 3rd, Govindarajan, A. and Tonegawa, S. (2004) Translational regulatory mechanisms in persistent forms of synaptic plasticity. *Neuron* **44**, 59–73.

Kemp, J. A. and McKernan, R. M. (2002) NMDA receptor pathways as drug targets. *Nature Neuroscience* **5**, 1039–1042.

Kemp, J. M. and Powell, T. P. (1971) The structure of the caudate nucleus of the cat: light and electron microscopy, *Philosophical Transactions of the Royal Society London B: Biological Sciences* **262**, 383–401.

Kempermann, G. (2002) Why new neurons? Possible functions for adult hippocampal neurogenesis. *Journal of Neuroscience* **22**, 635–638.

Kempermann, G., Kuhn, H. G. and Gage, F. H. (1997) More hippocampal neurons in adult mice living in an enriched environment. *Nature* **386**, 493–495.

Kempermann, G., Kuhn, H. G. and Gage, F. H. (1998) Experience-induced neurogenesis in the senescent dentate gyrus. *Journal of Neuroscience* **18**, 3206–3212.

Kempermann, G., Wiskott, L. and Gage, F. H. (2004b) Functional significance of adult neurogenesis. *Current Opinions in Neurobiology* **14**, 186–191.

Kempermann, G., Jessberger, S., Steiner, B. and Kronenberg, G. (2004a) Milestones of neuronal development in the adult hippocampus. *Trends in Neuroscience* **27**, 447–452.

Kerr, J. E., Allore, R. J., Beck, S. G. and Handa, R. J. (1995) Distribution and hormonal regulation of androgen receptor (AR) and AR messenger ribonucleic acid in the rat hippocampus. *Endocrinology* **136**, 3213–3221.

Kesner, R. P., Lee, I. and Gilbert, P. (2004) A behavioral assessment of hippocampal function based on a subregional analysis. *Neuroscience* **15**, 333–351.

Kesner, R. P., Gilbert, P. E. and Barua, L. A. (2002) The role of the hippocampus in memory for the temporal order of a sequence of odors. *Behavioral Neuroscience* **116**, 286–290.

Kesner, R. P., Gilbert, P. E. and Wallenstein, G. V. (2000) Testing neural network models of memory with behavioral experiments. *Current Opinion in Neurobiology* **10**, 260–265.

Kesner R. P., Hunsaker M. R. and Gilbert P. E. (2005) The role of CA1 in the acquisition of an object-trace-odor paired associate task. *Behavioral Neuroscience* **119**, 781–786.

Kessler, R. C., Berglund, P., Demler, O., Jin, R. and Walters, E. E. (2005) Lifetime prevalence and age-of-onset distributions of DSM-IV disorders in the National Comorbidity Survey Replication. *Archives of General Psychiatry* **62**, 593–602.

Kida, S., Josselyn, S. A., de Ortiz, S. P., Kogan, J. H., Chevere, I., Masushige, S. and Silva, A. J. (2002) CREB required for the stability of new and reactivated fear memories. *Nature Neuroscience* **5**, 348–355.

Kilman, V., van Rossum, M. C. and Turrigiano, G. G. (2002) Activity deprivation reduces miniature IPSC amplitude by decreasing the number of postsynaptic GABA(A) receptors clustered at neocortical synapses. *Journal of Neuroscience* **22**, 1328–1337.

Kim, A. J., Shi, Y., Austin, R. C. and Werstuck, G. H. (2005) Valproate protects cells from ER stress-induced lipid accumulation and apoptosis by inhibiting glycogen synthase kinase-3. *Journal of Cell Science* **118**, 89–99.

Kim, C. H., Chung, H. J., Lee, H. K. and Huganir, R. L. (2001a) Interaction of the AMPA receptor subunit GluR2/3 with PDZ domains regulates hippocampal long-term depression. *Proceedings of the National Academy of Science of the United States of America* **98**, 11,725–11,730.

Kim, I. J., Beck, H. N., Lein, P. J. and Higgins, D. (2002) Interferon gamma induces retrograde dendritic retraction and inhibits synapse formation. *Journal of Neuroscience* **22**, 4530–4539.

Kim, G. M., Xu, J., Song, S. K., Yan, P., Ku, G., Xu, X. M. and Hsu, C. Y. (2001b) Tumor necrosis factor receptor deletion reduces nuclear factor-kappaB activation, cellular inhibitor of apoptosis protein 2 expression, and functional recovery after traumatic spinal cord injury. *Journal of Neuroscience* **21**, 6617–6625.

Kim, P. M., Aizawa, H., Kim, P. S., Huang, A. S., Wickramasinghe, S. R., Kashani, A. H., Barrow, R. K., Huganir, R. L., Ghosh, A. and Snyder, S. H. (2005) Serine racemase: activation by glutamate neurotransmission via glutamate receptor interacting protein and mediation of neuronal migration. *Proceeding of the National Academy of Sciences of the United States of America* **102**, 2105–2110.

Kita, H. (1996) Glutamatergic and GABAergic postsynaptic responses of striatal spiny neurons to intrastriatal and cortical stimulation recorded in slice preparations. *Neuroscience* **70**, 925–940.

Kleene, R. and Schachner, M. (2004) Glycans and neural cell interactions. *Nature Reviews Neuroscience* **5**, 195–208.

Kleim, J. A., Barbay, S. and Nudo, R. J. (1998a) Functional reorganization of the rat motor cortex following motor skill learning. *Journal of Neurophysiology* **80**, 3321–3325.

Kleim, J. A., Vij, K., Ballard, D. H. and Greenough, W. T. (1997) Learning-dependent synaptic modifications in the cerebellar cortex of the adult rat persist for at least four weeks. *Journal of Neuroscience* **17**, 717–721.

Kleim, J. A., Lussnig, E., Schwarz, E. R., Comery, T. A. and Greenough, W. T. (1996) Synaptogenesis and Fos expression in the motor cortex of the adult rat after motor skill learning. *Journal of Neuroscience* **16**, 4529–4535.

Kleim, J. A., Hogg, T. M., VandenBerg, P. M., Cooper, N. R., Bruneau, R. and Remple, M. (2004) Cortical synaptogenesis and motor map reorganization occur during late, but not early, phase of motor skill learning. *Journal of Neuroscience* 24, 628–633.

Kleim, J. A., Swain, R. A., Armstrong, K. A., Napper, R. M., Jones, T. A. and Greenough, W. T. (1998) Selective synaptic plasticity within the cerebellar cortex following complex motor skill learning. *Neurobiology of Learning and Memory* 69, 274–289.

Kleim, J. A., Barbay, S., Cooper, N. R., Hogg, T. M., Reidel, C. N., Remple, M. S. and Nudo, R. J. (2002) Motor learning-dependent synaptogenesis is localized to functionally reorganized motor cortex. *Neurobiology of Learning and Memory* 77, 63–77.

Klein, P. S. and Melton, D. A. (1996) A molecular mechanism for the effect of lithium on development. *Proceedings of the National Academy of Science of the United States of America* 93, 8455–8459.

Klein, M. E., Impey, S. and Goodman, R. H. (2005) Role reversal: the regulation of neuronal gene expression by microRNAs. *Current Opinion in Neurobiology* 15, 507–513.

Klein, R., Naduri, V., Jing, S. A., Lamballe, F., Tapley, P., Bryant, S., Cordon-Cordo, C., Jones, K. R., Reichardt, L. F. and Barbacid, M. (1991) The trkB tyrosine protein kinase is a receptor for brain-derived neurotrophic factor and neurotrophin-3. *Cell* 66, 395–403.

Klintsova, A. Y., Dickson, E., Yoshida, R. and Greenough, W. T. (2004) Altered expression of BDNF and its high-affinity receptor TrkB in response to complex motor learning and moderate exercise. *Brain Research* 1028, 92–104.

Koechlin, E., Corrado, G., Pietrini, P. and Grafman, J. (2000) Dissociating the role of the medial and lateral anterior prefrontal cortex in human planning. *Proceedings of the National Academy of Science* 97, 7651–7656.

Koechlin, E., Danek, A., Burnod, Y. and Grafman, J. (2002) Medial prefrontal and subcortical mechanisms underlying the acquisition of motor and cognitive action sequences in humans. *Neuron* 35, 371–381.

Kohama, S. G., Goss, J. R., Finch, C. E., McNeill, T. H. (1995) Increases of glial fibrillary acidic protein in the aging female mouse brain. *Neurobiology of Aging* 16, 59–67.

Kohara, K., Kitamura, A., Morishima, M. and Tsumoto, T. (2001) Activity-dependent transfer of brain-derived neurotrophic factor to postsynaptic neurons. *Science* 291, 2419–2423.

Kojima, S., Wu, S. T., Parmley, W. W. and Wikman-Coffelt, J. (1994) Relationship between intracellular calcium and oxygen consumption: effects of perfusion pressure, extracellular calcium, dobutamine, and nifedipine. *American Heart Journal* 127, 386–391.

Kolb, B., Gorny, G., Soderpalm, A. H. and Robinson, T. E. (2003) Environmental complexity has different effects on the structure of neurons in the prefrontal cortex versus the parietal cortex or nucleus accumbens. *Synapse* 48, 149–153.

Komuro, H. and Rakic, P. (1993) Modulation of neuronal migration by NMDA receptors. *Science* 260, 95–97.

Koo, J. W., Han, J. S. and Kim, J. J. (2004) Selective neurotoxic lesions of basolateral and central nuclei of the amygdala produce differential effects on fear conditioning. *Journal of Neuroscience* 24, 7654–7662.

Koob, G. F. (2000) Neurobiology of addiction. Toward the development of new therapies, *Annals of the New York Academy of Sciences* 909, 170–185.

Kornack, D. R. and Rakic, P. (1999) Continuation of neurogenesis in the hippocampus of the adult macaque monkey. *Proceedings of the National Academy of Sciences of the United States of America* 96, 5768–5773.

Kornhuber, J., Weller, M., Schoppmeyer, K. and Riederer, P. (1994) Amantadine and memantine are NMDA receptor antagonists with neuroprotective properties. *Journal of Neural Transmission* 43(Suppl.), 91–104.

Korte, S. M., Bouws, G. A., Koolhaas, J. M. and Bohus, B. (1992a) Neuroendocrine and behavioral responses during conditioned active and passive behavior in the defensive burying/probe avoidance paradigm: effects of ipsapirone. *Physiology and Behavior* 52, 355–361.

Korte, M., Carroll, P., Wolf, E., Brem, G., Thoenen, H. and Bonhoeffer, T. (1995) Hippocampal long-term potentiation is impaired in mice lacking brain-derived neurotrophic factor. *Proceedings of the National Academy of Sciences of the United States of America* 92, 8856–8860.

Korte, S. M., Buwalda, B., Bouws, G. A., Koolhaas, J. M., Maes, F. W. and Bohus, B. (1992b) Conditioned neuroendocrine and cardiovascular stress responsiveness accompanying behavioral passivity and activity in aged and in young rats. *Physiology and Behavior* 51, 815–822.

Korz, V. and Frey, J. U. (2003a) Stress-related modulation of hippocampal long-term-potentiation in rats: involvement of adrenal steroid receptors. *Journal of Neuroscience* 23, 7281–7287.

Korz, V. and Frey, J. U. (2003b) Stress and novelty related moduation of rat hippocampal LTP. 2003 *Society for Neuroscience*, Abstract No 623.17

Korz, V., Ahmed, T. and Frey, J. U. (2004) Long-term effects of acute stress on hippocampal LTP and cellular signaling. 2004 *Society for Neuroscience*, Abstract No. 498.13.

Korzus, E. (2003) The relation of transcription to memory formation. *Acta Biochimica Polonica* **50**, 775–782.

Korzus, E., Rosenfeld, M. G. and Mayford, M. (2004) CBP histone acetyltransferase activity is a critical component of memory consolidation. *Neuron* **42**, 961–972.

Kovalchuk, Y., Hanse, E., Kafitz, K. W. and Konnerth, A. (2002) Postsynaptic Induction of BDNF-Mediated Long-Term Potentiation. *Science* **295**, 1729–1734.

Krakauer, J. W. and Shadmehr, R. (2006) Consolidation of motor memory. *Trends in Neuroscience* **29**, 58–64.

Krakauer, J. W., Ghez, C. and Ghilardi, M. F. (2005) Adaptation to visuomotor transformations: consolidation, interference, and forgetting. *Journal of Neuroscience* **25**, 473–478.

Kramar, E. A., Bernard, J. A., Gall, C. M. and Lynch, G. (2003) Integrins modulate fast excitatory transmission at hippocampal synapses. *Journal of Biological Chemistry* **278**, 10,722–10,730.

Krapivinksi, J., Medina, I., Krapivinski, L., Gapon, S. and Clapham, D. E. (2004) SynGAP-MUPP1-CaMKII synaptic complexes regulate p38 MAP kinase activity and NMDA receptor-dependent synaptic AMPA receptor potentiation. *Cell* **43**, 563–574.

Krebs, H. A. (1935) CXCVII. Metabolism of amino acids. III. Deamination of amino acids. *Biochemical Journal* **29**, 1620–1644.

Kreiman, K., Kock, C., and Fried, I. (2000a) Category-specific visual responses of single neurons in the human medial temporal lobe. Nature *Neuroscience* **3**, 946–953.

Kreiman, K., Kock, C., and Fried, I. (2000b). Imagery neurons in the human brain. *Nature* **408**, 357–361.

Kreitzer A. C. and Malenka R. C. (2007) Endocannabinoid-mediated rescue of striatal LTD and motor deficits in Parkinson's disease models. *Nature* **445**, 643–647.

Krichevsky, A. M. and Kosik, K. S. (2001) Neuronal RNA granules: A link between RNA localization and stimulation-dependent translation. *Neuron* **32**, 683–696.

Kriegler, S. and Chiu, S. Y. (1993) Calcium signaling of glial cells along mammalian axons. *Journal of Neuroscience* **13**, 4229–4245.

Krippeit-Drews, P., Kramer, C., Welker, S., Lang, F., Ammon, H. P. and Drews, G. (1999) Interference of H_2O_2 with stimulus-secretion coupling in mouse pancreatic beta-cells. *Journal of Physiology (London)* **514**, 471–481.

Kroner, S., Rosenkranz, J. A., Grace, A. A. and Barrionuevo, G. (2005) Dopamine modulates excitability of basolateral amygdala neurons in vitro. *Journal of Neurophysiology* **93**, 1598–1610.

Krug, M., Lossner, B. and Ott, T. (1984) Anisomycin blocks the late phase of long-term potentiation in the dentate gyrus of freely moving rats. *Brain Research Bulletin* **13**, 39–42.

Krystal, J. H., Karper, L. P., Seibyl, J. P., Freeman, G. K., Delaney, R., Bremner, J. D., Heninger, G. R., Bowers, M. B. Jr. and Charney, D. S. (1994) Subanesthetic effects of the noncompetitive NMDA antagonist, ketamine, in humans. Psychotomimetic, perceptual, cognitive, and neuroendocrine responses. *Archives of General Psychiatry* **51**, 199–214.

Kuhl, P. K., Tsao, F.-M. and Liu, H.-M. (2003) Foreign-language experience in infancy: effects of short-term exposure and social interactions on phonetic learning. *Proceedings of the National Academy of Science of the United States of America* **100**, 9096–9101.

Kuhn, H. G., Dickinson-Anson, H. and Gage, F. H. (1996) Neurogenesis in the dentate gyrus of the adult rat: age-related decrease of neuronal progenitor proliferation. *Journal of Neuroscience* **16**, 2027–2033.

Kujala, P., Portin, R. and Ruutiainen, J. (1997) The progress of cognitive decline in muyltiple sclerosis. A controlled 3-year follow-up. *Brain* **12**, 289–297.

Kukley, M, Capetillo-Zarate, E. and Dietrich, D. (2007) Vesicular glutamate release from axons in white matter. *Nature Neuroscience* **10**, 311–320.

Kullmann, D. M., Min, M. Y., Asztely, F. and Rusakov, D. A. (1999) Extracellular glutamate diffusion determines the occupancy of glutamate receptors at CA1 synapses in the hippocampus. *Philosophical Transactions of the Royal Society of London. Series B: Biological Sciences* **354**, 395–402.

Kupfermann, I., Castellucci, V., Pinsker, H. and Kandel, E. (1970) Neuronal correlates of habituation and dishabituation of the gill-withdrawal reflex in aplysia. *Science* **167**, 1743–1745.

Kuriyama, K., Stickgold, R. and Walker, M. P. (2004) Sleep-dependent learning and motor skill complexity. *Learning & Memory* 11, 705–713.

Kwok, R. P., Lundblad, J. R., Chrivia, J. C., Richards, J. P., Bachinger, H. P., Brennan, R. G., Roberts, S. G., Green, M. R. and Goodman, R. H. (1994) Nuclear protein CBP is a coactivator for the transcription factor CREB. *Nature* 370, 223–226.

Lachyankar, M. B., Condon, P. J., Quesenberry, P. J., Litofsky, N. S., Recht, L. D. and Ross, A. H. (1997) Embryonic precursor cells that express Trk receptors: induction of different cell fates by NGF, BDNF, NT-3, and CNTF. *Experimental Neurology* 144, 350–360.

Laggerbauer, B., Ostareck, D., Keidel, EM., Ostareck-Lederer, A. and Fischer, U. (2001) Evidence that fragile X mental retardation protein is a negative regulator of translation. *Human Molecular Genetics* 10, 329–338.

LaLumiere, R. T., Buen, T. V. and McGaugh, J. L. (2003) Post-training intra-basolateral amygdala infusions of norepinephrine enhance consolidation of memory for contextual fear conditioning. *Journal of Neuroscience* 23, 6754–6758.

LaLumiere, R. T., Nguyen, L. T. and McGaugh, J. L. (2004) Post-training intrabasolateral amygdala infusions of dopamine modulate consolidation of inhibitory avoidance memory: involvement of noradrenergic and cholinergic systems. *European Journal of Neuroscience* 20, 2804–2810.

Lambeth J. D. (2004) NOX enzymes and the biology of reactive oxygen. *Nature Reviews Immunology* 4, 181–189.

Lamprecht, R. (1999) CREB: a message to remember. *Cellular and Molecular Life Sciences* 55, 554–563.

Lamprecht R. and LeDoux, J. (2004) Structural plasticity and memory. *Nature Reviews. Neuroscience* 5, 45–54.

Lamprecht, R., Farb, C. R. and LeDoux, J. E. (2002) Fear memory formation involves p190 RhoGAP and ROCK proteins through a GRB2-mediated complex. *Neuron* 36, 727–738.

Lampson, L. A. (1995) Interpreting MHC class I expression and class I/class II reciprocity in the CNS: reconciling divergent findings. *Microsc Res Tech* 32, 267–285.

Lampson, L. A. and Fisher, C. A. (1984) Weak HLA and beta 2-microglobulin expression of neuronal cell lines can be modulated by interferon. *Proceedings of the National Academy of Science of the United States of America* 81, 6476–6480.

Lampson, L. A., Whelan, J. P. and Siegel, G. (1988) Functional implications of class I MHC modulation in neural tissue. *Annals of the New York Academy of Sciences* 540, 479–482.

Landfield, P. W. and Deadwyler, S. A. (1988) *Long-term Potentiation: From Biophysics to Behavior.* Alan R. Liss, Inc.

Landfield, P. W. and Lynch, G. (1977) Impaired monosynaptic potentiation in in vitro hippocampal slices from age, memory-deficient rats. *Journal of Gerontology* 32, 523–533.

Landfield, P.W., Pitler, T. A. and Applegate, M. D. (1986) The effects of high Mg^{2+}–Ca^{2+} ratios on frequency potentiation in hippocampal slices of young and aged rats. *Journal of Neurophysiology* 56, 797–811.

Landsteiner, K. (1900) Zur Kenntnis der antifermentativen, lytischen und agglutinierenden Wirkungen des Blutserums und der Lymphe. *Zbl. Bakt.* 27, 357–362.

Lattal, K. M. and Abel, T. (2004) Behavioral impairments caused by injections of the protein synthesis inhibitor anisomycin after contextual retrieval reverse with time. *Proceedings of the National Academy of Science of the United States of America* 101, 4667–4672.

Lauri, S. E., Rauvala, H., Kaila, K. and Taira, T. (1998) Effect of heparin-binding growth-associated molecule (HB-GAM) on synaptic transmission and early LTP in rat hippocampal slices. *European Journal of Neuroscience* 10, 188–194.

Lauri, S. E., Kaukinen, S., Kinnunen, T., Ylinen, A., Imai, S. and Kaila, K. (1999) Regulatory role and molecular interactions of a cell-surface heparan sulfate proteoglycan (N-syndecan) in hippocampal long-term potentiation. *Journal of Neuroscience* 19, 1226–1235.

Lauterborn, J. C., Isackson, P. J. and Gall, C. M. (1994) Cellular localization of NGF and NT-3 mRNAs in postnatal rat forebrain. *Molecular and Cellular Neuroscience* 5, 46–62.

Leach, M. J., Marden, C. M. and Miller, A. A. (1986) Pharmacological studies on lamotrigine, a novel potential antiepileptic drug: II. Neurochemical studies on the mechanism of action. *Epilepsia* 27, 490–497.

LeBaron, R. G., Hernandez, R. V., Orfila, J. E. and Martinez, J. L. Jr. (1999) An integrin binding peptide GRGDSP reduces LTP in area CA1 of the rat hippocampus. *Society for Neuroscience Abstracts* 25, 1495.

LeBaron, R. G., Orfila, J. E., Martinez, J. L. Jr. and Hernandez, R. V. (2003) An integrin binding peptide reduces rat CA1 hippocampal long-term potentiation during the first few minutes following theta burst stimulation. *Neuroscience Letters* 339, 199–202.

LeBaron, R. G., Esko, J. D., Woods, A., Johansson, S. and Hook, M. (1988) Adhesion of glycosaminoglycan-deficient Chinese hamster ovary cell mutants to fibronectin substrata. *Journal of Cell Biology* **106**, 945–952.

LeBaron, R. G., Hernandez, R. V., Navarro, M. M., Curry, L. R., Orfila, J. E. and Martinez, J. L. Jr. (2004) A focal adhesion-like process underlies induction of long-term potentiation in the Schaffer collateral-CA1 region of the hippocampus. *Neuron Glia Biology* **1**, 385–393.

Ledo, F., Kremer, L., Mellstrom, B. and Naranjo, J. R. (2002) Ca^{2+}-dependent block of CREB-CBP transcription by repressor DREAM. *EMBO Journal* **21**, 4583–4592.

LeDoux, J. E. (2000) Emotion circuits in the brain. *Annual Review of Neuroscience* **23**, 155–184.

LeDoux, J. E., Cicchetti, P., Xagoraris, A. and Romanski, L. M. (1990) The lateral amygdaloid nucleus: Sensory interface of the amygdala in fear conditioning. *Journal of Neuroscience* **10**, 1062–1069.

LeDoux, J. E., Iwata, J., Cicchetti, P. and Reis, D. J. (1988) Different projections of the central amygdaloid nucleus mediate autonomic and behavioral correlates of conditioned fear. *Journal of Neuroscience* **8**, 2517–2529.

Lee, H. and Kim, J. J. (1998) Amygdalar NMDA receptors are critical for new fear learning in previously fear-conditioned rats. *Journal of Neuroscience* **18**, 8444–8454.

Lee, R., Kermani, P., Teng, K. K. and Hempstead, B. L. (2001) Regulation of cell survival by secreted proneurotrophins. *Science* **294**, 1945–1948.

Lee, I. and Kesner, R. P. (2002) Differential contribution of NMDA receptors in hippocampal subregions to spatial working memory. *Nature Neuroscience* **5**, 162–168.

Lee, I. and Kesner, R. P. (2004) Differential contributions of hippocampal subregions to memory acquisition and retrieval in contextual fear conditioning. *Hippocampus* **14**, 301–310.

Lee, S. J. and McEwen, B. S. (2001) Neurotrophic and neuroprotective actions of estrogens and their therapeutic implications. *Annual Review of Pharmacology and Toxicology* **41**, 569–591.

Lee, J., Duan, W. and Mattson, M. P. (2002) Evidence that brain-derived neurotrophic factor is required for basal neurogenesis and mediates, in part, the enhancement of neurogenesis by dietary restriction in the hippocampus of adult mice. *Journal of Neurochemistry* **82**, 1367–1375.

Lee, J. L., Everitt, B. J. and Thomas, K. L. (2004) Independent cellular processes for hippocampal memory consolidation and reconsolidation. *Science* **304**, 839–843.

Lee, J. M., Zipfel, G. J. and Choi, D. W. (1999) The changing landscape of ischaemic brain injury mechanisms. *Nature* **399**, A7–14.

Lee, O., Lee, C. J. and Choi, S. (2002) Induction mechanisms for L-LTP at thalamic input synapses to the lateral amygdala: requirement of mGluR5 activation. *Neuroreport* **13**, 685–691.

Lee, P. R., Cohen, J. E., Becker, K. G. and Fields, R. D. (2005b) Gene expression in the conversion of early-phase to late-phase long-term potentiation. *Annals of the New York Academy of Sciences* **1048**, 259–271.

Lee, H. J., Berger, S. Y., Stiedl, O. and Kim, J. J. (2001b) Post-training injections of catecholaminergic drugs do not modulate fear conditioning in rats and mice. *Neuroscience Letters* **303**, 123–126.

Lee, H. J., Choi, J. S., Brown, T. H. and Kim, J. J. (2001a) Amygdalar nmda receptors are critical for the expression of multiple conditioned fear responses. *Journal of Neuroscience* **21**, 4116–4124.

Lee, J. L., Di Ciano, P., Thomas, K. L. and Everitt, B. J. (2005a) Disrupting reconsolidation of drug memories reduces cocaine-seeking behavior. *Neuron* **47**, 795–801.

Lee, H. K., Barbarosie, M., Kameyama, K., Bear, M. F. and Huganir, R. L. (2000) Regulation of distinct AMPA receptor phosphorylation sites during bidirectional synaptic plasticity. *Nature* **405**, 955–959.

Lee, S. J., Campomanes, C. R., Sikat, P. T., Greenfield, A. T., Allen, P. B. and McEwen, B. S. (2004) Estrogen induces phosphorylation of cyclic AMP response element binding (pCREB) in primary hippocampal cells in a time-dependent manner. *Neuroscience* **124**, 549–560.

Lee, S. J., Romeo, R. D., Svenningsson, P., Campomanes, C. R., Allen, P. B., Greengard, P. and McEwen, B. S. (2004) Estradiol affects spinophilin protein differently in gonadectomized males and females. *Neuroscience* **127**, 983–988.

Lenox, R. and Manji, H. (1998) Drugs for treatment of bipolar disorder: lithium. In Schatzberg, A. F. and Nemeroff, C. B. (eds) *Textbook of Psychopharmacology* 2nd ed. American Psychiatry Press, pp. 379–429.

Leonoudakis, D., Braithwaite, S. P., Beattie, M. S. and Beattie, E. C. (2004) TNFalpha-induced

AMPA-receptor trafficking in CNS neurons: relevance to excitotoxicity? *Neuron Glia Biology* 1, 263–273.

Leranth, C., Hajszan, T. and MacLusky, N. J. (2004) Androgens increase spine synapse density in the CA1 hippocampal subfield of ovariectomized female rats. *Journal of Neuroscience* 24, 495–499.

Leranth, C., Petnehazy, O. and MacLusky, N. J. (2003) Gonadal hormones affect spine synaptic density in the CA1 hippocampal subfield of male rats. *Journal of Neuroscience* 23, 1588–1592.

Leranth, C., Shanabrough, M. and Redmond, D. E. (2002) Gonadal hormones are responsible for maintaining the integrity of spine synapses in the CA1 hippocampal subfield of female nonhuman primates. *Journal of Comparative Neurology* 447, 34–42.

Lessmann, V., Gottmann, K. and Malcangio, M. (2003) Neurotrophin secretion: current facts and future prospects. *Progress in. Neurobiology* 69, 341–374.

Letellier, M., Willson, M. L., Gautheron, V., Mariani, J. and Lohof, A. M. (2008) Normal adult climbing fiber monoinnervation of cerebellar Purkinje cells in mice lacking MHC class I molecules. *Dev Neurobiol.*

Leviel, V., Gobert, A. and Guibert, B. (1990) The glutamate-mediated release of dopamine in the rat striatum: further characterization of the dual excitatory-inhibitory function. *Neuroscience* 39, 305–312.

Levine, S. (1957) Infantile experience and resistance to physiological stress. *Science* 126, 405.

Levine, S. (2005) Developmental determinants of sensitivity and resistance to stress. *Psychoneuroendocrinology* 30, 939–946.

Levine, E. S., Crozier, R. A., Black, I. B. and Plummer, M. R. (1998) Brain-derived neurotrophic factor modulates hippocampal synaptic transmission by increasing N-methyl-D-aspartic acid receptor activity. *Proceedings of the National Academy of Science of the United States of America* 95, 10,235–10,239.

Levine, E. S., Dreyfus, C. F., Black, I. B. and Plummer, M. R. (1995) Differential effects of NGF and BDNF on voltage-gated calcium currents in embryonic basal forebrain neurons. *Journal of Neuroscience* 15, 3084–3091.

Levine, E. S., Dreyfus, C. F., Black, I. B. and Plummer, M. R. (1995) Brain-derived neurotrophic factor rapidly enhances synaptic transmission in hippocampal neurons via postsynaptic tyrosine kinase receptors. *Proceedings of the National Academy of Sciences of the United States of America* 92, 8074–8077.

Levine, E. S., Dreyfus, C. F., Black, I. B. and Plummer, M. R. (1996) Selective role for trkB neurotrophin receptors in rapid modulation of hippocampal synaptic transmission. *Brain Research and Molecular Brain Research* 38, 300–303.

Levison, S. W. & McCarthy, K. D. (1991) Astroglia in cultures. In Banker, G. K. and Goslin, K. (eds) *Culturing Nerve Cells.* MIT Press, pp 309–336.

Lev-Ram, V. and Ellisman, M. H. (1995) Axonal activation-induced calcium transients in myelinating Schwann cells, sources, and mechanisms. *Journal of Neuroscience* 15, 2628–2637.

Lev-Ram, V. and Grinvald, A. (1986) Ca^{2+} and K^+-dependent communication between central nervous system myelinated axons and oligodendrocytes revealed by voltage-sensitive dyes. *Proceedings of the National Academy of Scienceof the United States of America* 83, 6651–6655.

Levy, W. B. (1989) A computational approach to hippocampal function. In Hawkins, R. D. and Bower, G. D. (eds), *Computational Models of Learning in Simple Systems.* Academic Press, pp. 243–305.

Levy, W. B. (1996) A sequence predicting CA3 is a flexible associator that learns and uses context to solve hippocampal-like tasks. *Hippocampus* 6, 579–590.

Lewin, G. R. and Barde, Y.-A. (1996) Physiology of the neurotrophins. *Annual Review in Neuroscience* 19, 289–317.

Lewis, D. J. and Bregman, N. J. (1973) Source of cues for cue-dependent amnesia in rats. *Journal of Comparative Physiological Psychology* 85, 421–426.

Lewis, C., McEwen, B. S. and Frankfurt, M. (1995) Estrogen-induction of dendritic spines in ventromedial hypothalamus and hippocampus: effects of neonatal aromatase blockade and adult GDX. *Developmental Brain Research* 87, 91–95.

Lewis, D. J., Bregman, N. J. and Mahan, J. J. Jr. (1972) Cue-dependent amnesia in rats. *Journal of Comparative Physiological Psychology* 81, 243–247.

Li, X., Phillips, R. G. and LeDoux, J. E. (1995) NMDA and non-NMDA receptors contribute to synaptic transmission between the medial geniculate body and the lateral nucleus of the amygdala. *Experimental Brain Research* 105, 87–100.

Li, X. F., Armony, J. L. and LeDoux, J. E. (1996) GABAa and GABAb receptors differentially regulate synaptic transmission in the auditory thalamo-amygdala pathway: an in vivo microiontophoretic study and a model. *Synapse* 24, 115–124.

Li, Z., Okamoto, K., Hayashi, Y. and Sheng, M. (2004) The importance of dendritic mitochondria in the morphogenesis and plasticity of spines and synapses. *Cell* 119, 873–887.

Li, R., Nishijo, H., Ono, T., Ohtani, Y. and Ohtani, O. (2002) Synapses on GABAergic neurons in the basolateral nucleus of the rat amygdala: double-labeling immunoelectron microscopy. *Synapse* 43, 42–50.

Li, W., Llopis, J., Whitney, M., Zlokarnik, G. and Tsien, R. Y. (1998b) Cell-permeant caged InsP3 ester shows that Ca^{2+} spike frequency can optimize gene expression. *Nature* 392, 936–941.

Li, Y., Vartanian, A. J., White, F. J., Xue, C. J. and Wolf, M. E. (1997) Effects of the AMPA receptor antagonist NBQX on the development and expression of behavioral sensitization to cocaine and amphetamine. *Psychopharmacology (Berl)* 134, 266–276.

Li, Y. X., Zhang, Y., Lester, H. A., Schuman, E. M. and Davidson, N. (1998a) Enhancement of neurotransmitter release induced by brain-derived neurotrophic factor in cultured hippocampal neurons. *Journal of Neuroscience* 18, 10,231–10,240.

Li, Z., Zhang, Y., Ku, L., Wilkinson, K. D., Warren, S. T. and Feng, Y. (2001). The fragile X mental retardation protein inhibits translation via interacting with mRNA. *Nucleic Acids Research* 29, 2276–2283.

Li, C., Brake, W. G., Romeo, R. D., Dunlop, J. C., Gordon, M., Buzescu, R., Margarinos, A. M., Allen, P., Greengard, P., Luine, V. and McEwen, B. S. (2004) Estrogen treatment alters hippocampal dendritic spine shape, enhances synaptic protein immunoreactivity and performance in a spatial working memory task in female C57BL/6J mice. *Proceedings of the National Academy of Sciences of the United States of America* 101, 2185–2190.

Liao, D., Hessler, N. A. and Malinow, R. (1995) Activation of postsynaptically silent synapses during pairing-induced LTP in CA1 region of hippocampal slice. *Nature* 375, 400–404.

Liao, D., Scannevin, R. H. and Huganir, R. (2001) Activation of silent synapses by rapid activity-dependent synaptic recruitment of AMPA receptors. *Journal of Neuroscience* 21, 6008–6017.

Lidman, O., Olsson, T. and Piehl, F. (1999) Expression of nonclassical MHC class I (RT1-U) in certain neuronal populations of the central nervous system. *European Journal of Neuroscience* 11, 4468–4472.

Lie, D. C., Song, H., Colamarino, S. A., Ming, G. L. and Gage, F. H. (2004) Neurogenesis in the adult brain: new strategies for central nervous system diseases. *Annual Review in Pharmacological Toxicology* 44, 399–421.

Ligon, L. A. and Steward, O. (2000) Role of microtubules and actin filaments in the movement of mitochondria in the axons and dendrites of cultured hippocampal neurons. *Journal of Comparative Neurology* 427, 351–361.

Lilienbaum, A. and Israel, A. (2003) From calcium to NF-kappa B signaling pathways in neurons. *Molecular and Cellular Biology* 23, 2680–2698.

Lilliehook, C., Chan, S., Choi, E. K., Zaidi, N. F., Wasco, W., Mattson, M. P. and Buxbaum, J. D. (2002) Calsenilin enhances apoptosis by altering endoplasmic reticulum calcium signaling. *Molecular and Cellular Neuroscience* 19, 552–559.

Lilliehook, C., Bozdagi, O., Yao, J., Gomez-Ramirez, M., Zaidi, N. F., Wasco, W., Gandy, S., Santucci, A. C., Haroutunian, V., Huntley, G. W. and Buxbaum, J. D. (2003) Altered Abeta formation and long-term potentiation in a calsenilin knock-out. *Journal of Neuroscience* 23, 9097–9106.

Lim, K. O., Adalsteinsson, E., Spielman, D., Sullivan, E. V., Rosenbloom, M. J. and Pfefferbaum, A. (1988) Proton magnetic resonance spectroscopic imaging of cortical gray and white matter in schizophrenia. *Archives of General Psychiatry* 55, 346–352.

Limback-Stokin, K., Korzus, E., Nagaoka-Yasuda, R. and Mayford, M. (2004). Nuclear calcium/calmodulin regulates memory consolidation. *Journal of Neuroscience* 24, 10,858–10,867.

Lin, B., Arai, A. C., Lynch, G. and Gall, C. M. (2003) Integrins regulate NMDA receptor-mediated synaptic currents. *Journal of Neurophysiology* 89, 2874–2878.

Lin, C. H., Yeh, S. H., Lu, K. T., Leu, T. H., Chang, W. C. and Gean, P. W. (2001) A role for the PI-3 kinase signaling pathway in fear conditioning and synaptic plasticity in the amygdala. *Neuron* 31, 841–851.

Lin, J. W., Ju, W., Foster, K., Lee, S. H., Ahmadian, G., Wyszynski, M., Wang, Y. T. and Sheng, M. (2000) Distinct molecular mechanisms and divergent endocytotic pathways of AMPA receptor internalization. *Nature Neuroscience* 3, 1282–1290.

Linda, H., Hammarberg, H., Piehl, F., Khademi, M. and Olsson, T. (1999) Expression of MHC class I heavy chain and beta2-microglobulin in rat brainstem motoneurons and nigral dopaminergic neurons. *Journal of Neuroimmunology* **101**, 76–86.

Linda, H., Hammarberg, H., Cullheim, S., Levinovitz, A., Khademi, M. and Olsson, T. (1998) Expression of MHC class I and beta2-microglobulin in rat spinal motoneurons: regulatory influences by IFN-gamma and axotomy. *Experimental Neurologty* **150**, 282–295.

Lindholm, D., Castren, E., Berzaghi, M., Blochl, A. and Thoenen, H. (1994) Activity-dependent and hormonal regulation of neurotrophin mRNA levels in the brain – implications for neuronal plasticity. *Journal of Neurobiology* **25**, 1362–1372.

Linnarsson, S., Bjorklund, A. and Ernfors, P. (1997) Learning deficit in BDNF mutant mice. *European Journal of Neuroscience* **9**, 2581–2587.

Lipp, P., Thomas, D., Berridge, M. J. and Bootman, M. D. (1997) Nuclear calcium signalling by individual cytoplasmic calcium puffs. *EMBO Journal* **16**, 7166–7173.

Lisman, J., Schulman, H. and Cline, H. (2002) The molecular basis of CaMKII function in synaptic and behavioural memory. *National Rev Neurosci* **3**, 175–190.

Lisman, J. E., Richter-Levin, G., Canevari, L., Bliss, T. V., Jarrard, L. E. and Jones, R. S. (1999) Relating hippocampal circuitry to function: recall of memory sequences by reciprocal dentate-CA3 interactions. *Neuron* **22**, 233–242.

Liss, B., Bruns, R. and Roeper, J. (1999) Alternative sulfonylurea receptor expression defines sensitivity to K-ATP channels in dopaminergic midbrain neurons, *EMBO Journal* **18**, 833–846.

Lissin, D. V., Carroll, R. C., Nicoll, R. A., Malenka, R. C. and von Zastrow, M. (1999) Rapid, activation-induced redistribution of ionotropic glutamate receptors in cultured hippocampal neurons. *Journal of Neuroscience* **19**, 1263–1272.

Liu, Y., Fiskum, G. and Schubert, D. (2002) Generation of reactive oxygen species by the mitochondrial electron transport chain. *Journal of Neurochemistry* **80**, 780–787.

Liu, Q. S., Xu, Q., Arcuino, G., Kang, J. and Nedergaard, M. (2004) Astrocyte-mediated activation of neuronal kainate receptors. *Proceedings of the National Academy of Science of the United States of America*.

Liu, Q. Y., Schaffner, A. E., Chang, Y. H., Vaszil, K. and Barker, J. L. (1997) Astrocytes regulate amino acid receptor current densities in embryonic rat hippocampal neurons. *Journal of Neurobiology* **33**, 848–864.

Liu, Q. Y., Schaffner, A. E., Li, Y. X., Dunlap, V. and Barker, J. L. (1996) Upregulation of GABAA current by astrocytes in cultured embryonic rat hippocampal neurons. *Journal of Neuroscience* **16**, 2912–2923.

Loconto, J., Papes, F., Chang, E., Stowers, L., Jones, E. P., Takada, T., Kumanovics, A., Fischer Lindahl, K. and Dulac, C. (2003) Functional expression of murine V2R pheromone receptors involves selective association with the M10 and M1 families of MHC class Ib molecules. *Cell* **112**, 607–618.

Loeb, J. A. and Fischbach, G. D. (1997) Neurotrophic factors increase neuregulin expression in embryonic ventral spinal cord neurons. *Journal of Neuroscience* **17**, 1416–1424.

Loeb, J. A., Hmadcha, A., Fischbach, G. D., Land, S. J. and Zakarian, V. L. (2002) Neuregulin expression at neuromuscular synapses is modulated by synaptic activity and neurotrophic factors. *Journal of Neuroscience* **22**, 2206–2214.

Lohof, A. M., Ip, N. Y. and Poo, M. M. (1993) Potentiation of developing neuromuscular synapses by the neurotrophins NT-3 and BDNF. *Nature* **363**, 350–353.

Lonstein, J. S., Quadros, P. S. and Wagner, C. K. (2001) Effects of neonatal RU486 on adult sexual, parental, and fearful behaviors in rats. *Behavioral Neuroscience* **115**, 58–70.

Loscertales, M., Rose, S. P. R. and Sandi, C. (1997) The corticosteroid synthesis inhibitors metyrapone and aminoglutethimide block long-term memory for a passive avoidance task in day-old chicks. *Brain Research* **769**, 357–361.

Losel, R. and Wehling, M. (2003) Nongenomic actions of steroid hormones. *Nature Reviews of Molecular and Cellular Biology* **4**, 46–56.

Lossner, B. and Rose, S. P. R. (1983) Passive avoidance training increases fucokinase activity in right forebrain base of day-old chicks. *Journal of Neurochemistry* **41**, 1357–1363.

Louie, K. and Wilson, M. A. (2001) Temporally structured replay of awake hippocampal ensemble activity during rapid eye movement sleep. *Neuron* **29**, 145–156.

Love, P. E., Shores, E. W., Johnson, M. D., Tremblay, M. L., Lee, E. J., Grinberg, A., Huang, S. P., Singer, A. and Westphal, H. (1993) T cell development in mice that lack the zeta chain of the T cell antigen receptor complex. *Science* **261**, 918–921.

Lozovaya, N. A., Grebenyuk, S. E., Tsinsadze, T. S., Feng, B., Monaghan, D. T. and Krishtal, O. A. (2004) Extrasynaptic NR2B and NR2D subunits of NMDA receptors shape "superslow" afterburst EPSC in rat hippocampus. *Journal of Physiology* **558**, 451–463.

Lu, B. (2003) BDNF and activity-dependent synaptic modulation. *Learning & Memory* **10**, 86–98.

Lu, B. and Chang, J. (2004) Regulation of neurogenesis by neurotrophins. *Neuron Glia Biology* **1**, 377–384.

Lu, B. and Chow, A. (1999) Neurotrophins and hippocampal synaptic transmission and plasticity. *Journal of Neuroscience Research* **58**, 76–87.

Lu, B. and Gottschalk, W. (2000) Modulation of hippocampal synaptic transmission and plasticity by neurotrophins. *Progress in Brain Research* **128**, 231–241.

Lu, Y. F. and Hawkins, R. D. (2002) Ryanodine receptors contribute to cGMP-induced late-phase LTP and CREB phosphorylation in the hippocampus. *Journal of Neurophysiology* **88**, 1270–1278.

Lu, W., Man, H., Ju, W., Trimble, W. S., MacDonald, J. F. and Wang, Y. T. (2001) Activation of synaptic NMDA receptors induces membrane insertion of new AMPA receptors and LTP in cultured hippocampal neurons. *Neuron* **29**, 243–254.

Lucas, F. R. and Salinas, P. C. (1997) WNT-7a induces axonal remodeling and increases synapsin I levels in cerebellar neurons. *Developmental Biology* **192**, 31–44.

Lucas, F. R., Goold, R. G., Gordon-Weeks, P. R. and Salinas, P. C. (1998) Inhibition of GSK-3beta leading to the loss of phosphorylated MAP-1B is an early event in axonal remodelling induced by WNT-7a or lithium. *Journal of Cell Science* **111**, 1351–1361.

Luckman, S. M., Dyball, R. E. and Leng, G. (1994) Induction of c-fos expression in hypothalamic magnocellular neurons requires synaptic activation and not simply increased spike activity. *Journal of Neuroscience* **14**, 4825–4830.

Luine, V. N. (1985) Estradiol increases choline acetyltransferase activity in specific basal forebrain nuclei and projection areas of female rats. *Experimental Neurology* **89**, 484–490.

Luine, V. N. (1997) Steroid hormone modulation of hippocampal dependent spatial memory. *Stress* **2**, 21–36.

Luine, V. N. and McEwen, B. S. (1983) Sex differences in cholinergic enzymes of diagonal band nuclei in the rat preoptic area. *Neuroendocrinology* **36**, 475–482.

Luine, V. N., Jacome, L. F. and MacLusky, N. J. (2003) Rapid enhancement of visual and place memory by estrogens in rats. *Endocrinology* **144**, 2836–2844.

Luine, V. N., McEwen, B. S. and Black, I. B. (1977) Effects of 17-beta estradiol on hypothalamic tyrosine hydroxylase activity. *Brain Research* **120**, 188–192.

Luine, V. N., Renner, K. J. and McEwen, B. S. (1986) Sex-dependent differences in estrogen regulation of choline acetyltransferase are altered by neonatal treatments. *Endocrinology* **119**, 874–878.

Luine, V. N., Richards, S. T., Wu, V. Y. and Beck, K. D. (1998) Estradiol enhances learning and memory in a spatial memory task and effects levels of monoaminergic neurotransmitters. *Hormones and Behavior* **34**, 149–162.

Lunn, M. P., Crawford, T. O., Hughes, R. A., Griffin, J. W. and Sheikh, K. A. (2002) Anti-myelin-associated glycoprotein antibodies alter neurofilament spacing. *Brain* **125**, 904–911.

Luskin, M. B., Parnavelas, J. G. and Barfield, J. A. (1993) Neurons, astrocytes, and oligodendrocytes of the rat cerebral cortex originate from separate progenitor cells: an ultrastructural analysis of clonally related cells. *Journal of Neuroscience* **13**, 1730–1750.

Luttges, M. W. and McGaugh, J. L. (1967) Permanence of retrograde amnesia produced by electroconvulsive shock. *Science* **156**, 408–410.

Madsen, T. M., Kristjansen, P. E., Bolwig, T. G. and Wortwein, G. (2003) Arrested neuronal proliferation and impaired hippocampal function following fractionated brain irradiation in the adult rat. *Neuroscience* **119**, 635–642.

Maehlen, J., Schroder, H. D., Klareskog, L., Olsson, T. and Kristensson, K. (1988) Axotomy induces MHC class I antigen expression on rat nerve cells. *Neuroscience Letters* **92**, 8–13.

Magarinos, A. M. and McEwen, B. S. (1995a) Stress-induced atrophy of apical dendrites of hippocampal CA3c neurons: comparison of stressors. *Neuroscience* **69**, 83–88.

Magarinos, A. M. and McEwen, B. S. (1995b) Stress-induced atrophy of apical dendrites of hippocampal CA3c neurons: involvement of glucocorticoid secretion and excitatory amino acid receptors. *Neuroscience* **69**, 89–98.

Magee, J. C. and Johnston, D. (1997) A synaptically controlled, associative signal for Hebbian plasticity in hippocampal neurons. *Science* 275, 209–213.

Maguire E. A. (2001) Neuroimaging studies of autobiographical events memory. *Philosophical Transactions of the Royal Society of London, Series B* 356, 1441–1452.

Mahanty, N. K. and Sah, P. (1998) Calcium-permeable AMPA receptors mediate long-term potentiation in interneurons in the amygdala. *Nature* 394, 683–687.

Maisonpierre, P. C., Belluscio, L., Friedman, B., Alderson, R. F., Wiegand, S. J., Furth, M. E., Lindsay, R. M. and Yancopoulos, G. D. (1990) NT-3, BDNF, and NGF in the developing rat nervous system: Parallel as well as reciprocal patterns of expression. *Neuron* 5, 501–509.

Makar, T. K., Nedergaard, M., Preuss, A., Gebard, A. S., Perumal, A. S. and Cooper, A. J. L. (1994) Vitamin E, ascorbate, glutathione, glutathione disulfide, and enzymes of oxidative metabolism in cultures of chick astrocytes and neurons: evidence that astrocytes play an important role in oxidative processes in the brain. *Journal of Neurochemistry* 62, 45–53.

Malberg, J. E., Eisch, A. J., Nestler, E. J. and Duman, R. S. (2000) Chronic antidepressant treatment increases neurogenesis in adult rat hippocampus. *Journal of Neuroscience* 20, 9104–9110.

Malenka, R. C. and Nicoll, R. A. (1999) Long-term potentiation – a decade of progress? *Science* 285, 1870–1874.

Malinow, R. (2003) AMPA receptor trafficking and long-term potentiation. *Philosophical Transactions of the Royal Society of London Series B Biological Science* 358, 707–714.

Malinow, R. and Malenka, R. C. (2002) AMPA receptor trafficking and synaptic plasticity. *Annual Review of Neuroscience* 25, 103–126.

Malkoski, S. P., Handanos, C. M. and Dorin, R. I. (1997) Localization of a negative glucocorticoid response element of the human corticotropin releasing hormone gene. *Molecular and Cellular Endocrinology* 127, 189–199.

Maltais, S., Cte, S., Drolet, G. and Falardeau, P. (2000) Cellular colocalization of dopamine D1 mRNA and D2 receptor in rat brain using a D2 dopamine receptor specific polyclonal antibody. *Progress in Neuro-Psychopharmacology and Biological Psychiatry* 24, 1127–1149.

Man, H. Y., Lin, J. W., Ju, W. H., Ahmadian, G., Liu, L., Becker, L. E., Sheng, M. and Wang, Y. T. (2000) Regulation of AMPA receptor-mediated synaptic transmission by clathrin-dependent receptor internalization. *Neuron* 25, 649–662.

Manabe, T., Togashi, H., Uchida, N., Suzuki, S. C., Hayakawa, Y., Yamamoto, M., Yoda, H., Miyakawa, T., Takeichi, M. and Chisaka, O. (2000) Loss of cadherin-11 adhesion receptor enhances plastic changes in hippocampal synapses and modifies behavioral responses. *Molecular and Cellular Neuroscience* 15, 534–546.

Mangiavacchi, S. and Wolf, M. E. (2004) D1 dopamine receptor stimulation increases the rate of AMPA receptor insertion onto the surface of cultured nucleus accumbens neurons through a pathway dependent on protein kinase A. *Journal of Neurochemistry* 88, 1261–1271.

Manji, H. K. (1992) G proteins: implications for psychiatry. *American Journal of Psychiatry* 149, 746–760.

Manji, H. K. and Lenox, R. H. (1999) Ziskind–Somerfeld Research Award. Protein kinase C signaling in the brain: molecular transduction of mood stabilization in the treatment of manic-depressive illness. *Biological Psychiatry* 46, 1328–1351.

Manji, H. K. and Lenox, R. H. (2000a) The nature of bipolar disorder. *Journal of Clinical Psychiatry* 61 (Supp. 13), 42–57.

Manji, H. K. and Lenox, R. H. (2000b) Signaling: cellular insights into the pathophysiology of bipolar disorder. *Biological Psychiatry* 48, 518–30.

Manji, H. K., Drevets, W. C. and Charney, D. S. (2001) The cellular neurobiology of depression. *Nature Medicine* 7, 541–547.

Manji, H. K., Chen, G., Shimon, H., Hsiao, J. K., Potter, W. Z. and Belmaker, R. H. (1995) Guanine nucleotide-binding proteins in bipolar affective disorder. Effects of long-term lithium treatment. *Archives of General Psychiatry* 52, 135–144.

Mansour, A. A., Babstock, D. M., Penney, J. H., Martin, G. M., McLean, J. H. and Harley, C. W. (2003) Novel objects in a holeboard probe the role of the locus coeruleus in curiosity: support for two modes of attention in the rat. *Behavioral Neuroscience* 117, 621–631.

Maquet, P., Schwartz, S., Passingham, R. and Frith, C. (2003) Sleep-related consolidation of a visuomotor skill: brain mechanisms as assessed by functional magnetic resonance imaging. *Journal of Neuroscience* **23**, 1432–1440.

Maquet, P., Laureys, S., Peigneux, P., Fuchs, S., Petiau, C., Phillips, C., Aerts, J., Del Fiore, G., Degueldre, C., Meulemans, T., Luxen, A., Franck, G., Van Der Linden, M., Smith, C. and Cleeremans, A. (2000) Experience-dependent changes in cerebral activation during human REM sleep. *Nature Neuroscience* **3**, 831–836.

Maren, S. (1999) Long-term potentiation in the amygdala: a mechanism for emotional learning and memory. *Trends in Neuroscience* **22**, 561–567.

Maren, S. & Fanselow, M. S. (1995) Synaptic plasticity in the basolateral amygdala induced by hippocampal formation stimulation in vivo. *Journal of Neuroscience* **15**, 7548–7564.

Maren, S. and Quirk, G. J. (2004) Neuronal signalling of fear memory. *Nat Rev Neurosci* **5**, 844–852.

Maren, S., De Oca, B. and Fanselow, M. S. (1994) Sex differences in hippocampal long-term potentiation (LTP) and Pavlovian fear conditioning in rats: positive correlation between LTP and contextual learning. *Brain Research* **661**, 25–34.

Maren, S., Yap, S. A. and Goosens, K. A. (2001) The amygdala is essential for the development of neuronal plasticity in the medial geniculate nucleus during auditory fear conditioning in rats. *Journal of Neuroscience* **21**, RC135.

Maren, S., Aharonov, G., Stote, D. L. and Fanselow, M. S. (1996) N-methyl-D-aspartate receptors in the basolateral amygdala are required for both acquisition and expression of conditional fear in rats. *Behavioral Neuroscience* **110**, 1365–1374.

Maren, S., Ferrario, C. R., Corcoran, K. A., Desmond, T. J. and Frey, K. A. (2003) Protein synthesis in the amygdala, but not the auditory thalamus, is required for consolidation of Pavlovian fear conditioning in rats. *European Journal of Neuroscience* **18**, 3080–3088.

Markham, M. A. and Greenough, W. T. (2004) Experience-driven brain plasticity: beyond the synapse. *Neuron Glia Biology* **1**, 351–364.

Markham, J. A., McKian, K. P., Stroup, T. S. and Juraska, J. M. (2004) Sexually dimorphic aging of dendritic morphology in CA1 of hippocampus. *Hippocampus* **15**, 97–103.

Markram, H., Lubke, J., Frotscher, M. and Sakmann, B. (1997) Regulation of synaptic efficacy by coincidence of postsynaptic APs and EPSPs. *Science* **275**, 213–215.

Martin, K. C. and Kosik, K. S. (2002). Synaptic tagging–who's it? *Nature Reviews Neuroscience* **3**, 813–820.

Martin, S. J., Grimwood, P. D. and Morris, R. G. M. (2000) Synaptic plasticity and memory, an evaluation of the hypothesis. *Annual Review of Neuroscience* **23**, 649–711.

Martin, K. C., Michael, D., Rose, J. C., Barad, M., Casadio, A., Zhu, H. and Kandel, E. R. (1997) MAP kinase translocates into the nucleus of the presynaptic cell and is required for long-term facilitation in Aplysia. *Neuron* **18**, 899–912.

Martinez, J. L. Jr. and Derrick, B. E. (1996) Long-term potentiation and learning. *Annual Reviews Psychology* **47**, 173–203.

Martinez, A., Alcantara, S., Borrell, V., Del Rio, J. A., Blasi, J., Otal, R., Campos, N., Boronat, A., Barbacid, M., Silos-Santiago, I. and Soriano, E. (1998) TrkB and TrkC signaling are required for maturation and synaptogenesis of hippocampal connections. *Journal of Neuroscience* **18**, 7336–7350.

Martinez-Turrillas, R., Frechilla, D. and Del Rio, J. (2002) Chronic antidepressant treatment increases the membrane expression of AMPA receptors in rat hippocampus. *Neuropharmacology* **43**, 1230–1237.

Marty, S., Wehrle, R. and Sotelo, C. (2000) Neuronal activity and brain-derived neurotrophic factor regulate the density of inhibitory synapses in organotypic slice cultures of postnatal hippocampus. *Journal of Neuroscience* **20**, 8087–8095.

Mascagni, F., McDonald, A. J. and Coleman, J. R. (1993) Corticoamygdaloid and corticocortical projections of the rat temporal cortex: a phaseolus vulgaris leucoagglutinin study. *Neuroscience* **57**, 697–715.

Mason, S. T. and Iversen, S. D. (1978) Reward, attention and the dorsal noradrenergic bundle. *Brain Research* **150**, 135–148.

Mason, J. W., Mangan, G., Brady, J. V., Conrad, D. and Rioch, D. M. (1961) Concurrent plasma epinephrine, norepinephrine and 17-hydroxycorticosteroid levels during conditioned emotional disturbances in monkeys. *Psychosomatic Medicine* **23**, 344–353.

Matsuda, S., Launey, T., Mikawa, S. and Hirai, H. (2000) Disruption of AMPA receptor GluR2 clusters following long-term depression induction in cerebellar Purkinje neurons. *Embo Journal* **19**, 2765–2774.

Matsumoto, A. and Arai, Y. (1980) **Sex dimorphism in "wiring pattern" in the hypothalamic arcuate nucleus** and its modification by neonatal hormonal environment. *Brain Research* **190**, 238–242.

Matsuzaki, M., Honkura, N., Ellis-Davies, G. C. and Kasai, H. (2004) Structural basis of long-term potentiation in single dendritic spines. *Nature* **429**, 761–766.

Matthews, K., Cauley, J., Yaffe, K. and Zmuda, J. M. (1999) Estrogen replacement therapy and cognitive decline in older community women. *Journal of the American Geriatric Society* **47**, 518–523.

Mattson, M. P. and Chan, S. L. (2003) Neuronal and glial calcium signaling in Alzheimer's disease. *Cell Calcium* **34**, 385–397.

Mattson, M. P., LaFerla, F. M., Chan, S. L., Leissring, M. A., Shepel, P. N. and Geiger, J. D. (2000) Calcium signaling in the ER: its role in neuronal plasticity and neurodegenerative disorders. *Trends in Neuroscience* **23**, 222–229.

Matute, C., Alberdi, E., Domercq, M., Perez-Cerda, F., Perez-Samartin, A. and Sanchez-Gomez, M. V. (2001) The link between excitotoxic oligodendroglial death and demyelinating diseases. *Trends in Neuroscience* **24**, 224–230.

Matyash, V., Filippov, V., Mohrhagen, K. and Kettenmann, H. (2001) Nitric oxide signals parallel fiber activity to Bergmann glial cells in the mouse cerebellar slice. *Molecular and Cellular Neuroscience* **18**, 664–670.

Mauch, D. H., Nägler, K., Schmacher, S., Göritz, C., Müller, E. C., Otto, A. and Pfrieger, F. W. (2001) CNS synaptogenesis promoted by glia-derived cholesterol. *Science* **294**, 1354–1357.

Maviel, T., Durkin, T. P., Menzaghi, F. and Bontempi, B. (2004) Sites of neocortical reorganization critical for remote spatial memory. *Science* **305**, 96–99.

Mayer, M. L., Westbrook, G. L. and Guthrie, P. B. (1984) Voltage-dependent block by Mg^{2+} of NMDA responses in spinal cord neurones. *Nature* **309**, 261–263.

Mayford, M., Bach, M. E., Huang, Y. Y., Wang, L., Hawkins, R. D. and Kandel, E. R. (1996) Control of memory formation through regulated expression of a CaMKII transgene. *Science* **274**, 1678–1683.

Mazzanti, M., Sul, J. Y. and Haydon, P. G. (2001) Glutamate on demand: astrocytes as a ready source. *Neuroscientist* **7**, 396–405.

Mazzucchelli, C. and Brambilla, R. (2000) Ras-related and MAPK signaling in neuronal plasticity and memory function. *Cell Molecular Life Science* **57**, 604–611.

McAllister, A. K. (2002) Spatial restricted actions of BDNF. *Neuron* **36**, 549–550.

McAllister, A. K., Katz, L. C. and Lo, D. C. (1996) Neurotrophin regulation of cortical dendritic growth requires activity. *Neuron* **17**, 1057–1064.

McAllister, A. K., Katz, L. C. and Lo, D. C. (1997) Opposing roles for endogenous BDNF and NT-3 in regulating cortical dendritic growth. *Neuron* **18**, 767–778.

McClelland, J. L., McNaughton, B. L. and O'Reilly, R. C. (1995) Why there are complementary learning systems in the hippocampus and neocortex: insights from the successes and failures of connectionist models of learning and memory. *Psychological Review* **102**, 419–457.

McCullumsmith, R. E., Gupta, D., Beneyto, M., Kreger, E., Haroutunian, V., Davis, K. L. and Meador-Woodruff, J. H. (2007) Expression of transcripts for myelination-related genes in the anterior cingulated cortex in schizophrenia. *Schizophrenia Research* **90**, 15–27.

McDonald, A. J. (1998) Cortical pathways to the mammalian amygdala. *Progress in Neurobiology* **55**, 257–332.

McDonald, A. J. and Augustine, J. R. (1993) Localization of GABA-like immunoreactivity in the monkey amygdala. *Neuroscience* **52**, 281–294.

McDonald, A. J. and Mascagni, F. (2001) Localization of the CB1 type cannabinoid receptor in the rat basolateral amygdala: high concentrations in a subpopulation of cholecystokinin-containing interneurons. *Neuroscience* **107**, 641–652.

McDonald, J. W., Althomsons, S. P., Hyrc, K. L., Choi, D. W. and Goldberg, M. P. (1998) Oligodendrocytes from forebrain are highly vulnerable to AMPA/kainate receptor-mediated excitotoxicity. *Nature Medicine* **4**, 291–297.

McEwen, B. S. (2001) Estrogen effects on the brain: multiple sites and molecular mechanisms. *Journal of Applied Physiology* **91**, 2785–2801.

McEwen, B. S. and Alves, S. E. (1999) Estrogen actions in the central nervous system. *Endocrine Reviews* **20**, 279–307.

McEwen, B. S., Gerlach, J. and Micco, D. (1975) Putative glucocorticoid receptors in hippocampus and other regions of the rat brain. In Pribram, R. I. K. (ed.) *The Hippocampus, Vol. 2: Neurophysiology and Behavior*. Plenum.

McEwen, B. S., Tanapat, P. and Weiland, N. G. (1999) Inhibition of dendritic spine induction on hippocampal

CA1 pyramidal neurons by a nonsteroidal estrogen antagonist in female rats. *Endocrinology* **140**, 1044–1047.

McGaugh, J. L. (1966) Time-dependent processes in memory storage. *Science* **153**, 1351–1358.

McGaugh, J. L. (1989) Involvement of hormonal and neuromodulatory systems in the regulation of memory storage. *Annual Review of Neuroscience* **12**, 255–287.

McGaugh, J. L. (2000) Memory – a century of consolidation. *Science* **287**, 248–251.

McGaugh J. L. (2002) Memory consolidation and the amygdala: a systems perspective. *Trends in Neurosciences* **25**, 456–461.

McGaugh, J. L. (2004) The amygdala modulates the consolidation of memories of emotionally arousing experiences. *Annual Review of Neuroscience* **27**, 1–28; doi: 10.1146/annurev.neuro27.070203.144157.

McGaugh, J. L. and Gold, P. E. (1989) Hormonal modulation of memory. In Brush, R. B. and Levine, S. (eds) *Psychoendocrinology*. Academic Press, pp 305–340.

McGaugh, J. L. and Introini-Collison, I. B. (1987) Hormonal and neurotransmitter interactions in the modulation of memory storage: involvement of the amygdala. *International Journal of Neurology* 21–22, 58–72.

McGaugh, J. L. and Roozendaal, B. (2002) Role of adrenal stress hormones in forming lasting memories in the brain. *Current Opinion in Neurobiology* **12**, 205–210.

McGaugh, J. L., McIntyre, C. K. and Power, A. E. (2002) Amygdala modulation of memory consolidation: interaction with other brain systems. *Neurobiology of Learning and Memory* **78**, 539–552.

McGaugh, J. L., Martinez, J. L. Jr., Messing, R. B., Liang, K. C., Jensen, R. A., Vasquez, B. J. and Rigter, H. (1982) Role of neurohormones as modulators of memory storage. *Advances in Biochemical Psychopharmacology* **33**, 123–130.

McGee, A. W., Yang, Y., Fischer, Q. S., Daw, N. W. and Strittmatter S. M. (2005) Experience-driven plasticity of visual cortex limited by myelin and Nogo receptor. *Science* **309**, 222–226.

McHugh, T. J., Blum, K. I., Tsien, J. Z., Tonegawa, S. and Wilson, M. A. (1996) Impaired hippocampal representation of space in CA1-specific NMDAR1 knockout mice. *Cell* **87**, 1339–1349.

McIlwain, H., Thomas, J. and Bell, J. L. (1956) The composition of isolated cerebral tissues: ascorbic acid and cozymase. *Biochemical Journal* **64**, 332–335.

McKernan, M. G. and Shinnick-Gallagher, P. (1997) Fear conditioning induces a lasting potentiation of synaptic currents in vitro. *Nature* **390**, 607–611.

McKerracher, L., David, S., Jackson, D. L., Kottis, V., Dunn, R. J. and Braun P. E. (1994) Identification of myelin-associated glycoprotein as a major myelin-derived inhibitor of neurite growth. *Neuron* **13**, 805–811.

McNaughton, B. L. and Morris, R. G. (1987) Hippocampal synaptic enhancement and information storatge within a distributed memory system. *Trends in Neurosciences* **10**, 408–415.

McTigue, D. M., Popovich, P. G., Morgan, T. E. and Stokes, B. T. (2000) Localization of transforming growth factor-beta1 and receptor mRNA after experimental spinal cord injury. *Experiments in Neurology* **163**, 220–230.

McTigue, D. M., Tani, M., Krivacic, K., Chernosky, A., Kelner, G. S., Maciejewski, D., Maki, R., Ransohoff, R. M. and Stokes, B. T. (1998) Selective chemokine mRNA accumulation in the rat spinal cord after contusion injury. *Journal of Neuroscience Research* **53**, 368–376.

Mead, A. N. and Stephens, D. N. (1998) AMPA-receptors are involved in the expression of amphetamine-induced behavioural sensitisation, but not in the expression of amphetamine-induced conditioned activity in mice. *Neuropharmacology* **37**, 1131–1138.

Meberg, P. J., Kinney, W. R., Valcourt, E. G. and Routtenberg, A. (1996) Gene expression of the transcription factor NF-kappa B in hippocampus: regulation by synaptic activity. *Brain Research. Molecular Brain Research* **38**, 179–190.

Medina, J. H. and Izquierdo, I. (1995) Retrograde messengers, long-term potentiation and memory. *Brain Research Brain Research Reviews* **21**, 185–194.

Medina, J. F., Repa, J., Mauk, M. D. and LeDoux, J. E. (2002) Parallels between cerebellum- and amygdala-dependent conditioning. *Nat Rev Neurosci* **3**, 122–131.

Meffert, M. K., Chang, J. M., Wiltgen, B. J., Fanselow, M. S. and Baltimore, D. (2003) NF-kappa B functions in synaptic signaling and behavior. *Nature Neuroscience* **6**, 1072–1078.

Meguro, H., Mori, H., Araki, K., Kushiya, E., Kutsuwada, T., Yamazaki, M., Kumanishi, T., Arakawa, M., Sakimura, K. and Mishina, M. (1992) Functional characterization of a heteromeric NMDA receptor channel expressed from cloned cDNAs. *Nature* **357**, 70–74.

Meister, A. (1994) Glutathione-ascorbic acid antioxidant system in animals. *Journal of Biological Chemistry* **269**, 9397–9400.

Merlo, E., Freudenthal, R., Maldonado, H. and Romano, A. (2005) Activation of the transcription factor NF-kappaB by retrieval is required for long-term memory reconsolidation. *Learning & Memory* **12**, 23–29.

Mermelstein, P. G., Becker, J. B. and Surmeier, D. J. (1996) Estradiol reduces calcium currents in rat neostriatal neurons via a membrane receptor. *Journal of Neuroscience* **16**, 595–604.

Mermelstein, P. G., Bito, H., Deisseroth, K. and Tsien, R. W. (2000) Critical dependence of cAMP response element-binding protein phosphorylation on L-type calcium channels supports a selective response to EPSPs in preference to action potentials. *Journal of Neuroscience* **20**, 266–273.

Mermelstein, P. G., Deisseroth, K., Dasgupta, N., Isaksen, A. L. and Tsien, R. W. (2001) Calmodulin priming: nuclear translocation of a calmodulin complex and the memory of prior neuronal activity. *Proceedings of the National Academy of Sciences of the United States of America* **98**, 15,342–15,347.

Meyer, U., van Kampen, M., Isovich, E., Flugge, G. and Fuchs, E. (2001) Chronic psychosocial stress regulates the expression of both GR and MR mRNA in the hippocampal formation of tree shrews. Hippocampus **11**, 329–336.

Meyer-Franke, A., Wilkinson, G. A., Kruttgen, A., Hu, M., Munro, E., Hanson, M. G., Jr., Reichardt, L. F. and Barres, B. A. (1998) Depolarization and cAMP elevation rapidly recruit TrkB to the plasma membrane of CNS neurons. *Neuron* **21**, 681–693.

Michailov, G. V., Sereda, M. W., Brinkmann, B. G., Fischer, T. M., Haug, B., Birchmeier, C., Role, L., Lai, C., Schwab, M. H. and Nave, K. A. (2004) Axonal neuregulin-1 regulates myelin sheath thickness. *Science* **304**, 700–703.

Migaud, M., Charlesworth, P., Dempster, M., Webster, L. C., Watabe, A. M., Makhinson, M., He, Y., Ramsay, M. F., Morris, R. G., Morrison, J. H., O'Dell, T. J. and Grant, S. G. (1998) Enhanced long-term potentiation and impaired learning in mice with mutant postsynaptic density-95 protein. *Nature* **396**, 433–439.

Migues, P. V., Johnston, A. N. B. and Rose, S. P. R. (2001) Dehydroepiandosterone and its sulphate enhance memory retention in day-old chicks. *Neuroscience* **109**, 243–251.

Milekic, M. H. and Alberini, C. M. (2002) Temporally graded requirement for protein synthesis following memory reactivation. *Neuron* **36**, 521–525.

Mileusnic, R., Anokhin, K. and Rose, S. P. R. (1996) Antisense oligodeoxynucleotides to c-fos are amnestic for passive avoidance in the chick *NeuroReport* **7**, 1269–1272.

Mileusnic, R., Lancashire, C. and Rose, S. P. R. (1999) Sequence specific impairment of memory formation by NCAM antisense oligonucleotides. *Learning & Memory* **6**, 120–127.

Mileusnic, R., Lancashire, C. and Rose, S. P. R. (2000) APP is required during an early phase of memory formation *European Journal of Neuroscience* **12**, 4487–4495.

Mileusnic, R., Lancashire, C. L. and Rose, S. P. R. (2004) The peptide sequence Arg-Glu-Arg, present in the amyloid precursor protein, protects against memory loss caused by Abeta and acts as a cognitive enhancer. *European Journal of Neuroscience* **19**, 1933–1938.

Mileusnic, R., Rose, S. P. R., Lancashire, C. and Bullock, S. (1995) Characterisation of antibodies specific for chick brain NCAM which cause amnesia for a passive avoidance task *Journal of Neurochemistry* **64**, 2598–2605.

Miller, R. F. (2001) The physiology and morphology of the vertebrate retina. In Ryan, S. J. (ed.) *Retina*, pp 138–170. Mosby.

Miller, R. F. (2004) D-Serine as a glial modulator of nerve cells. *Glia* **47**, 275–283.

Miller, S., Yasuda, M., Coats, J. K., Jones, Y., Martone, M. E. and Mayford, M. (2002) Disruption of dendritic translation of CaMKIIα impairs stabilization of synaptic plasticity and memory consolidation. *Neuron* **36**, 507–519.

Milner, B., Squire, L. R. and Kandel, E. R. (1998) Cognitive neuroscience and the study of memory. *Neuron* **20**, 445–468.

Milner, T. A., McEwen, B. S., Hayashi, S., Li, C. J., Reagan, L. and Alves, S. E. (2001) Ultrastructural evidence that hippocampal alpha estrogen receptors are located at extranuclear sites. *Journal of Comparative Neurology* **429**, 355–371.

Minichiello, L., Korte, M., Wolfer, D., Kuhn, R., Unsicker, K., Cestari, V., Rossi-Arnaud, C., Lipp, H.-P., Bonhoeffer, T. and Klein, R. (1999) Essential role for TrkB receptors in Hippocampus-mediated learning. *Neuron* **24**, 401–414.

Mirescu, C., Peters, J. D. and Gould, E. (2004) Early life experience alters response of adult neurogenesis to stress. *Nature Neuroscience* **7**, 841–846.

Misanin, J. R., Miller, R. R. and Lewis, D. J. (1968) Retrograde amnesia produced by electroconvulsive shock after reactivation of a consolidated memory trace. *Science* **160**, 554–555.

Miserendino, M. J., Sananes, C. B., Melia, K. R. and Davis, M. (1990) Blocking of acquisition but not expression of conditioned fear-potentiated startle by NMDA antagonists in the amygdala. *Nature* **345**, 716–718.

Mitchell, P. B., Manji, H. K., Chen, G., Jolkovsky, L., Smith-Jackson, E., Denicoff, K., Schmidt, M and Potter, W. Z. (1997) High levels of Gs alpha in platelets of euthymic patients with bipolar affective disorder. *American Journal of Psychiatry* **154**, 218–223.

Mitchell, S. J. and Silver, R. A. (2000a) GABA spillover from single inhibitory axons suppresses low-frequency excitatory transmission at the cerebellar glomerulus. *Journal of Neuroscience* **20**, 8651–8658.

Mitchell, S. J. and Silver, R. A. (2000b) Glutamate spillover suppresses inhibition by activating presynaptic mGluRs. *Nature* **404**, 498–502.

Mitra, S. K., Hanson, D. A. and Schlaepfer, D. D. (2005) Focal adhesion kinase: in command and control of cell motility. *Nature Review. Molecular Cell Biology* **6**, 56–68.

Moghaddam, B. and Gruen, R. J. (1991) Do endogenous excitatory amino acids influence striatal dopamine release? *Brain Research* **544**, 329–330.

Mohammed, A. K., Jonsson, G. and Archer, T. (1986) Selective lesioning of forebrain noradrenaline neurons at birth abolishes the improved maze learning performance induced by rearing in complex environment. *Brain Research* **398**, 6–10.

Mohammed, A. K., Winblad, B., Ebendal, T. and Larkfors, L. (1990) Environmental influence on behaviour and nerve growth factor in the brain. *Brain Research* **528**, 62–72.

Moita, M. A., Lamprecht, R., Nader, K. and LeDoux, J. E. (2002) A-kinase anchoring proteins in amygdala are involved in auditory fear memory. *Nature Neuroscience* **5**, 837–838.

Moita, M. A. P., Moisis, S., Zhou, Y., LeDoux, J. E., and Blair, H. T. (2003) Hippocampal place cells acquire location specific location specific responses to the conditioned stimulus during auditory fear conditioning. *Neuron* **37**, 485–497.

Monje, M. L., Mizumatsu, S., Fike, J. R. and Palmer, T. D. (2002) Irradiation induces neural precursor-cell dysfunction. *Nature Medicine* **8**, 955–962.

Moore, C. L., Kalil, R. and Richards, W. (1976) Development of myelination in optic tract of the cat. *Journal of Comparative Neurology* **165**, 125–136.

Moosmang, S., Haider, N., Klugbauer, N., Adelsberger, H., Langwieser, N., Muller, J., Stiess, M., Marais, E., Schulla, V., Lacinova, L., Goebbels, S., Nave, K. A., Storm, D. R., Hofmann, F. and Kleppisch, T. (2005) Role of hippocampal Cav1.2 Ca^{2+} channels in NMDA receptor-independent synaptic plasticity and spatial memory. *Journal of Neuroscience* **25**, 9883–9892.

Morgan, J. I. and Curran, T. (1991). Stimulus-transcription coupling in the nervous system: involvement of the inducible proto-oncogenes fos and jun. *Annual Review of Neuroscience* **14**, 421–451.

Morgan, S. L., Coussens, C. M. and Teyler, T. J. (2001) Depotentiation of vdccLTP requires NMDAR activation. *Neurobiology of Learning and Memory* **76**, 229–238.

Morgan, J. I., Cohen, D. R., Hempstead, J. L. and Curran, T. (1987). Mapping patterns of c-fos expression in the central nervous system after seizure. *Science* **237**, 192–197.

Morris, R. (1984).Developments of a water-maze procedure for studying spatial learning in the rat. *Journal of Neuroscience Methods* **11**, 47–60.

Morris, R. G., Moser, E. I., Riedel, G., Martin, S. J., Sandin, J., Day, M. and O'Carroll, C. (2003) Elements of a neurobiological theory of the hippocampus: the role of activity-dependent synaptic plasticity in memory. *Philosophical Transactions of the Royal Society of London: B Biological Science* **358**, 773–786.

Moser, M. B., Trommald, M., Egeland, T. and Andersen, P. (1997) Spatial training in a complex environment and isolation alter the spine distribution differently in rat CA1 pyramidal cells. *Journal of Comparative Neurology* **380**, 373–381.

Mothet, J. P., Pollegioni, L., Ouanounou, G., Martineau, M., Fossier, P., and Baux, G. (2005) Glutamate receptor activation triggers a calcium-dependent and SNARE protein-dependent release of the gliotransmitter D-serine. *Proceedings of the National Academy of Sciences, USA* **102**, 5606–5611.

Mothet, J. P., Parent, A. T., Wolosker, H., Brady Jr., R. O., Linden, D. J., Ferris, C. D., Rogawski, M. A. and Snyder, S. H. (2000) D-serine is an endogenous ligand for the glycine site of the N-methyl-D-aspartate receptor. *Proceedings of the National Academy of Sciences of the United States of America* **97**, 4926–4931.

Mourre, C., Smith, M. L., Siesjo, B. K. and Lazdunski, M. (1990) Brain ischemia alters the density of binding sites for glibenclamide, a specific blocker of ATP-sensitive K^+ channels. *Brain Research* **526**, 147–152.

Mu, J. S., Li, W. P., Yao, Z. B. and Zhao, X. F. (1999) Deprivation of endogenous brain-derived neurotrophic factor results in impairment of spatial learning and memory in adult rats. *Brain Research* **835**, 259–265.

Muellbacher, W., Ziemann, U., Wissel, J., Dang, N., Kofler, M., Facchini, S., Boroojerdi, B., Poewe, W. and Hallett, M. (2002) Early consolidation in human primary motor cortex. *Nature* **415**, 640–644.

Mulkey, R. M. and Malenka, R. C. (1992) Mechanisms underlying induction of homosynaptic long-term depression in area CA1 of the hippocampus. *Neuron* **9**, 967–975.

Muller, G. E. and Pilzecker, A. (1900) Experimentelle Beitrage zur Lehre von Gedachtnis. *Z Psychol Erganzungsband* **1**, 1–300.

Muller, D., Joly, M. and Lynch, G. (1988) Contributions of quisqualate and NMDA receptors to the induction and expression of LTP. *Science J1–s* **242**, 1694–1697.

Muller, J., Corodimas, K. P., Fridel, Z. LeDoux, J. E. (1997) Functional inactivation of the lateral and basal nuclei of the amygdala by muscimol infusion prevents fear conditioning to an explicit conditioned stimulus and to contextual stimuli. *Behavioral Neuroscience* **111**, 683–691.

Muller, R. U., Kubie, J. L., Ranck, J. B. Jr. (1987) Spatial firing patterns of hippocampal complex spike cells in a fixed environment. *Journal of Neuroscience* **7**, 1935–1950

Muller, R. U, Poucet, B., Fenton A. A. and Cressant, A. (1999) Is the hippocampus of the rat part of a specialized navigational system? *Hippocampus* **9**, 413–422.

Muller, R. A., Kleinhans, N., Pierce, K., Kemmotsu, N. and Courchesne, E. (2002) Functional MRI of motor sequence acquisition: effects of learning stage and performance. *Brain Research Cognitive Brain Research* **14**, 277–293.

Mulnard, R. A., Cotman, C. W., Kawas, C., van Dyck, C. H., Sano, M., Doody, R., Koss, E., Pfeiffer, E., Jin, S., Gamst, A., Grundman, M., Thomas, R. and Thal, L. J. (2000) Estrogen replacement therapy for treatment of mild to moderate Alzheimer's disease: a randomized controlled trial. *Alzheimer's Disease Cooperative Study. Journal of the American Medical Association* **283**, 1007–1015.

Mumby, D. G. (2001) Perspectives on object recognition memory following hippocampal damage: lessons from studies on rats. *Behavioural Brain Research* **127**, 159–181.

Mumby, D. G., Gaskin, S., Glenn, M. J., Scharamek, T. E. and Lehmann, H. (2002) Hippocampal damage and exploratory preferences in rats: memory for objects, place, and contexts. *Learning & Memory* **9**, 49–57.

Murai, K. K., Nguyen, L. N., Irie, F., Yamaguchi, Y. and Pasquale, E. B. (2003) Control of hippocampal dendritic spine morphology through ephrin-A3/EphA4 signaling. *Nature Neuroscience* **6**, 153–160.

Murphy, D. D. and Segal, M. (2000) Progesterone prevents estradiol-induced dendritic spine formation in cultured hippocampal neurons. *Neuroendocrinology* **72**, 133–143.

Murphy, G. G. and Glanzman, D. L. (1999) Cellular analog of differential classical conditioning in aplysia: disruption by the NMDA receptor antagonist DL-2-amino-5- phosphonovalerate. *Journal of Neuroscience* **19**, 10,595–10,602.

Murphy, S. J., Littleton-Kearney, M. T. and Hurn, P. D. (2002) Progesterone administration during reperfusion, but not preischemia alone, reduces injury in ovariectomized. *Journal of Cerebral Blood Flow and Metabolism* **22**, 1181–1188.

Murphy, D. D., Cole, N. B., Greenberger, V. and Segal, M. (1998) Estradiol increases dendritic spine density by reducing GABA neurotransmission in hippocampal neurons. *Journal of Neuroscience* **18**, 2550–2559.

Murray, N. and Steck, A. J. (1984) Impulse conduction regulates myelin basic protein phosphorylation in rat optic nerve. *Journal of Neurochemistry* **41**, 543–548.

Mustafa, A. K., Kim, P. M. and Snyder, S. H. (2004) D-serine as a putative glial neurotransmitter. *Neuron Glia Biology,* **1**, 275–281.

Myers, K. M. and Davis, M. (2002) Systems-level reconsolidation: reengagement of the hippocampus with memory reactivation. *Neuron* **36**, 340–343.

Nadel, L. and Bohbot, V. (2001) Consolidation of memory. *Hippocampus* **11**, 56–60.

Nader, K. (2003) Memory traces unbound. *Trends in Neuroscience* **26**, 65–72.

Nader, K., Ben Mamou, C. and Komorowski, B. (2004) *Double Dissociation of the Mechanisms Mediating the Induction of Reconsolidation from Those Mediating the Expression of a Conditioned Response.* Society for Neuroscience, Washington, DC.

Nader, K., Hardt, O. and Wang, S. H. (2005) Response to Alberini: right answer, wrong question. *Trends in Neuroscience* **28**, 346–347.

Nader, K., Schafe, G. E. and LeDoux, J. E. (2000) Fear memories require protein synthesis in the amygdala for reconsolidation after retrieval. *Nature* **406**, 722–726.

Nader, K., Schafe, G. E. and LeDoux, J. E. (2000b) The labile nature of consolidation theory. *Nature Reviews* **1**, 216–219.

Nader, K., Majidishad, P., Amorapanth, P. and LeDoux, J. E. (2001) Damage to the lateral and central, but not other, amygdaloid nuclei prevents the acquisition of auditory fear conditioning. *Learning & Memory* **8**, 156–163.

Nägler, K., Mauch, D. H. and Pfrieger, F. W. (2001) Glia-derived signals induce synapse formation in neurons of the rat central nervous system. *Journal of Physiology* **15**, 665–679.

Nakamura, T., Barbara, J. G., Nakamura, K. and Ross, W. N. (1999) Synergistic release of Ca^{2+} from IP3-sensitive stores evoked by synaptic activation of mGluRs paired with backpropagating action potentials. *Neuron* **24**, 727–737.

Nakazawa, K., McHugh, T. J., Wilson, M. A. and Tonegawa, S. (2004) NMDA receptors, place cells and hippocampal spatial memory. *Nature Reviews Neuroscience* **5**, 361–372.

Narisawa-Saito, M., Carnahan, J., Araki, K., Yamaguchi, T. and Nawa, H. (1999) Brain-derived neurotrophic factor regulates the expression of AMPA receptor proteins in neocortical neurons. *Neuroscience* **88**, 1009–1014.

Narisawa-Saito, M., Iwakura, Y., Kawamura, M., Araki, K., Kozaki, S., Takei, N. and Nawa, H. (2002) Brain-derived neurotrophic factor regulates surface expression of alpha-amino-3-hydroxy-5-methyl-4-isoxazoleproprionic acid receptors by enhancing the N-ethylmaleimide-sensitive factor/GluR2 interaction in developing neocortical neurons. *Journal of Biological Chemistry* **277**, 40,901–40,910.

Navarro, M. M., Orfila, J. E., Hernandez, R. V., Martinez, J. L. Jr. and LeBaron, R. G. (2000) The effect of a heparin sulfate proteoglycan binding peptide on long-term potentiation in area ca1 of the rat hippocampus. *Society for Neuroscience Abstracts* **26**, 1116.

Nawa, H. and Takei, N. (2001) BDNF as an anterophin; a novel neurotrophic relationship between brain neurons. *Trends in Neuroscience* **24**, 683–684.

Nedergaard, M. (1994) Direct signaling from astrocytes to neurons in cultures of mammalian brain cells. *Science* **263**, 1768–1771.

Nedergaard, M., Takano, T. and Hansen, A. J. (2002) Beyond the role of glutamate as a neurotransmitter. *Nature Reviews Neuroscience* **3**, 748–755.

Neeper, S. A., Gomez-Pinilla, F., Choi, J. and Cotman, C. W. (1996) Physical activity increases mRNA for brain-derived neurotrophic factor and nerve growth factor in rat brain. *Brain Research* **726**, 49–56.

Nestler, E. J., Barrot, M., DiLeone, R. J., Eisch, A. J., Gold, S. J. and Monteggia, L. M. (2002) Neurobiology of depression. *Neuron* **34**, 13–25.

Nestler, E. J., Gould, E. and Manji, H. (2002) Preclinical models: status of basic research in depression. *Biological Psychiatry* **52**, 503–528.

Neumann, H., Cavalie, A., Jenne, D. E. and Wekerle, H. (1995) Induction of MHC class I genes in neurons. *Science* **269**, 549–552.

Neumann, H., Schmidt, H., Cavalie, A., Jenne, D. and Wekerle, H. (1997) Major histocompatibility complex (MHC) class I gene expression in single neurons of the central nervous system: differential regulation by interferon (IFN)-gamma and tumor necrosis factor (TNF)-alpha. *Journal of Experimental Medixcine* **185**, 305–316.

Neve, K. A., Seamans, J. K. and Trantham-Davidson, H. (2004) Dopamine receptor signaling. *Journal of Receptor and Signal Transduction Research* **24**, 165–205.

Newman, E. A. (2001a) Glia of the retina. In Ryan, S. J. (ed.) *Retina*, pp 89–103. Mosby.

Newman, E. A. (2001b) Propagation of intercellular calcium waves in retinal astrocytes and Müller cells. *Journal of Neuroscience* **21**, 2215–2223.

Newman, E. A. (2003) Glial cell inhibition of neurons by release of ATP. *Journal of Neuroscience* **23**, 1659–1666.

Newman, E. A. (2003a) New roles for astrocytes: regulation of synaptic transmission. *TINS* **26**, 536–542.

Newman, E. A. (2004) Glial modulation of synaptic transmission in the retina. *Glia* **47**, 268–274.

Newman, E. A., (2004) Glia and synaptic transmission. In Kettenmann, H. and Ransom, B. R. (eds.) *Neuroglia*, 2nd edition. Oxford University Press, pp. 355–366.

Newman, E. A. (2005a) Glia and synaptic transmission. In Kettenmann, H. and Ransom, B. R. (eds) *Neuroglia*. Oxford University Press.

Newman, E. A. (2005b) Calcium increases in retinal glial cells evoked by light-induced neuronal activity. *Journal of Neuroscience* 25, 5502–5510.

Newman, E. A. and Zahs, K. R. (1997) Calcium waves in retinal glial cells. *Science* 275, 844–847.

Newman, E. A. and Zahs, K. R. (1998) Modulation of neuronal activity by glial cells in the retina. *Journal of Neuroscience* 18, 4022–4028.

Newton, A. C. (1995) Protein kinase C: structure, function, and regulation. *Journal of Biological Chemistry* 270, 28,495–28,498.

Ng, K. T., Gibbs, M. E., Crowe, S. F., Sedman, G. L., Hua, F., Zhao, W., O'Dowd, B., Rickard, N., Gibbs, C. L. and Sykova, E. et al. (1991) Molecular mechanisms of memory formation. *Molecular Neurobiology* 5, 333–350.

Nguyen, P. V., Abel, T. and Kandel, E. R. (1994) Requirement of a critical period of transcription for induction of a late phase of LTP. *Science* 265, 1104–1107.

Nibuya, M., Morinobu, S. and Duman, R. S. (1995) Regulation of BDNF and trkB mRNA in rat brain by chronic electroconvulsive seizure and antidepressant drug treatments. *Journal of Neuroscience* 15, 7539–7547.

Nichols, N. R., Zieba, M. and Bye, N. (2001) Do glucocorticoids contribute to brain aging? *Brain Research Brain Research Review* 37, 273–286.

Nicoll, R. A. and Malenka, R. C. (1999) Expression mechanisms underlying NMDA receptor-dependent long-term potentiation. *Annals of the New York Academy Sciences* 868, 515–525.

Nikolaev, E., Kaczmarek, L., Zhu, S. W., Winblad, B. and Mohammed, A. H. (2002) Environmental manipulation differentially alters c-Fos expression in amygdaloid nuclei following aversive conditioning. *Brain Research* 957, 91–98.

Nilsen, J. and Brinton, R. D. (2002a) Impact of progestins on estrogen-induced neuroprotection: synergy by progesterone and 19-norprogesterone and antagonism by medroxprogesterone acetate. *Endocrinology* 143, 205–212.

Nilsen, J. and Brinton, R. D. (2002b) Impact of progestins on estradiol potentiation of the glutamate calcium response. *Neuroreport* 13, 825–830.

Nilsen, J. and Brinton, R. D. (2003) Divergent impact of progesterone and medroxyprogesterone acetate (Provera) on nuclear mitogen-activated protein kinase signaling. *Proceedings of the National Academy of Sciences of the United States of America* 100, 10,506–10,511.

Nilsson, M., Perfilieva, E., Johansson, U., Orwar, O. and Eriksson, P. S. (1999) Enriched environment increases neurogenesis in the adult rat dentate gyrus and improves spatial memory. *Journal of Neurobiology* 39, 569–578.

Nirenberg, M. J., Chan, J., Liu, Y., Edwards, R. H. and Pickel, V. M. (1997) Vesicular monoamine transporter-2: immunogold localization in striatal axons and terminals. *Synapse* 26, 194–198.

Nishizuka, Y. (1992) Intracellular signaling by hydrolysis of phospholipids and activation of protein kinase C. *Science* 258, 607–614.

Nishizuka, Y. (1995) Protein kinase C and lipid signaling for sustained cellular responses. *Faseb Journal* 9, 484–496.

Noguchi, J., Matsuzaki, M., Ellis-Davies, G. C. and Kasai, H. (2005) Spine–neck geometry determines NMDA receptor-dependent Ca^{2+} signaling in dendrites. *Neuron* 46, 609–622.

Nonaka, S., Hough, C. J. and Chuang, D. M. (1998) Chronic lithium treatment robustly protects neurons in the central nervous system against excitotoxicity by inhibiting N-methyl-D-aspartate receptor-mediated calcium influx. *Proceedings of the National Academy of Science of the United States of America* 95, 2642–2647.

Nordeen, E. J., Nordeen, K. W., Sengelaub, D. R. and Arnold, A. P. (1985) Androgens prevent normally occurring cell death in a sexually dimorphic spinal nucleus. *Science* 229, 671–673.

Norris, C. M., Halpain, S. and Foster, T. C. (1998) Reversal of age-related alterations in synaptic plasticity by blockage of L-type Ca^{2+} channels. *Journal of Neuroscience* 18, 3171–3179.

Norris, C. M., Korol, D. L. and Foster, T. C. (1996) Increased susceptibility to induction of long-term depression and long-term potentiation reversal during aging. *Journal of Neuroscience* 16, 5382–5392.

Nottebohm, F. (1989) From bird song to neurogenesis. *Scientific American* 260, 74–79.

Novak, G., Kim, D., Seeman, P. and Tallerico, T. (2002) Schizophrenia and Nogo; elevated mRNA in cortex, and high prevalence of a homozygous CAA insert. *Brain Research, Molecular Brain Research* 107, 183–189.

Nowak, L., Bregestovski, P., Ascher, P., Herbet, A. and Prochiantz, A. (1984) Magnesium gates glutamate-activated channels in mouse central neurones. *Nature* 307, 462–465.

Nunez, J. L., Nelson, J., Pych, J. C., Kim, J. H. and Juraska, J. M. (2000) Myelination in the splenium of the corpus callosum in adult male and female rats. *Brain Research. Developmental Brain Research* 120, 87–90.

O'Brien, R. J., Lau, L. F. and Huganir, R. L. (1998) Molecular mechanisms of glutamate receptor clustering at excitatory synapses. *Current Opinion in Neurobiology* 8, 364–369.

O'Brien, R. J., Kamboj, S., Ehlers, M. D., Rosen, K. R., Fischbach, G. D. and Huganir, R. L. (1998) Activity-dependent modulation of synaptic AMPA receptor accumulation. *Neuron* 21, 1067–1078.

O'Dell, T. J., Huang, P. L., Dawson, T. M., Dinerman, J. L., Snyder, S. H., Kandel, E. R. and Fishman, M. C. (1991) Endothelial NOS and the blockade of LTP by NOS inhibitors in mice lacking neuronal NOS. *Science* 265, 542–546.

Ogoshi, F. and Weiss, J. H. (2003) Heterogeneity of Ca^{2+}-permeable AMPA/kainate channel expression in hippocampal pyramidal neurons: fluorescence imaging and immunocytochemical assessment. *Journal of Neuroscience* 23, 10,521–10,530.

Ohno, M., Frankland, P. W., Chen, A. P., Costa, R. M. and Silva, A. J. (2001) Inducible, pharmacogenetic approaches to the study of learning and memory. *Nature Neuroscience* 4, 1238–1243.

Ohyama, T., Nores, W. L., Murphy, M. and Mauk, M. D. (2003) What the cerebellum computes. *Trends in Neuroscience* 26, 222–227.

Olanow, C. W. and Tatton, W. G. (1999) Etiology and pathogenesis of Parkinson's disease. *Annual Review of Neuroscience* 22, 123–144.

Oleinick, N. L. (1977) Initiation and elongation of protein synthesis in growing cells: differential inhibition by cycloheximide and emetine. *Archives of Biochemical Biophysics* 182, 171–180.

Oliet, S. H., Piet, R. and Poulain, D. A. (2001) Control of glutamate clearance and synaptic efficacy by glial coverage of neurons. *Science* 292, 923–926.

Oliff, H. S., Berchtold, N. C., Isackson, P. and Cotman, C. W. (1998) Exercise-induced regulation of brain-derived neurotrophic factor (BDNF) transcripts in the rat hippocampus. *Brain Research Molecular Brain Research* 61, 147–153.

Oliveira, A. L., Thams, S., Lidman, O., Piehl, F., Hokfelt, T., Karre, K., Linda, H. and Cullheim, S. (2004) A role for MHC class I molecules in synaptic plasticity and regeneration of neurons after axotomy. *Proceedings of the National Academy of Science of the United States of America* 101, 17,843–17,848.

Olsson, T., Kristensson, K., Ljungdahl, A., Maehlen, J., Holmdahl, R. and Klareskog, L. (1989) Gamma-interferon-like immunoreactivity in axotomized rat motor neurons. *Journal of Neuroscience* 9, 3870–3875.

O'Mahony, A., Raber, J., Montano, M., Foehr, E., Han, V., Lu, S. M., Kwon, H., LeFevour, A., Chakraborty-Sett, S. and Greene, W. C. (2006) NF-kappaB/Rel regulates inhibitory and excitatory neuronal function and synaptic plasticity. *Molecular and Cellular Biology* 26, 7283–7298.

Orfila, J. E., Hernandez, R. V., LeBaron, R. G. and Martinez, J. L. (2000) The effects of an integrin binding peptide on theta burst and PTP in area CA1 of the rat hippocampus. *Society for Neuroscience Abstracts* 26, 879.

Orkand, R. K., Nicholls, J. G. and Kuffler, S. W. (1966) Effect of nerve impulses on the membrane potential of glial cells in the central nervous system of amphibia. *Journal of Neurophysiology* 29, 788–806.

Orth, M. and Schapira, A. H. (2002) Mitochondrial involvement in Parkinson's disease. *Neurochemistry International* 40, 533–541.

Osawa, M., Tong, K. I., Lilliehook, C., Wasco, W., Buxbaum, J. D., Cheng, H. Y., Penninger, J. M., Ikura, M. and Ames, J. B. (2001) Calcium-regulated DNA binding and oligomerization of the neuronal calcium-sensing protein, calsenilin/DREAM/KChIP3. *Journal of Biological Chemistry* 276, 41,005–41,013.

Ostroff, L. E., Fiala, J. C., Allwardt, B. and Harris, K. M. (2002) Polyribosomes redistribute from dendritic shafts into spines with enlarged synapses during LTP in developing rat hippocampal slices. *Neuron* 35, 535–545.

Otani, S. and Abraham, W. C. (1989) Inhibition of protein synthesis in the dentate gyrus, but not the entorhinal cortex, blocks maintenance of long-term potentiation in rats. *Neuroscience Letters* **106**, 175–80.

Ou, L. C. and Gean, P. W. (2006) Regulation of amygdala-dependent learning by brain-derived neurotrophic factor is mediated by extracellular signal-regulated kinase and phosphatidylinositol-3-kinase. *Neuropsychopharmacology* **31**, 287–296.

Packard, M. G. and Teather, L. A. (1997) Posttraining estradiol injections enhance memory in ovariectomized rats: cholinergic blockage and synergism. *Neurobiology of Learning and Memory* **68**, 172–188.

Paganini-Hill, A. and Henderson, V. W. (1996) Estrogen replacement therapy and risk of Alzheimer's disease. *Archives of Internal Medicine* **156**, 2213–2217.

Palmer, T. D., Ray, J. and Gage, F. H. (1995) FGF-2-responsive neuronal progenitors reside in proliferative and quiescent regions of the adult rodent brain. *Molecular Cellular Neuroscience* **6**, 474–486.

Palmer, T. D., Takahashi, J. and Gage, F. H. (1997) The adult rat hippocampus contains primordial neural stem cells. *Molecular and Cellular Neuroscience* **8**, 389–404.

Pan, W., Kastin, A. J., Bell, R. L. and Olson, R. D. (1999) Upregulation of tumor necrosis factor alpha transport across the blood-brain barrier after acute compressive spinal cord injury. *Journal of Neuroscience* **19**, 3649–3655.

Pang, P. T. and Lu, B. (2004) Regulation of late-phase LTP and long-term memory in normal and aging hippocampus: role of secreted proteins tPA and BDNF. *Ageing Research Reviews* **3**, 407–430.

Pantev, C., Ross, B., Fujioka, T., Trainor, L. J., Schulte, M. and Schulz, M. (2003) Music and learning-induced cortical plasticity. *Annals of the New York Academy of Sciences* **999**, 438–450.

Pantoni, L., Lamassa, M. and Inzitari, D. (2000) Transient global amnesia: a review emphasizing pathogenic aspects. *Acta Neurologica Scandinavica* **102**, 275–283.

Parducz, A. and Garcia-Segura, L. M. (1993) Sexual differences in the synaptic connectivity in the rat dentate gyrus. *Neuroscience Letters* **161**, 53–56.

Paré, D. and Smith, Y. (1998) Intrinsic circuitry of the amygdaloid complex: common principles of organization in rats and cats. *Trends in Neuroscience* **21**, 240–241.

Paré, D., Quirk, G. J. and LeDoux, J. E. (2004) New vistas on amygdala networks in conditioned fear. *Journal of Neurophysiology* **92**, 1–9.

Parent, J. M., Yu, T. W., Leibowitz, R. T., Geschwind, D. H., Sloviter, R. S. and Lowenstein, D. H. (1997) Dentate granule cell neurogenesis is increased by seizures and contributes to aberrant network reorganization in the adult rat hippocampus. *Journal of Neuroscience* **17**, 3727–3738.

Park, E., Liu, Y. and Fehlings, M. G. (2003) Changes in glial cell white matter AMPA receptor expression after spinal cord injury and relationship to apoptotic cell death. *Experimental Neurology* **182**, 35–48.

Park, C. S., Gong, R., Stuart, J. and Tang, S. J. (2006) Molecular network and chromosomal clustering of genes involved in synaptic plasticity in the hippocampus. *Journal of Biological Chemistry* **281**, 30,195–30,211.

Park, M., Penick, E. C., Edwards, J. G., Kauer, J. A. and Ehlers, M. D. (2004) Recycling endosomes supply AMPA receptors for LTP. *Science* **305**, 1972–1975.

Parkes, J. D., Calver, D. M., Zilkha, K. J. and Knill-Jones, R. P. (1970) Controlled trial of amantadine hydrochloride in Parkinson's disease. *Lancet* **1**, 259–262.

Parpura, V., Basarsky, T. A., Liu, F., Jeftinija, K., Jeftinija, S. and Haydon, P. G. (1994) Glutamate-mediated astrocyte-neuron signaling. *Nature* **369**, 744–747.

Parri, H. R., Gould, T. M. and Crunelli, V. (2001) Spontaneous astrocytic Ca^{2+} oscillations in situ drive NMDAR-mediated neuronal excitation. *Nature Neuroscience* **4**, 803–812.

Pascual-Leone, A. and Torres, F. (1993) Plasticity of the sensorimotor cortex representation of the reading finger in Braille readers. *Brain* **116 (Pt 1)**, 39–52.

Passafaro, M., Piech, V. and Sheng, M. (2001) Subunit-specific temporal and spatial patterns of AMPA receptor exocytosis in hippocampal neurons. *Nature Neuroscience* **4**, 917–926.

Pasti, L., Volterra, A., Pozzan, T. and Carmignoto, G. (1997) Intracellular calcium oscillations in astrocytes: a highly plastic, bidirectional form of communication between neurons and astrocytes in situ. *Journal of Neuroscience* **17**, 7817–7830.

Pasti, L., Zonta, M., Pozzan, T., Vicini, S. and Carmignoto, G. (2001) Cytosolic calcium oscillations in astrocytes may regulate exocytotic release of glutamate. *Journal of Neuroscience* **21**, 477–484.

Patapoutian, A. and Reichardt, L. F. (2001) Trk receptors: mediators of neurotrophin action. *Current Opinion in Neurobiology* **11**, 272–280.

Patel, J., Avshalumov, M. V. and Rice, M. E. (2004) Glutamate-dependent inhibition of striatal dopamine release is mediated by H_2O_2 generated in medium spiny neurons but not in dopamine terminals. *Abstract Viewer/Itinerary Planner*. Society for Neuroscience, 46.6.

Patel, S. N., Rose, S. P. R. and Stewart, M. G. (1988) Training induced dendritic spine density changes are specifically related to memory formation processes in the chick, gallus domesticus. *Brain Research* **463**, 168–173.

Patterson, T. A. and Rose, S. P. R. (1992) Memory in the chick: multiple cues, distinct brain locations. *Behavioral Neuroscienc*, **106**, 465–470.

Patterson, S. L., Grover, L. M., Schartzkroin, P. A. and Bothwell, M. (1992) Neurotophin expression in rat hippocampal slices: a stimulus paradigm inducing LTP in CA1 evokes increases in BDNF and NT-3 mRNAs. *Neuron* **9**, 1081–1088.

Patterson, T. A., Alvarado, M. C., Rosenzweig, M. R. and Bennett, E. L. (1988) Time courses of amnesia development in two different areas of the chick brain. *Neurochemical Resesearch* **10**, 1071–1081.

Patterson, S. L., Abel, T., Deuel, T. A., Martin, K. C., Rose, J. C. and Kandel, E. R. (1996) Recombinant BDNF rescues deficits in basal synaptic transmission and hippocampal LTP in BDNF knockout mice. *Neuron* **16**, 1137–1145.

Pavlides, C., Greenstein, Y. J., Grudman, M. and Winson, J. (1988) Long-term potentiation in the dentate gyrus is induced preferentially on the positive phase of theta-rhythm. *Brain Research* **439**, 383–387.

Paylor, R., Johnson, R. S., Papaioannou, V., Spiegelman, B. M. and Wehner, J. M. (1994) Behavioral assessment of c-fos mutant mice. *Brain Research* **651**, 275–282.

Payne, A. H. and Hales, D. B. (2004) Overview of steroidogenic enzymes in the pathway from cholesterol to active steroid hormones. *Endocrine Reviews* **25**, 947–970.

Payne, J. L., Quiroz, J. A., Gould, T. G., Zarate, C. A. J. and Manji, H. K. (2004) *The Cellular Neurobiology of Bipolar Disorder*, 2nd edition. Oxford University Press.

Pedreira, M. E., Perez-Cuesta, L. M. and Maldonado, H. (2002) Reactivation and reconsolidation of long-term memory in the crab Chasmagnathus: protein synthesis requirement and mediation by NMDA-type glutamatergic receptors. *Journal of Neuroscience* **22**, 8305–8311.

Peigneux, P., Laureys, S., Fuchs, S., Collette, F., Perrin, F., Reggers, J., Philips, C., Degueldre, C., Del Fiore, G., Aerts, J., Luxen, A. and Maquet, P. (2004) Are spatial memories strengthened in the human hippocampus during slow wave sleep? *Neuron* **44**, 535–545.

Peigneux, P., Laureys, S., Fuchs, S., Destrebecqz, A., Collette, F., Delbeuck, X., Phillips, C., Aerts, J., Del Fiore, G., Degueldre, C., Luxen, A., Cleeremans, A. and Maquet, P. (2003) Learned material content and acquisition level modulate cerebral reactivation during posttraining rapid-eye-movements sleep. *Neuroimage* **20**, 125–134.

Peirce, T. R., Bray, N. J., Williams, N. M., Norton, N., Moskvina, V., Preece, A., Haroutunian, V., Buxbaum, J. D., Owen, M. J. and O'Donovan M. C. (2006) Convergent evidence for 2′, 3′-cyclic nucleotide 3′-phosphodiesterase as a possible susceptibility gene for schizophrenia. *Archives in General Psychiatry* **63**, 18–24.

Peles, E. and Salzer, J. L. (2000) Molecular domains of myelinated axons. *Current Opinion in Neurobiology* **10**, 558–565.

Pellegrini-Giampietro, D. E., Gorter, J. A., Bennett, M. V. and Zukin, R. S. (1997) The GluR2 (GluR-B) hypothesis: $Ca^{(2+)}$-permeable AMPA receptors in neurological disorders. *Trends in Neuroscience* **20**, 464–470.

Pellmar, T. (1986) Electrophysiological correlates of peroxide damage in guinea pig hippocampus in vitro. *Brain Research* **364**, 377–381.

Pellmar, T. C. (1987) Peroxide alters neuronal excitability in the CA1 region of guinea-pig hippocampus in vitro. *Neuroscience* **23**, 447–456.

Pellmar, T. C. (1995) Use of brain slices in the study of free-radical actions. *Journal of Neuroscience Methods* **59**, 93–98.

Pencea, V., Bingaman, K. D., Wiegand, S. J. and Luskin, M. B. (2001) Infusion of brain-derived neurotrophic factor into the lateral ventricle of the adult rat leads to new neurons in the parenchyma of the striatum, septum, thalamus, and hypothalamus. *Journal of Neuroscience* **21**, 6706–6717.

Pereira, R. A., Tscharke, D. C. and Simmons, A. (1994) Upregulation of class I major histocompatibility complex gene expression in primary sensory neurons, satellite cells, and Schwann cells of mice in response to acute but not latent herpes simplex virus infection in vivo. *Journal of Experimental Medicine* **180**, 841–850.

Perez, J., Tardito, D., Mori, S., Racagni, G., Smeraldi, E. and Zanardi, R. (2000) Abnormalities of cAMP signaling in affective disorders: implication for pathophysiology and treatment. *Bipolar Disord* **2**, 27–36.

Perez, J. L., Khatri, L., Chang, C., Srivastava, S., Osten, P. and Ziff, E. B. (2001) PICK1 targets activated protein kinase Calpha to AMPA receptor clusters in spines of hippocampal neurons and reduces surface levels of the AMPA-type glutamate receptor subunit 2. *Journal of Neuroscience* **21**, 5417–5428.

Perrot-Sinal, T. S., Kostenuik, M. A., Ossenkopp, K. P. and Kavaliers, M. (1996) Sex differences in performance in the Morris water maze and the effects of initial nonstationary hidden platform training. *Behavioral Neuroscience* **110**, 1309–1320.

Perry, V. H., Bell, M. D., Brown, H. C. and Matyszak, M. K. (1995) Inflammation in the nervous system. *Current Opinion in Neurobiology* **5**, 636–641.

Peschon, J. J., Torrance, D. S., Stocking, K. L., Glaccum, M. B., Otten, C., Willis, C. R., Charrier, K., Morrissey, P. J., Ware, C. B. and Mohler, K. M. (1998) TNF receptor-deficient mice reveal divergent roles for p55 and p75 in several models of inflammation. *Journal of Immunology* **160**, 943–952.

Peters, O., Schipke, C. G., Hashimoto, Y. and Kettenmann, H. (2003) Different mechanisms promote astrocyte Ca^{2+} waves and spreading depression in the mouse neocortex. *Journal of Neuroscience* **23**, 9888–9896.

Peuchen, S., Bolanos, J. P., Heales, S. J., Almeida, A., Duchen, M. R. and Clark, J. B. (1997) Interrelationships between astrocyte function, oxidative stress and antioxidant status within the central nervous system. *Progress in Neurobiology* **52**, 261–281.

Pfaff, D. W. and McEwen, B. S. (1983) Actions of estrogens and progestins on nerve cells. *Science* **219**, 808–814.

Pfrieger, F. W. and Barres, B. A. (1997) Synaptic efficacy enhanced by glial cells in vitro. *Science* **277**, 1684–1687.

Pham, T. M., Winblad, B., Granholm, A.-C., and Mohammed, A. H. (2002) Environmental influences on brain neurotrophins in rats. *Pharmacology, Biochemistry, and Behavior* **73**, 167–175.

Phillips, A. G., Ahn, S. and Floresco, S. B. (2004) Magnitude of dopamine release in medial prefrontal cortex predicts accuracy of memory on a delayed response task. *Journal of Neuroscience* **24**, 547–553.

Pierce, T. R. et al. (2006) Convergent evidence for $2',3'$-cyclic nucleotide $3'$-phosphodiesterase as a possible susceptibility gene for schizophrenia. *Archives in General Psychiatry* **63**, 18–24.

Pierschbacher, M. D. and Ruoslahti, E. (1984) Cell attachment activity of fibronectin can be duplicated by small synthetic fragments of the molecule. *Nature* **309**, 30–33.

Pin, J. P. and Duvoisin, R. (1995) The metabotropic glutamate receptors: structure and functions. *Neuropharmacology* **34**, 1–26.

Pinard, C. R., Mascagni, F. and McDonald, A. J. (2005) Neuronal localization of Ca(v)1.2 L-type calcium channels in the rat basolateral amygdala. *Brain Research* **1064**, 52–55.

Pinsker, H., Kupfermann, I., Castellucci, V. and Kandel, E. (1970) Habituation and dishabituation of the gill-withdrawal reflex in Aplysia. *Science* **167**, 1740–1742.

Pisani, A., Bonsi, P., Martella, G. et al (2004) Intracellular calcium increase in epileptiform activity: modulation by levetiracetam and lamotrigine. *Epilepsia* **45**, 719–728.

Pitkanen, A., Savander, V. and LeDoux, J. E. (1997) Organization of intra-amygdaloid circuitries in the rat: an emerging framework for understanding functions of the amygdala. *Trends in Neuroscience* **20**, 517–523.

Pitman, R. K., Sanders, K. M., Zusman, R. M., Healy, A. R., Cheema, F., Lesko, M. B., Cahill, L. and Orr, S. P. (2002) Pilot study of secondary prevention of post-traumatic stress disorder with propranolol. *Biological Psychiatry* **51**, 241–242.

Platenik, J., Kuramoto, N. and Yoneda, Y. (2000) Molecular mechanisms associated with long-term consolidation of the NMDA signals. *Life Sciences* **67**, 335–364.

Poe, G. R., Nitz, D. A., McNaughton, B. L. and Barnes, C. A. (2000) Experience-dependent phase-reversal of hippocampal neuron firing during REM sleep. *Brain Research* **855**, 176–180.

Poldrack, R. A. and Rodriguez, P. (2003) Sequence learning: what's the hippocampus to do? *Neuron* **37**, 891–893.

Poo, M. M. (2001) Neurotrophins as synaptic modulators. *Nature Reviews Neuroscience* **2**, 24–32.

Popoli, P., Frank, C., Tebano, M. T., Potenza, R. L., Pintor, A., Domenici, M. R., Nazzicone, V., Pèzzola, A and Reggio, R. (2003) Modulation of glutamate release and excitotoxicity by adenosine A(2A) receptors. *Neurology* **61**, S69–S71.

Popovich, P. G., Guan, Z., McGaughy, V., Fisher, L., Hickey, W. F. and Basso, D. M. (2002) The neuropathological and behavioral consequences of intraspinal microglial/macrophage activation. *Journal of Neuropathology and Experimental Neurology* **61**, 623–633.

Porter, J. T. and McCarthy, K. D. (1996) Hippocampal astrocytes in situ respond to glutamate released from synaptic terminals. *Journal of Neuroscience* **16**, 5073–5081.

Porter, J. T. and McCarthy, K. D. (1997) Astrocytic neurotransmitter receptors in situ and in vivo. *Progress in Neurobiology* **51**, 439–455.

Pouliot, W. A., Handa, R. J. and Beck, S. G. (1996) Androgen modulates N-methyl-D-aspartate-mediated depolarization in CA1 hippocampal pyramidal cells. *Synapse* **23**, 10–19.

Power, J. M., Wu, W. W., Sametsky, E., Oh, M. M. and Disterhoft, J. F. (2002) Age-related enhancement of the slow outward calcium-activated potassium current in hippocampal CA1 pyramidal neurons in vitro. *Journal of Neuroscience* **22**, 7234–7243.

Pozzo-Miller, L. D., Gottschalk, W., Zhang, L., McDermott, K., Du, J., Gopalakrishnan, R., Oho, C., Sheng, Z. H. and Lu, B. (1999) Impairments in high-frequency transmission, synaptic vesicle docking, and synaptic protein distribution in the hippocampus of BDNF knockout mice. *Journal of Neuroscience* **19**, 4972–4983.

Preston, A., Shrager, Y., Dudukovic, N. M. and Gabrieli, J. D. E. (2004) Hippocampal contribution to the novel use of relational information in declarative memory. *Hippocampus* **14**, 148–152.

Presutti, C., Rosati, J., Vincenti, S. and Nasi, S. (2006) Non coding RNA and brain. *BMC Neuroscience* **7**, S5.

Przybyslawski, J. and Sara, S. J. (1997) Reconsolidation of memory after its reactivation. *Behavioural Brain Research* **84**, 241–246.

Przybyslawski, J., Roullet, P. and Sara, S. J. (1999) Attenuation of emotional and nonemotional memories after their reactivation: role of beta adrenergic receptors. *Journal of Neuroscience* **19**, 6623–6628.

Purves, D. (1988) *Body and Brain: A Trophic Theory of Neural Connections*. Harvard University Press.

Quartermain, D. and McEwen, B. S. (1970) Temporal characteristics of amnesia induced by protein synthesis inhibitor: determination by shock level. *Nature* **228**, 677–678.

Quevedo, J., Vianna, M., Daroit, D., Born, A. G., Kuyven, C. R., Roesler, R. and Quillfeldt, J. A. (1998) L-type voltage-dependent calcium channel blocker nifedipine enhances memory retention when infused into the hippocampus. *Neurobiology of Learning and Memory* **69**, 320–325.

Quevedo, J., Vianna, M. R., Martins, M. R., Barichello, T., Medina, J. H., Roesler, R. and Izquierdo, I. (2004) Protein synthesis, PKA, and MAP kinase are differentially involved in short- and long-term memory in rats. *Behavioral Brain Research* **154**, 339–343.

Quirk, G. J., Armony, J. L. and LeDoux, J. E. (1997) Fear conditioning enhances different temporal components of tone-evoked spike trains in auditory cortex and lateral amygdala. *Neuron* **19**, 613–624.

Quirk, G. J., Repa, C. and LeDoux, J. E. (1995) Fear conditioning enhances short-latency auditory responses of lateral amygdala neurons: parallel recordings in the freely behaving rat. *Neuron* **15**, 1029–1039.

Rainnie, D. G. (1999) Serotonergic modulation of neurotransmission in the rat basolateral amygdala. *Journal of Neurophysiology* **82**, 69–85.

Rainnie, D. G., Asprodini, E. K. and Shinnick-Gallagher, P. (1991) Inhibitory transmission in the basolateral amygdala. *Journal of Neurophysiology* **66**, 999–1009.

Raisman, G. and Field, P. M. (1973) Sexual dimorphism in the neuropil of the preoptic area of the rat and its dependence on neonatal androgen. *Brain Research* **54**, 1–29.

Rall, G. F., Mucke, L. and Oldstone, M. B. (1995) Consequences of cytotoxic T lymphocyte interaction with major histocompatibility complex class I-expressing neurons in vivo. *Journal of Experimental Medicine* **182**, 1201–1212.

Rampon, C. and Tsien, J. Z. (2000) Genetic analysis of learning behavior-induced structural plasticity. *Hippocampus* **10**, 605–609.

Rampon, C., Tang, Y. P., Goodhouse, J., Shimizu, E., Kyin, M. and Tsien, J. Z. (2000) Enrichment induces structural changes and recovery from nonspatial memory deficits in CA1 NMDAR1-knockout mice. *Nature Neuroscience* **3**, 238–244.

Ramsden, M., Shin, T. M. and Pike, C. J. (2003) Androgens modulate neuronal vulnerability to kainate lesion. *Neuroscience* **122**, 573–578.

Ramsden, M., Nyborg, A. C., Murphy, M. P., Chang, L., Stanczyk, F. Z., Golde, T. E. and Pike, C. J. (2003) Androgens modulate b-amyloid levels in male rat brain. *Journal of Neurochemistry* **87**, 1052–1055.

Ranganath, C., Yonelinas, A. P., Cohen, M. X., Dy, C. J., Tom, S. M. and D'Esposito, M. D. (2003) Dissociable correlates of recollection and familiarity with the medial temporal lobes. *Neuropsychologia* **42**, 2–13.

Ransom, B. R. (2000) Glial modulation of neural excitability mediated by extracellular pH: a hypothesis revisited. *Progress in Brain Research* **125**, 217–228.

Raps, S. P., Lai, J. C. K., Hertz, L. and Cooper, A. J. L. (1989) Glutathione is present in high concentrations in cultured astrocytes but not in cultured neurons. *Brain Research* **493**, 398–401.

Rattiner, L. M., Davis, M., French, C. T. and Ressler, K. J. (2004) Brain-derived neurotrophic factor and tyrosine kinase receptor B involvement in amygdala-dependent fear conditioning. *Journal of Neuroscience* **24**, 4796–4806.

Raymond, C. R. and Redman, S. J. (2006) Spatial segregation of neuronal calcium signals encodes different forms of LTP in rat hippocampus. *Journal of Physiology* **570**, 97–111.

Recanzone, G. H., Merzenich, M. M., Jenkins, W. M., Grajski, K. A. and Dinse, H. R. (1992) Topographic reorganization of the hand representation in cortical area 3b owl monkeys trained in a frequency-discrimination task. *Journal of Neurophysiology* **67**, 1031–1056.

Rechtschaffen, A. and Kales, A. (1968) *A Manual of Standardized Terminology, Techniques, and Scoring System for Sleep Stages of Human Subjects.* US Department of Health.

Redmond, L. and Ghosh, A. (2001) The role of Notch and Rho GTPase signaling in the control of dendritic development. *Current Opinion in Neurobiology* **11**, 111–117.

Redwine, J. M., Buchmeier, M. J. and Evans, C. F. (2001) In vivo expression of major histocompatibility complex molecules on oligodendrocytes and neurons during viral infection. *American Journal of Pathology* **159**, 1219–1224.

Rees, P. M. (2003) Contemporary issues in mild traumatic brain injury. *Archives of Physical Medicine and Rehabilitation* **84**, 1885–1894.

Reif, A., Schmitt, A., Fritzen, S., Chourbaji, S., Bartsch, C., Urani, A., Wycislo, M., Mossner, R., Sommer, C., Gass, P. and Lesch, K. P. (2004) Differential effect of endothelial nitric oxide synthase (NOS-III) on the regulation of adult neurogenesis and behaviour. *European Journal of Neuroscience* **20**, 885–895.

Reimann-Philipp, U., Ovase, R., Weigel, P. H. and Grammas, P. (2001) Mechanisms of cell death in primary cortical neurons and PC12 cells. *Journal of Neuroscience Research* **64**, 654–660.

Repa, J. and LeDoux, J. (2001) Conditioned neural activity in the dorsal subnucleus of the lateral amygdala predicts behavioral fear learning in freely moving rat. *Journal of Neuroscience*.

Repa, J. C., Muller, J., Apergis, J., Desrochers, T. M., Zhou, Y. and LeDoux, J. E. (2001) Two different lateral amygdala cell populations contribute to the initiation and storage of memory. *Nature Neuroscience* **4**, 724–731.

Rescorla, R. A. (1967) Pavlovian conditioning and its proper control procedures. *Psychological Reviews* **74**, 71–80.

Resnick, S. M. and Maki, P. M. (2001) Effects of hormone replacement therapy on cognitive and brain aging. *Annals of the New York Academy of Sciences* **949**, 203–314.

Reymann, K. G. and Frey, J. U. (2007) The late maintenance of hippocampal LTP: requirements, phases, "synaptic tagging," "late-associativity" and implications. *Neuropharmacology* **52**, 24–40.

Reynolds, B. A. and Weiss, S. (1992) Generation of neurons and astrocytes from isolated cells of the adult mammalian central nervous system. *Science* **255**: 1707–1710.

Reynolds, I. J. and Hastings, T. G. (1995) Glutamate induces the production of reactive oxygen species in cultured forebrain neurons following NMDA receptor activation. *Journal of Neuroscience* **15**, 3318–3327.

Reynolds, B. A., Tetzlaff, W. and Weiss, S. (1992) A multipotent EGF-responsive striatal embryonic progenitor cell produces neurons and astrocytes. *Journal of Neuroscience* **12**: 4565–4574.

Rhee, S. G. (2006) H_2O_2, a necessary evil for cell signaling. *Science* **312**, 1882–1883.

Rhee, S. G., Kang, S. W., Chang, T. S., Jeong, W. and Kim, K. (2001) Peroxiredoxins: a novel family of peroxidases. *IUMB Life* **52**, 35–41.

Rhee S. G., Kang S. W., Jeong W., Chang T. S., Yang K. S. and Woo H. A. (2005) Intracellular messenger function of hydrogen peroxide and its regulation by peroxiredoxins. *Current Opinion in Cell Biology* **17**, 183–189.

Rhyu, I. J., Boklewski, J., Ferguson, B., Lee, K. J., Lange, H., Bytheway, J., Lamb, J., McCormick, K., Williams, N., Cameron, J. and Greenough, W. T. (2003) Exercise training is associated with increased cortical vascularization in adult female cynomolgus monkeys. *Society for Neuroscience Abstracts*, 920–921.

Ribeiro, S., Goyal, V., Mello, C. V. and Pavlides, C. (1999) Brain gene expression during REM sleep depends on prior waking experience. *Learning & Memory* **6**, 500–508.

Ribeiro, C. S., Reis, M., Panizzutti, R., de Miranda, J. and Wolosker, H. (2002) Glial transport of the neuromodulator D-serine. *Brain Research* **929**, 202–209.

Ribeiro, S., Mello, C. V., Velho, T., Gardner, T. J., Jarvis, E. D. and Pavlides, C. (2002) Induction of hippocampal long-term potentiation during waking leads to increased extrahippocampal zif-268 expression during ensuing rapid-eye-movement sleep. *Journal of Neuroscience* **22**, 10,914–10,923.

Ribeiro, S., Gervasoni, D., Soares, E. S., Zhou, Y., Lin, S. C., Pantoja, J., Lavine, M., Nicolelis, M. A., Mello, C. V., Velho, T., Gardner, T. J., Jarvis, E. D. and Pavlides, C. (2004) Long-lasting novelty-induced neuronal reverberation during slow-wave sleep in multiple forebrain areas induction of hippocampal long-term potentiation during waking leads to increased extrahippocampal zif-268 expression during ensuing rapid-eye-movement sleep. *PLoS Biology* **2**, E24.

Riccio, D. C., Hodges, L. A. and Randall, P. K. (1968) Retrograde amnesia produced by hypothermia in rats. *Journal of Comparative Psychology* **66**, 618–622.

Rice, M. E. (2000) Ascorbate regulation and its neuroprotective role in the brain. *Trends in Neurosciences* **23**, 209–216.

Rice, M. E. and Russo-Menna, I. (1998) Differential compartmentalization of brain ascorbate and glutathione between neurons and glia. *Neuroscience* **82**, 1213–1223.

Rice, M. E., Pérez-Pinzón, M. A. and Lee, E. J. K. (1994) Ascorbic acid, but not glutathione, is taken up by brain slices and preserves cell morphology. *Journal of Neurophysiology* **71**, 1591–1596.

Rice, M. M., Graves, A. B., McCurry, S. M., Gibbons, L. E., Bowen, J. D., McCormick, W. C. and Larson, E. B. (2000) Postmenopausal estrogen and estrogen-progestin use and 2-year rate of cognitive change in a cohort of older Japanese American women: the Kame project. *Archives of Internal Medicine* **160**, 1641–1649.

Richardson, M. P., Strange, B. A. and Dolan, R. J. (2004) Encoding of emotional memories depends on amygdala and hippocampus and their interactions. *Nature Neuroscience* **7**, 278–285; doi: 10.1038/nn1190.

Richter-Levin, G. and Akirav, I. (2003). Emotional tagging of memory formation – in the search for neural mechanisms. *Brain Research Brain Research Reviews* **43**, 247–256.

Richter-Levin, G., Canevari, L., Bliss, T. V., Jarrard, L. E. and Jones, R. S. (1995) Long-term potentiation and glutamate release in the dentate gyrus: links to spatial learning. *Behavioral Brain Research* **66**, 37–40.

Rickard, N. S., Ng, K. T. and Gibbs, M. E. (1998) Further support for nitric oxide-dependent memory processing in the day-old chick. *Neurobiology of Learning and Memory* **69**, 79–86.

Rico, B., Xu, B. and Reichardt, L. F. (2002) TrkB receptor signaling is required for establishment of GABAergic synapses in the cerebellum. *Nature Neuroscience* **5**, 225–233.

Riedel, G., Platt, B. and Micheau, J. (2003) Glutamate receptor function in learning and memory. *Behavioral Brain Research* **140**, 1–47.

Rivard, B., Li, Y., Lenck-Santini, P.-P., Poucet, B. and Muller, R. U. (2004) Representation of objects in space by two classes of hippocampal pyramidal cells. *Journal of General Physiology* **124**, 9–25.

Rizzo, M. and Morselli, P. L. (1972) Amantadine-induced aggressiveness. *British Medical Journal* **3**, 50.

Robertson, E. M., Pascual-Leone, A. and Press, D. Z. (2004) Awareness modifies the skill-learning benefits of sleep. *Current Biology* **14**, 208–212.

Robinson, S. R., Hampson, E.C.G.M., Munro, M. N. and Vaney, D. I. (1993) Unidirectional coupling of gap junctions between neuroglia. *Science* **262**, 1072–1074.

Rodrigues, S. M., Schafe, G. E. and LeDoux, J. E. (2001) Intra-amygdala blockade of the NR2B subunit of the NMDA receptor disrupts the acquisition but not the expression of fear conditioning. *Journal of Neuroscience* **21**, 6889–6896.

Rodrigues, S. M., Schafe, G. E. and LeDoux, J. E. (2004a) Molecular mechanisms underlying emotional learning and memory in the lateral amygdala. *Neuron* **44**, 75–91.

Rodrigues, S. M., Bauer, E. P., Farb, C. R., Schafe, G. E. and LeDoux, J. E. (2002) The group I metabotropic glutamate receptor mGluR5 is required for fear memory formation and long-term potentiation in the lateral amygdala. *Journal of Neuroscience* **22**, 5219–5229.

Rodrigues, S. M., Farb, C. R., Bauer, E. P., LeDoux, J. E. and Schafe, G. E. (2004b) Pavlovian fear conditioning regulates Thr286 autophosphorylation of $Ca^{2+}/$

calmodulin-dependent protein kinase II at lateral amygdala synapses. *Journal of Neuroscience* **24**, 3281–3288.

Rodriguez-Moreno, A., Lopez-Garcia, J. C. and Lerma, J. (2000) Two populations of kainate receptors with separate signaling mechanisms in hippocampal interneurons. *Proceedings of the National Academy of Science of the United States of America* **97**, 1293–1298.

Rogan, M. T. and LeDoux, J. E. (1995) LTP is accompanied by commensurate enhancement of auditory-evoked responses in a fear conditioning circuit. *Neuron* **15**, 127–136.

Rogan, M. T., Staubli, U. V. and LeDoux, J. E. (1997) Fear conditioning induces associative long-term potentiation in the amygdala. *Nature* **390**, 604–607.

Rogers, L. J. and Deng, C. (1999) Light experience and the lateralisation of two visual pathways in the chick. *Behavioural Brain Research* **88**, 277–288.

Rola, R., Raber, J., Rizk, A., Otsuka, S., VandenBerg, S. R., Morhardt, D. R. and Fike, J. R. (2004) Radiation-induced impairment of hippocampal neurogenesis is associated with cognitive deficits in young mice. *Experimental Neurology* **188**, 316–330.

Romanski, L. M., LeDoux, J. E., Clugnet, M. C. and Bordi, F. (1993) Somatosensory and auditory convergence in the lateral nucleus of the amygdala. *Behavioral Neuroscience* **107**, 444–450.

Romeo, R. D. (2003) Puberty: a period of both organizational and activational effects of steroid hormones on neurobehavioral development. *Journal of Neuroendocrinology* **15**, 1185–1192.

Romeo, R. D., Waters, E. M. and McEwan, B. S. (2004) Steroid-induced hippocampal synaptic plasticity: sex differences and similarities. *Neuron Glia Biology* **1**, 219–229.

Romeo, R. D., Staub, D., Jasnow, A. M., Karatsoreos, I. N., Thronton, J. E. and McEwan, B. S. (2005) Dihydrotestosterone increases hippocampal NMDA binding but does not affect choline acetyltransferase cell number in the forebrain or choline transporter levels in the CA1 region of adult male rats. *Endocrinology* **146**, 2091–2097.

Roof, R. L. (1993) The dentate gyrus is sexually dimorphic in prepubescent rats: testosterone plays a significant role. *Brain Research* **610**, 148–151.

Roof, R. L. and Stein, D. G. (1999) Gender differences in Morris water maze performance depend on task parameters. *Physiology & Behavior* **68**, 81–86.

Rosario, E. R., Chang, L., Stanczyk, F. Z. and Pike, C. J. (2004) Age-related testosterone depletion and the development of Alzheimer disease. *Journal of American Medical Association* **292**, 1431–1432.

Rose, S. P. R. (2000) God's organism? The chick as a model system for memory studies. *Learning & Memory* **7**, 1–17.

Rose, S. P. R. (2004) *The Making of Memory* (2nd edition). Cape.

Rose, S. P. R. (2005) The 21st *Century Brain: Explaining, Mending and Manipulating the Mind*. Cape.

Rose, C. R., Blum, R., Pichler, B., Lepier, A., Kafitz, K. W. and Konnerth, A. (2003) Truncated TrkB-T1 mediates neurotrophin-evoked calcium signalling in glia cells. *Nature* **426**, 74–78.

Rosen, J. B. (2004) The neurobiology of conditioned and unconditioned fear: a neurobehavioral system analysis of the amygdala. *Behav Cogn Neurosci Rev* **3**, 23–41.

Rosenblum, K., Meiri, N. and Dudai, Y. (1993) Taste memory: the role of protein synthesis in gustatory cortex. *Behavioral and Neurological Biology* **59**, 49–56.

Rosenkranz, J. A. and Grace, A. A. (2002) Dopamine-mediated modulation of odour-evoked amygdala potentials during Pavlovian conditioning. *Nature* **417**, 282–287.

Rosenzweig, M. R., Bennett, E. L., Hebert, M. and Morimoto, H. (1978) Social grouping cannot account for cerebral effects of enriched environments. *Brain Research* **153**, 563–576.

Roullet, P. and Sara, S. (1998) Consolidation of memory after its reactivation: involvement of beta noradrenergic receptors in the late phase. *Neural Plast* **6**, 63–68.

Rowan, R. A. and Maxwell, D. S. (1981) Patterns of vascular sprouting in the postnatal development of the cerebral cortex of the rat. *American Journal of Anatomy* **160**, 247–255.

Rowe, M. H. and Stone, J. (1976) Conduction velocity groupings among axons of cat retinal ganglion cells, and their relationship to retinal topography. *Experimental Brain Research* **25**, 339–357.

Rudge, J. S., Alderson, R. F., Pasnikowski, E., McClain, J., Ip, N. Y. and Lindsay, R. M. (1992) Expression of ciliary neurotrophic factor and the neurotrophins-nerve growth factor, brain-derived neurotrophic factor and neurotrophin-3 in cultured rat hippocampal astrocytes. *European Journal of Neuroscience* **4**, 459–471.

Rudick, C. N. and Woolley, C. S. (2001) Estrogen regulates functional inhibition of hippocampal CA1 pyramidal cells in the adult female rat. *Journal of Neuroscience* **21**, 6532–6543.

Ruiz-Gomez, A., Mellstrom, B., Tornero, D., Morato, E., Savignac, M., Holguin, H., Aurrekoetxea, K., Gonzalez, P., Gonzalez-Garcia, C., Cena, V., Mayor, F. Jr. and Naranjo, J. R. (2007) G protein-coupled receptor kinase 2-mediated phosphorylation of downstream regulatory element antagonist modulator regulates membrane trafficking of Kv4.2 potassium channel. *Journal of Biological Chemistry* **282**, 1205–1215.

Rumbaugh, G., Sia, G. M., Garner, C. C. and Huganir, R. L. (2003) Synapse-associated protein-97 isoform-specific regulation of surface AMPA receptors and synaptic function in cultured neurons. *Journal of Neuroscience* **23**, 4567–4576.

Rumpel, S., LeDoux, J., Zador, A. and Malinow, R. (2005) Postsynaptic receptor trafficking underlying a form of associative learning. *Science* **308**, 83–88.

Rushton, W. A. H. (1951) A theory of the effects of fibre size in medullated nerve. *Journal of Physiology* **115**, 101–122.

Russo-Neustadt, A., Beard, R. C. and Cotman, C. W. (1999) Exercise, antidepressant medications, and enhanced brain derived neurotrophic factor expression. *Neuropsychopharmacology* **21**, 679–682.

Russo-Neustadt, A. A., Beard, R. C., Huang, Y. M. and Cotman, C. W. (2000) Physical activity and antidepressant treatment potentiate the expression of specific brain-derived neurotrophic factor transcripts in the rat hippocampus. *Neuroscience* **101**, 305–312.

Rutherford, L. C., Nelson, S. B. and Turrigiano, G. G. (1998) BDNF has opposite effects on the quantal amplitude of pyramidal neuron and interneuron excitatory synapses. *Neuron* **21**, 521–530.

Rutherford, L. C., DeWan, A., Lauer, H. M. and Turrigiano, G. G. (1997) Brain-derived neurotrophic factor mediates the activity-dependent regulation of inhibition in neocortical cultures. *Journal of Neuroscience* **17**, 4527–4535.

Ryff, C. D. and Singer, B. (2005) Social environments and the genetics of aging: advancing knowledge of protective health mechanisms. *Journals of Gerontology. Series B, Psychological Sciences and Social Sciences* **60 Spec No 1**, 12–23.

Saarelainen, T., Pussinen, R., Koponen, E., Alhonen, L., Wong, G., Sirvio, J. and Castren, E. (2000) Transgenic mice overexpressing truncated trkB neurotrophin receptors in neurons have impaired long-term spatial memory but normal hippocampal LTP. *Synapse* **38**, 102–104.

Sabha, M. Jr., Emirandetti, A., Cullheim, S. and A. L. De Oliveira (2008) MHC I expression and synaptic plasticity in different mice strains after axotomy. *Synapse* **62**, 137–148.

Sagara, J. I., Miura, K. and Bannai, S. (1993) Maintenance of neuronal glutathione by glial cells. *Journal of Neurochemistry* **61**, 1672–1676.

Sajikumar, S. and Frey, J. U. (2003) Anisomycin inhibits the late maintenance of long-term depression in rat hippocampal slices in vitro. *Neuroscience Letters* **338**, 147–150.

Sajikumar, S. and Frey, J. U. (2004) Late-associativity, synaptic tagging, and the role of dopamine during LTP and LTD. *Neurobiology of Learning and Memory* **82**, 12–25.

Sakimura, K., Kutsuwada, T., Ito, I., Manabe, T., Takayama, C., Kushiya, E., Yagi, T., Aizawa, S., Inoue, Y., Sugiyama, H. and Mishina, M. (1995) Reduced hippocampal LTP and spatial learning in mice lacking NMDA receptor 1 subunit. *Nature* **373**, 151–155.

Salinska, E., Bourne, R. C. and Rose, S. P. R. (2004) Reminder effects: the molecular cascade following a reminder in young chicks does not recapitulate that following training on a passive avoidance task. *European Journal of Neuroscience* **19**, 3042–3047.

Salinska, E. J., Bourne, R. C. and Rose, S. P. R. (2001) Long-term memory formation in the chick requires mobilisation of ryanodine sensitive intracellular calcium stores. *Neurobiology of Learning and Memory* **75**, 293–302.

Salinska, E. J., Chaudhury, D., Bourne, R. C. and Rose, S. P. R. (1999) Passive avoidance training results in increased responsiveness of synaptosomal voltage and ligand gated channels in chick brain. *Neuroscience* **93**, 1507–1514.

Samson, R. D., Dumont, E. C. and Pare, D. (2003) Feedback inhibition defines transverse processing modules in the lateral amygdala. *Journal of Neuroscience* **23**, 1966–1973.

Sanchez, I., Hassing L., Paskevich, P. A., Shine, H. D. and Nixon, R. A. (1996) Oligodendroglia regulate the regional expansion of axon caliber and local accumulation of neurofilaments during development independently of myelin formation. *Journal of Neuroscience* **16**, 5095–5105.

Sanchez, M. M., Hearn, E. F., Do, D., Rilling, J. K. and Herndon, J. G. (1998) Differential rearing affects corpus callosum size and cognitive function of rhesus monkeys. *Brain Research* **812**, 38–49.

Sanchez-Gomez, M. V., Alberdi, E., Ibarretxe, G., Torre, I. and Matute, C. (2003) Caspase-dependent and caspase-independent oligodendrocyte death mediated by AMPA and kainate receptors. *Journal of Neuroscience* **23**, 9519–9528.

Sandi, C., and Rose, S. P. R. (1994a) Corticosterone enhances long-term retention in one day old chicks trained in a weak passive avoidance learning paradigm. *Brain Research* **647**, 106–112.

Sandi, C. and Rose, S. P. R. (1994b) Corticosteroid receptor antagonists are amnestic for passive avoidance learning in day-old chicks *European Journal of Neuroscience* **6**, 1292–1297.

Sandi, C. and Rose, S. P. R. (1997) Training-dependent biphasic effects of corticosterone in memory formation for a passive avoidance tasks in chicks. *Psychopharmacology* **133**, 152–160.

Sandstrom, N. J. and Williams, C. L. (2001) Memory retention is modulated by acute estradiol and progesterone replacement. *Behavioral Neuroscience* **115**, 384–393.

Sandstrom, N. J. and Williams, C. L. (2004) Spatial memory retention is enhanced by acute and continuous estradiol replacement. *Hormones and Behavior* **45**, 128–135.

Sanford, L. D., Silvestri, A. J., Ross, R. J. and Morrison, A. R. (2001) Influence of fear conditioning on elicited ponto-geniculo-occipital waves and rapid eye movement sleep. *Archives italiennes de biologie* **139**, 169–183.

Sangha, S., Scheibenstock, A. and Lukowiak, K. (2003a) Reconsolidation of a long-term memory in Lymnaea requires new protein and RNA synthesis and the soma of right pedal dorsal 1. *Journal of Neuroscience* **23**, 8034–8040.

Sangha, S., Scheibenstock, A., McComb, C. and Lukowiak, K. (2003b) Intermediate and long-term memories of associative learning are differentially affected by transcription versus translation blockers in Lymnaea. *Journal of Experimental Biology* **206**, 1605–1613.

Santarelli, L., Saxe, M., Gross, C., Surget, A., Battaglia, F., Dulawa, S., Weisstaub, N., Lee, J., Duman, R., Arancio, O., Belzung, C. and Hen, R. (2003) Requirement of hippocampal neurogenesis for the behavioral effects of antidepressants. *Science* **301**, 805–809.

Sapolsky, R. M. (2003) Stress and plasticity in the limbic system. *Neurochemical Research* **28**, 1735–1742.

Sar, M., Lubahn, D. B., French, F. S. and Wilson, E. M. (1990) Immunohistochemical localization of the androgen receptor in rat and human tissues. *Endocrinology* **127**, 3180–3186.

Sara, S. J. (2000) Retrieval and reconsolidation: toward a neurobiology of remembering. *Learning & Memory* **7**, 73–84.

Saucier, D., Hargreaves, E. L., Boon, F., Vaderwolf, C. H. and Cain, D. P. (1996) Detailed behavioral analysis of water maze acquisition under systemic NMDA or muscarinic antagonism: Nonspatial pretraining eliminates spatial learning deficits. *Behavioral Neuroscience* **110**, 103–116.

Sauer, H., Wartenberg, M. and Hescheler, J. (2001) Reactive oxygen species as intracellular messengers during cell growth and differentiation. *Cellular Physiology and Biochemistry* **11**, 173–186.

Sawa, A. and Snyder, S. H. (2002) Schizophrenia: diverse approaches to a complex disease. *Science* **296**, 692–695.

Scanziani, M., Gahwiler, B. H. and Charpak, S. (1998) Target cell-specific modulation of transmitter release at terminals from a single axon. *Proceedings of the National Academy of Science of the United States of America* **95**, 12,004–12,009.

Schacter, D. and Tulving, E. (1994) *What are the Memory Systems of 1994?* MIT Press, viii, 1–38.

Schafe, G. E. and LeDoux, J. E. (2000) Memory consolidation of auditory Pavlovian fear conditioning requires protein synthesis and protein kinase A in the amygdala. *Journal of Neuroscience* **20**, RC96.

Schafe, G. E., Nader, K., Blair, H. T. and LeDoux, J. E. (2001) Memory consolidation of Pavlovian fear conditioning: a cellular and molecular perspective. *Trends in Neuroscience* **24**, 540–546.

Schafe, G. E., Nadel, N. V., Sullivan, G. M., Harris, A. and LeDoux, J. E. (1999) Memory consolidation for contextual and auditory fear conditioning is dependent on protein synthesis, PKA, and MAP kinase. *Learning & Memory* **6**, 97–110.

Schafe, G. E., Bauer, E. P., Rosis, S., Farb, C. R., Rodrigues, S. M. and LeDoux, J. E. (2005) Memory consolidation of Pavlovian fear conditioning requires nitric oxide signaling in the lateral amygdala. *European Journal of Neuroscience* **22**, 201–211.

Schaller, M. D., Otey, C. A., Hildebrand, J. D. and Parsons, J. T. (1995) Focal adhesion kinase and paxillin bind to peptides mimicking beta integrin cytoplasmic domains. *Journal of Cell Biology* **130**, 1181–1187.

Scharfman, H. E., Goodman, J. H. and Sollas, A. L. (2000) Granule-like neurons at the hilar/CA3 border after status epilepticus and their synchrony with area CA3 pyramidal cells: functional implications of seizure-induced neurogenesis. *Journal of Neuroscience* **20**, 6144–6158.

Schell, M. J., Cooper, O. B. and Snyder, S. H. (1997a) D-aspartate localizations imply neuronal and neuroendocrine roles. *Proceeding of the National Academy of Sciences of the United States of America* **94**, 2013–2018.

Schell, M. J., Molliver, M. E. and Snyder, S. H. (1995) D-serine, an endogenous synaptic modulator: localization to astrocytes and glutamate-stimulated release. *Proceeding of the National Academy of Sciences of the United States of America* **92**, 3948–3952.

Schell, M. J., Brady, R. O., Jr., Molliver, M. E. and Snyder, S. H. (1997b) D-serine as a neuromodulator: regional and developmental localizations in rat brain glia resemble NMDA receptors. *Journal of Neuroscience* **17**, 1604–1615.

Schendan, H. E., Searl, M. M., Melrose, R. J. and Stern, C. E. (2003) An FMRI study of the role of the medial temporal lobe in implicit and explicit sequence learning. *Neuron* **37**, 1013–1025.

Schipke, C. G. and Kettenmann, H. (2004) Astrocyte responses to neuronal activity. *Glia* **47**, 226–232.

Schmidt-Hieber, C., Jonas, P. and Bischofberger, J. (2004) Enhanced synaptic plasticity in newly generated granule cells of the adult hippocampus. *Nature* **429**, 184–187.

Schneider, A. M. and Sherman, W. (1968) Amnesia: a function of the temporal relation of footshock to electroconvulsive shock. *Science* **159**, 219–221.

Schneiderman, N., Francis, J., Sampson, L. D. and Schwaber, J. S. (1974) CNS integration of learned cardiovascular behavior. In DiCara, L. V. (ed.) *Limbic and Autonomic Nervous System Research*. Plenum, pp. 277–309.

Scholey, A. B., Mileusnic, R., Schachner, M. and Rose, S. P. R. (1995) A role for a chicken homolog of the neural cell adhesion molecule L1 in consolidation of memory for a passive avoidance task. *Learning & Memory* **2**, 17–25.

Scholey, A. B., Rose, S. P. R., Zamani, M. R., Bock, E. and Schachner, M. (1993) A role for the neural cell adhesion molecule in a late consolidating phase of glycoprotein synthesis 6 hr following passive avoidance training of the young chick. *Neuroscience* **55**, 499–509.

Schousboe, A., Westergaard, N., Sonnewald, U., Petersen, S. B., Yu, A. C. and Hertz, L. (1992) Regulatory role of astrocytes for neuronal biosynthesis and homeostasis of glutamate and GABA. *Progress in Brain Research* **94**, 199–211.

Schratt, G. M., Nigh, E. A., Chen, W. G., Hu, L. and Greenberg, M. (2004) BDNF regulates the translation of a select group of mRNAs by a mammalian target of rapamycin-phosphatidylinositol 3-kinase-dependent pathway during neuronal development. *Journal of Neuroscience* **24**, 9366–9377.

Schratt, G. M., Tuebing, F., Nigh, E. A., Kane, C. G., Sabatini, M. E., Kiebler, M. and Greenberg, M. E. (2006). A brain-specific microRNA regulates dendritic spine development. *Nature* **439**, 283–289.

Schroder, J. M., Bohl, J. and Brodda, K. (1978) Changes of the ratio between myelin thickness and axon diameter in the human developing sural nerve. *Acta Neuropathologica* **43**, 169–178.

Schubert, P., Ogata, T., Marchini, C., Ferroni, S. and Rudolphi, K. (1997) Protective mechanisms of adenosine in neurons and glial cells. *Annals of the New York Academy of Sciences* **825**, 1–10.

Schuman, E. and Chan, D. (2004) Fueling synapses. *Cell* **119**, 738–740.

Schwab, M. E. and Thoenen, H. (1985) Dissociated neurons regenerate into sciatic but not optic nerve explants in culture irrespective of neurotrophic factors. *Journal of Neuroscience* **5**, 2415–2423.

Schwanstecher, C. and Bassen, D. (1997) KATP-channel on the somata of spiny neurones in rat caudate nucleus: regulation by drugs and nucleotides. *British Journal of Pharmacology* **121**, 193–198.

Screaton, R. A., Conkright, M. D., Katoh, Y., Best, J. L., Canettieri, G., Jeffries, S., Guzman, E., Niessen, S., Yates, J. R 3rd, Takemori, H., Okamoto, M. and Montminy, M. (2004) The CREB coactivator TORC2 functions as a calcium- and cAMP-sensitive coincidence detector. *Cell* **119**, 61–74.

Sebaai, N., Lesage, J., Vieau, D., Alaoui, A., Dupouy, J. P. and Deloof, S. (2001). Altered control of the hypothalamo-pituitary-adrenal axis in adult male rats exposed perinatally to food deprivation and/or dehydration. *Neuroendocrinology* **76**, 243–253.

Seidenbecher, T., Reymann, K. G. and Balschun, D. (1997) A post-tetanic time window for the reinforcement of long-term potentiation by appetitive and aversive stimuli. *Proceedings of the National Academy of Sciences of the United States of America* **94**, 1494–1499.

Seidenbecher, T., Laxmi, T. R., Stork, O. and Pape, H. C. (2003) Amygdalar and hippocampal theta rhythm synchronization during fear memory retrieval. *Science* **301**, 846–850.

Seil, F. J. (2003) TrkB receptor signaling and activity-dependent inhibitory synaptogenesis. *Histology and Histopathology* **18**, 635–646.

Seil, F. J. and Drake-Baumann, R. (2000) TrkB receptor ligands promote activity-dependent inhibitory synaptogenesis. *Journal of Neuroscience* **20**, 5367–5373.

Seitz, R. J., Roland, E., Bohm, C., Greitz, T. and Stone-Elander, S. (1990) Motor learning in man: a positron emission tomographic study. *Neuroreport* **1**, 57–60.

Sejnowski, T. J. and Destexhe, A. (2000) Why do we sleep? *Brain Research* **886**, 208–223.

Selcher, J. C., Atkins, C. M., Trzaskos, J. M., Paylor, R. and Sweatt, J. D. (1999) A necessity for MAP kinase activation in mammalian spatial learning. *Learning & Memory* **6**, 478–490.

Selcherm, J. C., Weeber, E. J., Varga, A. W., Sweatt, J. D. and Swank, M. (2002) Protein kinase signal transduction cascades in mammalian associative conditioning. *Neuroscientist* **8**, 122–131.

Selden, N. R., Cole, B. J., Everitt, B. J. and Robbins, T. W. (1990) Damage to ceruleo-cortical noradrenergic projections impairs locally cued but enhances spatially cued water maze acquisition. *Behavioural Brain Research* **39**, 29–51.

Selye, H. (1950) *Stress. A Treatise based on the Concept of the General-Adaptation-Syndrome and the Diseases of Adaptation*. Acta, Inc. Medical Publishers.

Selye, H. (1956) *The Stress of Life*. New York: McGraw-Hill.

Semont, A., Fache, M.-P. Ouafik, L., Hery, M., Faudon, M. and Hery, F. (1999) Effect of serotonin inhibition on glucocorticoids and mineralocorticoid expression in various brain structures. *Neuroendocrinology* **69**, 121–128.

Semyanov, A. and Kullmann, D. M. (2000) Modulation of GABAergic signaling among interneurons by metabotropic glutamate receptors. *Neuron* **25**, 663–672.

Serrano, P. A., Rodriguez, W. A., Bennett, E. L. and Rosenzweig, M. R. (1995) Protein kinase inhibitors disrupt memory formation in two chick brain regions. *Pharmacology, Biochemistry and Behavior* **52**, 547–554.

Serrano, F., Kolluri, N. S., Wientjes, F. B., Card, J. P. and Klann, E. (2003) NADPH oxidase immunoreactivity in the mouse brain. *Brain Research* **988**, 193–188.

Serrano, P. E., Beniston, D. S., Oxonian, M. G., Rodriguez, W. A., Rosenzweig, M. R. and Eennett, E. L. (1994) Differential effects of protein kinase inhibitors and activators on memory formation in the 2-day old chick *Behavioral & Neural Biology* **61**, 60–72.

Seutin, V., Scuvee-Moreau, J., Masotte, L. and Dresse, A. (1995) Hydrogen peroxide hyperpolarizes rat CA1 pyramidal neurons by inducing an increase in potassium conductance. *Brain Research* **683**, 275–278.

Shaffery, J. P., Roffwarg, H. P., Speciale, S. G. and Marks, G. A. (1999) Ponto-geniculo-occipital-wave suppression amplifies lateral geniculate nucleus cell-size changes in monocularly deprived kittens. *Brain Research Developmental Brain Research* **114**, 109–119.

Shaffery, J. P., Sinton, C. M., Bissette, G., Roffwarg, H. P. and Marks, G. A. (2002) Rapid eye movement sleep deprivation modifies expression of long-term potentiation in visual cortex of immature rats. *Neuroscience* **110**, 431–443.

Shaffery, J. P., Oksenberg, A., Marks, G. A., Speciale, S. G., Mihailoff, G. and Roffwarg, H. P. (1998) REM sleep deprivation in monocularly occluded kittens reduces the size of cells in LGN monocular segment. *Sleep* **21**, 837–845.

Shaikh, A. A. and Shaikh, S. A. (1975) Adrenal and ovarian steroid secretion in the rat estrous cycle temporally related to gonadotropins and steroid levels found in peripheral plasma. *Endocrinology* **96**, 37–44.

Shaked, I., Porat, Z., Gersner, R., Kipnis, J. and Schwartz, M. (2004) Early activation of microglia as antigen-presenting cells correlates with T cell-mediated protection and repair of the injured central nervous system. *Journal of Neuroimmunology* **146**, 84–93.

Shapiro, M. L. and Eichenbaum, H. (1999) Hippocampus as a memory map, synaptic plasticity and memory encoding by hippocampal neurons. *Hippocampus* **9**, 365–384.

Shen, K., Teruel, M. N., Connor, J. H., Shenolikar, S. and Meyer, T. (2000) Molecular memory by reversible translocation of calcium/calmodulin-dependent protein kinase II. *Nature Neuroscience* **3**, 881–886.

Sheng, M. and Greenberg, M. E. (1990) The regulation and function of c-fos and other immediate early genes in the nervous system. *Neuron* **4**, 477–485.

Shi, S., Hayashi, Y., Esteban, J. A. and Malinow, R. (2001) Subunit-specific rules governing AMPA receptor trafficking to synapses in hippocampal pyramidal neurons. *Cell* **105**, 331–343.

Shi, S. H., Hayashi, Y., Petralia, R. S., Zaman, S. H., Wenthold, R. J., Svoboda, K. and Malinow, R. (1999) Rapid spine delivery and redistribution of AMPA receptors after synaptic NMDA receptor activation. *Science* **284**, 1811–1816.

Shieh, P. B., Hu, S. C., Bobb, K., Timmusk, T. and Ghosh, A. (1998) Identification of a signaling pathway involved in calcium regulation of BDNF expression. *Neuron* **20**, 727–740.

Shigemoto, R., Kulik, A., Roberts, J. D., Ohishi, H., Nusser, Z., Kaneko, T. and Somogyi, P. (1996) Target-cell-specific concentration of a metabotropic glutamate receptor in the presynaptic active zone. *Nature* **381**, 523–525.

Shigemoto, R., Kinoshita, A., Wada, E., Nomura, S., Ohishi, H., Takada, M., Flor, P. J., Neki, A., Abe, T., Nakanishi, S. and Mizuno, N. (1997) Differential presynaptic localization of metabotropic glutamate receptor subtypes in the rat hippocampus. *Journal of Neuroscience* **17**, 7503–7522.

Shimazu, K., Zhao, M., Sakata, K., Akbarian, S., Bates, B., Jaenisch, R. and Lu B. (2006) NT-3 facilitates hippocampal plasticity and learning and memory by regulating neurogenesis. *Learn Mem.* **13**, 307–315.

Shirayama, Y., Chen, A. C., Nakagawa, S., Russell, D. S. and Duman, R. S. (2002) Brain-derived neurotrophic factor produces antidepressant effects in behavioral models of depression. *Journal of Neuroscience* **22**, 3251–3261.

Shors, T. (2004) Learning during stressful times. *Learning & Memory* **11**, 137–144.

Shors, T. J. and Matzel, L. D. (1997) Long-term potentiation: what's learning got to do with it? *Brain and Behavioral Sciences* **20**, 597–614.

Shors, T. J., Chua, C. and Falduto, J. (2001) Sex differences and opposite effects of stress on dendritic spine density in the male versus female hippocampus. *Journal of Neuroscience* **21**, 6292–6297.

Shors, T. J., Falduto, J. and Leuner, B. (2004) The opposite effects of stress on dendritic spines in male vs. female rats are NMDA receptor-dependent. *European Journal of Neuroscience* **19**, 145–150.

Shors, T. J., Miesegaes, G., Beylin, A., Zhao, M., Rydel, T. and Gould, E. (2001) Neurogenesis in the adult is involved in the formation of trace memories. *Nature* **410**, 372–376.

Shumyatsky, G. P., Tsvetkov, E., Malleret, G., Vronskaya, S., Hatton, M., Hampton, L., Battey, J. F., Dulac, C., Kandel, E. R. and Bolshakov, V. Y. (2002) Identification of a signaling network in lateral nucleus of amygdala important for inhibiting memory specifically related to learned fear. *Cell* **111**, 905–918.

Shumyatsky, G. P., Malleret, G., Shin, R. M., Takizawa, S., Tully, K., Tsvetkov, E., Zakharenko, S. S., Joseph, J., Vronskaya, S., Yin, D., Schubart, U. K., Kandel, E. R. and Bolshakov, V. Y. (2005) stathmin, a gene enriched in the amygdala, controls both learned and innate fear. *Cell* **123**, 697–709.

Sidló, Z., Reggio, P. H. and Rice, M. E. (2008) Inhibition of striatal dopamine release by CB1 receptor activation requires nonsynaptic communication via GABA, H_2O_2, and KATP channels. *Neurochemistry International* **52**, 80–88.

Silva, A. J., Kogan, J. H., Frankland, P. W. and Kida, S. (1998) CREB and memory. *Annual Review of Neuroscience* **21**, 127–148.

Simerly, R. B. (2002) Wired for reproduction: organization and development of sexually dimorphic circuits in the mammalian forebrain. *Annual Review of Neuroscience* **25**, 507–536.

Simon, G. E. (2003) Social and economic burden of mood disorders. *Biological Psychiatry* **54**, 208–215.

Simoncini, T., Hafezi-Moghadam, A., Brazil, D. P., Ley, K., Chin, W. W. and Liao, J. K. (2000) Interaction of oestrogen receptor with the regulatory subunit of phosphatidylinositol-3-OH kinase. *Nature* **407**, 538–541.

Singer, D. S., Mozes, E., Kirshner, S. and Kohn, L. D. (1997) Role of MHC Class I molecules in autoimmune disease. *Critical Reviews in Immnology* **17**, 463–468.

Singer, C. A., Figueroa-Masot, X. A., Batchelor, R. H. and Dorsa, D. M. (1999) The mitogen-activated protein kinase pathway mediates estrogen neuroprotection after glutamate toxicity in primary cortical neurons. *Journal of Neuroscience* **19**, 2455–2463.

Singh, M., Setalo, G. Jr., Guan, X., Frail, D. E. and Toran-Allerand, C. D. (2000) Estrogen-induced activation of the mitogen-activated protein kinase cascade in the cerebral cortex of estrogen receptor-alpha knock-out mice. *Journal of Neuroscience* **20**, 1694–1700.

Singh, M., Meyer, E. M., Millard, W. J. and Simpkins, J. W. (1994) Ovarian steroid deprivation results in a reversible learning impairment and compromised cholinergic function in female Sprague-Dawley rats. *Brain Research* **664**, 305–312.

Sirevaag, A. M. and Greenough, W. T. (1985) Differential rearing effects on rat visual cortex synapses. *II. Synaptic morphometry. Brain Research* **351**, 215–226.

Sirevaag, A. M. and Greenough, W. T. (1987) Differential rearing effects on rat visual cortex synapses. III Neuronal and glial nuclei, boutons, dendrites, and capillaries. *Brain Research* **424**, 320–332.

Sirevaag, A. M. and Greenough, W. T. (1991) Plasticity of GFAP-immunoreactive astrocyte size and number in visual cortex of rats reared in complex environments. *Brain Research* **540**, 273–278.

Sirevaag, A. M., Black, J. E., Shafron, D. and Greenough, W. T. (1988) Direct evidence that complex experience increases capillary branching and surface area in visual cortex of young rats. *Brain Research* **471**, 299–304.

Siuciak, J. A., Lewis, D. R., Wiegand, S. J. and Lindsay, R. M. (1997) Antidepressant-like effect of brain-derived neurotrophic factor (BDNF). *Pharmacology, Biochemistry and Behavior* **56**, 131–137.

Skaggs, W. E. and McNaughton, B. L. (1996) Replay of neuronal firing sequences in rat hippocampus during sleep following spatial experience. *Science* **271**, 1870–1873.

Slivka, A., Mytilineou, C. and Cohen, G. (1987) Histochemical evaluation of glutathione in brain. *Brain Research* **409**, 275–284.

Small, S. A., Nava, A. S., Perera, G. M., DeLaPaz, R., Mayeux, R. and Stern, Y. (2001) Circuit mechanisms underlying memory encoding and retrieval in the long axis of the hippocampal formation. *Nature Neuroscience* **4**, 442–449.

Smith, M. S. (1975) The control of progesterone secretion during the estrous cycle and early pseudopregnancy in the rat: prolactin, gonadotropin and steroid levels associated with rescue of the corpus luteum of pseudopregnancy. *Endocrinology* **96**, 219–226.

Smith, A. D. and Bolam, J. P. (1990) The neural network of the basal ganglia as revealed by the study of synaptic connections of identified neurons. *Trends in Neurosciences* **13**, 259–265.

Smith, C. and MacNeill, C. (1994) Impaired motor memory for a pursuit rotor task following Stage 2 sleep loss in college students. *Journal of Sleep Research* **3**, 206–213.

Smith, K. M. and Larive, L.L.R.F. (2002) Club drugs: methylenedioxymethamphetamine, flunitrazepam, ketamine hydrochloride, and gamma-hydroxybutyrate. *American Journal of Health-System Pharmacy* **59**, 1067–1076.

Smith, R. S. and Koles, Z. J. (1970) Myelinated nerve fibers: computed effect of myelin thickness on conduction velocity. *American Journal of Physiology* **219**, 1256–1258.

Smith, M. D., Jones, L. S. and Wilson, M. A. (2002) Sex differences in hippocampal slice excitability: role of testosterone. *Neuroscience* **109**, 517–530.

Smith, C., Tenn, C. and Annett, R. (1991) Some biochemical and behavioural aspects of the paradoxical sleep window. *Canadian Journal of Psychology* **45**, 115–124.

Smith, Y., Pare, J. F. and Pare, D. (2000) Differential innervation of parvalbumin-immunoreactive interneurons of the basolateral amygdaloid complex by cortical and intrinsic inputs. *Journal of Comparative Neurology* **416**, 496–508.

Snyder, J. S., Hong, N. S., McDonald, R. J. and Wojtowicz, J. M. (2005) A role for adult neurogenesis in spatial long-term memory. *Neuroscience* **130**, 843–852.

Snyder, J. S., Kee, N. and Wojtowicz, J. M. (2001) Effects of adult neurogenesis on synaptic plasticity in the rat dentate gyrus. *Journal of Neurophysiology* **85**, 2423–2431.

Soderling, T. R., Chang, B. and Brickey, D. (2001) Cellular signaling through multifunctional Ca^{2+}/calmodulin-dependent protein kinase II. *Journal of Biological Chemistry* **276**, 3719–3722.

Sohal, V. S. and Hasselmo, M. E. (1998) Changes in GABAb modulation during a theta cycle may be analogous to the fall of temperature during annealing. *Neural Computation* **10**, 889–902.

Sokolova, T., Gutterer, J. M., Hirrlinger, J., Hamprecht, B. and Dringen, R. (2001) Catalase in astroglia-rich cultures from rat brain: immunocytochemical localization and inactivation during the disposal of hydrogen peroxide. *Neuroscience Letters* **297**, 129–132.

Song, I. and Huganir, R. L. (2002) Regulation of AMPA receptors during synaptic plasticity. *Trends in Neuroscience* **25**, 578–588.

Song, H. J., Stevens, C. F. and Gage, F. H. (2002) Neural stem cells from adult hippocampus develop essential properties of functional CNS neurons. *Nature Neuroscience* **5**, 438–445.

Sonsalla, P. K., Manzino, L., Sinton, C. M., Liang, C. L., German, D. C. and Zeevalk, G. D. (1997) Inhibition of striatal energy metabolism produces cell loss in the ipsilateral substantia nigra. *Brain Research* **773**, 223–226.

Soppet, D., Escandon, E., Maragos, J., Middlemas, D. S., Reid, S. W., Blair, J., Burton, L. E., Stanton, B. R., Kaplan, D. R. and Hunter, T. (1991) The neurotrophic factors brain-derived neurotrophic factor and neurotrophin-3 are ligands for the trkB tyrosine kinase receptor. *Cell* **65**, 895–903.

Sorra, K. E. and Harris, K. M. (2000) Overview on the structure, composition, function, development, and plasticity of hippocampal dendritic spines. *Hippocampus* **10**, 501–511.

Sorra, K. E., Fiala, J. C. and Harris, K. M. (1998) Critical assessment of the involvement of perforations, spinules, and spine branching in hippocampal synapse formation. *Journal of Comparative Neurology* **398**, 225–240.

Sotres-Bayon, F., Bush, D. E. and LeDoux, J. E. (2004) Emotional perseveration: an update on prefrontal-amygdala interactions in fear extinction. *Learning & Memory* **11**, 525–535.

Speigel, A. (1998) *Proteins, Receptors and Disease*. Humana Press.

Sperling, R., Chua, E., Cocchiarella, A., Rand-Giovannetti, E., Poldrack, R., Schacter, D. L. and Albert, M. (2003) Putting names to faces: successful encoding of associative memories activates the anterior hippocampal formation. *Neuroimage* **20**, 1400–1410.

Spleiss, O., van Calker, D., Scharer, L., Adamovic, K., Berger, M. and Gebicke-Haerter, P. J. (1998) Abnormal G protein alpha(s) – and alpha(i2)-subunit mRNA expression in bipolar affective disorder. *Molecular Psychiatry* **3**, 512–520.

Springer, J. E., Azbill, R. D. and Knapp, P. E. (1999) Activation of the caspase-3 apoptotic cascade in traumatic spinal cord injury. *Nature Medicine* **5**, 943–946.

Squinto, S. P., Stitt, T. N., Aldrich, T. H., Davis, S., Bianco, S. M., Radziejewski, C., Glass, D. J., Masiakowski, P., Furth, M. E. and Valenzuela, D. M. (1991) trkB encodes a functional receptor for brain-derived neurotrophic factor and neurotrophin-3 but not nerve growth factor. *Cell* **65**, 885–893.

Squire, L. R. and Alvarez, P. (1995) Retrograde amnesia and memory consolidation: a neurobiological perspective. *Current Opinions in Neurobiology* **5**, 169–177.

Squire, L. R. and Zola-Morgan, S. (1996) Structure and function of declarative and non-declarative memory systems. *Proceedings of the National Academy of Science* **93**, 13,515–13,522.

Squire, L. R., Stark, C. E. L. and Clark, R. E. (2004) The medial temporal lobe. *Annual Review of Neuroscience* **27**, 279–306.

Stabel, S. and Parker, P. J. (1991) Protein kinase C. *Pharmacology and Therapeutics* **51**, 71–95.

Stamler, J. S., Lamas, S. and Fang, F. C. (2001) Nitrosylation, the prototypic redox-based signaling mechanism. *Cell* **106**, 675–683.

Stanfield, B. B. and Trice, J. E. (1988) Evidence that granule cells generated in the dentate gyrus of adult rats extend axonal projections. *Experimental Brain Research* **72**, 399–406.

Stanford, L. R. (1987) Conduction velocity variations minimize conduction time differences among retinal ganglion cell axons. *Science* **238**, 358–360.

Starkov, A. A., Chinopoulos, C. and Fiskum, G. (2004) Mitochondrial calcium and oxidative stress as mediators of ischemic brain injury. *Cell Calcium* **36**, 257–264.

Staubli, U. V. (1995) Parallel properties of long-term potentiation and memory. In McGaugh, J. L., Weinburger, N. M. and Lynch, G. (eds) *Brain and Memory: Modulation and Mediation of Neuroplasticity*. Oxford University Press, pp. 303–318.

Staubli, U., Chun, D. and Lynch, G. (1998) Time-dependent reversal of long-term potentiation by an integrin antagonist. *Journal of Neuroscience* **18**, 3460–3469.

Staubli, U., Thibault, O., DiLorenzo, M. and Lynch, G. (1989) Antagonism of NMDA receptors impairs acquisition but not retention of olfactory memory. *Behavioral Neuroscience* **103**, 54–60.

Steele, R. J., Stewart, M. G. and Rose, S. P. R. (1995) Increases in NMDA receptor binding are specifically related to memory formation for a passive avoidance task in the chick: a quantitative autoradiographic study. *Brain Research* **674**, 352–356.

Stefani, A., Chen, Q., Flores-Hernandez, J., Jiao, Y., Reiner, A. and Surmeier, D. J. (1998) Physiological and molecular properties of AMPA/Kainate receptors expressed by striatal medium spiny neurons. *Developmental Neuroscience* **20**, 242–252.

Stein, V., House, D. R., Bredt, D. S. and Nicoll, R. A. (2003) Postsynaptic density-95 mimics and occludes hippocampal long-term potentiation and enhances long-term depression. *Journal of Neuroscience* **23**, 5503–5506.

Steriade, M. (2001) *The Intact and Sliced Brain*. MIT Press.

Steriade, M. (1997) Synchronized activities of coupled oscillators in the cerebral cortex and thalamus at different levels of vigilance. *Cerebral Cortex* **7**, 583–604.

Steriade, M. (1999) Coherent oscillations and short-term plasticity in corticothalamic networks. *Trends in Neuroscience* **22**, 337–345.

Stevens, B. *et al.* (2002) Adenosine: a neuron–glial transmitter promoting myelination in the CNS in response to action potentials. *Neuron* **36**, 855–868.

Stevens, C. F. (1994) CREB and memory consolidation. *Neuron* **13**, 769–770.

Stevens, B. and Fields, R. D. (2000) Response of Schwann cells to action potentials in development. *Science* **287**, 2267–2271.

Stevens, B., Tanner, S. and Fields, R. D. (1998) Control of myelination by specific patterns of neural impulses. *Journal of Neuroscience* **15**, 9303–9311.

Stevens, B., Porta, S., Haak, L. L., Gallo, V. and Fields, R. D. (2002) Adenosine: a neuron-glial transmitter promoting myelination in the CNS in response to action potentials. *Neuron* **36**, 855–868.

Stevens, E. R., Esquerra, M., Kim, P., Newman, E. A., Snyder, S. H., Zahs, K. R. and Miller, R. F. (2003) D-serine and serine racemase are present in the vertebrate retina and contribute to the functional expression of NMDA receptors. *Proceedings of the National Academy of Science of the United States of America* **100**, 6789–6794.

Steward, O. and Schuman, E. M. (2001) Protein synthesis at synaptic sites on dendrites. *Annual Review of Neuroscience* **24**, 299–325.

Steward, O. and Worley, P. F. (2001) Selective targeting of newly synthesized Arc mRNA to active synapses requires NMDA receptor activation. *Neuron* **30**, 227–240.

Stickgold, R. (2002) EMDR: a putative neurobiological mechanism of action. *Journal of Clinical Psychology* **58**, 61–75.

Stickgold, R., James, L. and Hobson, J. A. (2000) Visual discrimination learning requires post-training sleep. *Nature Neuroscience* **2**, 1237–1238.

Stickgold, R., Whidbee, D., Schirmer, B., Patel, V. and Hobson, J. A. (2000) Visual discrimination task improvement: a multi-step process occurring during sleep. *Journal of Cognitive Neuroscience* **12**, 246–254.

Stollhoff, N., Menzel, R. and Eisenhardt, D. (2005) Spontaneous recovery from extinction depends on the reconsolidation of the acquisition memory in an appetitive learning paradigm in the honeybee (Apis mellifera). *Journal of Neuroscience* **25**, 4485–4492.

Stoop, R. and Poo, M. M. (1995) Potentiation of transmitter release by ciliary neurotrophic factor requires somatic signaling. *Science* **267**, 695–699.

Stoop, R. and Poo, M. M. (1996) Synaptic modulation by neurotrophic factors. *Progress in Brain Research* **109**, 359–364.

Strack, S. and Colbran, R. J. (1998) Autophosphorylation-dependent targeting of calcium/calmodulin-dependent protein kinase II by the NR2B subunit of the N-methyl-D-aspartate receptor. *Journal of Biological Chemistry* **273**, 20,689–20,692.

Strack, S., Choi, S., Lovinger, D. M. and Colbran, R. J. (1997) Translocation of autophosphorylated calcium/calmodulin-dependent protein kinase II to the post-synaptic density. *Journal of Biological Chemistry* **272**, 13,467–13,470.

Straube, T., Korz, V., Balschun, D. and Frey, J. U. (2003) Novelty-exploration induces a protein synthesis-dependent late phase of long-term potentiation: involvement of β-adrenergic receptors. *Journal of Physiology* **552**, 953–960.

Streilein, J. W. (1993) Immune privilege as the result of local tissue barriers and immunosuppressive microenvironments. *Current Opinions in Immunology* **5**, 428–432.

Streit, W. J., Graeber, M. B. and Kreutzberg, G. W. (1989) Peripheral nerve lesion produces increased levels of major histocompatibility complex antigens in the central nervous system. *Journal of Neuroimmunology* **21**, 117–123.

Stuart, G. J. (2001) Determinants of spike timing-dependent synaptic plasticity. *Neuron* **32**, 966–968.

Stuart, G., Spruston, N., Sakmann, B. and Hausser, M. (1997) Action potential initiation and backpropagation in neurons of the mammalian CNS. *Trends in Neuroscience* **20**, 125–131.

Stutzmann, G. E. and LeDoux, J. E. (1999) GABAergic antagonists block the inhibitory effects of serotonin in the lateral amygdala: a mechanism for modulation of sensory inputs related to fear conditioning. *Journal of Neuroscience* **19**, RC8.

Sugihara, I., Lang, E. J. and Llinas, R. (1993) Uniform olivocerebellar conduction time underlies Purkinje cell complex spike synchronicity in the rat cerebellum. *Journal of Physiology* **470**, 243–271.

Summers, C. H. (2001) Mechanisms for quick and variable responses. *Brain and Behavioral Evolution* **57**, 283–292.

Sunanda, B. S., Rao, M. S. and Raju, T. R. (1995) Effect of chronic restraint stress on dendritic spines and excrescences of hippocampal CA3 pyramidal neurons – a quantitative study. *Brain Research* **694**, 312–317.

Suzuki, Y., Ikari, H., Hayashi, T. and Iguchi, A. (1996) Central administration of a nitric oxide synthase inhibitor impairs spatial memory in spontaneous hypertensive rats. *Neuroscience Letters* **207**, 105–108.

Suzuki, Y. J., Forman, H. J. and Sevanian, A. (1997) Oxidants as stimulators of signal transduction. *Free Radical Biology and Medicine* **22**, 269–285.

Suzuki, A., Josselyn, S. A., Frankland, P. W., Masushige, S., Silva, A. J. and Kida, S. (2004) Memory reconsolidation and extinction have distinct temporal and biochemical signatures. *Journal of Neuroscience* **24**, 4787–4795.

Svenningsson, P., Tzavara, E. T., Witkin, J. M., Fienberg, A. A., Nomikos, G. G. and Greengard, P. (2002) Involvement of striatal and extrastriatal DARPP-32 in biochemical and behavioral effects of fluoxetine (Prozac). *Proceedings of the National Academy of Science of the United States of America* **99**, 3182–3187.

Swadlow, H. A. (1985) Physiological properties of individual cerebral axons studied in vivo for as long as one year. *Journal of Neurophysiology* **54**, 1346–1362.

Swadlow, H. A., Waxman, S. G. and Geschwind, N. (1980) Small-diameter nonmyelinated axons in the primate corpus callosum. *Archives in Neurology* **37**, 114–115.

Swain, R. A., Harris, A. B., Wiener, E. C., Dutka, M. V., Morris, H. D., Theien, B. E., Konda, S., Engberg, K., Lauterbur, P. C. and Greenough, W. T. (2003) Prolonged exercise induces angiogenesis and increases cerebral blood volume in primary motor cortex of the rat. *Neuroscience* **117**, 1037–1046.

Sweatt, J. D. (2001) The neuronal MAP kinase cascade: a biochemical signal integration system subserving synaptic plasticity and memory. *Journal of Neurochemistry* **76**, 1–10.

Sweatt, J. D. (2004a) Mitogen-activated protein kinases in synaptic plasticity and memory. *Current Opinion in Neurobiology* **14**, 311–317.

Sweatt, J. D. (2004b) Hippocampal function in cognition. *Psychopharmacology* **174**, 99–110.

Syken, J., and Shatz, C. J. (2003a) Expression of T cell receptor beta locus in central nervous system neurons. *Proceedings of the National Academy of Sciences of the United States of America* **100**, 13,048–13,053.

Syken, J. and Shatz, C. J. (2003b) Neuronal expression of immunoreceptors. *Society for Neuroscience Abstracts*.

Syken, J., Grandpre, T., Kanold P. O. and Shatz C. J. (2006) PirB restricts ocular-dominance plasticity in visual cortex. *Science* **313**, 1795–1800.

Szabo, S. T., Gould, T. D. and Manji, H. K. (2003) Neurotransmitters, receptors, signal transduction, and second messengers in psychiatric disorders. In Schatzberg, A. and Nemeroff, C. B. (eds) *The American Psychiatric Publishing Textbook of Psychopharmacology*. American Psychiatric Publishing Inc., pp. 3–52.

Szabo, B., Dorner, L., Pfreundtner, C., Norenberg W. and Starke, K. (1998) Inhibition of GABAergic inhibitory postsynaptic currents by cannabinoids in rat corpus striatum. *Neuroscience* **85**, 395–403.

Szeligo, F. and Leblond, C. P. (1977) Response of the three main types of glial cells of cortex and corpus callosum in rats handled during suckling or exposed to enriched, control and impoverished environments following weaning. *Journal of Comparative Neurology* **172**, 247–263.

Szinyei, C., Heinbockel, T., Montagne, J. and Pape, H. C. (2000) Putative cortical and thalamic inputs elicit convergent excitation in a population of GABAergic interneurons of the lateral amygdala. *Journal of Neuroscience* **20**, 8909–8915.

Tabibnia, G., Cooke, B. M. and Breedlove, S. M. (1999) Sex difference and laterality in the volume of mouse dentate gyrus granule cell layer. *Brain Research* **827**, 41–45.

Tabori, N. E., Stewart, L. S., Znamensky, V., Romeo, R. D., Alves, S. E., McEwen, B. S. and Milner, T. A. (2005) Ultrastructural evidence that androgen receptors are located at extra nuclear sites in the rat hippocampal formation. *Neuroscience* **130**, 151–163.

Tailby, C., Wright, L. L., Metha, A. B. and Calford, M. B. (2005) Activity-dependent maintenance and growth of dendrites in adult cortex. *Proceedings of the National Academy of Sciences of the United States of America* **102**, 4631–4636.

Takahashi, J., Palmer, T. D. and Gage, F. H. (1999) Retinoic acid and neurotrophins collaborate to regulate neurogenesis in adult-derived neural stem cell cultures. *Journal of Neurobiology* **38**, 65–81.

Takahashi, T., Svoboda, K. and Malinow, R. (2003) Experience strengthening transmission by driving AMPA receptors into synapses. *Science* **299**, 1585–1588.

Takei, N., Sasaoka, K., Inoue, K., Takahashi, M., Endo, Y. and Hatanaka, H. (1997) Brain-derived neurotrophic factor increases the stimulation-evoked release of glutamate and the levels of exocytosis-associated proteins in cultured cortical neurons from embryonic rats. *Journal of Neurochemistry* **68**, 370–375.

Takimoto, K., Yang, E. K. and Conforti, L. (2002) Palmitoylation of KChIP splicing variants is required for efficient cell surface expression of Kv4.3 channels. *Journal of Biological Chemistry* **277**, 26,904–26,911.

Tamai, K., Ikari, K. and Hayashi, M. (1983) An ultrastructural change in developing rat cerebral cortex: a morphometrical study. *Folia Psychiatrica et Neurologica Japonica* **37**, 475–485.

Tanaka, H., Grooms, S. Y., Bennett, M. V. and Zukin, R. S. (2000) The AMPAR subunit GluR2: still front and center-stage. *Brain Research* **886**, 190–207.

Tanaka, J., Toku, K., Zhang, B., Ishihara, K., Sakanaka, M. and Maeda, N. (1999) Astrocytes prevent neuronal death induced by reactive oxygen and nitrogen species. *Glia* **28**, 85–96.

Tanapat, P., Galea, L. A. and Gould, E. (1998) Stress inhibits the proliferation of granule cell precursors in the developing dentate gyrus. *International Journal of Developmental Neuroscience* **16**, 235–239.

Tanapat, P., Hastings, N. B., Reeves, A. J. and Gould, E. (1999) Estrogen stimulates a transient increase in the number of new neurons in the dentate gyrus of the adult female rat. *Journal of Neuroscience* **19**, 5792–5801.

Tanapat, P., Hastings, N. B., Rydel, T. A., Galea, L. A. and Gould E. (2001) Exposure to fox odor inhibits cell proliferation in the hippocampus of adult rats via an adrenal hormone-dependent mechanism. *Journal of Comparative Neurology* **437**, 496–504.

Tang, X. D., Santarelli, L. C., Heinemann, S. H. and Hoshi, T. (2004) Metabolic regulation of potassium channels. *Annual Review of Physiology* **66**, 131–159.

Tang, M. X., Jacobs, D., Stern, Y., Marder, K., Schofield, P., Gurland, B., Andrews, H. and Mayeux, R. (1996) Effect of oestrogen during menopause on risk and age at onset of Alzheimer's disease. *Lancet* **348**, 429–432.

Tang, Y. P., Shimizu, E., Dube, G. R., Rampon, C., Kerchner, G. A., Zhuo, M., Liu, G. and Tsien, J. Z. (1999) Genetic enhancement of learning and memory in mice. *Nature* **401**, 63–69.

Tanzi, E. (1893/1909) *A Textbook of Mental Diseases*. Rebman.

Tao, X., Finkbeiner, S., Arnold, D. B., Shaywitz, A. J. and Greenberg, M. E. (1998) Ca^{2+} influx regulates BDNF transcription by a CREB family transcription factor-dependent mechanism. *Neuron* **20**, 709–726.

Tao, Y. X., Rumbaugh, G., Wang, G. D., Petralia, R. S., Zhao, C., Kauer, F. W., Tao, F., Zhuo, M., Wenthold, R. J., Raja, S. N., Huganir, R. L., Bredt, D. S. and Johns, R. A. (2003) Impaired NMDA receptor-mediated postsynaptic function and blunted NMDA receptor-dependent persistent pain in mice lacking postsynaptic density-93 protein. *Journal of Neuroscience* **23**, 6703–6712.

Tartaglia, N., Du, J., Tyler, W. J., Neale, E., Pozzo-Miller, L. and Lu, B. (2001) Protein synthesis-dependent and -independent regulation of hippocampal synapses by brain-derived neurotrophic factor. *Journal of Biological Chemistry* **276**, 37,585–37,593.

Taschenberger, H., Juttner, R. and Grantyn, R. (1999) Ca^{2+}-permeable P2X receptor channels in cultured rat retinal ganglion cells. *Journal of Neuroscience* **19**, 3353–3366.

Tassin, J. P., Studler, J. M., Herve, D., Blanc, G. and Glowinski, J. (1986) Contribution of noradrenergic neurons to the regulation of dopaminergic (D1) receptor denervation supersensitivity in rat prefrontal cortex. *Journal of Neurochemistry* **46**, 243–248.

Taubenfeld, S. M., Milekic, M. H., Monti, B. and Alberini, C. M. (2001) The consolidation of new but not reactivated memory requires hippocampal C/EBPbeta. *Nature Neuroscience* **4**, 813–818.

Tauber, H., Waehneldt, T. V. and Neuhoff, V. (1980) Myelination in rabbit optic nerves is accelerated by artificial eye opening. *Neuroscience Letters* **16**, 235–238.

Teicher, M. H., Dumont, N. L., Ito, Y., Vaituzis, C., Giedd, J. N. and Andersen S. L. (2004) Childhood neglect is associated with reduced corpus callosum area. *Biological Psychiatry* **56**, 80–85.

Temple, S. (2001) The development of neural stem cells. *Nature* **414**, 112–117.

Terasawa, E. and Timiras, P. S. (1968) Electrical activity during the estrous cycle of the rat: cyclic changes in limbic structures. *Endocrinology* **83**, 207–216.

Teyler, T. J., Vardaris, R. M., Lewis, D. and Rawitch, A. B. (1980) Gonadal steroids: effect on excitability of hippocampal pyramidal cells. *Science* **209**, 1017–1019.

Thams, S., Oliveira, A. and Cullheim, S. (2008) MHC class I expression and synaptic plasticity after nerve lesion. *Brain Res Rev.* **57**, 265–9.

Thomas, G. M. and Huganir, R. L. (2004) MAPK cascade signalling and synaptic plasticity. *Nat Rev Neurosci* **5**, 173–183.

Thompson, R. F., Bao, S., Chen, L., Cipriano, B. D., Grethe, J. S., Kim, J. J., Thompson, J. K., Tracy, J. A., Weninger, M. S. and Krupa, D. J. (1997) Associative learning. *International Review of Neurobiology* **41**, 151–189.

Tieman, S. B. (1984) Effects of monocular deprivation on geniculocortical synapses in the cat. *Journal of Comparative Neurology* **222**, 166–176.

Tieman, S. B. (1985) The anatomy of geniculocortical connections in monocularly deprived cats. *Cellular and Molecular Neurobiology* **5**, 35–45.

Tkachev, D., Mimmack, M. L., Ryan, M. M., Wayland, M., Freeman, T., Jones, P. B., Starkey, M., Webster, M. J., Yolken, R. H. and Bahn S. (2003) Oligodendrocyte dysfunction in schizophrenia and bipolar disorder. *Lancet* **362**, 798–805.

Tokube, K., Kiyosue, T. and Arita, M. (1998) Effects of hydroxyl radicals on KATP channels in guinea-pig ventricular myocytes. *Pflugers Archive* **437**, 155–157.

Tomida, T., Hirose, K., Takizawa, A., Shibasaki, F. and Iino, M. (2003) NFAT functions as a working memory of Ca^{2+} signals in decoding Ca^{2+} oscillation. *EMBO Journal* **22**, 3825–3832.

Tomita, S., Chen, L., Kawasaki, Y., Petralia, R. S., Wenthold, R. J., Nicoll, R. A. and Bredt, D. S. (2003) Functional studies and distribution define a family of transmembrane AMPA receptor regulatory proteins. *Journal of Cell Biology* **161**, 805–816.

Toms, N. J., Jane, D. E., Kemp, M. C., Bedingfield, J. S. and Roberts, P. J. (1996) The effects of (RS)-alpha-cyclopropyl-4-phosphonophenylglycine ((RS)-CPPG), a potent and selective metabotropic glutamate receptor antagonist. *British Journal of Pharmacology* **119**, 851–854.

Tonegawa, S., Tsien, J. Z., McHugh, T. J., Huerta, P., Blum, K. I. and Wilson, M. A. (1996) Hippocampal CA1-region-restricted knockout of NMDAR1 gene disrupts synaptic plasticity, place fields, and spatial learning. *Cold Spring Harbor Symposia Quantitative Biology* **61**, 225–238.

Tongiorgi, E., Domenici, L. and Simonato, M. (2006) What is the biological significance of BDNF mRNA targeting in the dendrites? Clues from epilepsy and cortical development. *Molecular Neurobiology* **33**, 17–32.

Tongiorgi, E., Righi, M. and Cattaneo, A. (1997) Activity-dependent dendritic targeting of BDNF and TrkB mRNAs in hippocampal neurons. *Journal of Neuroscience* **17**, 9492–9505.

Toni, I., Krams, M., Turner, R. and Passingham, R. E. (1998) The time course of changes during motor sequence learning: a whole-brain fMRI study. *Neuroimage* **8**, 50–61.

Toran-Allerand, C. D. (1976) Sex steroids and development of the newborn mouse hypothalamus and preoptic area in vitro: implications for sexual differentiation. *Brain Research* **106**, 407–412.

Torasdotter, M., Metsis, M., Henriksson, B. G., Winblad, B. and Mohammed, A. H. (1998) Environmental enrichment results in higher levels of nerve growth factor mRNA in the rat visual cortex and hippocampus. *Behavioral Brain Research* **93**, 83–90.

Torimoto, Y., Kinebuchi, M., Matsuura, A., Kikuchi, K. and Uede, T. (1990) A monoclonal antibody (8h3) that binds to rat T lineage cells and augments in vitro proliferative responses. *Journal of Experimental Medicine* **172**, 1315–1323.

Torras-Garcia, M., Lelong, J., Tronel, S. and Sara, S. J. (2005) Reconsolidation after remembering an odor-reward association requires NMDA receptors. *Learning & Memory* **12**, 18–22.

Towart, L. A., Alves, S. E., Znamensky, V., Hayashi, S., McEwen, B. S. and Milner, T. A. (2003) Subcellular relationships between cholinergic terminals and estrogen receptor-a in the dorsal hippocampus. *Journal of Comparative Neurology* **463**, 390–401.

Tower, D. B. and Elliott, K. A. C. (1952) Activity of the acetylcholine system in cerebral cortex of various unanesthetized animals. *American Journal of Physiology* **168**, 747–759.

Tower, D. B. and Young, O. M. (1973) The activities of butyrylcholinesterase and carbonic anhydrase, the rate of anaerobic glycolysis, and the question of a constant density of glial cells in the cerebral cortices of various mammalian species from mouse to whale. *Journal of Neurochemistry* **20**, 269–278.

Trachtenberg, J. T., Chen, B. E., Knott, G. W., Feng, G., Sanes, J. R. and Welker E. (2002) Long-term in vivo imaging of experience-dependent synaptic plasticity in adult cortex. *Nature* **420**, 788–794.

Traub, R. J. and Mendell, L. M. (1988) The spinal projection of individual identified A-delta- and C-fibers. *Journal of Neurophysiology* **59**, 41–55.

Trépanier, G., Furling, D., Puymirat, J. and Mirault, M. E. (1996) Immunocytochemical localization of seleno-glutathione peroxidase in the adult mouse brain. *Neuroscience* **75**, 231–243.

Treves A and Rolls E. T. (1994) Computational analysis of the role of the hippocampus in memory. *Hippocampus* **4**, 374–391.

Trombley, P. Q. and Shepherd, G. M. (1994) Glycine exerts potent inhibitory actions on mammalian olfactory bulb neurons. *Journal of Neurophysiology* **71**, 761–767.

Tronel, S., Milekic, M. H. and Alberini, C. M. (2005) Linking new information to a reactivated memory requires consolidation and not reconsolidation mechanisms. *PLoS Biol* **3**, e293.

Tsai, G. and Coyle, J. T. (2002) Glutamatergic mechanisms in schizophrenia. *Annual Review of Pharmacology and Toxicology* **42**, 165–179.

Tsai, M. J. and O'Malley, B. W. (1994) Molecular mechanisms of action of steroid/thyroid receptor superfamily members. *Annual Review of Biochemistry* **63**, 451–468.

Tsai, G., Yang, P., Chung, L. C., Lange, N. and Coyle, J. T. (1998) D-serine added to antipsychotics for the treatment of schizophrenia. *Biological Psychiatry* **44**, 1081–1089.

Tsay, D. and Yuste, R. (2004) On the electrical function of dendritic spines. *Trends in Neurosciences* **27**, 77–83.

Tsuang, M. (2000) Schizophrenia: genes and environment. *Biological Psychiatry* **47**, 210–220.

Tsukaguchi, H., Tokui, T., Mackenzie, B., Berger, U. V., Chen, X. Z., Wang, Y., Brubaker, R. F. and Hediger, M. A. (1999) A family of mammalian Na^+-dependent L-ascorbic acid transporters. *Nature* **399**, 70–75.

Tsvetkov, E., Carlezon, W. A., Benes, F. M., Kandel, E. R. and Bolshakov, V. Y. (2002) Fear conditioning occludes LTP-induced presynaptic enhancement of synaptic transmission in the cortical pathway to the lateral amygdala. *Neuron* **34**, 289–300.

Tully, T., Preat, T., Boynton, S. C. and Del Vecchio, M. (1994) Genetic dissection of consolidated memory in Drosophila. *Cell* **79**, 35–47.

Tulving, E. (1985) How many memory systems are there? *American Psychologist* **40**, 385–398.

Tulving, E. (2002) Episodic memory: From mind to brain. *Annual Review of Psychology* **53**, 1–25.

Turner, A. M. and Greenough, W. T. (1985) Differential rearing effects on rat visual cortex synapses. I. Synaptic and neuronal density and synapses per neuron. *Brain Research* **329**, 195–203.

Turner, B. and Herkenham, M. (1991) Thalamoamygdaloid projections in the rat: a test of the amygdala's role in sensory processing. *Journal of Comparative Neurology* **313**, 295–325.

Turrigiano, G. G. and Nelson, S. B. (2000) Hebb and homeostasis in neuronal plasticity. *Current Opinion in Neurobiology* **10**, 358–364.

Turrigiano, G. G. and Nelson, S. B. (2004) Homeostatic plasticity in the developing nervous system. *Nature Review Neuroscience* **5**, 97–107.

Tyler, W. J. and Pozzo-Miller, L. D. (2001) BDNF enhances quantal neurotransmitter release and increases the number of docked vesicles at the active zones of hippocampal excitatory synapses. *Journal of Neuroscience* **21**, 4249–4258.

Tyler, W. J., Alonso, M., Bramham, C. R. and Pozzo-Miller, L. D. (2002) From acquisition to consolidation: on the role of brain-derived neurotrophic factor signaling in hippocampal-dependent learning. *Learning & Memory* **9**, 224–237.

Tzingounis, A. V. and Nicoll, R. A. (2006) Arc/Arg3.1: linking gene expression to synaptic plasticity and memory. *Neuron* **52**, 403–407.

Tzschentke, T. M. (2003) Reassessment of buprenorphine in conditioned place preference: temporal and pharmacological considerations. *Psychopharmacology (Berl)*, published online November 13.

Tzschentke, T. M. and Schmidt, W. J. (2003) Glutamatergic mechanisms in addiction. *Molecular Psychiatry* **8**, 373–382.

Ullian, E. M., Harris, B. T., Wu, A., Chan, J. R. and Barres, B. A. (2004) Schwann cells and astrocytes induce synapse formation by spinal motor neurons in culture. *Molecular and Cellular Neuroscience* **25**, 241–251.

Ullian, E. M., Sapperstein, S. K., Christopherson, K. S. and Barres, B. A. (2001) Control of synapse number by glia. *Science* **291**, 657–661.

Ungerleider, L. G. and Haxby, J. V. (1994) "What" and "where" in the human brain. *Current Opinion in Neurobiology* **4**, 157–165.

Ungerleider, L. G., Doyon, J. and Karni, A. (2002) Imaging brain plasticity during motor skill learning. *Neurobiology of Learning and Memory* **78**, 553–564.

Uo, T., Yoshimura, T., Shimizu, S. and Esaki, N. (1998) Occurrence of pyridoxal 5'-phosphate-dependent serine racemase in silkworm, Bombyx mori. *Biochemical and Biophysical Research Communications* **246**, 31–34.

Uzakov, S., Frey, J. U. and Korz, V. (2005) Reinforcement of rat hippocampal LTP by holeboard training. *Learning & Memory* **12**, 165–171.

Uzakov, S., Behnisch, T., Frey, J. U. and Korz, V. (2003) Modulation of hippocampal long-term potentiation by holeboard learning in rats. 2003 Society for Neuroscience, Abstract No. 623.16.

Vaiva, G., Ducrocq, F., Jezequel, K., Averland, B., Lestavel, P., Brunet, A. and Marmar, C. R. (2003) Immediate treatment with propranolol decreases posttraumatic stress disorder two months after trauma. *Biological Psychiatry* **54**, 1471.

Vale, S., Espejel, M. A. and Dominguez, J. C. (1971) Amantadine in depression. *Lancet* **2**, 437.

Van der Kloot, W. (1991) The regulation of quantal size. *Progress in Neurobiology* **36**, 93–130.

van Praag, H., Kempermann, G. and Gage, F. H. (1999) Running increases cell proliferation and neurogenesis in the adult mouse dentate gyrus. *Nature Neuroscience* **2**, 266–270.

van Praag, H., Kempermann, G. and Gage, F. H. (2000) Neural consequences of environmental enrichment. *Nature Reviews Neuroscience* **1**, 191–198.

van Praag, H., Christie, B. R., Sejnowski, T. J. and Gage, F. H. (1999a) Running enhances neurogenesis, learning, and long-term potentiation in mice. *Proceedings of the National Academy of Sciences of the United States of America* **96**, 13,427–13,431.

van Praag, H., Schinder, A. F., Christie, B. R., Toni, N., Palmer, T. D. and Gage, F. H. (2002) Functional neurogenesis in the adult hippocampus. *Nature* **415**, 1030–1034.

van Steensel, B., van Binnendijk, E. P., Hornsby, C. D. van der Voort, H. T., Krozowski, Z. S., De Kloet, E. R. and van Driel, R. (1996) Partial colocalization of glucocorticoid and mineralocorticoid receptors in discrete compartments in nuclei of rat hippocampus neurons. *Journal of Cell Science* **109**, 787–792.

Vardaris, R. M. and Teyler, T. J. (1980) Sex differences in the response of hippocampal CA1 pyramids to gonadal steroids: effects of testosterone and estradiol on the in vitro slice preparation. *Society for Neuroscience Abstracts*, p. A153.159.

Ventura, R. and Harris, K. M. (1999) Three-dimensional relationships between hippocampal synapses and astrocytes. *Journal of Neuroscience* **19**, 6897–6906.

Vertes, R. P. (2004) Memory consolidation in sleep; dream or reality. *Neuron* **44**, 135–148.

Vicario-Abejon, C., Collin, C., McKay, R. D. and Segal, M. (1998) Neurotrophins induce formation of functional excitatory and inhibitory synapses between cultured hippocampal neurons. *Journal of Neuroscience* **18**, 7256–7271.

Vicario-Abejon, C., Owens, D., McKay, R. and Segal, M. (2002) Role of neurotrophins in central synapse formation and stabilization. *Nature Reviews Neuroscience* **3**, 965–974.

Vicario-Abejon, C., Johe, K. K., Hazel, T. G., Collazo, D. and McKay, R. D. (1995) Functions of basic fibroblast growth factor and neurotrophins in the differentiation of hippocampal neurons. *Neuron* **15**, 105–114.

Vickers, C. A., Dickson, K. S. and Wyllie, D. J. (2005) Induction and maintenance of late-phase long-term potentiation in isolated dendrites of rat hippocampal CA1 pyramidal neurones. *Journal of Physiology* **568**, 803–813.

Vincent, G. P., Paré, W. P., Prenatt, J. E. D. and Glavin, G. B. (1984) Aggression, body temperature, and stress ulcer. *Physiology & Behavior* **32**, 265–268.

Viola, H., Furman, M., Izquierdo, L. A., Alonso, M., Barros, D. M., de Souza, M. M., Izquierdo, I. and Medina, J. H. (2000) Phosphorylated cAMP response element-binding protein as a molecular marker of memory processing in rat hippocampus: effect of novelty. *Journal of Neuroscience* **20**, RC112.

Vitkovic, L., Bockaert, J. and Jacque, C. (2000) "Inflammatory" cytokines: neuromodulators in normal brain? *Journal of Neurochemistry* **74**, 457–471.

Vo, N., Klein, M. E., Varlamova, O., Keller, D. M., Yamamoto, T., Goodman, R. H. and Impey, S. (2005). A cAMP-response element binding protein-induced microRNA regulates neuronal morphogenesis. *Proceedings of the National Academy of Sciences of the United States of America* **102**, 16,426–16,431.

Volkmar, F. R. and Greenough, W. T. (1972) Rearing complexity affects branching of dendrites in the visual cortex of the rat. *Science* **176**, 1445–1447.

Volterra, A., Magistretti, P. and Haydon, P. G. (2002) *The Tripartite Synapse: Glia in Synaptic Transmission.* Oxford University Press.

von Hertzen, L. S. and Giese, K. P. (2005) Memory reconsolidation engages only a subset of immediate-early genes induced during consolidation. *Journal of Neuroscience* **25**, 1935–1942.

Votyakova, T. V. and Reynolds, I. J. (2001) $\Delta \Psi$m-Dependent and -independent production of reactive oxygen species by rat brain mitochondria. *Journal of Neurochemistry* **79**, 266–277.

Vouimba, R. M., Foy, M. R. and Thompson, R. F. (2000) 17B-estradiol suppresses expression of long-term depression in aged rats. *Brain Research Bulletin* **53**, 783–787.

Wagner, U., Gais, S. and Born, J. (2001) Emotional memory formation is enhanced across sleep intervals with high amounts of rapid eye movement sleep. *Learning & Memory* **8**, 112–119.

Walker, M. P. (2005) A refined model of sleep and the time course of memory formation. *Behavioural and Brain Science* **28**, 51–64.

Walker, D. L. and Davis, M. (2000) Involvement of NMDA receptors within the amygdala in short- versus long-term memory for fear conditioning as assessed with fear-potentiated startle. *Behavioral Neuroscience* **114**, 1019–1033.

Walker, M. P. and Stickgold, R. (2004) Sleep-dependent learning and memory consolidation. *Neuron* **44**, 121–133.

Walker, M., Brakefield, T., Hobson, J. A. and Stickgold, R. (2003) Dissociable stages of human memory consolidation and reconsolidation. *Nature* **425**, 616–620.

Walker, M. P., Stickgold, R., Jolesz, F. A. and Yoo, S.-S. (2005a) The functional anatomy of sleep-dependent visual skill learning. *Cerebral Cortex* **15**, 1666–1675.

Walker, M. P., Brakefield, T., Morgan, A., Hobson, J. A. and Stickgold, R. (2002) Practice with sleep makes perfect: sleep dependent motor skill learning. *Neuron* **35**, 205–211.

Walker, M. P., Stickgold, R., Alsop, D., Gaab, N. and Schlaug, G. (2005b) Sleep-dependent motor memory plasticity in the human brain. *Neuroscience* **133**, 911–917.

Wallace, C. S., Kilman, V. L., Withers, G. S. and Greenough, W. T. (1992) Increases in dendritic length in occipital cortex after 4 days of differential housing in weanling rats. *Behavioral and Neural Biology* **58**, 64–68.

Wallace, T. L., Stellitano, K. E., Neve, R. L. and Duman, R. S. (2004) Effects of cyclic adenosine monophosphate response element binding protein overexpression in the basolateral amygdala on behavioral models of depression and anxiety. *Biological Psychiatry* **56**, 151–160.

Wallenstein, G. V. and Hasselmo, M. E. (1997) GABAergic modulation of hippocampal population activity: sequence learning, place field development, and the phase precession effect. *Journal of Neurophysiology* **78**, 393–408.

Wallenstein, G. V., Eichenbaum, H. and Hasselmo, M. E. (1998) The hippocampus as an associator of discontiguous events. *Trends in Neurosciences* **21**, 315–365.

Walling, S. G. and Harley, C. W. (2004) Locus ceruleus activation initiates delayed synaptic potentiation of perforant path input to the dentate gyrus in awake rats: a novel beta-adrenergic- and protein synthesis dependent mammalian plasticity mechanism. *Journal of Neuroscience* **24**, 598–604.

Waltereit, R. and Weller, M. (2003) Signaling from cAMP/PKA to MAPK and synaptic plasticity. *Molecular Neurobiology* **27**, 99–106.

Wampler, M. S. and McAllister, K. A. (2004) MHC class I protein is present and regulated by activity in dissociated cortical neurons during synaptogenesis. *Society for Neuroscience Abstracts*.

Wan C., Yang Y., Feng G., Gu, N., Liu, H., Zhu, S., He, L. and Wang, L. (2005) Polymorphisms of myelin-associated glycoprotein gene are associated with schizophrenia in the Chinese Han population. *Neuroscience Letters* **18**, 126–131.

Wang, K. C., Koprivica, V., Kim, J. A., Sivasankaran, R., Guo, Y., Neve, R. L. and He, Z. (2002) Oligodendrocyte-myelin glycoprotein is a Nogo receptor ligand that inhibits neurite outgrowth. *Nature* **417**, 941–944.

Wang, H. Y. and Friedman, E. (1996) Enhanced protein kinase C activity and translocation in bipolar affective disorder brains. *Biological Psychiatry* **40**, 568–575.

Wang, H. Y. and Friedman, E. (1999) Effects of lithium on receptor-mediated activation of G proteins in rat brain cortical membranes. *Neuropharmacology* **38**, 403–414.

Wang, X. F. and Cynader, M. S. (2000) Astrocytes provide cysteine to neurons by releasing glutathione. *Journal of Neurochemistry* **74**, 1434–1442.

Wang, R., Dineley, K. T., Sweatt, J. D. and Zheng, H. (2004) Presenilin 1 familial Alzheimer's disease mutation leads to defective associative learning and impaired adult neurogenesis. *Neuroscience* **126**, 305–312.

Wang, S. H., Ostlund, S. B., Nader, K. and Balleine, B. W. (2005) Consolidation and reconsolidation of incentive learning in the amygdala. *Journal of Neuroscience* **25**, 830–835.

Wang, K. C., Kim, J. A., Sivasankaran, R., Segal, R. and He, Z. (2002) P75 interacts with the Nogo receptor as a co-receptor for Nogo, MAG and OMgp. *Nature* **420**, 74–78.

Wang, S. J., Huang, C. C., Hsu, K. S., Tsai, J. J. and Gean, P. W. (1996) Presynaptic inhibition of excitatory neurotransmission by lamotrigine in the rat amygdalar neurons. *Synapse* **24**, 248–255.

Wang, J. Q., Tang, Q., Parelkar, N. K., Liu, Z., Samdani, S., Choe, E. S., Yang, L., Mao, L. (2004) Glutamate signaling to Ras-MAPK in striatal neurons: mechanisms for inducible gene expression and Plasticity. *Molecular Neurobiology* **29**, 1–14.

Wang, W. L., Yeh, S. F., Chang, Y. I., Hsiao, S. F., Lian, W. N., Lin, C. H., Huang, C. Y. and Lin, W. J. (2003b) PICK1, an anchoring protein that specifically targets protein kinase C-alpha to mitochondria selectively upon serum stimulation in NIH 3T3 cells. *Journal of Biological Chemistry* **278**, 37,705–37,712.

Wang, H., Shimizu, E., Tang, Y. P., Cho, M., Kyin, M., Zuo, W., Robinson, D. A., Alaimo, P. J., Zhang, C., Morimoto, H., Zhuo, M., Feng, R., Shokat, K. M. and Tsien, J. Z. (2003a) Inducible protein knockout reveals temporal requirement of CaMKII reactivation for memory consolidation in the brain. *Proceedings of the National Academy of Science of the United States of America* **100**, 4287–4292.

Warren, S. G. and Juraska, J. M. (1997) Spatial and non-spatial learning across the rat estrous cycle. *Behavioral Neuroscience* **111**, 259–266.

Warren, S. G., Humphreys, A. G., Juraska, J. M. and Greenough, W. T. (1995) LTP varies across the estrous cycle: enhanced synaptic plasticity in proestrus rats. *Brain Research* **703**, 26–30.

Warsh, J., Young, L. and Li, P. (2000) Guanine nucleotide binding (G) proteindisturbances. In Manji, H., Bowden, C. and Belmaker, R. (eds) *Bipolar Affective Disorder in Bipolar Medications: Mechanisms of Action*. American Psychiatric Press, pp. 299–329.

Warsh, J. J., Young, L. T. and Li, P. P. (2000) Guanine nucleotide binding (G) protein disturbances. In Belmaker, R. (ed.) *Bipolar Affective Disorder in Bipolar Medications: Mechanisms of Action*. American Psychiatric Press, pp. 299–329.

Waxman, S. G. (1972) Regional differentiation of the axon: a review with special reference to the concept of the multiplex neuron. *Brain Research* **47**, 269–288.

Waxman, S. G., Pappas, G. D. and Bennett, M. V. L. (1972) Morphological correlates of functional differentiation of nodes of Ranvier along single fibers in the neurogenic electric organ of the knife fish Sternarchus. *Journal of Cell Biology* **53**, 210–224.

Weick, J. P., Kuo, S. P., and Mermelstein, P. G. (2005) L-type calcium channel regulation of gene expression. *Cell Science Reviews* **1**, 44–49.

Weiland, N. G. (1992a) Estradiol selectively regulates agonist binding sites on the N-methyl-D-aspartate receptor complex in the CA1 region of the hippocampus. *Endocrinology* **131**, 662–668.

Weiland, N. G. (1992b) Glutamic acid decarboxylase messenger ribonucleic acid is regulated by estradiol and progesterone in the hippocampus. *Endocrinology* **131**, 2697–2702.

Weiland, N. G., Orisaka, C., Hayashi, S. and McEwen, B. S. (1997) Distribution and hormone regulation of estrogen receptor-immunoreactive cells in the hippocampus of male and female rats. *Journal of Comparative Neurology* **388**, 603–612.

Weinberger, N. M. (1995) Retuning the brain by fear conditioning. In Gazzaniga, M. S. (ed.) *The Cognitive Neurosciences*. MIT Press, pp. 1071–1090.

Weiss, J. H. and Sensi, S. L. (2000) Ca^{2+}-Zn^{2+} permeable AMPA or kainate receptors: possible key factors in selective neurodegeneration. *Trends in Neuroscience* **23**, 365–371.

Weisskopf, M. G. and LeDoux, J. E. (1999) Distinct populations of NMDA receptors at subcortical and cortical inputs to principal cells of the lateral amygdala. *Journal of Neurophysiology* **81**, 930–934.

Weisskopf, M. G., Bauer, E. P. and LeDoux, J. E. (1999) L-type voltage-gated calcium channels mediate NMDA-independent associative long-term potentiation at thalamic input synapses to the amygdala. *Journal of Neuroscience* **19**, 10,512–10,519.

Weissman, A. (1967) Drugs and retrograde amnesia. *International Review of Neurobiology* **10**, 167–198.

Weng, G., Bhalla, U. S. and Iyengar, R. (1999) Complexity in biological signaling systems. *Science* **284**, 92–96.

Wesa, J. M., Chang, F. L., Greenough, W. T. and West, R. W. (1982) Synaptic contact curvature: effects of differential rearing on rat occipital cortex. *Brain Research* **256**, 253–257.

West, A. E., Griffith, E. C. and Greenberg, M. E. (2002) Regulation of transcription factors by neuronal activity. *Nature Reviews Neuroscience* **3**, 921–931.

West, R. W. and Greenough, W. T. (1972) Effect of environmental complexity on cortical synapses of rats: preliminary results. *Behavioral Biology* **7**, 279–284.

West, A. E., Chen, W. G., Dalva, M. B., Dolmetsch, R. E., Kornhauser, J. M., Shaywitz, A. J., Takasu, M. A., Tao, X. and Greenberg, M. E. (2001) Calcium regulation of neuronal gene expression. *Proceedings of the National Academy of Sciences of the United States of America* **98**, 11,024–11,031.

Westerink, B. H., Santiago, M. and De Vries, J. B. (1992) In vivo evidence for a concordant response of terminal and dendritic dopamine release during intranigral infusion of drugs. *Naunyn-Schmiedeberg's Archives of Pharmacology* 345, 523–529.

Whishaw, I. Q., Zaborowski, J. A. and Kolb, B. (1984) Postsurgical enrichment aids adult hemidecorticate rats on a spatial navigation task. *Behavioral and Neural Biology* 42, 183–190.

White, L. A., Keane, R. W. and Whittemore, S. R. (1994) Differentiation of an immortalized CNS neuronal cell line decreases their susceptibility to cytotoxic T cell lysis in vitro. *Journal of Neuroimmunology* 49, 135–143.

Wilde, G. J., Pringle, A. K., Sundstrom, L. E., Mann, D. A. and Iannotti, F. (2000) Attenuation and augmentation of ischaemia-related neuronal death by tumour necrosis factor-alpha in vitro. *European Journal of Neuroscience* 12, 3863–3870.

Wilensky, A. E., Schafe, G. E. and LeDoux, J. E. (1999) Functional inactivation of the amygdala before but not after auditory fear conditioning prevents memory formation. *Journal of Neuroscience* 19, RC48.

Will, B. and Kelche, C. (1992) Environmental approaches to recovery of function from brain damage: a review of animal studies (1981 to 1991). *Advances in Experimental Medicine and Biology* 325, 79–103.

Williams, J. H., Li, Y. G., Nayak, A., Errington, M. L., Murphy, K. P. and Bliss, T. V. (1993) The suppression of long-term potentiation in rat hippocampus by inhibitors of nitric oxide synthase is temperature and age dependent. *Neuron* 11, 877–884.

Wilquet, V. and De Strooper, B. (2004) Amyloid-beta precursor protein processing in neurodegeneration. *Current Opinion in Neurobiology* 14, 582–588.

Wilson, M. A. and McNaughton, B. L. (1994) Reactivation of hippocampal ensemble memories during sleep. *Science* 265, 676–679.

Wirth, S., Yanike, M., Frank, L. M., Smith, A. C., Brown, E. N. and Suzuki, W. A. (2003) Single neurons in the monkey hippocampus and learning of new associations. *Science* 300, 1578–1581.

Withers, G. S. and Greenough, W. T. (1989) Reach training selectively alters dendritic branching in subpopulations of layer II–III pyramids in rat motor-somatosensory forelimb cortex. *Neuropsychologia* 27, 61–69.

Wo, Z. G., Bian, Z. C. and Oswald, R. E. (1995) Asn-265 of frog kainate binding protein is a functional glycosylation site: implications for the transmembrane topology of glutamate receptors. *FEBS Letters* 368, 230–234.

Wolosker, H., Blackshaw, S. and Snyder, S. H. (1999) Serine racemase: a glial enzyme synthesizing D-serine to regulate glutamate-N-methyl-D-aspartate neurotransmission. *Proceedings of the National Academy of Science of the United States of America* 96, 13,409–13,414.

Wolosker, H., Blackshaw, S. and Snyder, S. H. (1999a) Serine racemase: a glial enzyme synthesizing D-serine to regulate glutamate-N-methyl-D-aspartate neurotransmission. *Proceeding of the National Academy of Sciences of the United States of America* 96, 13,409–13,414.

Wolosker, H., Sheth, K. N., Takahashi, M., Mothet, J. P., Brady, R. O., Jr., Ferris, C. D. and Snyder, S. H. (1999b) Purification of serine racemase: biosynthesis of the neuromodulator D-serine. *Proceeding of the National Academy of Sciences of the United States of America* 96, 721–725.

Wong, M. and Moss, R. L. (1991) Electrophysiological evidence for a rapid membrane action of the gonadal steroid, 17b-estradiol, on CA1 pyramidal neurons of the rat hippocampus. *Brain Research Bulletin* 543, 148–152.

Wong, M. and Moss, R. L. (1992) Long-term and short-term electrophysiological effects of estrogen on the synaptic properties of hippocampal CA1 neurons. *Journal of Neuroscience* 12, 3217–3225.

Wong, M. and Moss, R. L. (1994) Patch-clamp analysis of direct steroidal modulation of glutamate receptor-channels. *Journal of Neuroendocrinology* 6, 347–355.

Wong, M., Thompson, T. L. and Moss, R. L. (1996) Nongenomic actions of estrogens in the brain: physiological significance and cellular mechanisms. *Critical Reviews in Neurobiology* 10, 189–203.

Wong, C. S., Cherng, C. H., Luk, H. N., Ho, S. T. and Tung, C. S. (1996) Effects of NMDA receptor antagonists on inhibition of morphine tolerance in rats: binding at mu-opioid receptors. *European Journal of Pharmacology* 297, 27–33.

Wong, G. H., Bartlett, P. F., Clark-Lewis, I., Battye, F. and Schrader, J. W. (1984) Inducible expression of H-2 and Ia antigens on brain cells. *Nature* 310, 688–691.

Wood, P. L. (1995) The co-agonist concept: is the NMDA-associated glycine receptor saturated in vivo? *Life Sciences* 57, 301–310.

Wood, E., Dudchenko, P. A. and Eichenbaum, H. (1999) The global record of memory in hippocampal neuronal activity. *Nature* 397, 613–616.

Wood, E., Dudchenko, P., Robitsek, J. R. and Eichenbaum, H. (2000) Hippocampal neurons encode information about different types of memory episodes occurring in the same location. *Neuron* 27, 623–633.

Wood-Kaczmar A., Gandhi S. and Wood N. W. (2006) Understanding the molecular causes of Parkinson's disease. *Trends in Molecular Medicine* 12, 521–528.

Woods, A. and Couchman, J. R. (1994) Syndecan 4 heparan sulfate proteoglycan is a selectively enriched and widespread focal adhesion component. *Molecular Biology of the Cell* 5, 183–192.

Woods, A., Couchman, J. R., Johansson, S. and Hook, M. (1986) Adhesion and cytoskeletal organization of fibroblasts in response to fibronectin fragments. *EMBO Journal* 5, 665–670.

Woods, A., McCarthy, J. B., Furcht, L. T. and Couchman, J. R. (1993) A synthetic peptide from the COOH-terminal heparin-binding domain of fibronectin promotes focal adhesion formation. *Molecular Biology of the Cell* 4, 605–613.

Woodson, W., Farb, C. R. and LeDoux, J. E. (2000) Afferents from the auditory thalamus synapse on inhibitory interneurons in the lateral nucleus of the amygdala. *Synapse* 38, 124–137.

Woolf, C. J. (2003) No Nogo: now where to go? *Neuron* 38, 153–156.

Woolf, N. J. (1998) A structural basis for memory storage in mammals. *Progress in Neurobiology* 55, 59–77.

Woolley, C. S. (1998) Estrogen-mediated structural and functional synaptic plasticity in the female rat hippocampus. *Hormones and Behavior* 34, 140–148.

Woolley, C. S. and McEwen, B. S. (1992) Estradiol mediates fluctuation in hippocampal synapse density during the estrous cycle in the adult rat. *Journal of Neuroscience* 12, 2549–2554.

Woolley, C. S. and McEwen, B. S. (1993) Roles of estradiol and progesterone in regulation of hippocampal dendritic spine density during the estrous cycle in the rat. *Journal of Comparative Neurology* 336, 293–306.

Woolley, C. S. and McEwen, B. S. (1994) Estradiol regulates hippocampal dendritic spine density via an N-methyl-D-aspartate receptor-dependent mechanism. *Journal of Neuroscience* 14, 7680–7687.

Woolley, C. S., Wenzel, H. J. and Schwartzkroin, P. A. (1996) Estradiol increases the frequency of multiple synapse boutons in the hippocampal CA1 region of the adult female rat. *Journal of Comparative Neurology* 373, 108–117.

Woolley, C. S., Gould, E., Frankfurt, M. and McEwen, B. S. (1990) Naturally occurring fluctuations in dendritic spine density on adult hippocampal pyramidal neurons. *Journal of Neuroscience* 10, 4035–4039.

Woolley, C. S., Weiland, N. G., McEwen, B. S. and Schwartzkroin, P. A. (1997) Estradiol increases the sensitivity of hippocampal CA1 pyramidal cells to NMDA receptor-mediated synaptic input: correlation with dendritic spine density. *Journal of Neuroscience* 17, 1848–1859.

Worley, P. F., Bhat, R. V., Baraban, J. M., Erickson, C. A., McNaughton, B. L. and Barnes, C. A. (1993) Thresholds for synaptic activation of transcription factors in hippocampus: correlation with long-term enhancement. *Journal of Neuroscience* 13, 4776–4786.

Wrathall, J. R., Teng, Y. D. and Marriott, R. (1997) Delayed antagonism of AMPA/kainate receptors reduces long-term functional deficits resulting from spinal cord trauma. *Experimental Neurology* 145, 565–573.

Wu, S. Z., Bodles, A. M., Porter, M. M., Griffin, W. S., Basile, A. S. and Barger, S. W. (2004) Induction of serine racemase expression and D-serine release from microglia by amyloid beta-peptide. *Journal of Neuroinflammation* 1, 2.

Wu, Y., Pearl, S. M., Zigmond, M. J. and Michael, A. C. (2000) Inhibitory glutamatergic regulation of evoked dopamine release in striatum. *Neuroscience* 96, 65–72.

Wurtz, C. C. and Ellisman, M. H. (1986) Alterations in the ultrastructure of peripheral nodes of Ranvier associated with repetitive action potential propagation. *Journal of Neuroscience* 6, 3133–3145.

Xia, Y. and Haddad, G. G. (1991) Major differences in CNS sulfonylurea receptor distribution between the rat (newborn, adult) and turtle. *Journal of Comparative Neurology* 314, 278–289.

Xia, S. Z., Feng, C. H. and Guo, A. K. (1998) Multiple-phase model of memory consolidation confirmed by behavioral and pharmacological analyses of operant conditioning in Drosophila. *Pharmacology, Biochemistry and Behavior* 60, 809–816.

Xia, J., Chung, H. J., Wihler, C., Huganir, R. L. and Linden, D. J. (2000) Cerebellar long-term depression requires PKC-regulated interactions between GluR2/3 and PDZ domain-containing proteins. *Neuron* 28, 499–510.

Xia, S., Miyashita, T., Fu, T. F., Lin, W. Y., Wu, C. L., Pyzocha, L., Lin, I. R., Saitoe, M., Tully, T. and Chiang, A. S. (2005) NMDA receptors mediate olfactory learning and memory in Drosophila. *Current Biology* 15, 603–615.

Xiao, L. and Jordan, C. L. (2002) Sex differences, laterality, and hormonal regulation of androgen receptor immunoreactivity in rat hippocampus. *Hormones and Behavior* **42**, 327–336.

Xie, X., Berger, T. W. and Barrionuevo, G. (1992) Isolated NMDA receptor-mediated synaptic responses express both LTP and LTD. *Journal of Neurophysiology* **67**, 1009–1013.

Xu, H., Luo, C., Richardson, J. S. and Li, X. M. (2004) Recovery of hippocampal cell proliferation and BDNF levels, both of which are reduced by repeated restraint stress, is accelerated by chronic venlafaxine. *Pharmacogenomics J* **4**, 322–331.

Xu, J., Kao, S. Y., Lee, F. J., Song, W., Jin, L. W. and Yankner, B. A. (2002) Dopamine-dependent neurotoxicity of α-synuclein: A mechanism for selective neurodegeneration in Parkinson disease. *Nature Medicine* **8**, 600–606.

Yakovlev, P. a. L., A. (1967) The myelinogenetic cycles of regional maturation of the brain. In Minkowski, A. (ed.) *Rational Development of the Brain Early in Life.* Blackwell Scientific Publications, Inc., pp. 3–70.

Yamada, K. and Nabeshima, T. (2003) Brain-derived neurotrophic factor/TrkB signaling in memory processes. *Journal of Pharmacological Science* **91**, 267–270.

Yamada, M. K., Nakanishi, K., Ohba, S., Nakamura, T., Ikegaya, Y., Nishiyama, N. and Matsuki, N. (2002) Brain-derived neurotrophic factor promotes the maturation of GABAergic mechanisms in cultured hippocampal neurons. *Journal of Neuroscience* **22**, 7580–7585.

Yamada, T., McGeer, P. L., Baimbridge, K. G. and McGeer, E. G. (1990) Relative sparing in Parkinson's disease of substantia nigra dopamine neurons containing calbindin-D28K. *Brain Research* **526**, 303–307.

Yamamoto, F., Clausen, H., White, T., Marken, J. and Hakomori, S. (1990) Molecular genetic basis of the histo-blood group ABO system. *Nature* **345**, 229–233.

Yang, D. W., Pan, B., Han, T. Z. and Xie, W. (2004) Sexual dimorphism in the induction of LTP: critical role of tentanizing stimulation. *Life Sciences* **75**, 119–127.

Yang, L., Lindholm, K., Konishi, Y., Li, R. and Shen, Y. (2002) Target depletion of distinct tumor necrosis factor receptor subtypes reveals hippocampal neuron death and survival through different signal transduction pathways. *Journal of Neuroscience* **22**, 3025–3032.

Yang, Y., Ge, W., Chen, Y., Zhang, Z., Shen, W., Wu, C., Poo, M. and Duan, S. (2003) Contribution of astrocytes to hippocampal long-term potentiation through release of D-serine. *Proceeding of the National Academy of Sciences of the United States of America* **100**, 15,194–15,199.

Yanike, M., Wirth, S. and Suzuki, W. A. (2004) Representations of well-learned information in the monkey hippocampus. *Neuron* **42**, 477–487.

Yao, J. K., Reddy, R. D. and van Kammen, D. P. (2001) Oxidative damage and schizophrenia: an overview of the evidence and its therapeutic implications. *CNS Drugs* **15**, 287–310.

Yasuda, R., Sabatini, B. L. and Svoboda, K. (2003) Plasticity of calcium channels in dendritic spines. *Nature Neuroscience* **6**, 948–955.

Yatham, L. N., Lecrubier, Y., Fieve, R. R., Davis, K. H., Harris, S. D. and Krishnan, A. A. (2004) Quality of life in patients with bipolar I depression: data from 920 patients. *Bipolar Disord* **6**, 379–385.

Ye, Z. C., Wyeth, M. S., Baltan-Tekkok, S. and Ransom, B. R. (2003) Functional hemichannels in astrocytes: a novel mechanism of glutamate release. *Journal of Neuroscience* **23**, 3588–3596.

Yeh, S. H., Lin, C. H., Lee, C. F. and Gean, P. W. (2002) A requirement of nuclear factor-kappaB activation in fear-potentiated startle. *Journal of Biological Chemistry* **277**, 46,720–46,729.

Yin, J. C. and Tully, T. (1996) CREB and the formation of long-term memory. *Current Opinion in Neurobiology* **6**, 264–268.

Yin, X., Crawford, T. O., Griffin, J. W., Tu, P., Lee, V. M., Li, C., Roder, J. and Trapp, B. D. (1998) Myelin-associated glycoprotein is a myelin signal that modulates the caliber of myelinated axons. *Journal of Neuroscience* **18**, 1953–1962.

Yoon, I. S., Li, P. P., Siu, K. P., Kennedy, J. L., Cooke, R. G. Parikh, S. V. (2001a) Altered IMPA2 gene expression and calcium homeostasis in bipolar disorder. *Molecular Psychiatry* **6**, 678–683.

Yoon, I. S., Li, P. P., Siu, K. P., Kennedy, J. L., Macciardi, F. and Cooke, R. G. (2001b) Altered TRPC7 gene expression in bipolar-I disorder. *Biological Psychiatry* **50**, 620–626.

Yoshikawa, M., Kobayashi, T., Oka, T., Kawaguchi, M. and Hashimoto, A. (2004) Distribution and MK-801-induced expression of serine racemase mRNA in rat brain by real-time quantitative PCR. *Brain Research: Molecular Brain Research* **128**, 90–94.

Young, L. T. (2002) Neuroprotective effects of antidepressant and mood stabilizing drugs. *Journal of Psychiatry and Neuroscience* **27**, 8–9.

Young, D., Lawlor, P. A., Leone, P., Dragunow, M. and During, M. J. (1999) Environmental enrichment inhibits spontaneous apoptosis, prevents seizures and is neuroprotective. *Nature Medicine* **5**, 448–453.

Young, L. T., Li, P. P., Kish, S. J., Siu, K. P., Kamble, A., Hornykiewicz, O. and Warsh, J. J. (1993) Cerebral cortex Gs alpha protein levels and forskolin-stimulated cyclic AMP formation are increased in bipolar affective disorder. *Journal of Neurochemistry* **61**, 890–898.

Yu, B. P. (1994) Cellular defenses against damage from reactive oxygen species. *Physiological Reviews* **74**, 139–162.

Yuste, R. and Tank, D. W. (1996) Dendritic integration in mammalian neurons, a century after Cajal. *Neuron* **16**, 701–716.

Zafra, F., Castren, E., Thoenen, H. and Lindholm, D. (1991) Interplay between glutamate and gamma-aminobutyric acid transmitter systems in the physiological regulation of brain-derived neurotrophic factor and nerve growth factor synthesis in hippocampal neurons. *Proceedings of the National Academy of Science of the United States of America* **88**, 10,037–10,041.

Zahs, K. R. and Newman, E. A. (1997) Asymmetric gap junctional coupling between glial cells in the rat retina. *Glia* **20**, 10–22.

Zai, G., King, N., Wigg, K., Couto, J., Wong, G. W., Honer, W. G., Barr, C. L. and Kennedy, J. L. (2005) Genetic study of the myelin oligodendrocyte glycoprotein (MOG) gene in schizophrenia. *Genes, Brain and Behavior* **4**, 2–9.

Zarate, C. A., Jr., Du, J., Quiroz, J., Gray, N. A., Denicoff, K. D., Singh, J., Charney, D. S. and Manji, H. K. (2003) Regulation of cellular plasticity cascades in the pathophysiology and treatment of mood disorders: role of the glutamatergic system. *Annals of the New York Academy of Sciences* **1003**, 273–291.

Zarate, C. A., Jr., Quiroz, J. A., Singh, J. B., Denicoff, K. D., De Jesus, G., Luckenbaugh, D. A., Charney, D. S. and Manji, H. K. (2005) An open-label trial of the glutamate-modulating agent riluzole in combination with lithium for the treatment of bipolar depression. *BiologicalPsychiatry* **57**, 430–432.

Zarate, C. A. J., Singh, J., Quiroz, J., De Jesus, G., Denicoff, K. K., Luckenbaugh, D. A., Manji, H. K. and Charney, D. S. (2006) A double-blind placebo-controlled study of memantine in major depression. *American Journal of Psychiatry* **163**, 153–155.

Zeineh, M. M., Engel, S. A., Thompson, P. M. and Brookheimer, S. Y. (2003) Dynamics of the hippocampus during encoding and retrieval of face-name pairs. *Science* **299**, 577–580.

Zeman, A. Z., Boniface, S. J. and Hodges, J. R. (1998) Transient epileptic amnesia: a description of the clinical and neuropsychological features in 10 cases and a review of the literature. *Journal of Neurology, Neurosurgery and Psychiatry* **64**, 435–443.

Zeng, Y., Foy, M. R., Teyler, T. J. and Thompson, R. F. (2004) 17B-estradiol enhancement of nmda LTP in hippocampus. *Society for Neuroscience Abstracts*.

Zenisek, D. and Matthews, G. (2000) The role of mitochondria in presynaptic calcium handling at a ribbon synapse. *Neuron* **25**, 229–237.

Zhang, Q. and Haydon, P. G. (2005) Roles for gliotransmission in the nervous system. *Journal of Neural Transmission* **112**, 121–125.

Zhang, Y., Dawson, V. L. and Dawson, T. M. (2000) Oxidative stress and genetics in the pathogenesis of Parkinson's disease. *Neurobiology of Disease* **7**, 240–250.

Zhang, J. M., Wang, H. K., Ye, C. Q., Ge, W., Chen, Y., Jiang, Z. L., Wu, C. P., Poo, M. M. and Duan, S. (2003) ATP released by astrocytes mediates glutamatergic activity-dependent heterosynaptic suppression. *Neuron* **40**, 971–982.

Zhao, M., Li, D., Shimazu, K., Zhou, X., Deng, C.-X. and Lu, B. (2007) Fibroblast growth factor receptor 1 is required for long-term potentiation, memory consolidation, and neurogenesis. *Biological Psychiatry* **62**, 381–390.

Zhao, M., Li, D., Shimazu, K., Zhou, Y., Deng, C.-X. and Lu, B. (2005) FGF Receptor 1-mediated proliferative neurogenesis is critical for hippocampus-dependent synaptic plasticity and memory consolidation.

Zhao, X., Bausano, B., Pike, B. R., Newcomb-Fernandez, J. K., Wang, K. K., Shohami, E., Ringger, N. C., DeFord, S. M., Anderson, D. K. and Hayes, R. L. (2001) TNF-alpha stimulates caspase-3 activation and apoptotic cell death in primary septo-hippocampal cultures. *Journal of Neuroscience Research* **64**, 121–131.

Zhao, X., Ueba, T., Christie, B. R., Barkho, B., McConnell, M. J., Nakashima, K., Lein, E. S., Eadie, B. D., Willhoite, A. R., Muotri, A. R., Summers, R. G., Chun, J., Lee, K. F. and

Gage, F. H. (2003) Mice lacking methyl-CpG binding protein 1 have deficits in adult neurogenesis and hippocampal function. *Proceedings of the National Academy of Sciences of the United States of America* **100**, 6777–6782.

Zhou, Y., Wu, H., Li, S., Chen, Q., Cheng, X. W., Zheng, J., Takemori, H. and Xiong, Z. Q. (2006) Requirement of TORC1 for late-phase long-term potentiation in the hippocampus. *PLoS ONE* **1**, e16.

Zhu, J. J., Qin, Y., Zhao, M., Van Aelst, L. and Malinow, R. (2002) Ras and Rap control AMPA receptor trafficking during synaptic plasticity. *Cell* **110**, 443–455.

Zigova, T., Pencea, V., Wiegand, S. J. and Luskin, M. B. (1998) Intraventricular administration of BDNF increases the number of newly generated neurons in the adult olfactory bulb. *Molecular and Cellular Neuroscience* **11**, 234–245.

Zijlstra, M., Bix, M., Simister, N. E., Loring, J. M., Raulet, D. H. and Jaenisch, R. (1990) Beta 2-microglobulin deficient mice lack CD 4–8+ cytolytic T cells. *Nature* **344**, 742–746.

Ziskin, J. L., Nishiyama, A., Rubio, M., Fukaya, M. Bergles, D. E. (2007) Vesicular release of glutamate from unmyelinated axons in white matter. *Nature Neuroscience* **10**, 321–330.

Zohar, O., Reiter, Y., Bennink, J. R., Lev, A., Cavallaro, S., Paratore, S., Pick, C. G., Brooker, G. and Yewdell, J. W. (2008) Cutting edge: MHC class I-ly49 interaction regulates neuronal function. *Journal of Immunology* **180**, 6447–6451.

Zonta, M., Angulo, M. C., Gobbo, S., Rosengarten, B., Hossmann, K. A., Pozzan, T. and Carmignoto, G. (2003) Neuron-to-astrocyte signaling is central to the dynamic control of brain microcirculation. *Nature Neuroscience* **6**, 43–50.

Zou, L. B., Yamada, K., Tanaka, T., Kameyama, T. and Nabeshima, T. (1998) Nitric oxide synthase inhibitors impair reference memory formation in a radial arm maze task in rats. *Neuropharmacology* **37**, 323–330.

Zucker, R. S. (1989) Short-term synaptic plasticity. *Annual Review of Neuroscience* **12**, 13–31.

Zukin, S. Y., Grooms, K. M., Noh, R., Regis, M. K., Bryan, G. J. and Carroll, R. C. (2004) Activity bidirectionally regulates AMPA receptor mRNAs in dendrites of hippocampal neurons. Program No. 260.11. *Abstract Viewer/Itinerary Planner. Society for Neuroscience*.

Zuo, Y., Yang, G., Kwon, E. and Gan, W. B. (2005) Long-term sensory deprivation prevents dendritic spine loss in primary somatosensory cortex. *Nature* **436**, 261–265.

Index

Abe, K. 94
acetylcholine receptor (AChR) 148
AChR *see* acetylcholine receptor
action potentials 39, 178
activator protein 1 (AP-1) 112
activity-dependent genes 170
activity-dependent myelination 36–42
activity-dependent neurotrophic factor (ADNF) 149
activity-dependent neurotrophins 143–6
activity-dependent plasticity
    MHC class I 123–9
activity-dependent remodeling 126–7
Adams, J. P. 116
addiction 133, 192
adenosine potentiation 202–3
adenosine triphosphate (ATP) 195, 208, 217
ADNF *see* activity-dependent neurotrophic factor
Aggleton 14
aging
    cerebrovasculature 33
    estrogen 110
    LTP 105
Akama, K. T. 116
Akopian, G. 108
Alberini, C. M. 72
Albin, R. L. 186
Alderson, R. F. 143
Alford, S. 173
alpha Ca^{2+}/calmodulin-dependent protein kinase II (αCaMKII) 80
Alzheimer's disease 8, 13, 97, 120, 133
Amaral, D. G. 16
amino-3-hydroxy-5-methylisoxazole-4-propionic acid (AMPA)
    neurotrophin signaling 142
    receptor activation 100, 109, 164
    receptor trafficking 47–8, 51–2
    sex differences 114
amino-3-hydroxy-5-methylisoxazole-4-propionic acid (AMPA) type glutamate receptor (AMPARs) 130, 131, 137
    abnormal trafficking 133
    surface localization 135
    TNFα 134–5, 140
γ-aminobutyric acid (GABA)-dependent modulation
    dopamine release 183–4
    uptake 201
amnesia 12
AMPA *see* amino-3-hydroxy-5-methylisoxazole-4-propionic acid
AMPARs *see* amino-3-hydroxy-5-methylisoxazole-4-propionic acid-(AMPA) type glutamate receptors
amygdala
    fear conditioning 75–6
amyloid precursor protein (APP) 8, 10, 13
An, W. F. 174
Andrew, R. J. 9
androgen-induced hippocampal synaptic plasticity 118–19
antidepressants 28, 156
    signaling networks 43
Antonny, B. 48
AP-1 *see* activator protein 1
apoptosis 133
APP *see* amyloid precursor protein
Araque, A. 212, 213, 216, 217
ascorbate 191
associative learning 75
associative representation 14, 15, 16, 17–18
astrocytes 201
    calcium signaling 216
    D-serine 194
    glutamate 218
    metabotropic glutamate receptors 210–18
    neuronal modulation 216–17
    plasticity 29–31
    synapse formation 142, 145, 150
    TNFα 136
ATP *see* adenosine triphosphate
auditory fear conditioning 19
Auerbach, J. M. 182
autoimmunity 123
Avshalumov, Marat V. 181–92

axon diameter 37, 41
axonal conduction time 36
axotomy 127

Backsai, B. J. 170
Bading, H. 171, 172
Bahar, A. 88
Bailey, C. H. 85
Balice-Gordon, Rita J. 142–52
Bao, Li 181–92
Bao, S. 145
Barco, A. 176
basal nucleus 75
basolateral amygdala (BLA) 94
Baudry, Michael 97–110
BDNF *see* brain-derived neurotrophic factor
Bear, M. F. 97
Beattie, Eric C. 130–42
Beattie, Michael S. 130–42
behavior
    inheritance 22
Bekkers, J. M. 215
Benavides, J. 55
Bengtsson, S. L. 42
Benington, J. H. 68
Bennett, E. L. 23
Berman, R. M. 54
Bernard, V. 191
Berridge, M. J. 173
Bezzi, P. 135
Bi, R. 101, 102
bidirectional communication 208, 210
Biedenkapp, J. C. 88
bipolar disorder (BPD) 43–58
Birnbaum, S. G. 46
BLA *see* basolateral amygdala
Black, J. E. 25, 33
Bliss, T. V. P. 8, 74, 172
Boehning, D. 197
Boulanger, Lisa M. 123–9
Bourne, H. R. 45
Bourtchouladze, R. 81
BPD *see* bipolar disorder
Brahma, B. 190

brain
 antioxidant network 188
 morphological plasticity 23
brain injury 181
brain plasticity
 sleep 66–70
brain-derived neurotrophic factor (BDNF)
 bipolar disorder 44
 calcium signaling 143
 LTP 145, 176
 neurogenesis 28, 153, 155
 synaptic plasticity 83–4
Braithwaite, Steven P. 130–42
Bredt, D. S. 197
Brinley, F. J. 32
Brinton, R. D. 100
Briones, T. L. 24
Bruce, A. J. 135
Bukalo, Olena 169–78
Bunsey, M. 20
Butt, A. M. 32

Cain, Christopher 72–88
Caithness, G. 61
calcium channels
 estrogen 107–8
calcium signaling 47, 172–3, 175–6, 201, 210
calcium
 fear-conditioning 80
calmodulin-dependent kinase II pathways 50
alphaCaMKII see alpha Ca^{2+}/calmodulin-dependent protein kinase II
cAMP response element (CRE) 112, 118
cAMP response element-binding protein (CREB) 170, 173, 176, 178
CAMs see cell adhesion molecules
Cao, W. 181
capillary volume 33
carbon monoxide (CO) 197, 199
Carmona, M. A. 144
Carrion, A. M. 174
Carroll, R. C. 131, 132
Castren, E. 144, 155
catalase 190
catamenial epilepsy 98
catecholamine modulation 84–5
CBP see CREB binding protein
CE see central nucleus
cell adhesion 166
cell adhesion molecules (CAMs) 10, 42
cell excitability 174
cellular neuroprotection
 estrogen 108

cellular signalling
 LTP 89
central nervous system (CNS)
 steroid hormones 111
 TNFα 130, 133
central nucleus (CE) 75
cerebrovasculature
 plasticity 33–4
Chang, Jay H. 153–9
Chapman, P. F. 84
Chen, B. T. 182
Chenard, B. L. 80
Cheng, B. 133
Cheng, H. Y. 174
Cheremy, A. 182
childhood neglect 38
Choi, D. W. 133, 134
cholesterol 111
Cirelli, C. 69
CNS see central nervous system
CO see carbon monoxide
cognition
 estrogen 97, 120
 LTP 89–96
 myelin 37
cognitive reinforcement 95
Cohen, N. J. 15, 17
Collin, C. 144
concanamycin A 198
conditioned fear responses 75
conduction velocity 39, 40–1, 42
Conn, P. 212
conscious memory 15
convergent afferents 15
corpus callosum
 childhood neglect 38
Corriveau, R. A. 123, 124, 125, 126, 128
corticosterone 90, 94
Cotman, C. W. 28
Crane, G. 53
CRE see cAMP response element
CREB see cAMP response element-binding protein
CREB binding protein (CBP) 174
critical periods 40
Cummins, R. A. 154
Curry, Lisa R. 160–8
cytokine induction
 neuronal receptor trafficking 130–42
cytoplasmic signaling proteins 168

D-amino acid oxidase (DAO) 193
D-aspartate 193, 199–200
D-serine 193–200, 203

localisation 193–4
Müller cells 207
neuronal migration 198–9
NMDAR regulation 196–7
storage 197–8
DA see dopamine
Daisley, J. N. 9
Dan, Y. 173
DAO see D-amino acid oxidase
Davachi, L. 18
Dave, A. S. 68
Davies, H. P. 10
Davis, H. P. 72
De Miranda, J. 195
Debiec, Jacek 72–88
Deisseroth, K. 157
delay conditioning 156
dendrites
 MHC class I 124
dendritic action potentials 170
dendritic length 24
dendritic spines 26, 113, 115, 119
dentate gyrus (DG)
 plasticity 153
 sex differences 114
Desahger, S. 188
DeZazzo, T. 75
DG see dentate gyrus
DHT see 5α-dihydrotestosterone
Diamond, M. C. 29
5α-dihydrotestosterone (DHT) 118
Dolan, R. 93
dopamine (DA) release 181–92
 dorsal striatum 186–8
 GABA-dependent modulation 183–4
 glutamate-dependent modulation 182–3
dopamine (DA) transmission 84, 96
dorsal striatum (DS)
 dopamine release 186–8
downstream regulatory element antagonistic modulator (DREAM) 174
downstream signaling 129
DREAM see downstream regulatory element antagonistic modulator
Drevets, W. C. 48
Dringen, R. 190
Drukarch, B. 188
DS see dorsal striatum
Du, Jing 43–58
Dudek, S. M. 169, 170, 171, 172, 176, 178
Dugan, L. L. 183
Dunlop, D. S. 193, 195
Dusek, J. A. 20
Dyson, S. E. 26

Ebinu, J. O. 175
EC *see* enriched condition
ECS *see* electroconvulsive shock
Edstrom, E. 126
Eichenbaum, H. 14–21, 111, 112
Einat, H. 47
Ekstrom, A. D. 17, 19
electrical activity 125–6
electroconvulsive shock (ECS) 61
Elmariah, Sarina B. 142–52
Emamghoreishi, M. 47
emotion 89–96
emotional reinforcement 93–4, 95
endocannabinoids 129
endocytosis 132
English, J. D. 81
enriched condition (EC) paradigm 22, 29
epilepsy 133
episodic memory 14–17
EPSPs *see* excitatory post-synaptic potentials
Eriksson, P. S. 27
ERK *see* extracellular signal-regulated kinase
estrogen
  activation 97–8
  aging 110
  AMPA receptor activation 100, 109
  calcium channels 107–8
  cellular neuroprotection 108
  cognition 97, 120
  dendritic spines 115
  hippocampal electrophysiology 98
  hippocampal excitability 98
  hippocampal LTP 101–5
  hippocampal synaptic plasticity 97–110
  LTP 105–8
  memory 109–10, 120
  NMDA receptor activation 100, 109
  NMDA receptor-mediated EPSPs 100–1
  nongenomic action 98–100
  synaptic plasticity 108–9
estrogen-induced hippocampal synaptic plasticity 115–17
estrous cycle
  synaptic plasticity 102–5
N-ethylmaleimide-sensitive fusion protein (NSF) 132
evoked IPSCs 210–12
  astrocyte-induced depression 213–14
excitatory post-synaptic potentials (EPSPs)
  estrogen 100
  ganglion cell activity 204–5
  synapse-to-nucleus signaling 171
excitotoxicity 130–42

exercise
  adult neurogenesis 28, 154
  angiogenesis 33
experience-driven brain plasticity 22–36
extracellular modulation
  Pavlovian fear conditioning 72–88
extracellular signal-regulated kinase (ERK) MAP kinase pathway 47, 50, 81
Eyre, J. A. 38

FAK *see* focal adhesion kinase
Falkenberg, T. 153
Fallon, J. H. 84
Farb, C. R. 78
Fendt, M. 80
Feng, R. 157
Fields, R. Douglas 36–42, 169–78
fight or flight syndrome 94
Fink, C. C. 50
fluoxetine 118
focal adhesion kinase (FAK) 168
focal adhesion-like processes 160–8
Foltyn, V. N. 195
forskolin 81
Foster, C. M. 107
Foster, J. A. 127
Foy, Michael 97–110
fragile X syndrome 177
Frank, M. G. 69
Freeman, F. M. 10
Frey, Julietta U. 89–96
Frey, U. 75, 169, 170, 171
Friedman, E. 45, 46
Frye, C. A. 118
functional plasticity
  MHC class I 123–9
Furukawa, S. 149

G protein-coupled receptors (GPCRs) 44
G protein/cAMP signaling pathway 45
GABA *see* γ-aminobutyric acid
GABAergic inhibition 82
Gage, F. H. 153, 155
Gais, S. 64
Gale, G. D. 72
Ganeshina, O. 27
ganglion cell activity
  post-synaptic modulation 205–6
  pre-synaptic modulation 204–5
Garber, K. 177
gastrin-releasing peptide (GRP) 82
gene expression 178
gene transcription
  learning 169–70
geniculocortical projections 128
Ghosh, A. 153, 155, 157

Gibbs, R. B. 97
Giedd, J. N. 38
glia
  myelinating 39, 41
  neuronal excitability 201–9
  neurotrophin signaling 142–52
  oxidative stress 188
glia-to-glia signaling 207–8
glial microdomains 29
glial neurotransmitters 193–200
Globus, A. 24
glucocorticoid receptors (GR) 95
glucocorticoid-induced hippocampal synaptic plasticity 119–20
glutamate 47–8, 49, 130, 201, 203, 216–17
glutamate receptor interacting protein (GRIP) 132, 198
glutamate receptor trafficking 131
glutamate synapses 187
glutamate transporters 217
glutamate-dependent modulation
  dopamine release 182–4
glutamatergic system
  mood disorders 52–5
glutathione (GSH) peroxidase 181, 188, 190
GMP *see* guanosine monophosphate
Goddard, C. A. 128, 129
Gold, P. E. 84
Gonzalez, M. 148
Gothard, K. M. 17
Gould 115, 154, 156
GPCRs *see* G protein-coupled receptors
GR *see* glucocorticoid receptors
Graef, I. A. 174
granule cells 194, 199
Greenough, William T. 22–35
GRIP *see* glutamate receptor interacting protein
Grosche, J. 30
Grossman, A. W. 22–35
Grover, L. M. 107
GRP *see* gastrin-releasing peptide
GSH *see* glutathione
GSK-3 46–7
guanosine monophosphate (GMP) 197
Gyllensten, L. 31, 38, 42

habituation 73
Hakak, Y. 37
Hampson 19
Hara, Y. 41
Harik, S. I. 34
Hayashi, Y. 212
Hebb, Donald 7, 22, 73, 169
Heckers, S. 20
Henderson 110, 120

Hermann, V. W. 140
Hernandez, Ruben V. 160–8
hippocampal long-term potentiation
  estrogen 101–5
  reinforcement 89–96
hippocampal neurons
  AMPAR surface localization 135–6, 138–9
hippocampal slice preparation 99
hippocampal synaptic plasticity
  androgen-induced 118–19
  estrogen-induced 115–17
  glucocorticoid-induced 119–20
  progesterone-induced 117–18
  steroid-induced 111–20
hippocampus 14–21
  amygdalar activation 94
  androgens 118
  damage 20
  estrogen 98
  excitability 98
  LTP 160–8
  metabotropic glutamate receptors 210–18
  MHC class I 124
  plasticity 15, 97–110, 157
  progesterone 118
  sex differences 112–15
  subgranular zone 153
hippocampus-dependent memory 153–9
Hirbec, H. 131
Hollman, M. 130
Honey, R. C. 18
hormone-response elements (HREs) 112
hormones
  signaling pathways 43
HPA see hypothalamo-pituitary-adrenal axis
HREs see hormone-response elements
Huang, Y. Y. 81
Huber 170
Hughes, Ethan G. 142–52
Huh, G. S. 124, 126, 127, 128, 129
Hyde, T. M. 37
hydrogen peroxide 181–92, 197
hyperpolarization 205
hypothalamo-pituitary-adrenal (HPA) axis 94
hypothalamus
  sex differences 115
hypoxia 34

IEGs see immediate early genes
ifenprodil 80
iGluRs see ionotropic receptors
immediate early genes (IEGs) 69, 175, 178

IMMP see intermediate medial mesopallium
impulse activity 42
Inagaki, N. 185
infection 127
information-processing
  myelin 37
inheritance
  behavior 22
inhibitory post-synaptic currents (IPSCs) 210–18
inhibitory synapses 145
integrins 160, 164, 168
intermediate medial mesopallium (IMMP) 9, 12
interneurons 82, 212
intracellular signaling
  Pavlovian fear conditioning 72–88
ionotropic receptors (iGluRs) 130, 133
IPSCs see inhibitory post-synaptic currents
Itoh, K. 39
Izquierdo, I. 85

Janeway, C. A. 124
Javitt, D. C. 199
Jiang, Z. G. 186
Johnson, J. W. 196
Jones, T. A. 29
Jope, R. S. 47
Ju, W. 135
Juraska, J. M. 24

kainate receptors 130
Kandel, E. R. 62
Kang, H. 142
Kang, Jian 210–19
Karschin, C. 185
Kawas, C. 110
Kemp, J. M. 186, 199
Kempermann, G. 28, 153, 154, 156
ketamine 54
Kim, A. J. 198
Kim, Paul M. 193–200
Kleim, J. A. 25, 26
Koechlin, E. 66
Kohara, K. 144
Komuro, H. 199
Korz, Volker 89–96
Korzus, E. 174
Kramar, E. A. 162, 166
Kreiman, K. 19
Krichevsky, A. M. 177
Krug, M. 89, 170

LA see lateral nucleus
lamotrigine 54

Lampson, L. A. 123
Landfield 105
lateral geniculate nucleus (LGN)
  development 125
lateral nucleus (LA) 75–6, 78–9
lateral perforant path (LPP) synapses 158
learning 7, 8–9, 72
  dopamine 84
  gene transcription 169–70
  LTP 89–96
  memory 88
  neurogenesis 153, 156–7
  NMDA receptors 78
  ponto-geniculo-occipital waves 68
  sleep-dependent 65
  white matter development 39
learning paradigms
  LTP-reinforcement 93
LeBaron, Richard G. 160–8
LeDoux, Joseph E. 72–88
Lee, H. J. 85, 143
Lee, I. 21
Lee, P. R. 177
Lee, S. J. 112
Leonoudakis, Dmitri 130–42
Levine, E. S. 146
Levy, W. B. 16
Lewin, G. R. 153
Lewis, C. 116, 117
LGN see lateral geniculate nucleus
Li, C. 32
Li, X. 79
light-evoked activity 203
Ligon, L. A. 32
Lim, K. O. 37
Liss, B. 185
lithium 46, 51, 54
Liu, Q. S. 214, 216, 217
lobus parolfactorius see medial striatum (MS)
Loeb, J. A. 148
long-term depression (LTD) 105, 127, 132, 170
long-term memory (LTM) 72, 81
  nucleus signaling 169–78
long-term potentiation (LTP) 15, 74, 81, 97
  AMPAR trafficking 132
  BDNF 145
  estrogen 105–8
  focal adhesion-like processes 160–8
  neurogenesis 154
  NMDARs 199
  reinforcement 89–96
  RGD 160–4, 167–8
  sex differences 114
LPP see lateral perforant path synapses

LTD *see* long-term depression
LTM *see* long-term memory
LTP *see* long-term potentiation
Lu, B. 83, 153–9
Lu, W. 136
Lu, Y. F. 173
Lucas, F. R. 47
Luine, V. N. 115, 116, 117

MacGregor, Duncan G. 181–92
Magarinos, A. M. 119, 120
major histocompatibility complex (MHC) class I 123–9
Malenka, R. C. 132
Malinmow, R. 131
Manji, Husseini K. 43–58
MAP kinase pathway
 synaptic plasticity 109
MAPK *see* mitogen-activated protein kinase
Maquet, P. 66, 68
Maren, S. 94, 114
Markham, Julie A. 22–35
Martin, K. C. 171
Martinez, A. 144
Martinez Jr, Joe L. 160–8
Matsuda, S. 50
Mauch, D. H. 135
Mazzanti, M. 216
Mazzucchelli, C. 109
McAllister, A. K. 176
McDonald, A. J. 82
McDonald, J. W. 134
McEwen, B. S. 98, 111–20
McGaugh, J. L. 85, 94
medial striatum (MS) 9
medium spiny neurons 186–8
Meguro, H. 196
memantine 55
memory 60, 72
 BDNF signaling 83
 consolidation 59, 60–2, 72, 73, 85
 degradation 63
 dopamine 84
 enhancement 62, 64
 erasure 62
 estrogen 109–10, 120
 formation 10, 11, 13
 hippocampus-dependent 153–9
 learning 88
 neurogenesis 153, 156–7
 Pavlovian fear conditioning 72–88
 protein synthesis 73
 reconsolidation 59, 62–3, 70–1, 73, 85–8
 stabilization 61, 64
memory deficit 12

memory trace 12
metabotropic glutamate receptors (mGLuRs) 210–18
 fear-conditioning 79–80
N-methyl-D-aspartate (NMDA) receptor-dependent plasticity 20
N-methyl-D-aspartate (NMDA) receptors (NMDARs) 48, 130
 activity-dependent plasticity 127
 D-serine 193, 194, 196–7
 estrogen 100, 109
 learning 78
 LTP 199
 sex differences 114
mGLuRs *see* metabotropic glutamate receptors
MHC *see* major histocompatibility complex
mineralocorticoid receptors (MR) 89, 95
miniature IPSCs 210, 214–16
Mitchell, S. J. 218
mitochondria 185–6
mitochondrial plasticity 32–3
mitogen-activated protein kinase (MAPK)
 progesterone 118
mitogens 155
Moita, M. A. 19
mood disorders
 glutamatergic system 52–5
mood stabilizers
 signaling networks 43
 synaptic plasticity 51–2
Mothet, J. P. 197, 207
motor connections
 remodeling 126
motor learning 26, 28
Mourre, C. 185
MR *see* mineralocorticoid receptors
mRNA translation 73
MS *see* medial striatum
MSBs *see* multiple synaptic boutons
Mulkey, R. M. 132
Müller cells
 adenosine potentiation 202–3
 Ca^{2+} transients 201–2
 D-serine 207
Muller, D. 78
Muller, G. E. 59
Muller, R. U. 17
multiple sclerosis (MS) 134
multiple synaptic boutons (MSBs) 27
Mumby, D. G. 18
Mustafa, Asif K. 193–200
myelin 36–7
 thickness 41
myelin plasticity 38–9, 40

myclination 31–2, 34
 activity-driven 36–42

N-ethylmaleimide-sensitive fusion protein (NSF) 132
N-methyl-D-aspartate (NMDA) receptor-dependent plasticity 20
N-methyl-D-aspartate (NMDA) receptors (NMDARs) 48, 130
 activity-dependent plasticity 127
 D-serine 193, 194, 196–7
 estrogen 100, 109
 learning 78
 LTP 199
 sex differences 114
Nader, K. 59, 62, 63, 85, 87
Narisawa-Saito, M. 146
Navarro, Mary M. 160–8
NCAM *see* neural cell adhesion molecule
NE *see* norepinephrine
Nedergaard, Maiken 210–19
Neeper, S. A. 154
nerve growth factor (NGF) 143, 155
Neumann, H. 125
neural cell adhesion molecule (NCAM) 10
neural connectivity 7
neural progenitor cells (NPCs) 153, 155
neural trauma 134–5
neuregulin 31
neurogenesis 27–9
 regulation 153–9
neuromodulation 182
neuromuscular synapses 148
neuronal dysfunction
 AMPARs 133
neuronal excitability 201–9
neuronal migration
 D-serine 198–9
neuronal modulation
 astrocytes 216–17
neuronal plasticity 23–9
 AMPA receptor trafficking 47–8
neuronal receptor trafficking
 cytokine induction 130–42
neurons
 major histocompatibility complex (MHC) class I 123–5
 neurotrophin signaling 142–52
neuroplasticity 46
neurosteroids 11
neurotoxicity
 iGluRs 133
 NMDARs 196, 199
neurotransmission 181
neurotransmitter release 197–8
neurotrophin-3 (NT-3) 153

neurotrophins
    hippocampus-dependent memory
        153–9
    signaling 142–52
Newman, Eric A. 201–9
NFAT *see* nuclear factor of activated
    T cells
NGF *see* nerve growth factor
Nilsen, J. 108, 118
Nirenberg, M. J. 191
nitric oxide (NO) 83–4, 129, 197, 199
NMDA *see* N-methyl-D-aspartate
NO *see* nitric oxide
nodes of Ranvier 37, 41
norepinephrine (NE) signaling 85
novelty
    LTP 89, 90–1
NPCs *see* neural progenitor cells
NSF *see* N-ethylmaleimide-sensitive
    fusion protein
NT-3 *see* neurotrophin-3
nuclear factor of activated T cells
    (NFAT) 174
nucleus
    calcium signaling 47, 172–3, 175–6
    transcription 173–5
nucleus signaling
    long-term memory 169–78

Oh, Eun Joo 142–52
Ohyama, T. 66
Olanow, C. W. 192
olfactory bulb 194
Oliet, S. H. 29
oligodendrocyte precursor cells (OPC)
    194
oligodendrocytes
    AMPARs 134
    myelination 39
    plasticity 31–2
Olsson, T. 127
one-trial passive avoidance task 8
OPC *see* oligodendrocyte precursor
    cells
optic nerve 31
    glia 39
Orfila, James E. 160–8
oxidative stress 181, 188, 192

paired-pulse facilitation (PPF) 84, 158
Pang, P. T. 176
Parkinson's disease 53, 181, 192
Patel, Jyoti 181–92
Patterson, S. L. 83, 144
Pavlovian fear conditioning 72–88
PDEs *see* phosphodiesterases

Peigneux, P. 64, 66
Pellmar, T. 188
perforated synapses 27
Perry, V. H. 133, 135
Pfrieger 135, 142
PGO *see* ponto-geniculo-occipital waves
Phillips, A. G. 96
phosphodiesterases (PDEs) 44
phosphorylation 132
PICK *see* protein interacting with
    C-kinase
Pierschbacher, M. D. 166
PKA pathway 48–9
PKC *see* protein kinase C
plasticity 34
    MHC class I 123–9
ponto-geniculo-occipital (PGO) waves
    68
Porter, J. T. 211
potassiumAPT channels 184–5, 192
Pozzo-Miller, L. D. 145
PPF *see* paired-pulse facilitation
Preston, A. 20
progesterone 120
progesterone-induced hippocampal
    synaptic plasticity 117–18
proneurotrophins 143
protein interacting with C-kinase (PICK)
    132, 198
protein kinase C (PKC) 44, 80
protein kinase C (PKC) signaling 45–6,
    50, 142
protein kinases 80–1
protein synthesis
    learning 169–70
    LTP 89, 93
    memory 73, 81
proteoglycans 160
Przybyslawski, J. 87
psychostimulants 52
purinergic signaling 42
Purves, D. 8

Quirk, G. J. 76
Quiroz, Jorge 43–58

Raisman, G. 115
Rattiner, L. M. 83
Raymond 172
reactive oxygen species (ROS) 181
Recanzone, G. H. 34
recurrent connections 15, 16
regulated intra-membrane proteolysis
    (RIP) 175
Reif, A. 157
reinforcement

training 93
relational networking 14, 15, 19–20
retina
    electrical activity 125
    neuronal excitability 201–9
retinogeniculate projections 128
retrograde amnesia 72
rewarding 95
Ribeiro, S. 69, 70, 197
Rice, Margaret E. 181–92
riluzole 55
RIP *see* regulated intra-membrane
    proteolysis
Rodrigues, S. M. 78, 79
Romeo, Russell D. 111–20
ROS *see* reactive oxygen species
Rose, S. 143
Rose, S. P. R. 7–13
Rosenkranz, J. A. 85
rotenone 185
Rutherford, L. C. 146

Salinska, E. 9
Sandi, C. 11
Sara, S. J. 12
Schafe, G. E. 84, 87
Schaffer collateral-CA1 pathway
    LTP 160–8
Schaller, M. D. 168
Schell, M. J. 193, 194, 199, 200
Schipke, C. G. 201
schizophrenia 37, 133, 192, 199
Schwann cells 39
second-messengers 80–1
Selcher, J. C. 81
Semont, A. 94
sequential organization 14, 15, 16, 18–19
serine racemase (SR) 195–6
Seutin, V. 184
SGZ *see* subgranular zone
Shaffery, J. P. 69
Shaikh, A. A. 117
Shigemoto, R. 210
Shimazu, K. 158
Shors, T. 113
short-term memory (STM) 72
Shumyatsky, G. P. 82
Sidló, Zsuzsanna 181–92
signaling cascades
    bipolar disorder 43–58
signaling factors 11
signaling networks 43–8
silent synapses 132
Singer, D. S. 108
sleep 60
    brain plasticity 66–70

sleep deprivation 68, 69
sleep-dependent memory consolidation 59–71
sleep-dependent motor memory reorganization 67
Smith, A. D. 192
Smith, C. 69
Snyder, Solomon H. 193–200
Sokolova, T. 190
spatial training 91–2
spinal cord injury 40, 134, 135
spinogenesis
  estrogen 116
spontaneous IPSCs 210, 211
SR *see* serine racemase
Staubli, U. 164, 167, 168
stereospecificity 193
Steriade, M. 68
steroid hormones 111–12
  action 112, 113
steroid-induced hippocampal synaptic plasticity 111–20
Stevens, B. 39, 42
Stevens, E. R. 207
Steward, O. 177
Stickgold, Robert 59–71
STM *see* short-term memory
Strack, S. 80
stress 11, 27
  LTP 89, 90–1
  neurotrophins 154
structural plasticity
  MHC class I 123–9
subgranular zone (SGZ) 153
Swadlow, H. A. 40
Syken, J. 129
synapse 7–13
  astrocytes 30
  cytokine induction 130–41
  formation 149
  hydrogen peroxide 181–92
  morphology 26–7
  specificity 171–2
synapse-to-nucleus signaling 169
synaptic connectivity 11
synaptic modulators 146–51
synaptic plasticity 8, 13, 16, 20–1, 36, 165
  estrogen 108–9
  estrous cycle 102–5
  fear conditioning 77–82
  MAP kinase pathway 109
  memory consolidation 73–5
  mood stabilizers 51–2
  myelination 40

signaling networks 43–8
  steroid-induced 115
synaptic strengthening
  learning 169–70
synaptic tagging 169, 171–2
synaptic transmission
  glial cells 204
  integrins 164
synaptogenesis 23, 31, 144
Szabo, B. 187

Tailby, C. 25
Tamai, K. 32
Tanzi, E. 7
TARPs *see* transmembrane AMPAR regulatory proteins
Taschenberger, H. 203, 206
Taubenfeld, S. M. 88
temporal cascade 9–11
temporal order 18
testosterone 111, 120
tetrodotoxin (TTX) 125
Teyler, T. J. 98
Thompson, Richard F. 97–110
TNFα *see* tumor necrosis factor alpha
TORCs *see* transducers of regulated CREB activity
trace conditioning 156
Trachtenberg, J. T. 24, 25
training 9, 12, 25
  reinforcement 93
transcription factor activation 173
transcription factors 10
transducers of regulated CREB activity (TORCs) 174
translational regulatory mechanisms 177–8
transmembrane AMPAR regulatory proteins (TARPs) 132
Traub, R. J. 41
Trk receptors 143
Tsai, G. 112
TTX *see* tetrodotoxin
Tulving, E. 14
tumor necrosis factor alpha (TNFα) 130, 133
  AMPAR trafficking 134–5
  glial-derived 135–8
  surface localized AMPARs 138, 139, 140
Tyler, W. J. 153
tyrosine kinase 83
Tzingounis, A. V. 176

Ungerleider, L. G. 66
Uzakov, S. 92, 93, 96

valproic acid (VPA) 46, 51, 54
vascular stroke 196, 199
Ventura, R. 210, 217
vesicle trafficking 48
vitamin C *see* ascorbate
VNO *see* vomeronasal organ
Vo, N. 177, 178
voltage-gated calcium channels 80
voltage-sensitive calcium channels (VSCCs) 171
Volterra, A. 30, 201
vomeronasal organ (VNO) 124
Votyankova, T. V. 186
VPA *see* valproic acid
VSCCs *see* voltage-sensitive calcium channels

Wagner, U. 64
Walker, Mathew P. 59–71
Wang, H. Y. 46
Wang, J. Q. 157
Wang, W. L. 198
water maze 92–3
Waters, Elizabeth M. 111–20
Waxman, S. G. 41
Weick, J. P. 172
Weiland, N. G. 115, 116
Weinberger, N. M. 77
white matter abnormalities 37
Wilson, M. A. 68
Wing-song Liu 210–19
Wolosker, H. 195, 203, 207
Wong, M. 100
Wood, E. 19
Woods, A. 165, 168
Woolley, C. S. 115, 117

Xie, X. 101
Xu, Qiwu 210–19

Yang, Y. 199
Yasuda, R. 80
Young, L. T. 45
Yuan, Peixiong 43–58

Zarate Jr., Carlos 43–58
Zeng, Y. 108
Zhao, X. 133, 156, 158
Zhu, J. J. 50
Zigova, T. 155
zinc 55
Zuo, L. B. 25

For EU product safety concerns, contact us at Calle de José Abascal, 56–1°,
28003 Madrid, Spain or eugpsr@cambridge.org.

www.ingramcontent.com/pod-product-compliance
Ingram Content Group UK Ltd.
Pitfield, Milton Keynes, MK11 3LW, UK
UKHW050109230326
469255UK00020B/466